Structural Geology

and

Personal Computers

COMPUTER METHODS IN THE GEOSCIENCES

Daniel F. Merriam, Series Editor

Volumes in the series published by Elsevier Science Ltd
Geological Problem Solving with Lotus 1-2-3 for Exploration and Mining Geology: G.S. Koch Jr. (With program on diskette)
Exploration with a Computer: Geoscience Data Analysis Applications: W.R. Green
Contouring: A Guide to the Analysis and Display of Spatial Data: D.F. Watson (with program on diskette)
Management of Geological Data Bases: J. Frizado (Editor)
Simulating Nearshore Environments: P.A. Martinez and J.W. Harbaugh
Geographic Information Systems for Geoscientists: Modelling with GIS: G.F. Bonham-Carter
Computing Risk for Oil Prospects:
Principles and Programs: J.W. Harbaugh, J.C. Davis and J. Wendebourg.

\+ **Simulating Oil Migration and Stratigraphic Traps:** J. Wendebourg and J.W. Harbaugh

* Volumes published by Van Nostrand Reinhold Co. Inc.:
Computer Applications in Petroleum Geology: J.E. Robinson
Graphic Display of Two- and Three-Dimensional Markov
Computer Models in Geology: C. Lin and J.W. Harbaugh
Image Processing of Geological Data: A.G. Fabbri
Contouring Geologic Surfaces with the Computer: T.A. Jones, D.E. Hamilton, and C.R. Johnson
Exploration-Geochemical Data Analysis with the IBM PC: G.S. Koch, Jr. (With programs on diskettes)
Geostatistics and Petroleum Geology: M.E. Hohn
Simulating Clastic Sedimentation: D.M. Tetzlaff and J.W. Harbaugh

*Orders to: Van Nostrand Reinhold Co. Inc. 7625 Empire Drive, Florence, KY 41042, USA.

Related Elsevier Science Ltd Publications

Books
MACEACHREN & TAYLOR (Editors): Visualization in Modern Cartography
TAYLOR (Editor): Geographic Information Systems (The Microcomputer and Modern Cartography)

Journals
Computers & Geosciences
Journal of Geodynamics
Journal of Structural Geology
Tectonophysics

Full details of all Elsevier publications available on request from your nearest Elsevier office. See also http://www.elsevier.nl

\+ In preparation

Structural Geology

and

Personal Computers

Edited by

Declan G De Paor
Department of Earth and Planetary Sciences
Harvard University

Pergamon

U.K.	Elsevier Science Ltd, The Boulevard, Langford Lane, Kidlington, Oxford OX5 lGB, U.K.
U.S.A.	Elsevier Science Inc., 660 White Plains Road, Tarrytown, New York 10591-5153. U.S.A.
JAPAN	Elsevier Science Japan, Tsunashima Building Annex, 3-20-12 Yushima, Bunkyo-ku, Tokyo 113, Japan

---

Copyright ©1996 Elsevier Science Ltd

All Rights Reserved. No part of this publication may be reproduced, stored in a retrieval system or transmitted in any form or by any means electronic, electrostatic, magnetic tape, mechanical, photocopying, recording or otherwise, without permission in writing from the publishers.

First edition 1996

Library of Congress Cataloging in Publication Data
A catalog record for this book is available from the Library of Congress.

British Library Cataloguing in Publication Data
Structural geology and personal computers. – (Computer
methods in the geosciences; v. 15)
1. Geology, Structural – Data processing
I. De Paor, Declan G.
551.8'0285

ISBN 0 08 042430 9 (cased version)
ISBN 0 08 043110 0 (flexi version)

Printed in Great Britain by BPC Wheatons Ltd, Exeter

Series Editor's Foreword

Structural Geology has been revolutionized by the computer, as have all subdisciplines in the earth sciences. Early uses in the 1960s included statistical analysis and the handling of routine, time-consuming chores of the geologist such as the graphic display of β- and π-diagrams, stereonets, *etc.* (Whitten 1969) . Later in the 1970s multivariate statistical techniques were applied to the mapping and analysis of spatial variability of structural elements (Whitten 1981). Such techniques as trend analysis, Fourier analysis, and segmentation and partitioning were applied to structural data. These successful applications were followed in the 1980s by more sophisticated modeling, tectonic analysis, and three dimensional graphic displays. All of these applications went through a series of computer types and sizes from mainframes and minis to work stations, micros, and laptops. Declan De Paor has brought together in this book a series of papers by experts in fields relating to the involvement of PCs in structural geology as of the 1990s.

The book is organized into six parts: I Computer-Aided Learning; II Microstructural Analysis; III Analysis of Orientation Data; IV Strain and Kinematic Analysis; V Mathematical and Physical Modeling; and VI Structural Mapping and GIS. The 34 papers are by 45 authors and cover well the widespread use of PCs by structural geologists.

As the editor says "...This volume presents...some of the new directions and new possibilities that personal computers are opening up for structural geologists." Thus, this book fits well into the series - *Computer Methods in the Geosciences*. It is timely and of considerable interest with the rapid developments taking place in the computer world today. The papers presented here will help in this area, and prove a helpful background for others. May all readers discover something new and of interest to them!

References

Whitten, E. H. T. 1969. Trends in computer applications in structural geology. In: Merriam, D.F. (ed.). *Computer Applications in the Earth Sciences*. Plenum Press, New York, pp. 223-249.

Whitten, E. H. T. 1981. Trends in computer applications in structural geology: 1969-1979. In: Merriam, D.F. (ed.). *Computer Applications in the Earth Sciences: an Update of the 70s*. Plenum Press, New York, pp. 323-368.

Daniel F. Merriam

Preface

Some thirty years ago, structural geology underwent a revolution that fundamentally changed how we think about the deformation of rocks. Regional field observations and laboratory data were given profound new meaning in terms of the global model of lithospheric behavior called "Plate Tectonics". Today structural geologists are witnessing a second revolution and although it is of a very different, less fundamental, type it will clearly have a profound and lasting effect on our field. This new revolution has been fuelled by the widespread use of the personal computer which has become the principal tool of scientists worldwide for data storage or retrieval, teaching, communication with colleagues, number crunching, modeling, and preparation of publications.

The current exponential growth in the power of personal computers is radically changing how we work and how we study. Calculations that were intractable only a few years ago are now within the capabilities of geologists with only basic math skills. Data that would have taken months or years to gather from various sources are retrievable in an instant by electronic means. Fabrics and textures that were previously described only qualitatively can now be quantified with the aid of novel mathematical devices such as fractals (which themselves resulted from the effects of the personal computer revolution on the field of mathematics). Interpretations can be more rigorously tested and results more meaningfully and forcefully presented.

The radical change in work practices brought on by the personal computer revolution is redirecting our research efforts. Computers will ultimately influence what research problems we decide to tackle and how we think about geological structures. Hopefully, the results will include both a higher standard of research work, as new insights are obtained through new means of visualization, and also a greater democratization of research and education, as the wisdom of scientists working in the world's best-endowed universities is transmitted to the desktops of brilliant students who are unfortunate enough to be located in the world's most disadvantaged regions.

This volume presents just some of the new directions and new possibilities that personal computers are opening up for structural geologists. An attempt was made to include material of interest to a wide range of geologists, not just those who already know a great deal about computers. In addition to articles involving sophisticated modelling programs such as *AutoCAD*™ or *C*++ compilers, there are articles for those whose programming skills are limited to *HyperCard*™ or who still use *DOS*™. The aim is to present a view of the current state of personal computer use throughout structural geology, not just the "leading edge". Initially, a two-part book was planned, with part 1 written by the editor and part 2 by the contributing authors. However, the range of subjects submitted was such that it seemed more appropriate to distribute the editor's contributions through the book; hence the unusually large number of such contributions!

Camera-ready copy for this book was prepared by the editor on a Power Macintosh 9500/132 running *System* 7.5.2 with 16 MB RAM and a 2GB hard drive. Authors were given a choice of English or American usage. Text and images were submitted by contributors either by e-mail, on diskette, or on paper. The latter were scanned using *Read-It*™ optical character recognition and *Photoshop*™ 3.0 image access software on a LaCie Silverscanner III. Output was printed at a resolution of 600 dpi on 24lb. high-gloss bond with an Apple LaserWriter 4/600 PS laser printer and an Apple Color StyleWriter 2400 inkjet printer. Page proofs were prepared using *Nisus*™ 4.0 for word processor, *Pagemaker*™ 6.0 for page layout, *Canvas*™ 3.5 for drawing, and *Expressionist*™ 2.0 for mathematical typesetting. Some graphics were incorporated from hard copy by camera. Interestingly, no products manufactured by Microsoft Inc. were required! The principal fonts used were New Century Schoolbook and Times for the main text, Symbol for mathematical expressions, and Courier for computer code and command words. Names of items of computer hardware are capitalized; software product names are set in italics. All trademarks are the property of their respective owners.

The editor now understands why authors and editors commonly thank their spouses for not divorcing them during book preparation. In addition to lending moral support, Carol Simpson did more than her share of reviewing and proof-reading. I am grateful to her and to all the contributors and reviewers who made this publication possible. As a condition for consideration of their own manuscripts, all contributing authors agreed to serve as reviewers and most carried out this commitment punctually. In addition, I am very grateful to Bob Burger, Mike Ellis, Eric Erslev, Dave Gray, Laura Goodwin, Mary Beth Gray, Renée Panozzo Heilbronner, Rick Law, Stephen Marshak, Ken McCaffrey, Win Means, Dan Merriam, Katsuyoshi Michibayashi, Brendan Murphy, Paul Ryan, Dave Sanderson, and Robert Twiss, some of whom reviewed more than one manuscript. Julie Bartley compiled the subject index. Thanks are also due to Peter Henn of Elsevier Science Ltd. for his encouragement and patience. Grant support from the National Science Foundation (EAR-9304879) is gratefully acknowledged.

Contents

I: COMPUTER-AIDED LEARNING

A Computer Laboratory for Structural Geologists 3
Declan G. De Paor

GeologiCAL Structures – Multimedia Presentation and Modelling Software 13
Patrick R. James and Ian Clark

Courseware: Rock Deformation and Geological Structures 39
Dave Byron and Bill Sowerbutts

A Structural Study of the North Sea Petroleum Traps Using HyperCard 43
Iain Allison

Visualization of Basic Structural Geometries with Structure Lab 1 51
Declan G. De Paor and Carol Simpson

Using Graphics Programs to Help Students Understand Strain 57
Barbara J. Tewksbury

Visualization of Deformation: Computer Applications for Teaching 75
Basil Tikoff and Haakon Fossen.

Computer-aided Understanding of Deformation Microstructures 97
Carol Simpson and Declan G. De Paor

II: MICROSTRUCTURAL ANALYSIS

Image Analysis in Structural Geology Using NIH Image 105
Marcia G. Bjørnerud and Brian Boyer

Synkinematic Microscopic Analysis Using NIH Image 123
Youngdo Park

Image Analysis of Microstructures in Natural and Experimental Samples 135
Paul Bons and Mark W. Jessell

Calculation of Rock Properties from Pole Figures Using LabView 167
Johann Lapierre, David Mainprice, and Walid Ben Ismail

III: ANALYSIS OF ORIENTATION DATA

SpheriCAD: An AutoCAD Program for Analysis of Structural Orientation Data 181
Carl E. Jacobson

A Computer Program to Print Inclined Spherical Projections 195
John Starkey

Presentation of Orientation Data in Spherical Projection 203
John Starkey

Microcomputers and the Optical Universal Stage 217
John Starkey

Stereonet Applications for Windows and Macintosh 233
Declan G. De Paor

Manipulation of Orientation Data Using Spreadsheet Software 237
Gustavo Tolson and Francisco Correa-Mora

IV: STRAIN AND KINEMATIC ANALYSIS

Modeling Growth and Rotation of Porphyroblasts and Inclusion Trails 247
Eric Beam

Simulated Pressure Fringes, Vorticity, and Progressive Deformation 259
Kyuichi Kanagawa

Flinn Diagram Construction on Macintosh Computers 285
Jay Zimmerman

A Modified Data Input Procedure for the Fry 5.8 Strain Analysis Application 293
Jay Zimmerman

V: MATHEMATICAL AND PHYSICAL MODELING

Review of Theorist: a Symbolic Mathematics and Graphics Application 299
Andy R. Bobyarchick

Structural Geophysics: Integrated Structural and Geophysical Modelling 303
Mark W. Jessell and R. K. Valenta

Principal Stress Orientations from Faults: a C++ Program 325
Bruno Ciscato

A Spring-Network Model of Fault-System Evolution 343
Norihiro Nakamura, Kenshiro Otsuki, and Hiroyuki Nagahama

Linear-Elastic Crack Models of Jointing and Faulting 359
Juliet G. Crider, Michele Cooke, Emanuel J. M.Willemse, and J. Ramón Arrowsmith

Bézier Curves and Geological Design 389
Declan G. De Paor

VI: STRUCTURAL MAPPING AND GIS

Digital Terrain Models and the Visualization of Structural Geology 421
Robert N. Spark and Paul F. Williams

Computation of Orientations for GIS – the 'Roll' of Quaternions 447
Declan G. De Paor

Computerized Geologic Map Compilation 457
Mark G. Adams, Laura D. Mallard, Charles H. Trupe, and Kevin G. Stewart

Fieldlog: GIS Software as a Mapping Aid for Structural Geologists 471
Mohamed I. Matsah and Timothy Kusky

Computerized Cross Section Balance and Restoration 477
Richard H. Groshong Jr. and Jean-Luc Epard

Bitmap Rotation, Raster Shear, and Block Diagram Construction 499
Declan G. De Paor

Subject Index 513

Contributing Authors

Mark G. Adams
Department of Geology,
University of North Carolina,
Chapel Hill NC 27599-3315,
U.S.A.
mga4470@email.unc.edu

Iain Allison
Department of Geology and
Applied Geology,
University of Glasgow,
Glasgow G12 8QQ,
Scotland.
iallison@geology.gla.ac.uk

Ramón Arrowsmith
Department of Geological and
Environmental Sciences,
Stanford University,
Stanford CA 94305-2115,
U.S.A.
crider@pangea.stanford.edu

Eric C. Beam
Department of Geological Sciences,
University of Texas at Austin,
Austin TX 78712,
U.S.A.
eric@maestro.geo.utexas.edu

(Current address:
Exxon,
P.O. Box 4778,
Houston TX 77210-4778,
U.S.A.)

Marcia G. Bjørnerud
Geology Department,
Lawrence University,
Appleton WI 54912, U.S.A.
Marcia.Bjornerud@Lawrence.edu

Andy R. Bobyarchick
Department of Geography and
Earth Sciences, University of
North Carolina at Charlotte,
Charlotte NC 28223, U.S.A.
fgg00arb@email.uncc.edu
http://anb-mac.uncc.edu

Paul Bons
Victorian Institute of Earth and
Planetary Sciences,
Monash University,
Clayton, Victoria 3168,
Australia.
Paul@artemis.earth.monash.edu.au

Brian Boyer
Geology Department,
Miami University,
Oxford Ohio 45056,
U.S.A.

Dave Byron
UKES Courseware Consortium,
Department of Earth Sciences,
University of Manchester,
Manchester M13 9PL,
U.K.
ukescc@man.ac.uk

Bruno Ciscato
Dipartimento di Scienze della Terra,
Via La Pira 4,
50127 Firenze,
Italy.

Ian Clark
Centre for Environmental and
Recreation Management, University
of South Australia, Smith Road,
Salisbury East 5109, Australia.

Juliet G. Crider
Department of Geological and
Environmental Sciences,
Stanford University,
Stanford CA 94305-2115,
U.S.A.
crider@pangea.stanford.edu

Francisco Correa-Mora
Instituto de Geofísica,
UNAM,
Ciudad Universitaria,
México D.F. 04510.
pancho@tonatiuh.igeofcu.unam.mx

Michele L. Cooke
Department of Geological and Environmental Sciences,
Stanford University,
Stanford CA 94305-2115,
U.S.A.
crider@pangea.stanford.edu

Declan G. De Paor
Department of Earth and Planetary Sciences,
Harvard University,
20 Oxford Street,
Cambridge MA 02138,
U.S.A. EarthnMail@aol.com,
depaor@eps.harvard.edu

Haakon Fossen
Department of Geology,
University of Bergen,
N-5007 Bergen,
Norway.

Jean-Luc Epard
Institut de Géologie,
Université de Lausanne BFSH2,
CH-1015 Lausanne,
Switzerland.

Richard H. Groshong, Jr.
Department of Geology,
University of Alabama Box 870338,
Tuscaloosa AL 35487-0338,
U.S.A.
rgroshon@wgs.geo.ua.edu

Walid Ben Ismail
Laboratoire de Tectonophysique,
Université Montpellier II,
Place E. Bataillon,
34095 Montpellier cédex 05,
France.
walid@dstu.univ-montp2.fr

Carl. E. Jacobson
Department of Geological and Atmospheric Sciences,
Iowa State University,
Ames IA 50011-3210,
U.S.A.
cejac@iastate.edu

Patrick R. James
Department of Geology and Geophysics,
University of Adelaide, Box 498,
Adelaide, South Australia 5005.
pjames@geology.adelaide.edu.au

Mark W. Jessell
Victorian Institute of Earth and Planetary Sciences,
Monash University,
Clayton, Victoria 3168,
Australia.
mark@artemis.earth.monash.edu.au

Kyuichi Kanagawa
Department of Earth Sciences,
Chiba University,
Chiba 263,
Japan.
kyu@earth.s.chiba-u.ac.jp

Timothy Kusky
Department of Earth Sciences,
Boston University,
675 Commonwealth Avenue,
Boston MA 02215,
U.S.A.
kusky@bu.edu

Johann Lapierre
Laboratoire de Tectonophysique,
Université Montpellier II,
Place E. Bataillon,
34095 Montpellier cédex 05,
France.
johann@dstu.univ-montp2.fr

David Mainprice
Laboratoire de Tectonophysique,
Université Montpellier II,
Place E. Bataillon,
34095 Montpellier cédex 05,
France.
david@dstu.univ-montp2.fr

Laura D. Mallard
Department of Geology,
University of North Carolina,
Chapel Hill NC 27599-3315,
U.S.A.

Mohamed I. Matsah
Department of Earth Sciences,
Boston University,
675 Commonwealth Avenue,
Boston MA 02215,
U.S.A.

Hiroyuki Nagahama
Institute of Geology and
Paleontology,
Graduate School of Sciencc,
Tohoku University,
Sendai 980-77,
Japan

Norihiro Nakamura
Institute of Geology and
Paleontology,
Graduate School of Science,
Tohoku University,
Sendai 980-77,
Japan
nakamura@dges.tohoku.ac.jp

Kenshiro Otsuki,
Institute of Geology and
Paleontology,
Graduate School of Science,
Tohoku University,
Sendai 980-77,
Japan

Youngdo Park
Department of Geological Sciences,
University at Albany,
Albany NY 12222,
U.S.A.
ydpark@kuccnx.korea.ac.kr

Carol Simpson
Department of Earth Sciences,
Boston University,
675 Commonwealth Avenue,
Boston MA 02215,
U.S.A.
csimpson@bu.edu

Bill Sowerbutts
UKES Courseware Consortium,
Department of Earth Sciences,
University of Manchester,
Manchester M13 9PL,
U.K.
ukescc@man.ac.uk

Robert N. Spark
Institute of Geological and Nuclear
Sciences,
Private Bag 1930,
Dunedin,
New Zealand.
rspark@gns.cri.nz

John Starkey
Department of Earth Sciences,
University of Western Ontario,
London, Ontario N6A 3B7,
Canada.
jstarkey@julian.uwo.ca

Kevin G. Stewart
Department of Geology,
University of North Carolina,
Chapel Hill NC 27599-3315,
U.S.A.

Barbara J. Tewksbury
Department of Geology,
Hamilton College,
Clinton NY 13323,
U.S.A.
btewksbu@hamilton.edu

Basil Tikoff
Department of Geology and
Geophysics,
University of Minnesota,
Minneapolis MN 55455,
U.S.A.
teyssier@vx.cis.umn.edu

Gustavo Tolson
Instituto de Geología,
UNAM,
Ciudad Universitaria,
México D.F. 04510.
tolson@servidor.dgsca.unam.mx

Charles H. Trupe
Department of Geology,
University of North Carolina,
Chapel Hill NC 27599-3315, U.S.A.

Rick K.Valenta
Australian Geodynamics Cooperative
Research Centre,
Monash University,
Clayton,
Victoria, 3168,
Australia.

J. M. Willemse
Department of Geological and
Environmental Sciences,
Stanford University,
Stanford CA 94305-2115,
U.S.A.

Paul F. Williams
Department of Geology,
University of New Brunswick,
Fredericton NB,
Canada E3B 5A3.
pfw@unb.ca

Jay Zimmerman
Department of Geology,
Southern Illinois University at
Carbondale,
Carbondale IL 62901-4324,
U.S.A.
zimmerman@geo.siu.edu

I: COMPUTER-AIDED LEARNING

A Computer Laboratory for Structural Geologists

Declan G. De Paor
Department of Earth and Planetary Sciences,
Harvard University, 20 Oxford Street, Cambridge
MA 02138, U.S.A. depaor@eps.harvard.edu

Abstract– The ability to visualize and analyze complex natural patterns and processes is essential for a proper understanding of geologic structures. Personal computers can help students improve their understanding by aiding visual presentations, process simulations, and complex calculations. However, most academic institutions lack the funds necessary to set up state-of-the art facilities geared to the needs of structure students. Nevertheless, great improvements in facilities may be achieved relatively cheaply by careful selection of hardware and software that is considered obsolete by those in pursuit of the current leading edge. This contribution offers advice on the design of computer laboratory facilities for the impoverished professor.

Introduction

Most academic structural geologists would like to modernize their laboratory teaching facilities by purchasing an array of personal computers and peripheral devices but many simply do not have the financial resources to buy lots of equipment for a class that traditionally required only a minuscule budget to cover the costs of pencils and tracing paper. To them, my advice is this: buy obsolete equipment! There is absolutely no need to spend thousands of dollars on the latest PentiumPro PC or PowerTower Macintosh for undergraduate structure laboratory use. Apart from the fact that students will quickly trash such equipment, they won't necessarily learn more using faster, more sophisticated machines. In fact, the fancier computer may be a distraction; students may waste time investigating its new features instead of concentrating on their assigned task. Rather than opting to buy one top-end computer for, say, $6,000 in the U.S., you could get ten perfectly good second hand computers for $600 each (or equivalent sums in other currencies). You may recall that the processor aboard the Voyager mission to the outer planets had only the power of an Apple][computer, yet it probably provided more benefit to planetary science than any Cray supercomputer. I would not recommend that you buy an IBM 286 PC or the Mac Plus because software compiled using new systems may not run at all on such equipment and their built-in floppy diskette drives may not be able to read modern, high density diskettes. But even an IBM 386 or a Macintosh II will run a lot of teaching software perfectly adequately.

The secondhand equipment advertised for sale in trade magazines such as PC World or MacWorld is commonly supported by a warranty. Most often, these used computers were discarded because owners decided to buy bigger, faster computers, and not because of any operational defects. Right now, a Quadra or an IBM 486 is probably the optimal choice because such machines are being sold in large numbers by purchasers of PowerMacs and Pentium Pros. By deciding to purchase such obsolete equipment, you solve the problem faced by those who try to get all the latest features – every time they feel ready to purchase, they hear a rumor of a bigger, better machine in the pipeline and so are unable to commit themselves to whatever is already available!

Providing a basic computer for each student or between each two students in your class will be much more effective than gathering a dozen students around to watch passively as you demonstrate software on a single computer no matter how powerful the latter may be. If students do have to share computers, be aware that some males will tend to hog the keyboard while females watch over their shoulders, unless you intervene. If it is at all possible, try to provide each student with an individual computer, even if the power of each CPU is thereby lessened. Most of what you do today on your latest computer could have been done five years ago, albeit more slowly and with fewer fancy graphics. Microcomputers have never been fully exploited in academic use; their current level of operation is analogous to the use of Formula-1 racing cars as school buses! So concentrate on quantity first; you can always upgrade later.

If you follow the above advice, you'll be spending less than $1,000 per unit in the U.S. on your hardware. At that price, some sophisticated software packages are going to cost more than the hardware they run on. So it simply doesn't make sense to confine your laboratory to all PCs or all Macintoshes and thus restrict your software purchases. I've heard quite extraordinary reasons for restricting purchases to one system or the other. One academic told me that their geologists all used PCs (or was it Macs?) because the geographers with whom they shared a building all used the other type, and geographers weren't to be considered real scientists. Another said their University President allowed only PCs on campus so that every computer could talk to every other one - this despite the fact that PCs and Macs can exchange data over a network much more easily than Geology and German professors can in oral conversation. Ironically, the PCs in the structure laboratory in question weren't linked to each other or to any outside network. A third colleague didn't want to buy a Macintosh because they heard that the Apple Corporation would be out of business by the end of the year. The year in question was 1992. Even if Apple were to fold this week, Macintosh computers would be useful for years to come. In fact, you will probably change your current Macintosh sooner if Apple stays in business than you will if it doesn't. Any potential buyer of the Apple Corporation would doubtless seek to profit from Mac customer loyalty, especially in the area of education.

Some people also argue illogically against the purchase of IBM-compatibles. I've heard many Macintosh enthusiasts argue that the latest Windows operating system is still inferior to the Macintosh's last operating system. This is hardly the point, however. There was an case for choosing only Macintoshes when the alternative was *DOS*™, but even the most loyal fans of the Macintosh must admit that with the *Windows 95*™ operating system, IBM PC applications are almost as easy to run as Mac ones. Students shouldn't be confined to their professors' choice of best operating system; rather, they need exposure to the greatest variety of systems possible. Thus, the arguments commonly used against purchase of either Macs or IBM PCs are of questionable validity.

Hopefully, in the not-too-distant future, all personal computers, like automobiles, will share a common operating system (thankfully, when driving a car, one doesn't have to think: "...this is a Volkswagon, therefore the brake is the center pedal..."). In the meantime, don't limit your students' horizons by denying them access to one sector of available teaching materials. There is little enough available even when both main operating systems are considered. Indeed, given sufficient funds, you should consider adding a secondhand Sun Workstation, or equivalent, to your collection of PCs and Macs, if only to expose your students to the type of equipment they will likely encounter in their professional lives.

Hardware

After you have obtained a set of CPUs, the effectiveness of your computer laboratory as a teaching and research facility will be greatly enhanced by the addition of peripheral input / output devices. The list of such devices is surprisingly long and can consume a sizeable portion of your budget, so shop carefully.

Floppy Diskette Drives

Obviously, a basic necessity. When buying older equipment, consider the capacity of the built-in diskette drives. Can they read high density diskettes?

Hard Disk Drives

Another basic necessity; don't even think about buying a used computer without a hard drive. Most programs simply cannot run off floppy diskettes and those that can will run exceedingly slowly. Don't spend too much money on capacity as your students will fill any amount of hard disk space with garbage in no time. Just get enough to store the system and the software you intend to use (remember that, in general, IBM PC systems require larger hard disks than Macintoshes) and consider buying an external drive so that you can prepare your teaching materials on your own computer and transfer them for class use without having to duplicate lots of diskettes. If you are networked, such file transfer can be achieved using software, of course.

Optical Drives

With their removable cartridges, optical drives are handy for protecting disk contents between laboratory teaching sessions. However, most students won't buy their own cartridges and therefore machines equipped with optical drives may be under-utilized between class times unless they also have adequate hard drives.

CD ROM Drives

At least a couple of CD ROM drives per laboratory are essential. Already, there are a number of CDs geared towards the structural geologist and more are sure to follow in the near future. Expect to find CD ROM discs attached to standard textbooks in coming years. If CD drives are not built into at least some of the computers in your laboratory, consider purchasing cheap external drives (current U.S. prices are as low as $200). As with CPUs, there's no need to buy the latest, hex-speed drive on the market. What harm if students have to wait 10-15 seconds for data to load? These seconds are certainly not the rate-determining step in their acquisition of knowledge!

If you do have cash to spare, you may wish to consider a facility to manufacture your own CD ROMs. You will need the CD mastering hardware, a 1-2 Gigabyte hard drive and two computers - a PC and a Macintosh - if you want your discs to be readable on either platform. It is very important to get a drive that can transfer information to the mastering software sufficiently quickly. When mastering a CD, interruptions in data flow are a major problem. Talk to an agent before committing funds.

Scanners, Digitizers, and Whiteboards

A scanner is a near-essential device for inputting graphical data in pixel (bitmap) format. Most student laboratory needs are well served by a black & white scanner which can be purchased very cheaply these days. Color scanners are also becoming quite inexpensive, but they generate very large data files requiring lots of hard disk space. Also, large format scanners are still high-price items and even the largest can't handle an average-sized geological map in one scan. Therefore, you should consider purchasing a large format digitizing tablet in addition to a scanner. A digitizing tablet captures data in vector rather than raster mode and is therefore suited to many mapping applications but not to texture or fabric analysis.

If I had to choose between a scanner and a digitizer, I'd choose the scanner first. It can be made to do everything a digitizer does (you can always vectorize scanned data using a drawing program) but the opposite is not true - there are some images a digitizer just cannot capture. Try digitizing a photomicrograph of crenulation cleavage!

A novel input and display device is the Softboard™, a whiteboard on which you, the teacher, write with special markers and which transmits your scribblings to each student's desktop. Students may draw over your notes and save them in a graphics file format.

Microphones

A microphone can be useful, for example, for adding your own voice commands to hypertext course materials. Students may respond better to the sound of their own professor's voice than to robotic instructions. However, do not leave microphones attached to computers in your lab unless you wish to hear "feedback" from your students in no uncertain terms!

Video

Any video camera can be hooked up to your personal computer using a video card installed in one of the computer's extension slots. Some computers have video input/output built in, which should be a consideration at time of purchase if you plan to use video teaching or research materials. The advantage of inputting video to the computer rather than sending it directly to a monitor using a videotape player is that each student may proceed at their own pace, backtracking or fast-forwarding as they wish and linking video images to other data sources. Also, you may show video recordings of experimental deformation rigs that would be too tedious (or dangerous) to set up and run during class time, or that are located in other universities - even on other continents. Videos of recent field trips may be linked to textbook information about the structures seen in the field in order to reinforce the learning process. Microscope attachments enable you to view rock analog microstructure development in real time, or to demonstrate microstructures such as undulose extinction which require a rotation of the microscope stage. Electronic journals such as the American Geophysical Union's *Earth Interactions* will accept animations and visualizations for publication. Video output to VHS tape may be useful if you wish to distribute the results of numerical models or simulations to users who may not have the necessary equipment for digital viewing.

Fortunately, there is growing compatibility between the IBM and Macintosh platforms; newer Macs use PCI connections that are standard in the PC world and the latest PCs use Apple's *QuickTime*™ technology. Of course, this means that PC owners benefit from the ease of use of *QuickTime* whereas Macintosh users must bear with the frequent incompatibilities associated with PCI technology! Video-editing certainly requires a modern computer such as a PowerMac or Pentium Pro but the final, edited product can often be viewed on lesser computers. When selecting hardware, remember that the majority of geological applications do not require fast image grabbing (in fact, the problem may be that frames cannot be captured sufficiently infrequently). Don't be talked into buying expensive equipment whose only merit is its ability to grab and store dozens of frames per second unless you need to analyze neotectonic movements.

Monitors and LCDs

For effective teaching, it is important that your class can see what you do on your monitor. Do not try to gather more than three or four students around a standard monitor. Any more and the runt of the litter will starve, intellectually speaking! *LCDs* (panels that sit on an overhead projector) have fallen in price dramatically in recent years. It is now possible to purchase a color LCD in the U.S. for about $1,000-$2,000 (plus $500 for an extra-bright overhead projector). However, image quality varies widely and is not always correlated with price. Be sure that the system you buy allows you to duplicate the contents of the main monitor. You do not want to be groping in the dark with a system where the menubar appears only on the display behind your back. Also check that the computer you intend to use has a slot available for the LCD control card.

Direct projection devices are certainly preferable in terms of image quality but they are sometimes built in as a fixture in one classroom and they are more expensive (currently $5,000-$6,000 minimum in the U.S.). The monitors you buy for student computers should display 256 colors or more; it is just not worth cutting costs in this respect as black and white monitors are not that much cheaper.

Keyboard and Mouse

A keyboard and mouse for each computer are taken for granted. Budget for repairs and replacements of broken or stolen items in each case. Mice that reside in computer labs have a half-life about equal to that of their organic counterparts. Inspect mice frequently for a buildup of dirt on the rollers. Students will quickly tire of the task of selecting text or clicking on buttons if the mouse refuses to move as directed. You can make an interesting two dimensional input device by disassembling a dead mouse and attaching its rotation sensors to gear wheels using Technical Lego™ or Mechano™ parts.

Fancy mice (trackballs or optical sensors) are probably a waste of money in the classroom setting but if you plan to lecture directly from your own computer, consider getting a gyroscopic mouse. It will make mouse operations in real-time demonstrations much easier as you'll be able to control the cursor on a big screen while walking about in the traditional lecturer's fashion. Mice are available that are operated by foot-pedals; they may be useful if you are using an on-line microscope, for example. If your class includes a severely physically disabled person, check whether your university will pay for the latest brain-wave or ocular I/O controls. These will allow the student to control the computer by staring at the appropriate part of the monitor.

Joy Sticks, Light Pens, Touch Screens, and Foot Pedals

These are rarely if ever needed. Joy sticks should be avoided as they only invite students to play computer games. Touch screens are more often seen in museums and restaurants than on personal computers and light pens are becoming a rarity. I've never come across a structurally oriented program that required them. Foot pedals are useful for computer-aided microscopy because the user's hands are occupied with the microscope stage and users will not want to take their eyes from the microscope in order to locate a mouse button. For example, two foot pedals are used to record optical orientations in the author's application *U4* , a device driver for Futron™ digital universal stage hardware designed by E. de Graaf at Utrecht.

Control Cards

Most personal computers are equipped with expansion slots into which one may plug a control card for real-time experimental control. Examples include motion control of motors that drive a deformation rig or temperature control of rock analog materials. If you buy a new computer and plan to include computer controlled experimental deformation in your course, make sure that your chosen computer has PCI expansion slots. This will ensure maximum compatibility with control cards manufactured both for the IBM PC and Macintosh (PowerWave Mac clones have both Nubus™ and PCI slots). Leading producers of control cards in the U.S. include National Instruments Inc. and NuLogic Inc. Not all real-time motion control requires an internal card in your computer. Sometimes data is transmitted from an external control box to the computer (PC or Macintosh) via its RS-232-C serial port.

Printers

If you're waiting for the long-promised arrival of the paperless office, don't hold your breath! Hard copy output is an on-going necessity of life. Even students who are committed to saving the Earth will work their way through an immense amount of paper and toner if allowed. It would probably be wise to provide an old, slow, obsolete printer for general use and arrange for laser or color printer output by special request. If you still have money left over after all other purchases are made, a color laser printer will quickly solve your solvency. If this is not your problem, consider either inkjet or wax transfer. The choice is a matter of personal taste so go to your local computer store and ask for a demonstration.

Ethernet Cards, Modems, and Networks

You may wish to provide internet and World Wide Web access to your students via a modem or hardwired connection. However, you should plan your internet exercises carefully; cruising the web is a great time sink and it is difficult to monitor students' progress in the classroom setting. The best approach is to set your students an assignment to gather research information from on-line sources such as government agencies and professional companies that you have already located prior to class-time. Try to choose a classroom that is already hardwired for internet access; modems can be used to dial up university services but even their highest transmission rates feel relatively slow because of the amount of graphical data in the latest web pages. High-speed PIP connections will be costly. Internet access via commercial vendors such as *Compuserve* or *MyCountry-on-Line* (*America-On-Line*, *Ireland-On-Line*, *etc*.) are generally extremely expensive, and service can be painstakingly slow (it is not for naught that *America-on-Line* is nicknamed America-on-Hold) so check who is paying the phone bill! Currently, the best web access software is undoubtedly *Netscape*.

It is perfectly OK to assemble a room full of computers with either no local area network or with a simple link to shared printers and/or digitizers. However, a much more desirable arrangement, given the necessary funds, is to set up a LAN including a control computer, ideally located in a separate control room to which students do not have open access. The control computer acts as a file server and network manager. Software such as *Timbuktu*™ will permit you to transfer files among various campus locations without any hassles. Students can submit reports and projects using a "drop-box" - a folder which accepts file transfers but does not allow viewing or editing of its contents except by the instructor.

Laptops and GPS Devices

All field structural geologists need to learn how to use a laptop or notepad computer as a field notebook and how to transfer data from their laptop to a remote university computer. You should plan to bring some laptops on a field trip and set up a modem connection from a hotel room or airport office facility in order to demonstrate to your students the kind of work practice they will encounter in industry. Students also need to be familiar with the workings of a global positioning system. Hand-held GPS devices are currently available in the U.S. for less than $200. However, you'll pay over $1,000 for a high-end model such as the Magellan FieldPRO V™ that links directly into a personal computer via a 9-pin connector.

Software

This topic could fill the entire book on its own. Obviously, your choice of hardware and operating system will determine the specific software you need to buy. However some generalities can be stated.

First, it is necessary to consider the legal aspects of licensing software for a set of computers. The most common shady practice is for one person out of ten or twenty users in a department to buy a single user license in order to qualify for upgrades and technical support. Apart from being unethical, this practice may get you into trouble. In the U.S., an agency called the Software Protection Association regularly swoops on companies in breach of Federal law regarding software licenses. Universities have not come under quite the same pressure as corporations to date but many Deans nowadays take a dim view of software piracy by faculty or students.

Despite reductions for bulk purchases, multiple licenses can be prohibitively expensive. So the first thing you should do is to check with your institution's computer services or purchasing department to determine what site licenses are already freely available to you. Secondly, you should buy software from corporations that offer academic discounts. Many expensive products such as *AutoCAD* or *ArcInfo*™ are sold much, much cheaper to academics (on the assumption that student users will go on to purchase software with which they are familiar when they get a job). Thirdly, don't purchase unnecessarily sophisticated software for student use. The word processor on classroom computers need not be *MS Word*™; shareware products featuring styled text and in-line graphics are available for as little as $20. Between these end-members, you will find a word processor called *Nisus*™ for the Macintosh, for example, which costs significantly less money, occupies considerably less space, and works a lot better than *Word* (it also inputs and outputs *Word* file format, among others). Fourthly, note that whilst obtaining duplicate programs from other users is considered piracy, it is perfectly legal for you to purchase licenses, manuals, and diskettes second hand, provided the seller does not keep a copy of the software. If you ask around, you may find that colleagues have slightly dated software packages that they don't use anymore, having moved on to bigger and better products. Fifthly, if you are buying hardware and software at the same time, try to find a company that 'bundles' the latter with the former. (For example, I got my copy of *Photoshop*™ free with my scanner).

You should plan to place a spreadsheet program on each computer. Though most people associate spreadsheets with economics applications, they are also extremely useful for solving scientific equations. After the word processor and spreadsheet, the next essential for each CPU is a drawing application. Again, shopping around for the less well known brand names will save you perhaps a third to two thirds of total costs. Just make sure that you can save files in standard formats such as GIF, PICT, EPS, or TIFF. For mapping applications, your drawing program should support layered documents. A number of custom mapping software packages are available, notably *Fieldlog* from the Canadian Geological Survey. Look for a package that includes UTM to Latitude & Longitude conversion. A final essential for every computer is a stereonet program. In my personal evaluation, the top ranking products are *StereoNet* for *Windows* from Geological Software Inc. of Varden 94, N-9018 Tromsø and *StereoPlot* for the Macintosh from Neil Mancktelow of ETH Zentrum, CH-8092 Zürich. Runners-up include *StereoPC* for *Windows* from www.Rockware.com and *Stereonet* for the Macintosh from Rick Allmendinger of Cornell University, New York (the latter is free to academics only).

A significant volume of custom software for structural geologists is now available from commercial, semi-commercial, and academic sources. Noteworthy are Rockware Inc., the UK Software Consortium, and Earth'nWare Inc. Being the founder of the latter corporation, I will not attempt an unbiased comparison. You should also check the faculty-authored software listed in Robert Hatcher's structural geology textbook, that listed by Art Busbey on the internet (busbey@gamma.is.tcu.edu), and the worldwide web page maintained by Ben van der Pluijm (www-personal.umich.edu/~vdpluijm/gsasgtpage.htm). Other useful sources include the Tectonic Studies Group (www.dur.ac.uk/~dg10www/TSG/tsghome.html) and the Canadian Tectonics Group (http://craton-geol.brocku.ca/ctg.html). Whenever software is faculty authored and/or free, be prepared for a relatively rough user interface; you get what you pay for. New products are regularly reviewed in *GeoTimes*, *Terra*, *Journal of Geoscience Education,* and, more rarely, *Journal of Structural Geology*. For in-depth coverage, often including source code, check past issues of *Computers and Geoscience*.

Another category of software you should examine consists of commercial applications that are not specific to structural geology. You might consider loading these onto one of your lab's computers, in which case it makes sense to buy a high-end workstation. Single user licenses for such packages permit use by any number of different individuals provided the software is loaded onto one CPU only. Relevant categories include image editing (*Photoshop*™ is a clear leader in this field), image presentation (*e.g.*, *Premiere*™, *Persuasion*™, and *Portfolio*™), image analysis (if you're a Mac user, check out *NIH Image* which is free from the U.S. National Institutes of Health), GIS (here *ArcInfo* is the standard), Finite Element Analysis (*e.g.*, *Meshmaker*™ for the *MacOS, Windows*, or *Unix* from Argus.gdy@applelink.apple.com), CAD/CAM (*e.g.*, *AutoCAD*™ from Autodesk Inc.), experiment or instrument control (don't think of getting anything other than *LabView*™ – it's amazing!), system analysis (*e.g.*, *Stella*™), mathematics / statistics (*e.g.*, *Mathematica*™, *HiQ*™, *Maple*™, *MathLab*™, *Theorist*™), and landscape rendering and digital terrain modeling (*VistaPro*™, *Scenery Animator*™). Finally, even if you don't have any programming skills yourself, you should consider making a compiler available to your students. Don't automatically invest in a *C* / *C*++ compiler even though it is touted by advisors from the computing center. Students who know how to use *C* will have access to it elsewhere whilst those who don't are unlikely to teach themselves in the breaks between structural laboratory exercises. Instead, I recommend *Visual BASIC*™ for Windows or *FutureBASIC*™ for the Macintosh, both of which are more sophisticated than the name *BASIC* might suggest, yet fairly easily self-taught. Alternatively there are many decent *FORTRAN* and *Pascal* compilers available for both operating systems.

Having collected together even a portion of the above shopping list, the hard drive on your good computer will be pretty full. You can buy software that doubles your storage and/or RAM memory but it sometimes causes compatibility problems with applications. When clearing space on your hard drive, start with the junk that gets loaded automatically when you install an application. You probably don't need your word processor's help file or tutorial after you've gotten used to it (there's always the hard-copy manual). If you need still more space, don't make the mistake of trashing documents but keeping applications (your documents are probably not backed up because you're overworked and underpaid). Instead, throw out your least-used applications. One application will probably free up more disk space than a score of long documents and it can always be reinstalled in a matter of minutes when needed.

If you can't afford the more expensive items mentioned above, at least write to the manufacturers for a demo disk. Many demo versions are only slightly crippled; for example, you may be free to perform all operations but not be able to save or print data. These limitations would prevent use of the software for serious research applications but the demo version may be perfectly adequate for demonstrating modern analytical methods in class. Finally, don't forget to write to the contributing authors in this volume. Most academic authors will gladly distribute their 'wares free of charge.

GeologiCAL Structure – Multimedia Presentation and Modelling Software

Patrick R. James
Department of Geology and Geophysics, University of Adelaide, Box 498
Adelaide, South Australia 5005, Australia. pjames@geology.adelaide.edu.au

Ian Clark
Centre for Environmental and Recreation Management, University of South Australia, Smith Road, Salisbury East 5109, Australia.

Abstract– We have developed Computer Aided Learning (CAL) courseware for use in undergraduate structural geology subjects. The courseware covers the whole structural curriculum, is based on presentation and other interactive multimedia software, and is integrated throughout the lectures, laboratory classes, workshops, and even within the fieldwork programme. With the progressive establishment of a digital teaching and learning environment equipped with fully networked personal computers, mass storage devices, computer projection, and computer suites, we have now developed and exclusively use a fully integrated electronic classroom for our structure courses.

Introduction

We have taught introductory through advanced structural geology courses to undergraduate students using a variety of educational methodologies for the last twenty years. Innovative teaching practices have included the introduction of concepts maps (Clark & James 1993a), detailed field mapping techniques (James & Clark 1993, Clark & James 1993b), and also frequent use of analogs, models, simulations, and demonstrations as well as other techniques to encourage active and collaborative learning (Gibbs *et al.* 1988a,b).

With the advent and rapid proliferation of microcomputers over the last few years we have adapted new educational technologies, developed in a structured curriculum, to further innovate the teaching of our courses using a range of commercially available software in an essentially Macintosh-based environment (James & Clark 1991, James 1994). We started these developments with simple text editors and word processors to produce student notes and printed overhead transparencies in the mid 1980's. Through the early 1990's all lectures were replaced with full colour multimedia presentations. Laboratory exercises included paper-based and computer-based exercises. The courses we teach now comprise multimedia presentations, including digital images and videoclips, animations, interactions, on-line internet data, and a range of other software. The latest developments involve students producing their own interactive multimedia assignments in place of traditional essays and reports. Our innovative change to an electronic classroom environment has been followed by many of our colleagues in other departments. Thus much of the undergraduate curricula with which we are involved is also beginning to utilise similar innovative educational and learning technologies.

The move to develop a fully electronic classroom environment has involved the use of leading-edge technologies, including networked hardware and integrated software, available to us as lecturers (information servers) and to our students (learning clients). Sophisticated, cheap and powerful, but most importantly networked and integrated, microcomputer-based systems form the basis of the multimedia development, storage, and delivery systems which are installed within most of our current teaching areas (academic offices, lecture rooms, and teaching laboratories). Commercially available presentation and authoring software has allowed the development of multimedia delivery packages as teaching aids. The use of the electronic courseware has been combined with consistent pedagogic principles (including clear outlines of course aims and objectives) thorough use of a range of teaching models (Gibbs *et al.* 1988a, Habeshaw *et al.* 1989, Habeshaw & Steeds 1987), defined learning strategies such as concept mapping (Clark & James 1993a), and realistic outcome and assessment criteria (Gibbs *et al.* 1988b).

Electronic courseware has a number of advantages in teaching and learning making it a useful adjunct to traditional lecture and demonstration techniques. These advantages include simple updating of courseware and visual aid (digital slide) manipulation, high quality professional visual presentations, sophisticated three dimensional graphics and animations, interactivity, availability of handout material, and access and rapid review of the courseware by students at any time via floppy disc or the electronic network. This paper describes our efforts to integrate new learning technologies into our courses and hopefully confirms that in spite of the many technological, pedagogical, and financial difficulties involved, these concerns and difficulties need not deter a determined proponent of the electronic classroom and may provide a successful outcome for students and educators.

The Hardware Environment

To create and effectively use digital courseware throughout our courses we have been involved not only in the production of electronic teaching and presentation modules, but also in the development of a very comprehensive hardware system to develop, manage, and deliver the material. This material has been designed and implemented for use by many colleagues as well as for our own courseware production. The computer based electronic teaching and learning environment includes an array of three separate, but linked, areas (Fig. 1). Flexibility is essential as individual academics, students, and electronic network managers enter and leave the environment regularly, while changes and improvements to potential hardware and software outpace all of those personnel movements. The development area is where we and our colleagues (teaching staff), as content

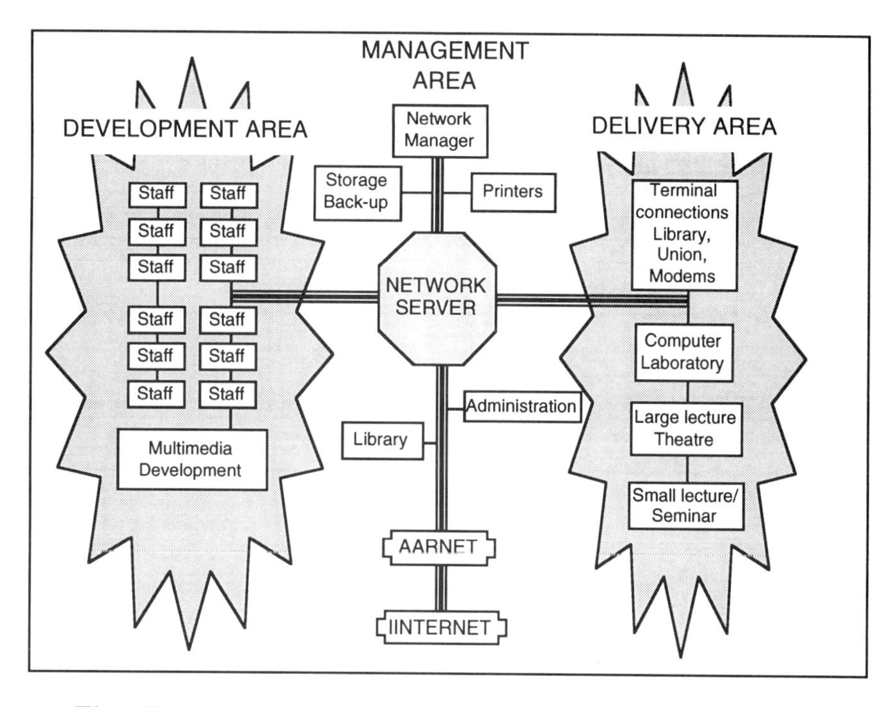

Fig. 1. Essential hardware component areas of the electronic classroom.

specialists, have developed individual courseware modules and integrated information using a variety of input sources and multimedia software. The delivery area is highly variable and includes large lecture theatres and small classrooms, laboratories of networked computers, and, most importantly, a variety of remote and isolated terminal access nodes. The electronic management area provides the links between the development and delivery areas and also links to external and peripheral areas (printers, mass storage, the University Libraries, student information systems, AARNet – Australian Academic Research Network and Internet–, *etc*.). Thus the three essential elements of the electronic environment are: the academic development area, the delivery area, and the management area.

(i) The Development Area.

This area comprises two main components: firstly, small networked personal desktop computers for academic staff, and, secondly, special multimedia hardware units set up in a development laboratory with technical and research assistance. Individual desktop computers are linked to a network which allows remote printing, graphics, and text scanning, local and external email, and file transfer, electronic file sharing and contact with academic colleagues, clerical assistance, and library and student information systems.

Within the development area the major multimedia development facility (Fig. 2) comprises a high-end microcomputer with internal accelerator, video

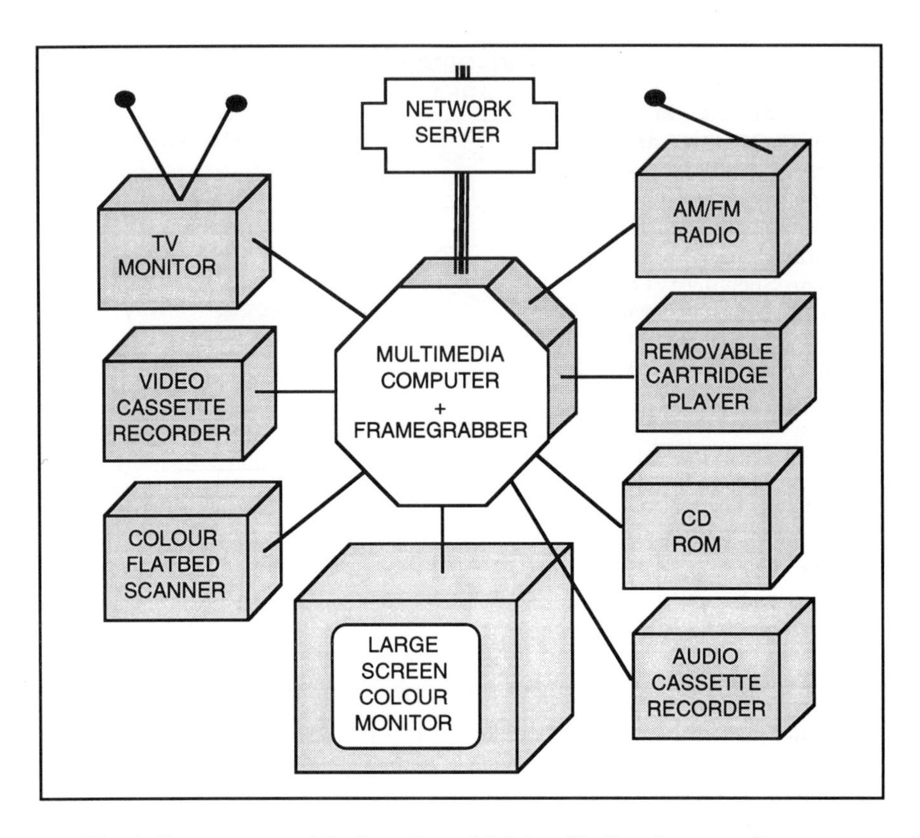

Fig. 2. Components of Geology Dept. Multimedia Development System

frame grabbing card, large internal memory (minimum 20MB RAM), and large internal disc storage (currently PowerMac 8100 valued at about A$7000). This computer has an array of dedicated adjacent peripheral equipment including a large-screen high-resolution colour monitor (A$2000), internal CD-ROM drive, removable cartridge drive for local storage (A$250-300), colour flatbed scanner (A$2000), video cassette recorder (VCR) (A$700), television monitor (TV), AM-FM radio, and audio cassette. The large-screen high-resolution colour monitor is essential for detailed drawing of figures, maps, sketches, *etc.*, for inclusion in the multimedia courseware and for the concurrent running of drawing, word processing, animation, and presentation software, which speeds-up courseware module development and of course necessitates the requirement for large internal machine RAM. The VCR/TV and radio/cassette devices allow the input and capture of video- and audio-clips from proprietary and locally produced media for further integration into the courseware. The flatbed colour scanner and CD-ROM allow capture and integration of further multimedia components (35mm slides and prints from field excursions, textbook illustrations, *etc.*), with appropriate copyright authorisations. Within our institution most of the funding for equipment was derived from special teaching grants, but Departmental, Faculty, and other grant funds were used as much of the equipment also fulfills a research function.

(ii) The Delivery Area

There are a number of hardware requirements for electronic classrooms. Small lecture classes, together with laboratory and tutorial classes, are located in small-class teaching areas which are unlikely to warrant full equipping with multimedia delivery hardware. We have therefore made use of a portable array of computer display and presentation hardware securely fixed to a mobile trolley. The system includes a portable notebook computer attached to a computer presentation panel and high luminosity overhead projector or direct computer projector. Such arrays can be purchased and assembled for between A$10000-12000 and can be timetabled for effective use in a variety of locations though are most effective if used only within one building. An essential planning item in the use of such systems is the early network cabling of all potential small teaching areas, such that large courseware files and external network databases can be readily accessed within these areas as they become part of the electronic learning environment.

Larger classes (50+) in large lecture theatres are controlled by University central administration. Such areas, due to their size and capacity, often warrant the installation of fixed high quality computer and video projection devices. Again, as with small group teaching areas, the network cabling of the large areas is essential.

Timetabling, course options, and course requirements currently require student learning to occur by traditional 50 minute lecture and tutorial timeslots. However, there is an increasing need for students to be able to access course material outside of normal scheduled times. In the recent past this has led to the addition of an increasing number of "computer laboratories" to regular work areas. This has not only been costly but has caused difficulties in finding the extra accommodation. Security and after-hours access are further associated problems. A traditional computer laboratory, with banks or rows of PC's on desks all facing in the same direction like the classrooms of old, can provide an efficient avenue for the delivery of electronic courseware. Such computer suites must obviously be networked to access the courseware together with the usual peripherals (printers, *etc.*).

A more recent addition to the delivery area is that of separate and isolated terminal nodes for accessing computers located in non-traditional areas (libraries, student union buildings, and other departments). Network sockets are located to allow students to plug in their own notebook computers in order to access the courseware through the network. Policy on computer standards, ownership, security, software, *etc.*, must all be considered when these concepts are added to the delivery area. An extension of this last facet is the idea of modem links to the network which would allow the students to access material from their own computers, which they are increasingly likely to possess at home. This has not yet been implemented in our teaching environment.

(iii) The Management Area.

The third area of hardware essential to the introduction of an electronic learning environment is the area of network management. All individual computers, multimedia facilities, and computer suites are linked and accessible through a managed computer network. The base of the network is an array of network servers or processors. These are served by bulk storage devices and further peripherals and must allow simple access to printing facilities, links to AARNet, student information systems, libraries, *etc*.

The Software Environment

We have consistently utilised network or site licences of simple, ready to use, commercially available Macintosh software. The basic software tools for creating material to be incorporated in the multimedia courseware are standard word processing (Microsoft *Word*™), drawing (Aldus *Freehand*™), scanning (*Applescan*™), optical character recognition (OCR - *Omnipage*™), and image manipulation (Adobe *Photoshop*™). The principal presentation mode is Microsoft *Powerpoint*™ for lecture delivery, with more sophisticated animation and interactivity developed using *Authorware Professional*™ and *Macromedia Director*™. For multimedia storage on removable cartridges and file back-up on remote high density servers we use Aldus *Fetch*™. More recent specialist software has included animation (*Elastic Reality*™) and personal low-end interactive multimedia packages (*Hyperstudio*™, *Digital Chisel*™).

Using Presentation Software in Structural Teaching and Learning

The technical/scientific lecture delivery programme described here uses direct projection of computer generated presentation images and is aimed to significantly improve the quality and effectiveness of lectures and other presentations to student classes. This is being carried out by raising the quality of illustrations, by increasing the flexibility of time-usage during the lecture period, and by providing greater student access to material after-hours. The project has been repeatedly tested in our university lecture programmes at all undergraduate and honours levels. When surveyed, students said the new teaching methods significantly enhanced their perception abilities.

The majority of the courseware that we have developed so far is in the form of comprehensive visual illustrations to accompany lectures or laboratory classes and is displayed using *Powerpoint. Powerpoint* is rapidly being recognised as a most significant influence in the improvement of the quality of lectures in many areas of teaching (Burton & Wynn 1994). At the tertiary level, structural geology lectures or other classes usually consist of the presentation of a series of factual statements, frequently illustrated with quantitative data displayed as graphs, tables, or figures. These data are used to erect models of natural/mechanical behaviour and to substantiate reasoned assertions or hypotheses. Three fundamental problems which have consistently hampered the lecturer's delivery of a professional presentation have been the poor quality of display material, the inflexibility of the lecture time-slot, and the student audience's restriction of access to valuable lecture material out-of the normal lecture period.

The quality of the display or illustration of the scientific data has varied historically depending on the currently available technology and the individual whim of the lecturer. Blackboard and chalk – or whiteboard and felt/marker pen – thumbnail sketches, with associated (often illegible) text, have been the most common display method, with illustration added by 35mm transparencies. Overhead projectors have allowed enlargement of written text and figures. Photocopies onto overhead transparencies from reference books and other publications have gradually improved the overall quality of technical presentation material.

It is well known that the optimum concentration of most students during a lecture presentation deteriorates rapidly from early on in a lecture reaching a very low level after about 20 minutes (Gibbs *et al.* 1988, p101). The standard 50 minute lecture, illustrated with poor quality visual images, is thus not an ideal mechanism for the distribution of geological information. As well as improving illustration quality, we have used many other methods to break-up the standard lecture into a number of smaller more digestible components (short question/ answer sessions, active demonstrations, or slide shows/film clips).

For the teaching of structural geology, copious field data are usually collected and presented as 35mm photographic slides, plates, or images which may or may not be embellished or synthesised by technical drawing and annotation. These images may range in scale from remotely sensed satellite scenes of large swathes of the earth's surface to the smallest petrographic image of natural rock microstructures. 35mm slides are expensive to acquire, store, and curate, time consuming to prepare for presentation (which often restricts their use), otherwise not available to students, prone to degradation after prolonged use, inflexible in the order of delivery, and impossible to annotate. However, 35mm slides as an illustration of a field excursion (transporting the field location to the lecture room) are significantly less expensive than the cost of transporting the students to the field area and thus provide a readily accessible and often exciting practical adjunct to theoretical geological presentations. An alternative mechanism to the acquisition, storage, retrieval, and display of field generated images would greatly aid the geology lecturer.

Overhead Projection Transparencies and Presentation Software

The initial development and trial of the programmes that we developed began in 1989 following the purchase of a Mac Plus and the *Powerpoint* presentation and image management program. *Powerpoint* produced professional style overhead projection transparencies (and/or 35mm slides) which could be copied onto transparent sheets from a high-quality encapsulated postscript black-and-white laser printer output. *Powerpoint* also allowed the production and mixing of text titles and bulleted or paragraph text with a range of fonts and formats to produce excellent quality transparencies. In-house graphics ranging from simple figures or diagrams to complex tables and graphs, or imported graphic images, were used to enhance the visual impression of the transparencies. Layout was aided by a significant range of professionally prepared templates provided. The program was cheap, simple to learn, and rapid to use, and was readily accepted by a variety of academic staff and students.

During the 1989 teaching year hard-copy overhead projector transparencies for a series of second and third year undergraduate introductory structural geology lectures were produced. Some of the text for the transparencies was transferred from an earlier DOS system into *Powerpoint* using a file transfer program. However, it proved faster in most cases to retype text and reproduce new transparency images for most presentations. The text facilities allowed the complete range of Macintosh fonts and styles (**bold**, *italic*, underline, *etc.*), while a sophisticated text ruler provided a range of paragraph alignments and graphic bullets. Of further important benefit to the lectures were the Greek symbol fonts which allowed the presentation of complex mathematical algorithms. The graphic drawing tools in the simple *Powerpoint* palette (line, circle, square box, rounded box, and ellipse) allowed the preparation of many unsophisticated figures showing for example the range of structures found in most introductory structural texts (see Figs. 3-38).

For most of the 35-40 50-minute lectures in the program, about 10-15 overhead projector transparencies were sufficient to illustrate the individual structural topics. Most lectures were further illustrated with 35mm field photographs and transparencies photocopied from reference papers. *Powerpoint* transparencies were produced at a cost of about $1.00- $1.20 (including printing, photocopying, and transparency cost). Additional costs included replacement transparencies where errors were found or changes required. The overall inflexibility of the transparency "hard copy" once produced, proved a negative inducement to produce more transparencies or alter those already copied.

Student perception of lectures, as gauged by anonymous student evaluation of teaching surveys, was significantly in favour of the improved quality of transparencies. The transparencies could be coloured for visual effect, altered to provide more in-depth explanations, or further annotated using marker pens, though this effectively reduced the lifetime of each transparency to a single lecture. The need to modify lecture content annually also severely reduced the likely lifetime of each individual transparency. Students were particularly keen on the availability of reduced size photocopies of all of the most pertinent transparencies which could be prepared to a professional standard by the software and distributed for each lecture (Figs. 3-38 provide an example of this type of student handout).

Direct Display of Computer-generated Images

A major new development in our teaching programmes began in 1990 with the acquisition of a Datashow portable computer image projector manufactured by Kodak Inc. Other equipment acquisitions included a faster and more powerful Macintosh SE30 with 4 MB RAM (allowing concomitant multiple software use under Multifinder), a 40 MB hard disk storage, a flat-bed black and white scanner, and a site licence for *Powerpoint*. The inexpensive multiple-copy site licence cost (~A$550/10 copy Ednet pack) provided a significant additional advantage to allow for multiple use by staff for development and by students for delivery via microcomputer teaching suites.

This stage of the programme provided a number of significant improvements to the presentation of lectures. These were mostly found in parts of *Powerpoint*'s slide/transparency lecture manipulation facilities, which could not be previously used. *Powerpoint* contained a sophisticated series of ways to store, arrange, rearrange, and view slides in each individual computer file, which thus became an individual lecture or presentation. Multiple slides could be viewed at a reduced size, in sequential order as either small slide images, or as listed by title. Either way, single images or groups of images could be moved, cut, copied, pasted, or deleted via single instructions or keystrokes.

The program was also designed to display the sequence of images in order via single instructions or as an automatic (time-variable) or manual "slide show". All of these features were ideal for the production of full length (50 minute), fully illustrated lecture presentations. These features were also ideal for alteration of the order or sequence of slides in a presentation with great ease and speed. The full sequence of slide images produced in the computer by the program were delivered in the lecture theatre directly to the lecture screen using the Datashow device and a standard overhead projector. The aim of this stage was the elimination of the preparation and presentation of hard-copy overhead transparencies (thus also eliminating the cost of their production), together with an increase in the flexibility of the lecture illustration system. This occurred with the change to a wholly electronic system of preparation, storage, and display of illustrations.

Each of the computer files containing a series of transparencies produced as single lectures in 1989 was re-edited and developed for direct presentation using the Datashow for the same series of lectures previously prepared in *Powerpoint* in 1990. The particularly advantageous features of the lecture procedure included the following points:

a) Sequential Display of Text and Graphic Images

Bulletted text may often be used to greatest effect if individual bulletted statements are revealed sequentially to illustrate the development of a time variable concept or a set of increasingly more complex statements. Standard lecturing practice is to use an opaque overlay mask (paper or card) as a simple solution while complex annotated figures may be similarly exposed using complex cut-out paper and opaque sticky-tape overlays. *Powerpoint* proved to be an ideal vehicle for this type of display using the multiple slide copy facility. Each single complex slide or figure from the previous array was copied typically 6-10 times (taking 1-2 seconds). The sequence of identical slides was then progressively edited by removing less and less information from each sequential slide. So for example with bulleted text the first slide contained the first bulleted statement, the second contained the first and second statements and so on. A similar process was carried out for figures containing complex and multiple components. It was estimated that turning a single complex image into a sequence of slides each containing gradually more information took from 5-10 minutes. Subsequent versions of *Powerpoint* (versions 3.0 and 4.0) contained a "build" facility, which allowed bulleted text to be sequentially added to each screen, but the process described here for building up complex figures is still one of the main advantages of this direct presentation technique.

b) Kinematic Modelling of Time-dependant Processes (Animation)

Many repeatable quantitative experimental studies in structural geology analyse physical material parameters that change states of matter over time. Such changes are observed and described in a variety of ways illustrated by models, graphical representations, algorithms, *etc*. Typically such procedures are illustrated in lectures with single (or occasionally multiple) images revealing one instance of the process under study, or as graphs showing how different parameters vary with time. Animation as a display technique, especially where the rate of progress may be slowed, halted, reversed, *etc.*, has the ability both to increase audience understanding of how such processes operate in real-time, and to significantly enhance the ability of the lecturer to demonstrate this in a visually enhanced way.

The *Powerpoint* program with direct display of computer slides was used as a crude animator of a variety of structural processes. As an example, the experimental deformation of an elastic-brittle theoretical rock analog (which is a typical behavioural response of rock at low temperatures and pressures) proved to be an excellent illustrator of the process (see Figs. 29-33). Such an empirical deformation process may be demonstrated in a number of ways in a lecture environment. A 2-D figure with a shaded box representing the material, arrows representing deviatoric stresses, and lines representing resultant fractures, is a simple illustration of sophisticated experiments carried out in high-pressure material testing laboratories. The time-dependent response may be indicated by varying the size of arrows (for increased stress), by varying the dimensions of the box (for strain or dilation), and by the presence or absence of fractures. The typical rheological analog to the process is a coiled spring stressed with varying weights and showing variable extension or shortening. Quantitative parametric relationships may be displayed on orthogonal graphical axes of stress versus strain magnitudes. For stress variation, the Mohr Circle representation is used.

All of these figures were prepared as a number of illustrations on a series of slides showing the progressive behavioural change during deformation. Thus a complete visual representation of the experiment was carried out by showing the sequence of about 6-8 slides. This technique was the forerunner for the further development of more sophisticated animations using more powerful software.

c) Modelling of Kinematic Processes

Another time-dependent process important in structural geology is the gross incremental change in mechanical state of rocks during ductile deformation. Such variations are observed as changes of position, shape, size, and orientation, which are quantitatively specified as translations, distortions, dilations, and rotations, respectively (Bjørnerud 1991, Tickoff *et al.* 1993). With the ability to draw simple or complex 3-dimensional models of undeformed rock bodies containing identifiable original features (*e.g.* clasts, fossils, layers, or earlier tectonic structures) using the graphic tools available in *Powerpoint*, the effects of the different types of deformation on such bodies can be illustrated in real-time visual experiments. All of the different styles of deformation mentioned above are beyond the transformation capabilities of *Powerpoint*, but these may alternatively be carried out using other drawing software (*Freehand*, *Superpaint*, *Illustrator*, *etc.*) and transferred as graphic images to *Powerpoint* for display.

d) Electronic File Storage and Lecture Linking

Individual lectures should ideally stand independently as complete presentations. However, in a series of lectures with a uniform theme, later lectures often depend on concepts and data introduced earlier. Remembering to bring along overhead transparencies or slides from earlier lectures often leads to weighty manilla folders and slide boxes ready to burst onto the floor in the lecture theatre just prior to the start of the lecture. The advantages of the electronic storage and retrieval of a whole series of lectures on a single floppy disc or even better accessed across the Local Area Network (LAN) reduces the risk of such embarrassing and inefficient occupational hazards for the lecturer.

Using Apple's *System 7*, it is possible not only to open and interrogate earlier presented lectures at will and at random, but also to access other software (*e.g.*, the *Freehand* program with all of the kinematic transformations just described) for demonstration.

These techniques thus indicate the clear advantages so far utilised in the presentation of text and graphic illustrations in formal lectures. Within each lecture the addition of further slides, including introduction, revision, linking, reference, and summary material, has led to the creation of presentations/ lectures comprising from 50 to 120 or more slides which were each prepared within a few hours. Each of these *Powerpoint* presentations/lectures, together with 35mm field illustrations, other figures, *etc*., comprises a series of lectures which have been delivered to classes of more than 50 undergraduate students each year since 1990.

The main disadvantage of the initial system was the need to transport and set up the Datashow and computer hardware in a lecture theatre for every lecture. This portability allowed the system to be used in whichever room was available (or even transported to other institutions). However, risk of damage in transit or theft was a constant concern. Few lecture rooms yet had permanently installed computer projection facilities, which would reduce such problems. Although clear, the black and white image suffered from the lack of colour for visual impact, and the dullness of the image required severely dimmed lecture room lights (conducive to audience drowsiness but not to note taking).

Over the last few years, the refurbishment of major lecture theatres has included the addition of electronic multimedia presentation systems with full colour computer projection, video, slide, and sound facilities. This, together with the more ready availability of high-quality, high-luminosity portable colour computer projectors, has allowed a significant improvement in the overall visual impact of the courseware. This has also been aided by the continuing introduction of network access within lecture theatres and suites of computers in laboratories available to students, any time.

The Structural Geology Courseware

Since 1989 an array of geological courseware has gradually been produced for all teaching levels but is mainly used for 2nd. and 3rd. year undergraduates. The courseware production and delivery began with a program of electronic lectures in introductory Structural Geology for classes of 30-50 students each (James & Clark 1991). The basic structural geology course (Table 1) now comprises a series of 20 fifty minute lecture periods which includes the use of *Powerpoint* slides, varying from 50 to 135 slides per lecture. Total disc memory used to store these slides is 7.5 MB uncompressed (about 5 * 1.4 MB floppy discs). At this introductory level, the content was developed from a few standard structural geology texts (*e.g.*, Park 1983, Hobbs *et al.* 1976, Means 1976, Davis 1984, McClay 1987) and all lectures contained a list of page numbers and chapter references to the appropriate sections in these texts. All lectures also contained a glossary of 20-30 terms to be introduced and defined in the lecture (a printed list is also given to students during the course). Finally a printed handout book with 6 *Powerpoint* images per page of summaries of illustrations was provided as hard copy to students for all of the lectures.

Basic Structural Geology Lecture Modules

The basic structural geology subject is part of a whole semester subject on structural geology and geophysics, which assumes no background in structural geology. It is generally separated into two approximately equally sized areas, the first half introducing the wide terminology involved with the description and morphology of field-scale natural structures (fractures, folds, and fabrics in that order). The second half deals with the more quantitative aspects of deformation processes (stress, strain, rheology, and experimental deformation). The course is introduced with comprehensive coverage of the brittle regime with 5 lectures, one lecture on joints, three on faults, and one on veins (Table 1). Fractures are introduced with a description of all aspects of joint formation, where joint terminology is illustrated with simple 3D graphic drawings. In the lectures on faults two complex 3D figures showing fault geometry and orientation (fault-strike, -dip, -hade, -trace, -hanging wall, -footwall) and displacement (fault-heave, throw, dip-slip, strike-slip, net-slip, oblique-slip) and their associated stereographic representations are built up with a series of frames which sequentially add more components and their names (Figs. 3 & 4). Three dimensional figures are also constructed (and 'built') to illustrate the concepts of slip, separation, and displacement and the important differences between them. Anderson's classification of the three major fault types and their stereographic representation is similarly displayed. A further innovation here is the real-time animation of the simulation of flow in mylonites and the development of asymmetric winged porphyroclasts using the *Powerpoint* automatic slide-show display option (Fig. 5). This "pseudo" animation can be presented, reversed, or cycled and the rate can be controlled by varying the time gap between the display of each screen. In describing the geometry and kinematics of different styles of faults, the animation facilities of *Powerpoint* have also been used to show the production of accommodation folds in association with faults, such as roll-over structures, synthetic and antithetic graben, pull apart basins associated with extensional faults, and hangingwall 'snakes head' fault-bend folds associated with thrusts (Fig. 6). The classic hindward-dipping duplexes of Boyer & Elliott (1982) have also been animated (Fig. 7). The final component of the brittle fracture section includes the description of veins and vein systems. The animation capabilities of *Powerpoint* are well displayed in the production of 'actively' dilating veins, and the addition of syntectonic growth fibres provides very realistic simulation to

Table 1 - Level 2 "basic" structural geology courseware example slides (Copies of these modules are available free on request to the authors)

Lecture Title	Keywords	Size	Slides
01. Joints	sets, systems, scale, surfaces, measurement & orientation, origin	638kb	102
02. Faults-A	nomenclature, geometry & kinematics, slip, separation	220kb	62
03. Faults-B	Anderson's classn., -scale, lineaments, scarps, roll-over, mylonite	275kb	70
04. Faults-C	extensional-, contractional-, wrench-, accommodation folds	484kb	84
05. Veins.	origin, types, growth fibres, stresses and shear fractures	308kb	103
06. Folds-A	geometry, nomenclature, families, attitude, coords, systems	319kb	129
07. Folds-B	classification, tightness, shape, style, generation, refolding	253kb	93
08. Foliations	fabric anisotropy, fabric components/arrangement, cleavage, *etc.*	1S7kb	98
09. Lineations	L-tectonites, striations, crenulation, intersection, mineral, stretch	209kb	68
10. Rock Mech.	defs., stress, strain, rheology, foundations slopes & excavations	583kb	67
11. Displacement	continuum mech., position, velocity, force, changing material state	715kb	109
12. Stress-A	gen. of force, body / surface force, measurement, Mohr stress	422kb	119
13. Stress-B	Mohr stress (cont), classes of stress, deviatoric & lithostatic stress	375kb	135
14. Strain-A	deformation, displ., strain, rotation, (in)homogen., measures	253kb	65
15. Strain-B	pure & simple shear, finite strain ellipsoids, Mohr circle	297kb	80
16. Elastic Rheol.	failure, isotrop., homogen., continuity, Hooke, stiffness, strength	462kb	102
17. Brittle Rocks	failure, stress, int. friction, cohesion, Navier-Coulomb, Mohr	297kb	103
18. Ductile Rocks	test rig, rheol., ductility, Newtonian visc., plasticity, creep, flow	352kb	106
19. Tectonics-A	Plate motion/orientation, convergence, divergence, transcurrence	413kb	68
20. Tectonics-B	Extens./contract. orogeny, fold-fault association, wrenching	428kb	50

natural examples of curved syntaxial and antitaxial fibrous veins (Fig. 8). The complex geometries of conjugate "tension" gash systems in brittle-ductile shear zones (Ramsay & Huber 1983), their stereographic representation, and their significance for paleostress analysis is also presented here (Fig. 9).

The geometry and nomenclature of single surface folds, multilayer folds, various fold hinge varieties, and their three dimensional morphology (Fig. 10) is explained in the next two lectures, as are the differences between plane, non-plane, cylindrical, non-cylindrical (Fig. 11), upward facing, and downward facing folds. Sander's (1970) fold coordinate system is used in conjunction with stereographic projections to illustrate the naming of fold orientation (as part of this exercise the students concurrently engage in a game to pick the correct model) (Fig. 12). The concepts of order of folding and fold vergence are also described. Fold classification is defined and described in terms of five parameters: tightness, shape, style, orientation, and origin. Simple 2D and 3D figures are built up to illustrate the different fold geometries (Figs. 13 & 14).

The development of natural rock fabrics by the preferred alignment of planar, linear, or inequidimensional minerals or mineral aggregates is very effectively illustrated by an array of 3D block diagrams (Fig. 15). Recently, as well as showing this to students, the concepts have been simulated with in-class modelling exercises using Philadelphia cheese blocks as the 3D space substrate and cheese biscuits and pretzels as the planar and linear elements, respectively. This combination of constructed, good-quality, computer generated, visual illustrations and tangible experiments appears to be a most effective, enjoyable, and nourishing learning exercise! The internal geometry and possible origin of a variety of cleavages and other foliations can also be modelled (Figs 16 & 17). Combined text and simple graphics have further been produced to illustrate the various components of fabrics producing tectonic lineations such as striations (Fig. 18), crenulations, intersections, mineral lineations, and stretching lineations.

The second half of this course is more process oriented, containing quite different content and thus the computer generated images are used in different ways. As a technique to emphasis the difference and also to provide immediate stimulation, the fundamental aspects of rock mechanics are introduced via the application of the concepts of stress, strain, and material behaviour as illustrated by seismic and other risks to building foundations, slopes, and excavations. Short videoclips, along with animations, are very effectively incorporated into the computer presentations to show the tangible effects of earthquake shock on major population centres. The fundamentals of continuum mechanics are then introduced from first principles by defining the material states of matter within a Cartesian reference frame and coordinate system in terms of the position, velocity, and force acting upon single and multiparticle systems, and their evolution through time, as described by Means (1976). The gradual construction of complex diagrams (Figs. 19 & 20) provides a simple but powerful technique for demonstrating this topic.

Stress and strain are theoretical concepts which require sequential development of mathematical algorithms and proofs to adequately define quantitatively. This can and has been utilised in the *Powerpoint* presentations (*e.g.* for the derivation of Mohr stress circles) (Figs. 21 & 22). The graphical nature of *Powerpoint* however has also allowed these concepts to be more visually illustrated and, for example, normal forces and stresses can be effectively demonstrated using the gravitational effects of known masses (bricks) as they impact on bodies of limited resistance (tomatoes) (Fig 23). One particular aspect of stress which can be very clearly demonstrated by sequential *Powerpoint* slides, and is also useful for the description of the mechanical behaviour of materials, is the simulation of axial loading of a specimen in tensile or compressional tests (Fig. 24). A graphical display of the experiment, the stress-strain curve, and a Mohr representation can all be shown with a simulation/animation of the change in all parameters with time (note - if a real demonstration of the experiment is conducted on the laboratory bench at the same time with, for example, a banana representing the material, the overall effect can be significantly enhanced). The geometrical aspects of strain can be most effectively demonstrated by the real-time graphical manipulation and deformation of images. These can be produced and then imported into *Powerpoint* for sequential display (Fig. 25). However, it is more effective to use the *Powerpoint* drawing tools to take an image of a square or other regular shape and illustrate the homogeneous transformations of simple and pure shear, plane strain, area change, rigid body translation, rotation, and the various combinations of these that are allowed by the programs' algorithms and drawing tools (see also Bjørnerud 1991 , Tickoff *et al.* 1993). These tools also allow the real-time manipulation of finite homogeneous strain ellipses, and the derivations of Mohr circles for strain simply following Ramsay 1967 (Figs. 26 & 27).

Rheological principles have proven to be the most stimulating concepts to develop and illustrate using our computer simulations. After introducing the concepts of homogeneity, continuity, and isotropy and the effects of material properties and scale (Fig. 28), the simulation of deformation experiments using

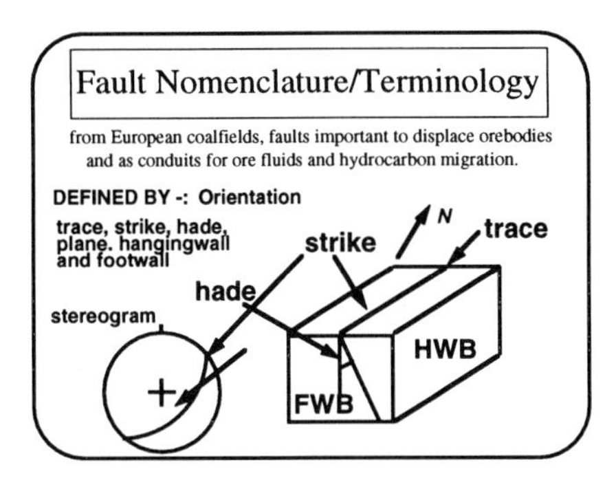

Fig. 3. Fault orientation.

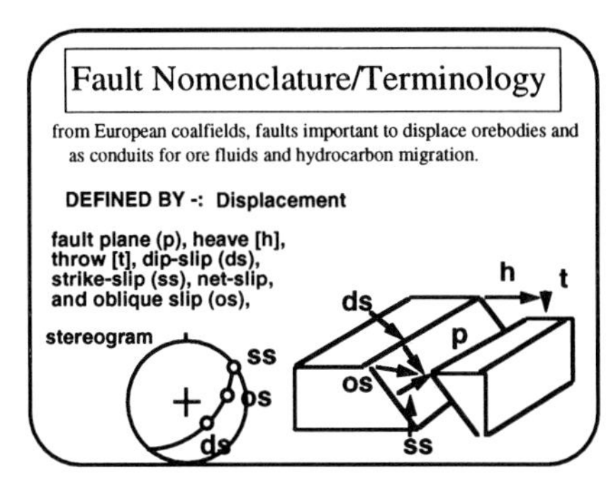

Fig. 4. Fault displacement.

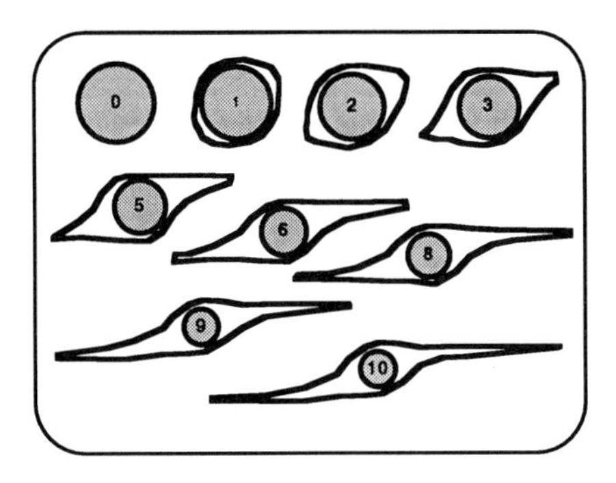

Fig. 5. Porphyroclast rotation.

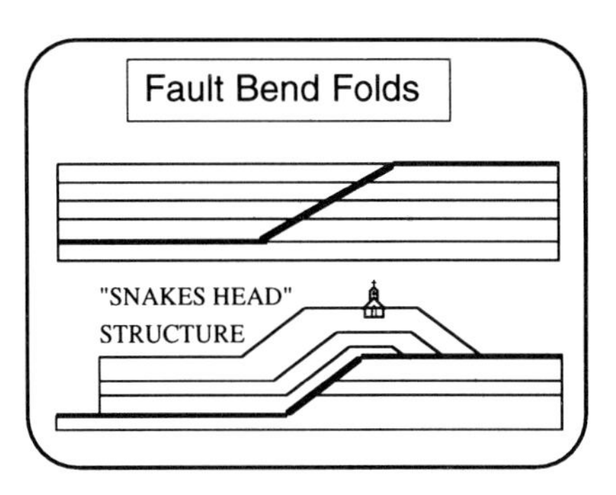

Fig. 6. Fault-bend folds.

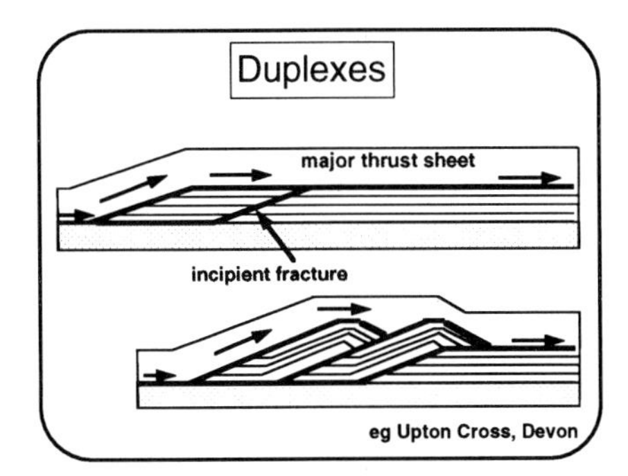

Fig. 7. Duplexes.

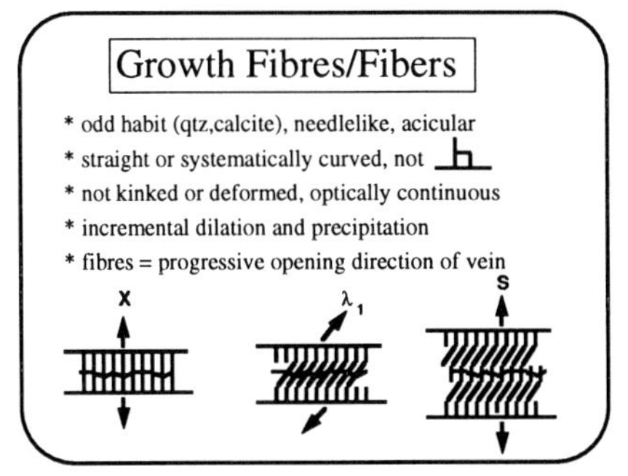

Fig. 8. Growth fibers.

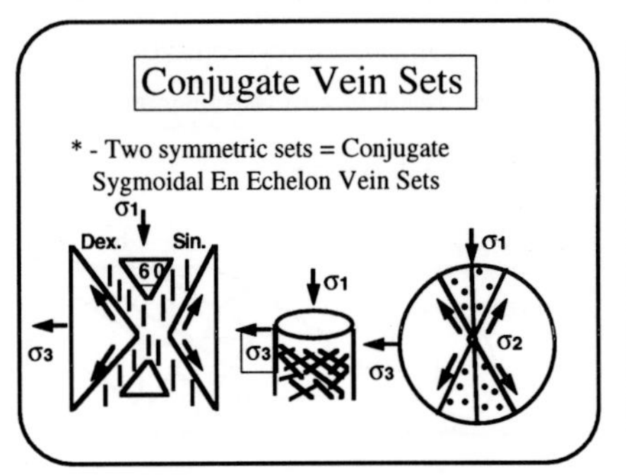

Fig. 9. Conjugate vein sets.

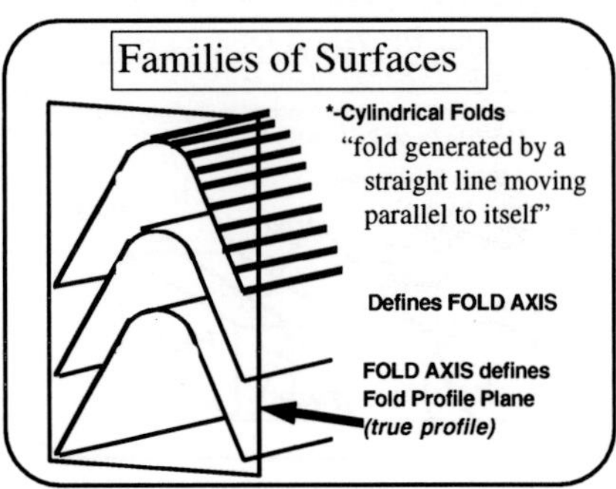

Fig. 10. Folded geometry.

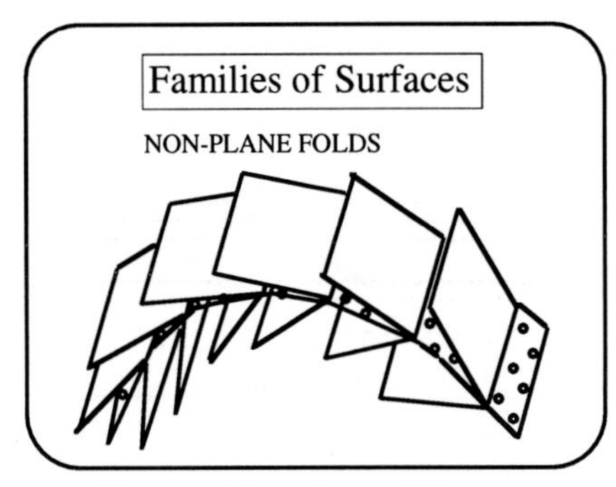

Fig. 11. Non-plane folds.

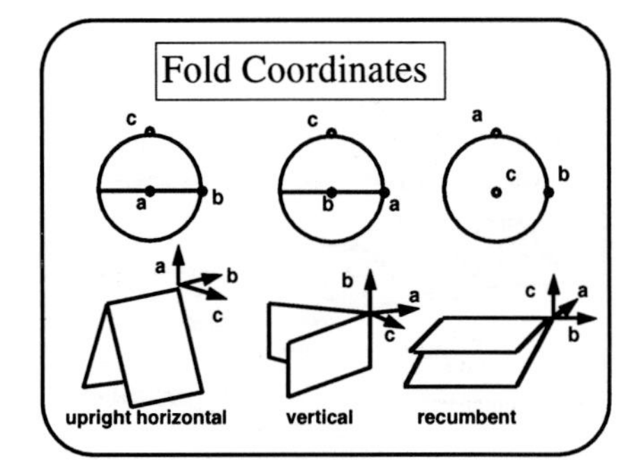

Fig. 12. Sander's coordinate system.

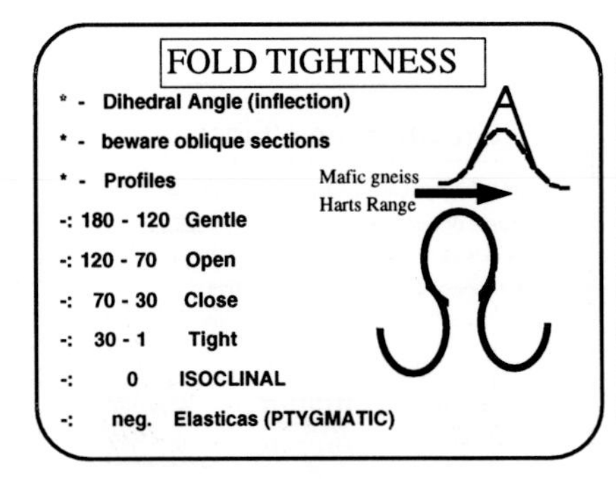

Fig. 13. Fold classification.

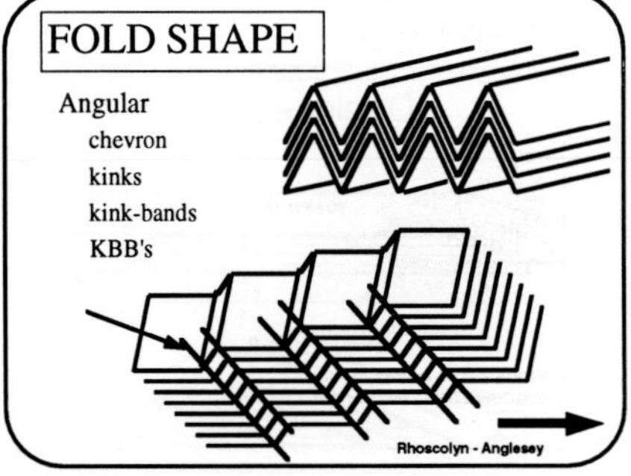

Fig. 14. Angular folds.

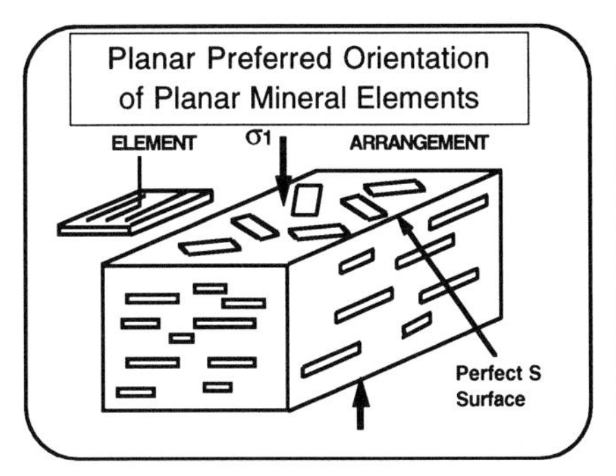

Fig. 15. Planar fabrics.

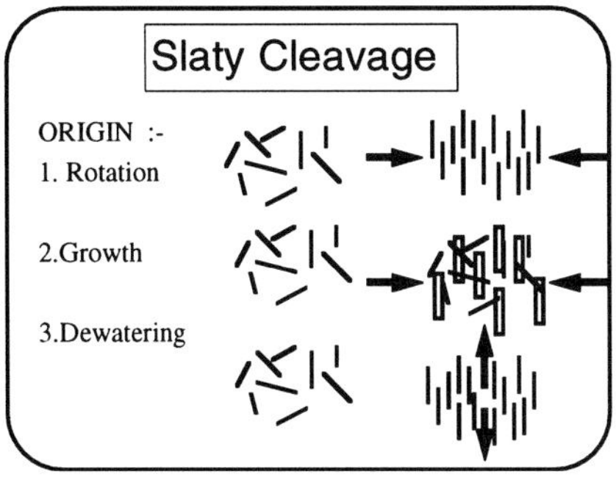

Slaty Cleavage

ORIGIN :-
1. Rotation
2.Growth
3.Dewatering

Fig. 16. Slaty cleavage.

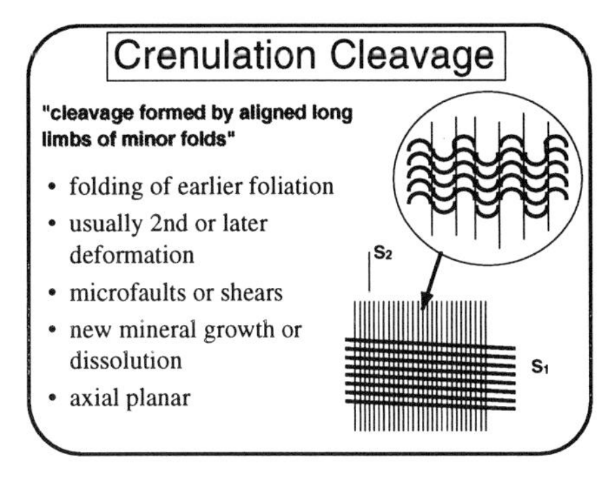

Fig. 17. Crenulation cleavage.

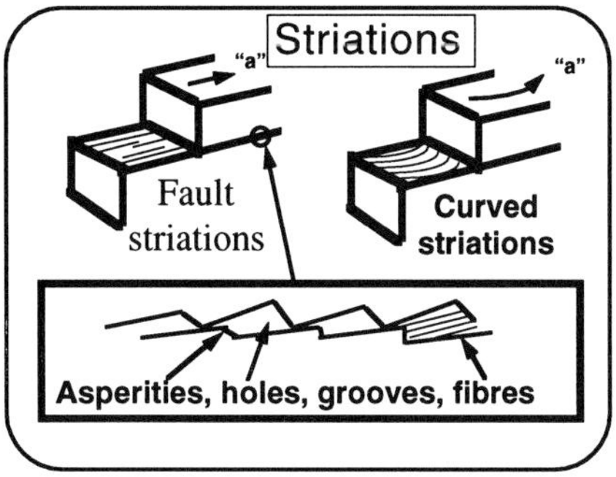

Striations

"a"
"a"
Fault
striations
Curved
striations
Asperities, holes, grooves, fibres

Fig.18. Fault striations.

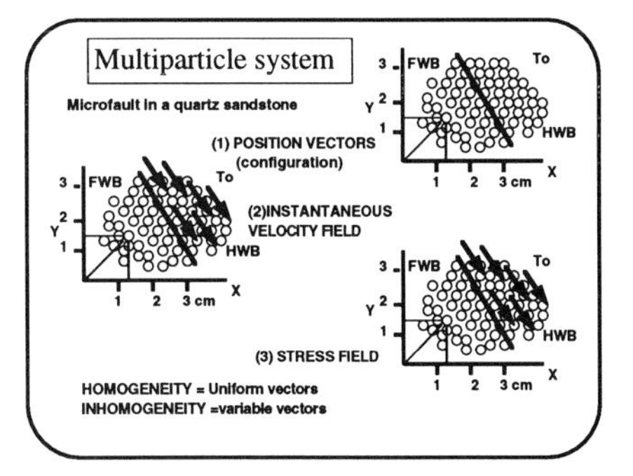

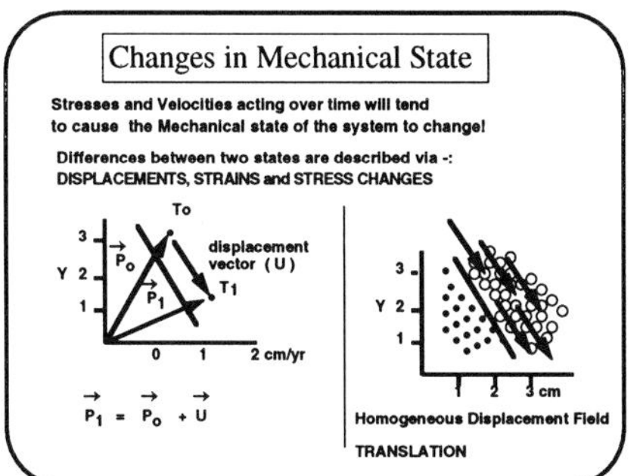

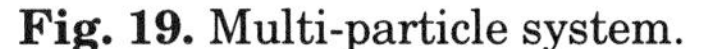

Fig. 19. Multi-particle system.

Fig. 20. Mechanical state.

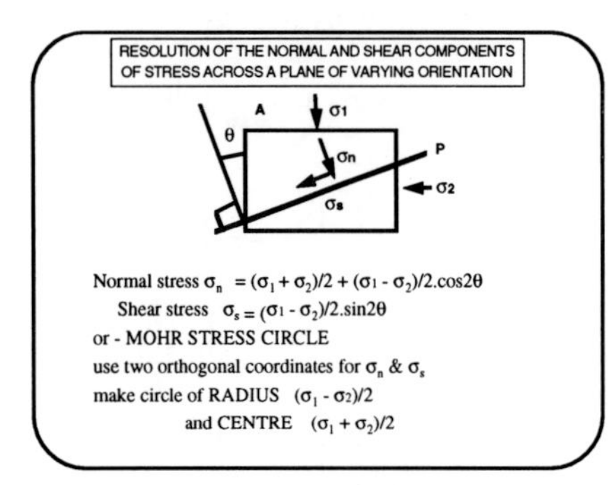

Fig. 21. Stress on a plane.

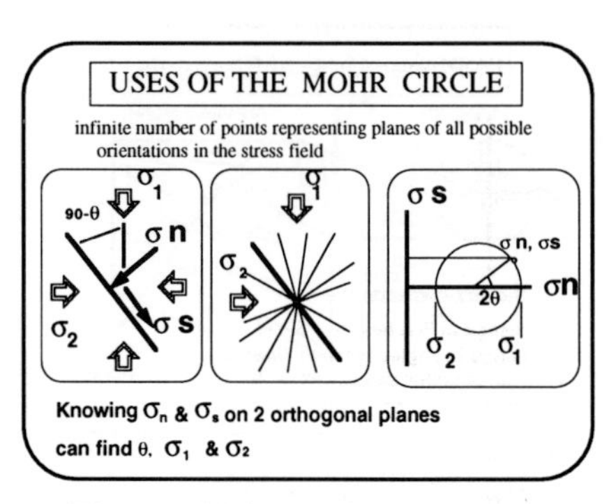

Fig. 22. Mohr circle for stress.

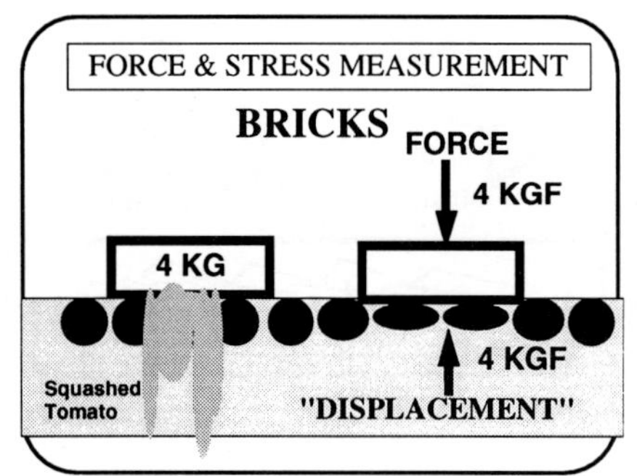

Fig. 23. Force and displacement.

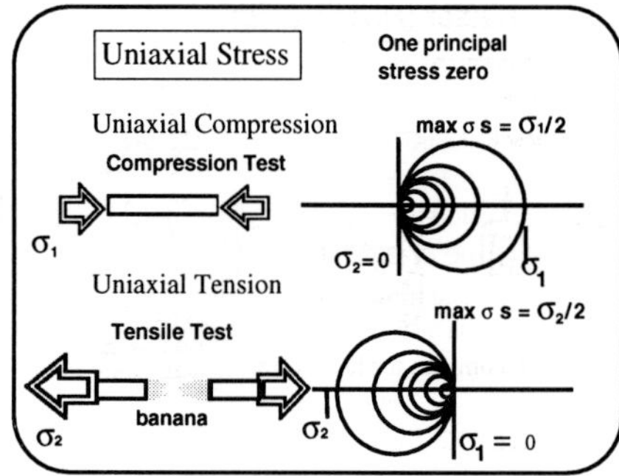

Fig. 24. Experimental deformation.

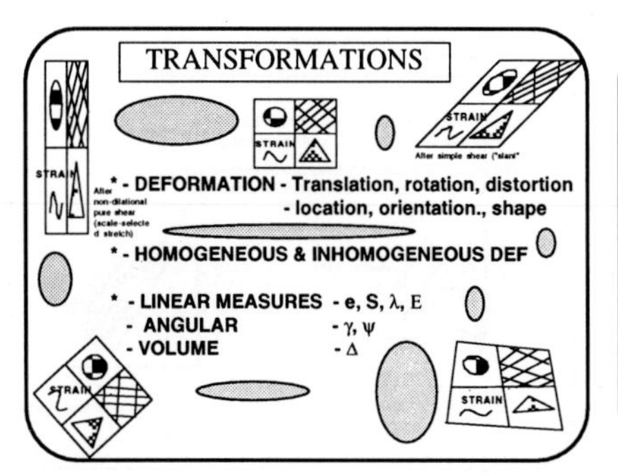

Fig. 25. Transformations and strain.

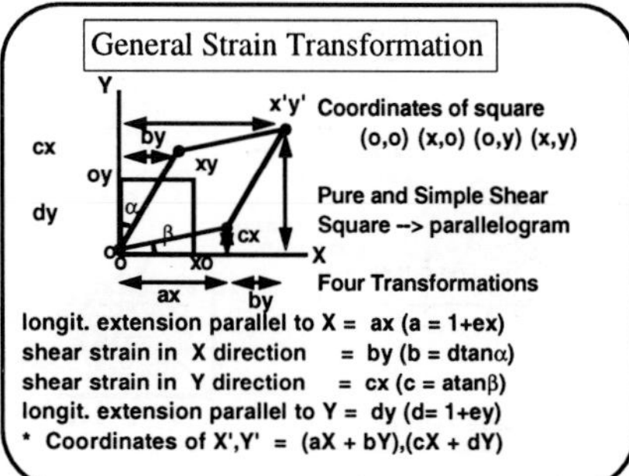

Fig. 26. General strain.

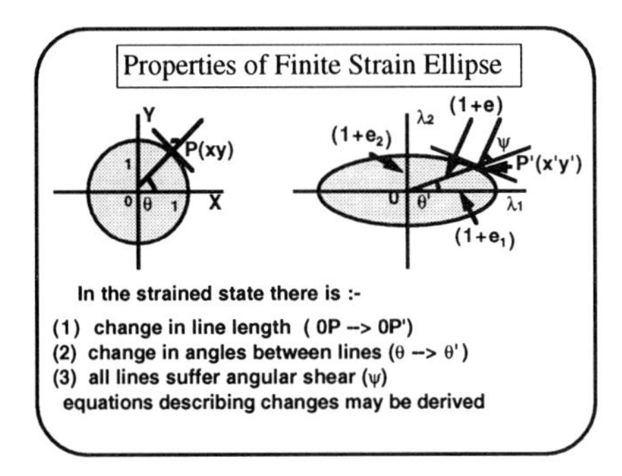

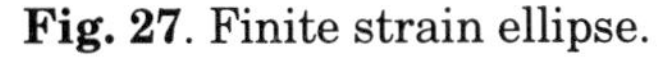

Fig. 27. Finite strain ellipse.

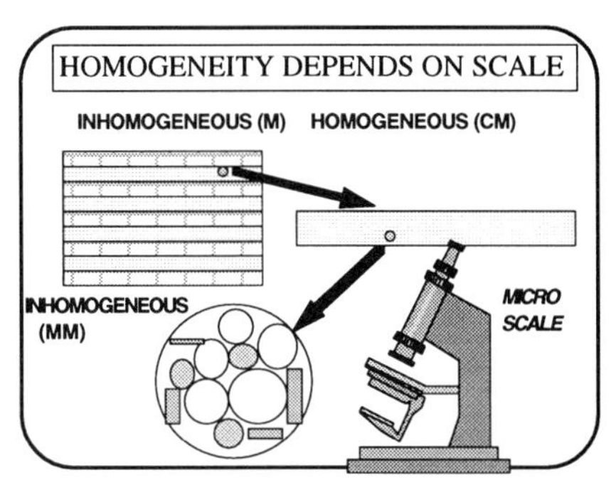

Fig. 28. Homogeneity & scale.

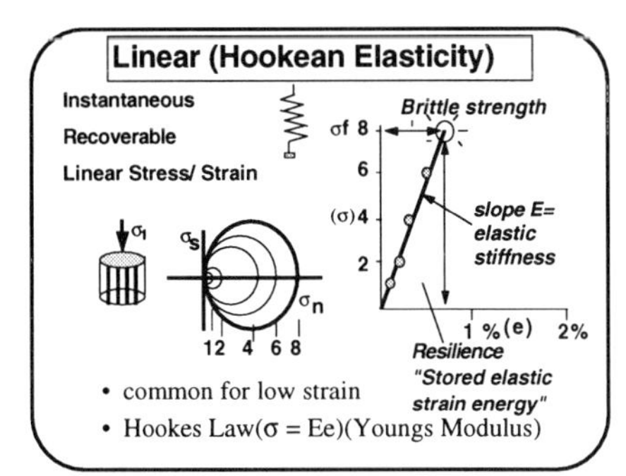

Fig. 29. Linear elasticity model.

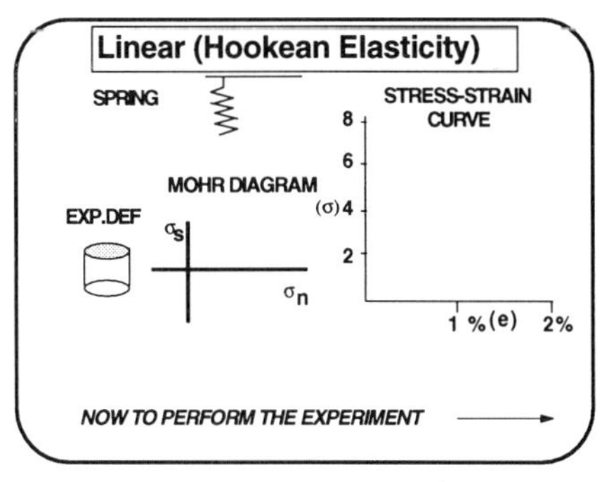

Fig. 30. Linear elasticity - expt. 1.

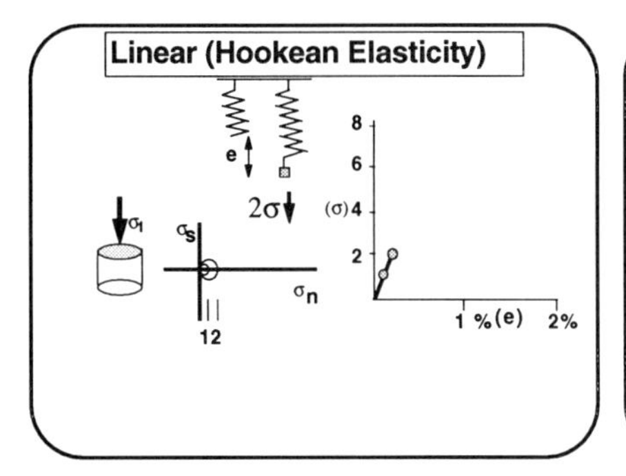

Fig. 31. Linear elasticity - expt. 2

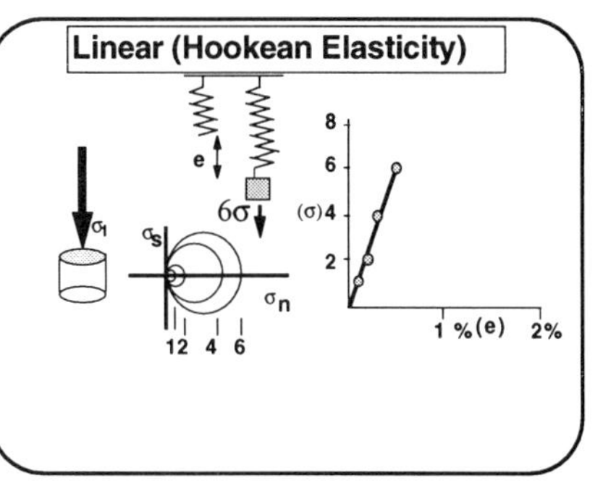

Fig. 32. Linear elasticity - expt. 3.

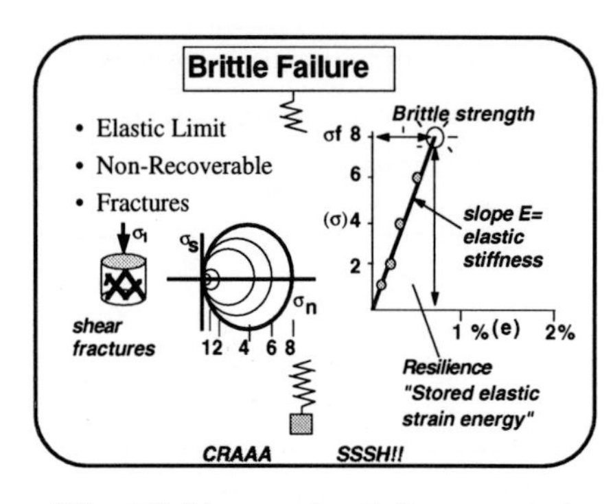

Fig. 33. Linear elasticity - expt. 4.

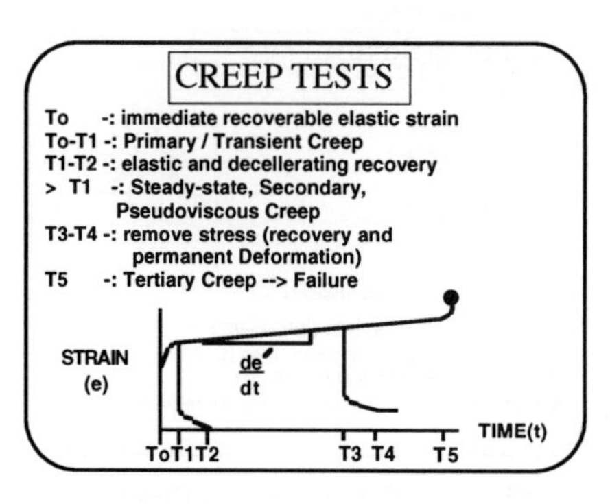

Fig. 34. Creep curves.

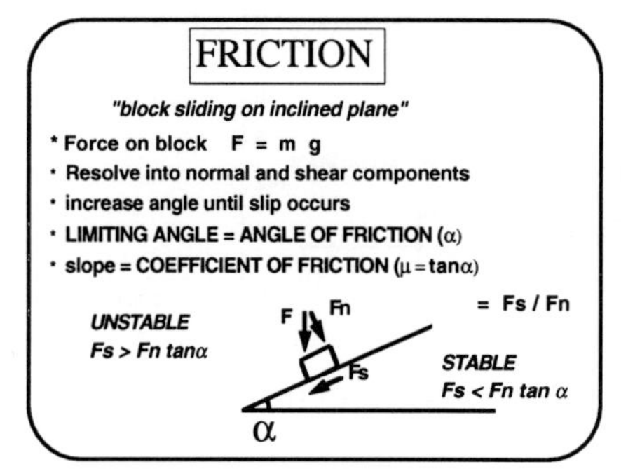

Fig. 35. Friction models.

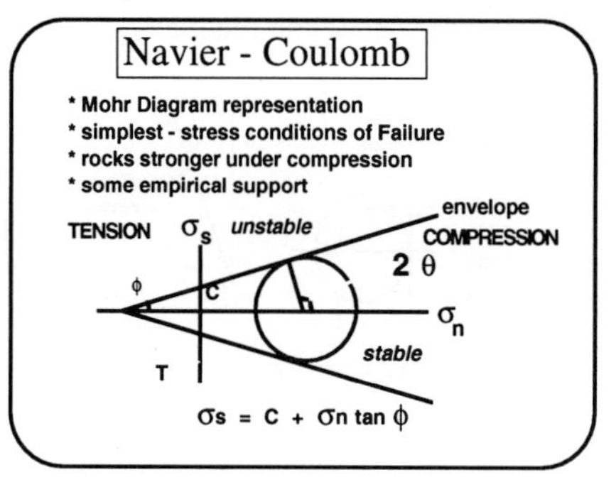

Fig. 36. Navier-Coulomb criteria.

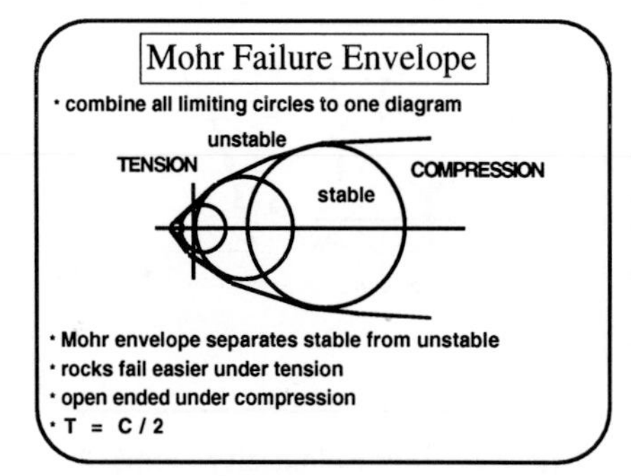

Fig. 37. Mohr failure.

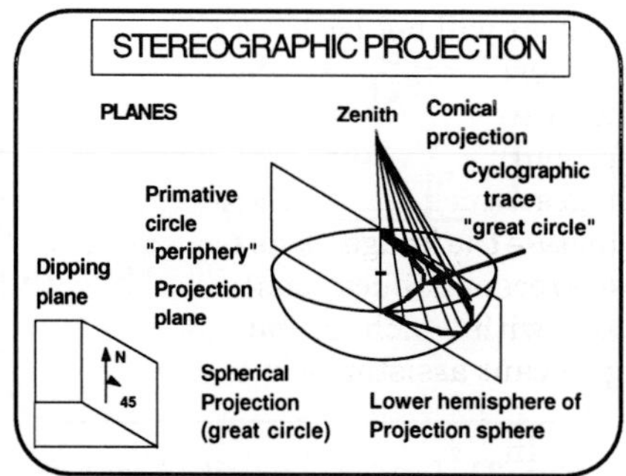

Fig. 38. Stereographic projections.

the pseudo animation capabilities of the *Powerpoint* slide-show have been used (along with analog model demonstration experiments) to show a range of different behaviour of materials and rock analogs. Figures 29-33 show a complex deformation experiment created to demonstrate short duration elastic and brittle responses to flexural and uniaxial deformation. These include an analog sample ('rock' cylinder), a deformation situation analog (spring, dashpot, or friction block), a stress strain graph, and a Mohr circle. The concepts of elastic rheology, anelasticity, strength, stiffness, resilience are all demonstrated for elastic materials, and more complex time-dependant responses of viscosity, plasticity, visco-elasticity, and even pseudoviscous creep (Fig. 34) are also produced for ductile materials. (*n.b.* along with the computer representations, real simulations using a variety of wood and metal bars, plasticine, rubber, fruit, and vegetables help to make the experiments more vivid, stimulating, and enjoyable). The values of many of these parameters for everyday materials are very adequately provided by Gordon (1968, 1978) and add further reality. Continued use of the experimental deformation graphics and analogs also allows the investigation of brittle deformation, including aspects of brittle extensional and shear failure, frictional sliding (Fig. 35) and the coefficient of internal friction and cohesion of rocks which defines the Navier Coulomb failure criteria (Fig. 36). The sequential development of Mohr failure envelopes also allows the illustration of real mechanisms of brittle failure (Fig. 37).

As well as the lecture topics described, the basic structural geology course also contains a laboratory schedule. Within this, one module on stereographic projection has been developed as a *Powerpoint* presentation. Over 6 or 7 two hour classes, all aspects of basic stereographic projection including an introduction and plotting of planes, poles, lines, pitches, plunges, intersections, angles, folds, rotations, and drill holes are delivered via 93 slides (187k) with accompanying student exercises (Fig. 38).

Advanced Structural Geology and Other Lecture Modules

An advanced course in structural geology as a component of a structural geology and exploration geophysics subject comprises 13 lectures, using a total of 9.2MB memory space and made up of 730 (in 1995) *Powerpoint* presentation slides (Table 2). These will not be described in detail here as they use many of the techniques previously mentioned. These presentations use more memory and less slides per lecture as they contain many more images and graphics scanned from textbooks and journal publications. They also contain more scanned field 35mm slide images which therefore takes advantage of the ability of the programme to annotate the images. The need to significantly modify the content each year (from more recent publications) at this advanced level has also meant that the ease and speed with which an individual presentation can be changed provides further significant assistance in lecture preparation.

In 1994 a larger, more ambitious programme involving the conversion of 60 lectures to electronic *Powerpoint* format for 120 introductory Geology students was carried out, with considerable success. This course was taught, and the content prepared, by six different subject lecturers. We understand that other earth science and other disciplines are also currently following similar pathways (CAUT 1993, Sowerbutts 1993, Burton & Wynn 1994).

Table 2: Level 3, "Advanced " structural geology courseware.

Lecture Title	Keywords	Size	Slides
01. Introduction	definitons. of structures, scale, deformation states, movement history	792kb	48
02. Contract. faults	ramps and flats, 3-D geometry, thrust propagation, thrust systems	792kb	91
03. Extension Faults	rifting, basin formation, listric and planar geometry, roll-over	1034k	82
04. Wrench Faults	transcurrent faults, en échelon folds, riedel shears, expts, oblique faults	1111k	58
05. Strain-A	displacement fields, transformations, ellipsoids, Flinn/Hsu diagrams	462kb	66
06. Strain-B	strain estimation, markers, Rf/ϕ, Fry plots, deformed fossils	858kb	37
07. Shear Zones-A	progressive simple shear, ductile fabrics, strain measurement	440kb	37
08. Shear Zones-B	sheath folds, S-C mylonites, lineations, ECC's, kinematics, veins	891kb	48
09. Folding-A	classification, thickness plots, isogons, parallel, similar, concentric	374kb	35
10. Folding-B	tangential and flexural strain, flow folds, buckling, wavelengths	616kb	61
11. Complex Defn.	transposition, differentiation, folds & strain, non-plane, -cylindrical	605kb	49
12. Refolded folds	superimposed fold interference pattems, classes, minor vergences	539kb	48
13. Complex Defn.	facing, structural analysis, scale, style, overprinting, generation	759kb	70

Further Developments

As well as *Powerpoint* presentations, we have used a number of other software options to teach a variety of topics and concepts in the electronic classroom. *Freehand* has been used as a direct presentation package to illustrate the visual aspect of physical transformations in geological situations. With the ability to draw simple or complex three dimensional models of undeformed rock bodies containing identifiable original features, the effects of the different types of deformation on such bodies can be illustrated in real-time visual experiments (Bjørnerud 1991). Similarly, *Mathematica* can be used to display a great variety of mathematical phenomena (*e.g.*, R_f/ϕ plots), *Endnote*™ can be used to show the power of current referencing and bibliographic search and storage facilities, and a variety of software is appearing on the worldwide web. There are also a number of recent developments in innovative specialist geological research and educational software (*e.g.*, Bjørnerud 1991, Tickoff *et al.* 1993; see also Learning Curve 1994 and Rockware's software catalog) that can be illustrated and examined in real time in the electronic classroom.

For the move into more realistic animation and more powerful interactions, multimedia presentation software packages, *Macromedia Director* and *Authorware* have allowed the development of fully interactive tutorials. These much more sophisticated programs allow multiple preparation of images "scripted" into full animations. However, expense and complex programming requirements have led to limited availability of useful teaching modules (Sowerbutts 1993, James *et al.* 1995). We are now beginning to investigate the development of simple interactive multimedia modules using less complex and less expensive software (*Hyperstudio* and *Digital Chisel*). We are introducing these packages to undergraduate students to develop their own multimedia presentations as an alternative to more traditional text-based assignments.

Conclusions

Electronic learning materials have had significant advantages in teaching and learning for us as lecturers and for our students, when used in conjunction with traditional teaching methods. For us, the advantages have been ease of acquisition, storage, arrangement, and display of visual aids in lectures and laboratory classes, plus the bonus of a professional and uniform presentation style. The ability to interact with a larger proportion of students and the possibilities of automatic revision and assessment have been a further advantage. The digital revolution has allowed almost instantaneous access to all material previously presented in a course. As later lectures often depend on concepts and data introduced in earlier lectures, electronic-file/lecture linking has allowed repeated return to material previously covered which offers considerable advantages to the learning process (Gibbs *et al.* 1988, p.79) The problems encountered by remembering and physically carrying the array of overhead transparencies, slides, manilla folders, and lecture notes into classes has been made redundant by the ability of the electronic storage and retrieval of a whole series of lectures on a single floppy disk or through a network link. With electronic notes attached to files and screens, there is no longer even the need to remember quantities of course notes nor to have them on hand for reference. Further, the ability to simply update a course year after year without the need to rewrite a whole series of notes or reproduce new overhead transparencies, together with the ability to link these with field slides and videos has created a vastly more pleasant and efficient working environment. Adding detailed graphics and animations to simplify understanding of complex natural phenomena has further enriched the learning environment.

For our students, a new, separate, and alternative learning environment is significant, as is the ability to access the teaching materials via floppy disk or electronic network which leads to greater impact and efficiency in essentially more "self-paced" learning. Students learn best in a variety of flexible learning environments. The flexibility of the electronic classroom, with the ability to repeatedly view lecture material out of normal lecture times or to interactively interrogate a subject, to view professional visual presentations, and also, if necessary, produce and distribute hard copy, is increasingly demanded by students who pay significant fees and expect excellent learning services.

The main disadvantages of the electronic learning system we have been developing are still largely a result of lack of financial and managerial support. There are insufficient computers and peripherals and the computers themselves lack sufficient power, speed, and memory to make electronic teaching and learning uniformly comfortable. Not enough lecture theatres have computer projectors. Portable projection systems are still not portable enough and require darkened rooms. Not enough large (Gigabyte) disk storage space is available. Risk of damage of equipment in transit, or theft, is a constant concern. Networks are not always pervasive in the academic environment and file transfer is often difficult and slow. Software is expensive, even for networked copies, often has compatibility problems for file transfer, and is commonly difficult to master. Fortunately, electronic data storage, integration, and presentation systems such as RISC-based PowerPC computers are becoming smaller, cheaper, and more available.

With increased student numbers in higher education, and with the increasing demand by students for equity and access in teaching and learning, new and innovative methods of teaching and learning are required. New media integration technologies have kept pace with the desire for new teaching methods, but authored packages for use in lectures, laboratory classes, and tutorials, have not. The system described in this paper is one attempt to move geological education into the electronic classroom and forward with the advanced technologies now available.

References

Bjørnerud, M. G. 1991. Conveying principles of finite strain with standard graphics software. *J. Geol. Ed.* **39**: 23-27.

Burton, A. & Wynn, S. 1994. Making the most of electronic media for teaching and learning. In: Steele, J.R. & Hedberg, J.G. (eds), *Learning Environment Technology Australia.* AJET publ., Canberra. pp. 27-32.

Boyer, S. E. & Elliott, D. 1982. Thrust systems. *Am. Assoc. Petrol. Geol. Bull.* **66**:1196-1230.

C. A. U. T., 1993. *Improving University Teaching. 1994 National Teaching Development Projects.* Comm. Adv. Univ. Teaching (CAUT), Canberra. 84 pp.

Clark, I. & James, P. R. 1993a. The Use of Concept Maps in the Teaching of Structural Geology. In: *Promoting Teaching in Higher Education (Reports from the National Teaching Workshop)*, Bain, Lietzow & Ross (eds.). Griffiths Univ. Press. pp. 291-304.

Clark, I. & James, P. R. 1993b. *Port Victoria Geology Trail, "The ancient volcanoes of Port Victoria".* Techsearch Inc., Univ. S. Australia Pubs. 24 pp.

Davis G. H. 1984. *Structural Geology of Rocks and Regions.* Wiley, New York. 492 pp.

Gibbs, G., Habeshaw, S. & Habeshaw, T. 1988a. *Fifty Three Interesting Things to Do in your Lectures.* Tech. & Ed. Services Ltd., Bristol. 156 pp.

Gibbs, G., Habeshaw, S., & Habeshaw, T. 1988b. *Fifty Three Interesting Ways to Assess Your Students.* Tech. & Ed. Services Ltd., Bristol. 176 pp.

Gordon, J. E. 1968. *The New Science of Strong Materials, or Why You Don't Fall through the Floor.* Penguin, Harmondsworth. 287 pp.

Gordon, J. E. 1978. *Structures, or why things don't fall down.* Penguin, Harmondsworth. 395 pp.

Habeshaw, T. Habeshaw, S. & Gibbs, G. 1989. *Fifty Three Interesting Ways of Helping Your Students to Study.* Tech. & Ed. Services Ltd., Bristol. 197 pp.

Habeshaw, S. & Steeds, G. 1987. *Fifty Three Interesting Communication Exercises for Science Students.* Tech. & Ed. Services Ltd., Bristol, 235 pp.

Hobbs, B. E., Means, W. D. & Williams P. F., 1967. *An Outline of Structural Geology.* Wiley International, New York, 571 pp.

James, P. R. 1994. Media Integration in Teaching & Learning: Best Practice in the Electronic Classroom. In: Steele, J. R. & Hedberg, J. G. (eds), *Learning Environment Technology Australia.* AJET publications, Canberra, pp. 120-127.

James, P. R. & Clark, I. 1991. Computer animation and multimedia presentation software as an aid to tertiary geological education in the Earth Sciences. In: *Simulation and Academic Gaming in Tertiary Education.* Godfrey, R. (ed), Proc. 8th ASCILITE Conf., Launceston, p339-348.

James, P. R. & Clark, I. 1993. Grid sketching to aid teaching geological mapping in an area of complcx polydcformation. *J. Gcol. Ed.* **41**: 433-437.

McClay, K. 1987. The Mapping of Geological Structures. *Geol. Soc. Lond. Handb. Ser.* Open Univ. Press., Milfort Haven. 161 pp.

Means, W. D. 1976. *Stress and Strain. Basic Concepts of Continuum Mechanics for Geologists.* Springer Verlag, New York. 339 pp.

Park, R. G. 1983. *Foundations of Structural Geology.* Blackie, Glasgow. 135 pp.

Ramsay, J. G. 1967. *Folding and Fracturing of Rocks.* McGraw Hill, New York. 568 pp.

Ramsay, J. G. & Huber, M.I. 1983. *Techniques of Modern Structural Geology, Volume 1: Strain Analysis.* Academic Press, London. 307 pp.

Sander, B. 1970. *An Introduction to the Study of Fabrics of Geological Bodies.* (Translated by F.C. Phillips & G. Windsor). Pergammon Press, New York. 641 pp.

Sowerbutts, W. 1993. Earth Science Info. *U. K. Earth Science Courseware Cons. Newsl.* **1**: 1-8.

Tickoff, B., Teyssier, C. & Fossen, H. 1993. Computer applications for teaching general two-dimensional deformation. *J. Geol. Ed.* **41**: 425-432.

Courseware: Rock Deformation and Geological Structures

Dave Byron and Bill Sowerbutts
UKES Courseware Consortium, Department of Earth Sciences,
University of Manchester, Manchester M13 9PL, U.K.
ukescc@man.ac.uk

Abstract– This communication describes a courseware module produced and distributed by the U.K. Earth Science Courseware Consortium. The module, entitled "Rock Deformation and Geological Structures" is a hypertext-based teaching aid designed primarily for undergraduates. *Windows*™ and *MacOS*™ versions are available.

Introduction

This interactive module deals with the key areas of a standard structural geology course. It is aimed at second year students, and assumes a basic knowledge of structural geology at first year level (fault and fold classification, use of stereonets, *etc.*). It may be of use to more advanced students who wish to revise various aspects of structural geology. The module concentrates on the concepts of stress and strain, the structures that may result, and how they relate to deformed rocks in the field.

The module is designed to be equivalent to about 4 hours of conventional teaching and occupies under 7MB of memory. To run on a PC it requires *Windows* 3.1 or later, 4MB RAM, VGA color, and a 386 processor or better. To run on the *MacOS*, it requires 4MB RAM and 256 colors. The module consists of six sections which are accessed via a main menu. One section is a tutorial that demonstrates the layout of the menu bars and how to navigate through the module. The other five sections (Fig. 1) are subdivided into topics that can be accessed at any time from sub-menus.

Information is presented in a variety of ways, including multiple-choice questions. With these, if an incorrect answer is selected, the reasons why it is incorrect are explained, clues about the correct answer given, and the user is asked to try again. When the correct answer is given, a confirming explanation is given and the user allowed to proceed.

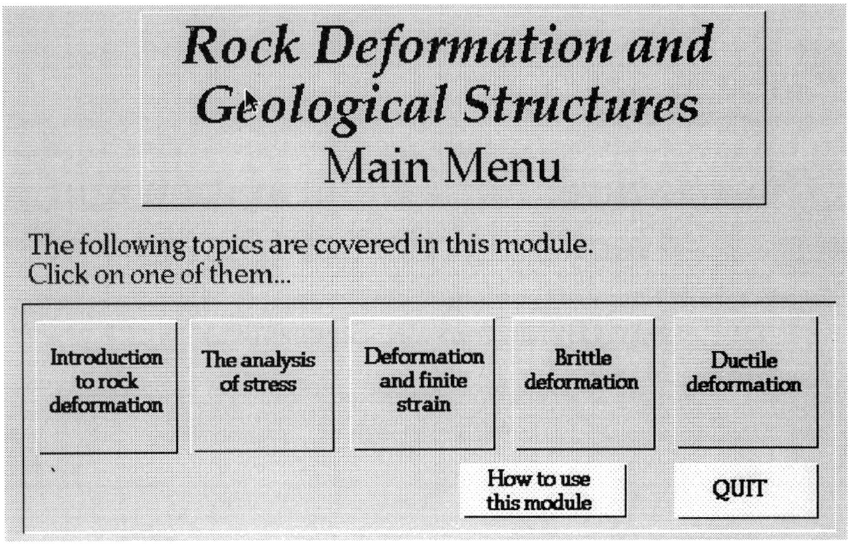

Fig. 1. The module's main menu offers a choice of five subject areas.

Introduction to Rock Deformation

This section (Fig. 2) introduces the rationale behind structural studies in geology and hence why it is important to understand forces, stresses, and strains. These concepts are outlined and differences between them explained. The other sections deal in more detail with the topics introduced here.

Stress

This section defines stress and explains how it arises. The distinction is made between force and stress and the different types of stress are described. Normal and shear stresses, stress components, principal stresses, and stress axes are covered, as well as hydrostatic and deviatoric stresses. The student is introduced to stress ellipses, stress ellipsoids, and the uses of Mohr Circle diagrams in solving structural problems.

Strain

This section discusses the different types of strain, units of strain, and how strain is measured (Fig. 3) . Homogeneous and heterogeneous strain are covered as well as strain ellipses and ellipsoids, and how these differ from stress ellipsoids. Also dealt with are pure shear and simple shear, Flinn diagrams and k values, deformation fields, deformation paths, and geological strain markers.

Brittle Deformation

Structures formed through tensile fracture and shear failure are discussed. The relationship between fault types (thrust, reverse, normal, and strike-slip) and the

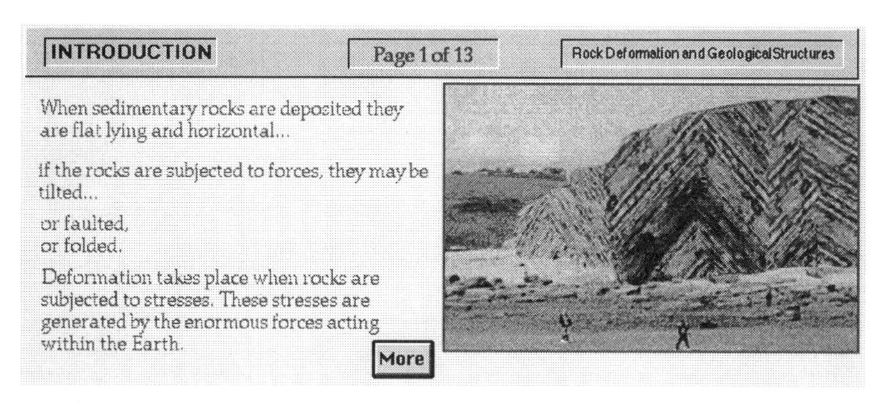
INTRODUCTION
Page 1 of 13
Rock Deformation and Geological Structures
When sedimentary rocks are deposited they are flat lying and horizontal...
If the rocks are subjected to forces, they may be tilted...
or faulted,
or folded.
Deformation takes place when rocks are subjected to stresses. These stresses are generated by the enormous forces acting within the Earth.
More

a)

Body Forces

Body forces are forces which can work on an object from a distance. The strength of the force depends upon the amount of material affected.

Examples include gravity and magnetism.

Surface forces

Surface forces (pushing and pulling) are so called because they operate across surfaces of contact. Such surfaces can be within rocks, as well as at the contacts between different rock types. The amount of a surface force depends upon the area over which it acts.

b)

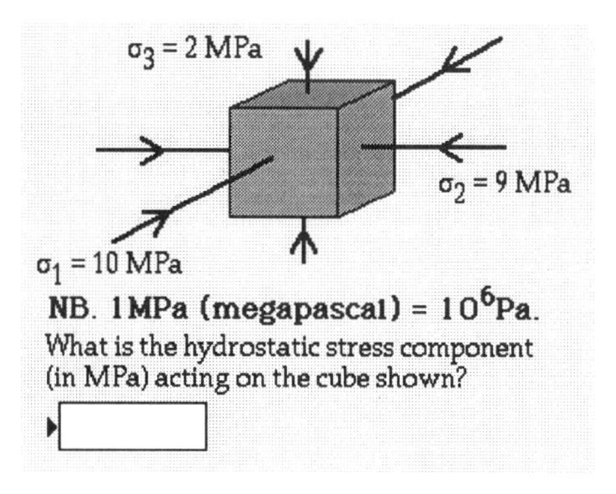
σ_3 = 2 MPa
σ_2 = 9 MPa
σ_1 = 10 MPa
NB. 1MPa (megapascal) = 10^6Pa.
What is the hydrostatic stress component (in MPa) acting on the cube shown?

c)

Fig. 2. Pages from the Introduction to Rock Deformation section. a) General, b) Body and surface forces, c) Stress.

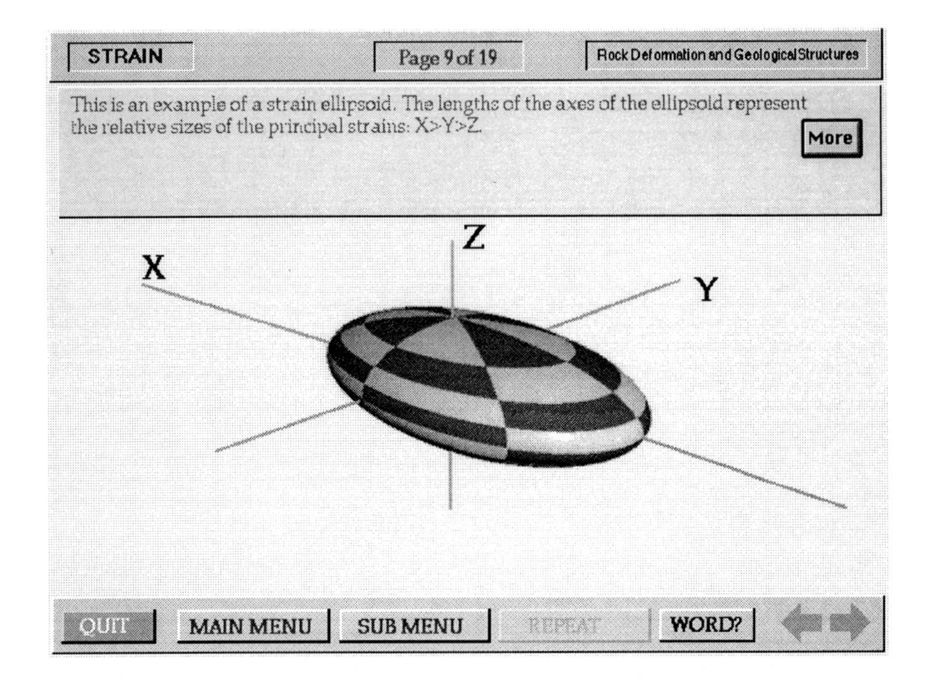

Fig. 3. Page from the Strain section.

state of stress are dealt with. The section describes the factors affecting failure using Mohr circle diagrams and considers ductile-brittle shear zones.

Ductile Deformation

This section covers the rock fabrics that form during ductile deformation. This includes folding and boudinage in layered rocks, folding and boudinage in 3D, ductile shear zones, and kinematic indicators. Fold mechanisms, the factors controlling fold wavelength, dip isogons, and fold classification are dealt with. Flexural slip and neutral surface folding are considered, together with slaty, crenulation, and spaced cleavage.

Availability

The module is offered for sale by the U.K. Earth Science Courseware Consortium, Department of Earth Sciences, University of Manchester, Manchester M13 9PL. The Consortium also distributes other modules of potential interest to structural geologists: "Geological Map Skills", "Visualising Geology in Three Dimensions", and "Preparing for Fieldwork 1: Using the Compass-Clinometer". See world wide web page: http://info.mec.ac.uk/geology/CAL/index.html

A Structural Study of North Sea Petroleum Traps Using HyperCard

Iain Allison
Department of Geology and Applied Geology,
University of Glasgow, Glasgow G12 8QQ, Scotland.
iallison@geology.gla.ac.uk

Abstract – Graduates of today may expect to change careers a number of times during their working life and companies are placing more emphasis than previously on acquiring personal transferable skills. Prime among these are the abilities to communicate effectively and to manage time and information. With such a visual subject as structural geology it is not just text that requires storage. Graphs, diagrams, maps, sections, photomicrographs and field photographs are just as important and need to be as easily accessible. Aspects of the structure of the North Sea basin are used to illustrate the power of *HyperCard* ™ to develop and maintain a comprehensive database of linked information. The *HyperCard* programming environment for Macintosh computers is capable of generating sophisticated applications yet is easy to use by those with no programming experience. The data are held in a 'stack' (the analogy is with a stack of file cards) and there may be text, graphics, sounds, or video sequences. Navigation through stacks is by means of 'buttons' which make the links between cards. These buttons may be visible and pass one to a specific card, the next in the sequence, for example, or invisible and sit over a feature or a word and bring up a card with more information on it. Two examples show not only how information may be linked but also how the visual appearance of a stack may be increased by the use of special effects and animation.

Introduction

Knowledge in all fields is increasing exponentially and there is a great need to keep track of information. Nowadays most undergraduate students have ready access to personal computers, some of which will be networked to allow communication across campuses and beyond, while most postgraduates will spend a considerable amount of their time on data analysis, modelling processes, entering text, *etc.*, on a variety of computers. With access to many networks, research workers can conduct literature searches from their offices, download the results to their own machine, and copy them straight into a bibliographic manager. Information, while not necessarily being power, needs to be readily accessible and the 'tip of the tongue' feeling of knowing that some piece of information exists, but is not immediately retrievable, leads to frustration. The stuff of our research includes our own data, laboriously gathered in field or laboratory, and articles in the scientific literature. Our personal collections of reprints may be filed according to accession number or by author, by subject, or in other ways. No single way is entirely satisfactory; cross referencing helps. Personal computers have the potential to make accessing information fast and easy. Here I describe how information may be stored in a structure like a neural network where there may be links from one piece of information or concept to many others leading to a three dimensional network or web of linked data. The database is generated using *HyperCard*, an object-based programming environment for Macintosh computers. The structural geology of the North Sea has been selected to show the versatility and ease of use of *HyperCard* for information management. For PC users, similar applications may be created using *Toolbook*™.

The North Sea is an important petroleum province and continues to provide a significant amount of oil, with production in early 1995 almost reaching the highest output during the peak of activity some 15 years ago. The geology of the basin is easily accessible in a number of compilations – for example, Abbotts (1991), Glennie (1984), and Parker (1993) – as well as in many journal articles.

HyperCard allows the integration of text, sound, and images (including video). It provides a simple, friendly, easily learned environment in which to generate databases of information to suit one's own interests and personal ways of linking facts, ideas, and concepts. Applications may be created for performing calculations as well as for information handling (Bjerg *et al.* 1992, Mogessie *et al.* 1990). *HyperCard* may also be used to control actions of other applications.

The analogy used in *HyperCard* is of a stack of file cards, especially that of a circular 'Rolodex' file which need not have a specific first nor last card. One can choose the level of complexity one wants in a stack. A stack for personal use may have few frills while one developed for others to browse can be made more sophisticated. With *HyperCard*, simple and effective applications can be generated quickly and easily. Interesting effects, such as visual wipes or iris motions between cards, may be added later as experience grows. *HyperCard* is, however, a very sophisticated and structured programming environment and, with experience, challenging multimedia applications may be produced.

Stacks may be made for other users, such as students. Such stacks can be visually interesting, easy to use and allow students to digest information at their own pace and in their own way as they navigate through the stacks. As examples, two stacks have been developed. One, based on the discovery of the Beatrice oil field, shows some of the visual effects that can be achieved in a stand-alone application for others to use. The other gives examples of a database of information which may be added to, and is intended for personal use. Copies of the demonstration stacks are available from the author on receipt of a disk.

The *HyperCard* Environment

Each card in a *HyperCard* stack contains certain information – text, graphics, sounds, *etc*. A stack is usually a collection of related information, in this case on the geological structures of the North Sea as they relate to hydrocarbon traps. Not only can the data be linked from card to card in this stack but they may be linked to those in others stacks; for example, on the stratigraphic evolution of the North Sea basin if there were one.

Each stack may contain a variety of different types of information – in this case, maps, sections, textual descriptions, definitions, bibliography, *etc.* – and cards for each of these types may be designed with the same style to differentiate visually the types of information. In other words, all cards containing maps, for instance, would share the same background. In a visual science such as geology, information is of both graphic and textual form and *HyperCard* can manipulate both easily.

Navigating Through the Stacks using Buttons

Stacks may be created in which there is a logical progression from one card to the next but in an information management system there will normally be no order in the acquisition of information. New cards therefore require to be linked to existing cards in the stack. This is accomplished using buttons. These may be transparent and sit atop words in the text or parts of a map or section or they may be visible in a "navigation panel" at the foot of the cards. Buttons perform actions and a click of the mouse on a button takes the user to another card with information on that topic or, perhaps, a more detailed part of a map. At their simplest the buttons allow one to navigate through the stack, directing one's view to the next card in the sequence, or the previous card, *etc.* They may also initiate an animation, play a sound, or run a video clip. Buttons allow the user to go to particular cards such as the "home" card and then into a different stack, or to quit. Other buttons bring up the next or the previous card in the stack or return to the last card viewed. It is also possible to see miniature versions of the previous 42 different cards viewed for instant access back in time. A search of all text on all cards in the stack may be made and, once again, a click on the relevant button on the navigation panel opens the search box for typing in the chosen term. *HyperCard* buttons are rectangular but duplicates may be overlain like shingles in order to outline an irregularly shaped area.

Although text can be painted as bitmapped characters in a graphics layer, it will more commonly be entered into text fields. These fields are generally of the 'fill in the blank' type. Textual information is entered into a field which may have a fixed size on the card or may be a scrolling field capable of accepting up to 30,000 characteristics, *i.e.* twice the length of this article.

Graphics may be created within *HyperCard* using painting tools which are in the *MacPaint*™ tradition (*i.e.* bitmapped graphics). Pictures in *MacPaint* format may be imported directly from file onto a card. Graphics in other formats can be copied and pasted via the clipboard and *HyperCard* can access PICT resources. *HyperCard* is an on-screen information resource and resolution is 72 pixels per inch. Diagrams from books or journals can be scanned, copied to the clipboard, and pasted into stacks as has been done with the example stacks. Beware of copyright requirements if you plan to distribute your stack.

A variety of visual effects can add impact to the stack; for example, a checkerboard fade out to the next card makes the transition more interesting.

These effects may be used consistently to indicate a difference between the various parts of the stack and are useful psychological aids when navigating through the information. The same or a complementary effect should be used when going to a particular card and returning to the original. For example, a scroll left effect could be used when going to the glossary part of the stack and a scroll right used to hide the glossary card on returning. Advanced techniques include animation, video, and sound and *Quicktime*™ movies can be played. Colour is now available for *HyperCard* (versions ≥ 2.0) and with version 2.3 *HyperCard* has colour fully integrated into its core. Default objects (buttons, edit fields, *etc.*) may be copied, renamed, resized, and pasted into place on any card, or custom controls may be created using a built-in scripting language. Users may write their own external commands in a language such as C and use *HyperCard* to run external applications.

The "Structural Geology of the North Sea" Stack

The stack consists of text, maps, and sections (Figs. 1 through 5). Certain card characteristics are shared; for example, all cards containing maps have a particular background image, a north arrow, and buttons for navigating through the stack. This is the background and the maps are added onto cards with that background (Fig. 2). New cards with that background are easily added and more text, say, typed onto them or pictures pasted from other applications or from scanned images. The title card (Fig. 1) shares the same background as the reference cards (Fig. 5). The cards containing cross sections (Fig. 3) are intended to give the impression of vertical sheets pinned up on a board to contrast with the plan view of maps. The glossary entries (Fig 4) look like pages from a notebook.

The "Beatrice" Stack

Developed initially as a resource for use by visiting school leavers on a University open day, the Beatrice stack considers aspects of the exploration for an offshore oilfield. It demonstrates the use of both visual effects and animation in a stand-alone stack of 43 cards.

Strengths and Weaknesses

HyperCard is an on-screen information manager which is easy to use and easy to learn yet is so comprehensive that with time and experience very sophisticated applications may be created. The prime advantage is for readily storing text and graphics in a form which is easy to search. All text in text fields may be searched using the `Find` command. Card information may be further linked using buttons. Cards with both text and diagrams may be printed singly or in groups at reduced size.

HyperCard is not intended as a report generating application. The report printing options are limited to text although other software is available to overcome this limitation and incorporate the graphics on cards; an example being *Reports*™ which enhances *HyperCard* into a true database manager.

Fig. 1. Title card from 'Structural Geology of the North Sea'.

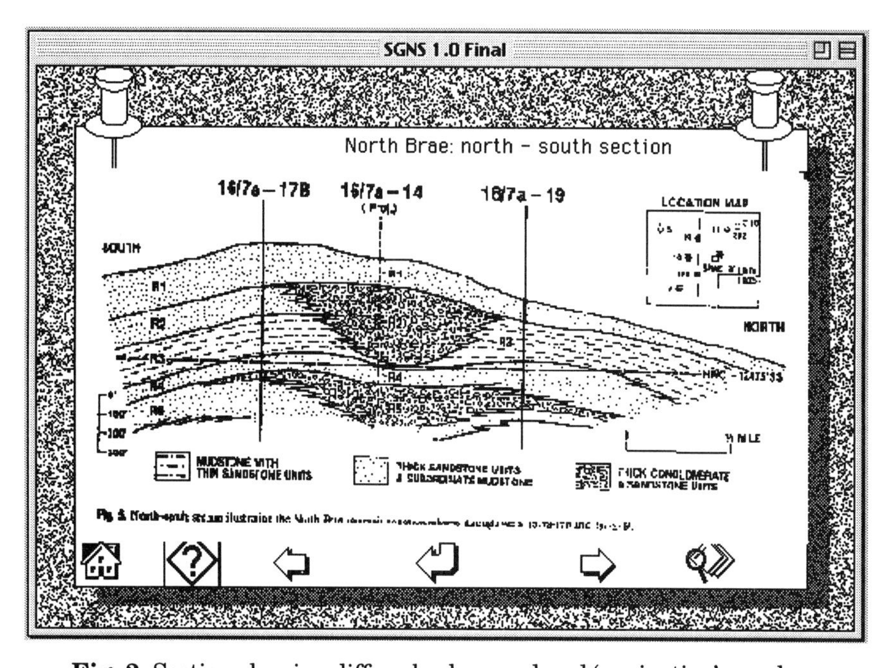

Fig. 2. Section showing diffuse background and 'navigation' panel.

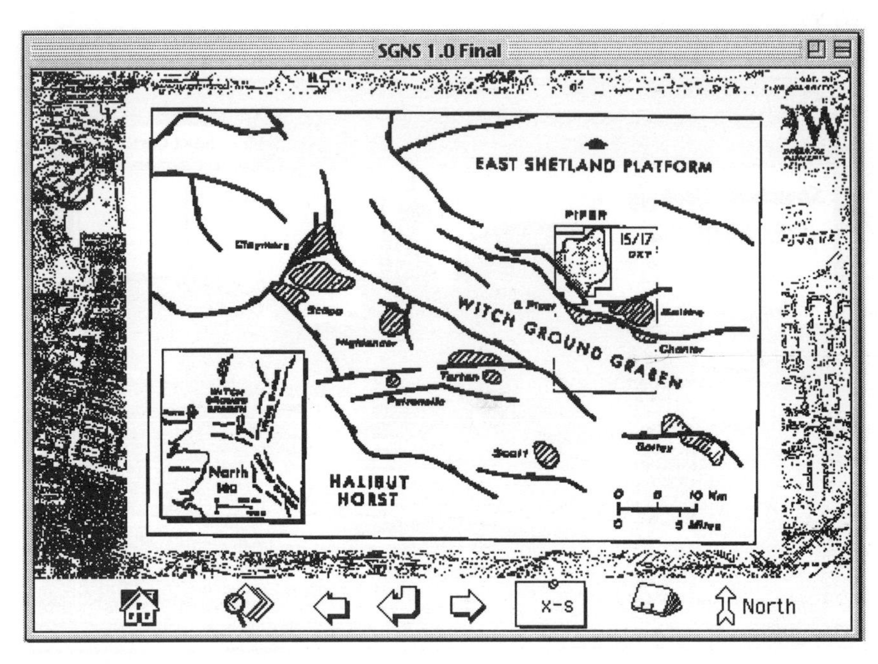

Fig. 3. Card containing cross sections giving impression of a vertical sheet pinned on a board.

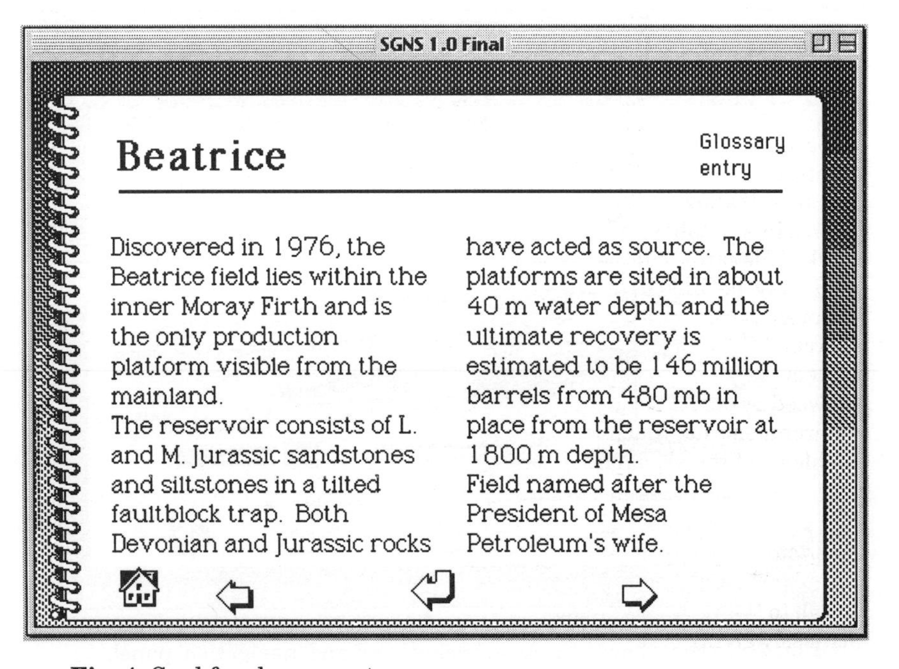

Fig. 4. Card for glossary entry.

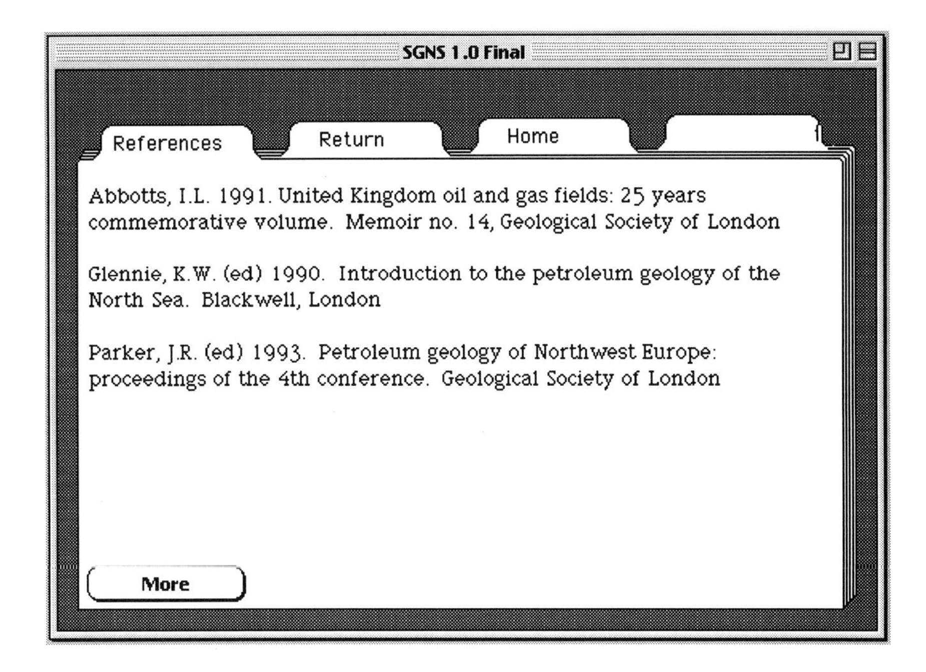

Fig. 5. References may be entered on typical file cards as shown here.

Conclusion

HyperCard for Macintosh computers is a versatile yet powerful environment for creating an information management system, incorporating both text and graphics. It allows researchers with no programming experience to generate and maintain a database of textual and graphical information. The analogy is of a stack of file cards and the power of this application is the way information on any card may be linked in ways of one's choice to other cards so creating an apparently three dimensional lattice of information. In addition all text may be searched by keyword. As well as providing an easy to use personal information system, *HyperCard* may be used to generate stacks which may be simply browsed by others or used as the basis for further development. The *HyperCard* environment for Macintosh computers provides one method of solving the problems of the exponential growth of information.

Acknowledgements

I wish to thank John Rennox for designing the first stack and George Bowes for selflessly giving of his time and advice on all aspects of computing.

References

Abbotts, I. L. 1991. *United Kingdom oil and gas fields: 25 years commemorative volume.* Memoir no. 14, Geol. Soc. Lond.

Bjerg, S. C. de, Mogessie, A. & Bjerg, E. 1992. Hyper-form - a *HyperCard*™ program for Macintosh microcomputers to calculate mineral formulae from electron microprobe and wet chemical analysis. *Comp. & Geosci.* **18**: 717-745.

Glennie, K. W. (ed.) 1990. *Introduction to the petroleum geology of the North Sea.* Blackwell, London.

Goodman, D. 1993. *The complete Hypercard 2.2 handbook.* Random House, New York.

Mogessie, A., Tessandri, R. & Veltman. C. B. 1990. EMP-AMPH - a *HyperCard*™ program to determine the name of an amphibole from electron microprobe analysis, according to the International Mineralogical Association scheme. *Comp. & Geosci.* **16**: 309-330.

Parker, J. R. (ed.) 1993. Petroleum Geology of Northwest Europe: *Proc. 4th Conf. Geol. Soc. Lond.*

Rennox, J. A. 1990. *The use of HyperCard™ in creating an application to be used in schools liaison work.* M.Sc. thesis, University of Glasgow.

Visualization of Basic Structural Geometries with *Structure Lab 1*

Declan G. De Paor
Department of Earth and Planetary Sciences,
Harvard University, 20 Oxford Street,
Cambridge MA 02138, U.S.A. depaor@eps.harvard.edu

Carol Simpson
Department of Earth Sciences,
Boston University, 675 Commonwealth Avenue,
Boston MA 02215, U.S.A. csimpson@bu.edu

Abstract– *Structure Lab 1* is a Macintosh application designed for use in the first few laboratory classes of an introductory structural geology course. The program permits students to relate the orientations of planar structures on a map, stereonet, and cross section. It automatically solves strike line, apparent *vs*. true dip, and 3-point problems.

Introduction

The first hurdle that most students of structural geology face is the vizualization of complex (or even simple) structures in three dimensions. Angles such as strike, dip, pitch (alias rake), trend, and plunge are particularly difficult for teachers to explain because they are distorted by projection onto the teacher's whiteboard or the students' textbook page. Strike line or structure contour problems are laborious – usually occupying a whole lab session – and students often make mistakes in construction which prevent them from seeing the structure's three dimensional attitude. To help students overcome problems of visualization in their first two or three structure laboratory classes, we have developed a Macintosh application called *Structure Lab 1.* Program code was written by the first author and student exercises were developed by the second. During the past three years, *Structure Lab 1* has been distributed commercially by Earth'nWare Inc. to about 100 universities and colleges worldwide.

Program Operation

Using *Structure Lab 1*, students learn the fundamentals of map interpretation, including cross section construction and outcrop pattern recognition for simple planar structures such as beds, dikes, or faults. By interactively changing

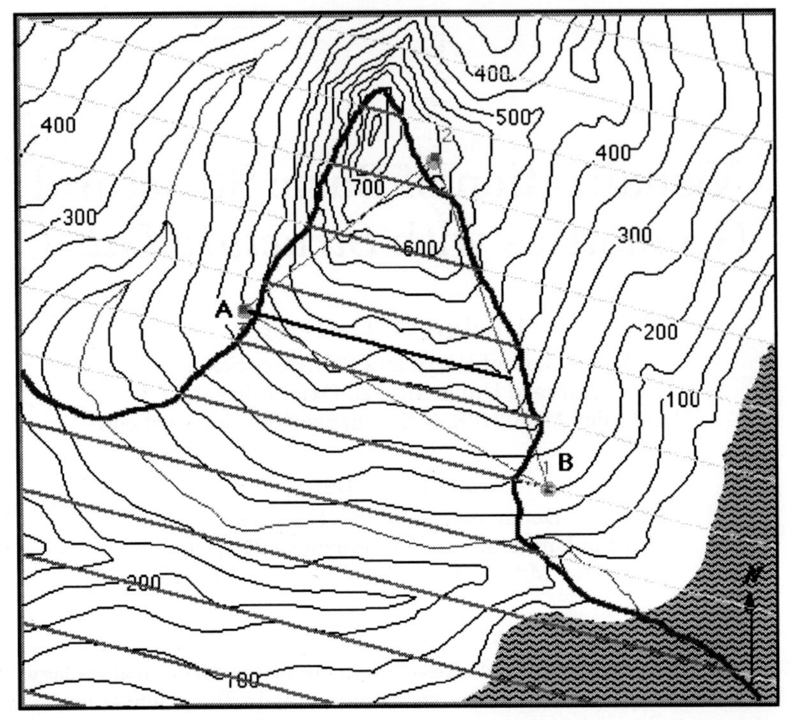

Fig. 1. Problem map based on a real topographic map. A,B,C are map locations. 1,2,3 are triangle vertices. Dark gray lines are strikes lines of plane below surface, light gray lines represent eroded portion.

parameter values with the computer mouse or keyboard, they see how dip and strike angles relate to strike lines on a map, apparent dips on a cross section, and great circles on a stereonet.

The program permits the user to open Macintosh files of two types – PICT or CNTR. PICT is a standard image file type with which most readers will be familiar. All Macintosh drawing and scanning programs permit the user to save files in this format. CNTR files are a special type created for this program. A CNTR file incorporates a PICT image of a problem map and data structures defining contour locations and elevations. Thus the computer 'knows' the elevation of every point on the map; it can echo the mouse's location in all three dimensions and it can change the color or gray shade of strike lines where they intersect the topography. The program comes with a set of ready-made CNTR files (*e.g.,* Fig. 1). Teachers may convert their own scanned problem map PICT files into CNTR files using a separate utility program called *Contour File Maker*.

As students move the mouse around on the map after opening a CNTR file, the computer shows the elevation of a planar structure at the current point. The plane is defined by three vertices of a triangle which can be edited by a click and drag of the mouse or by keyboard entry of dip and strike data into a dialog box. By changing the plane's location and orientation, students may discover the significance of Veeing upstream and Veeing downstream patterns. Features such

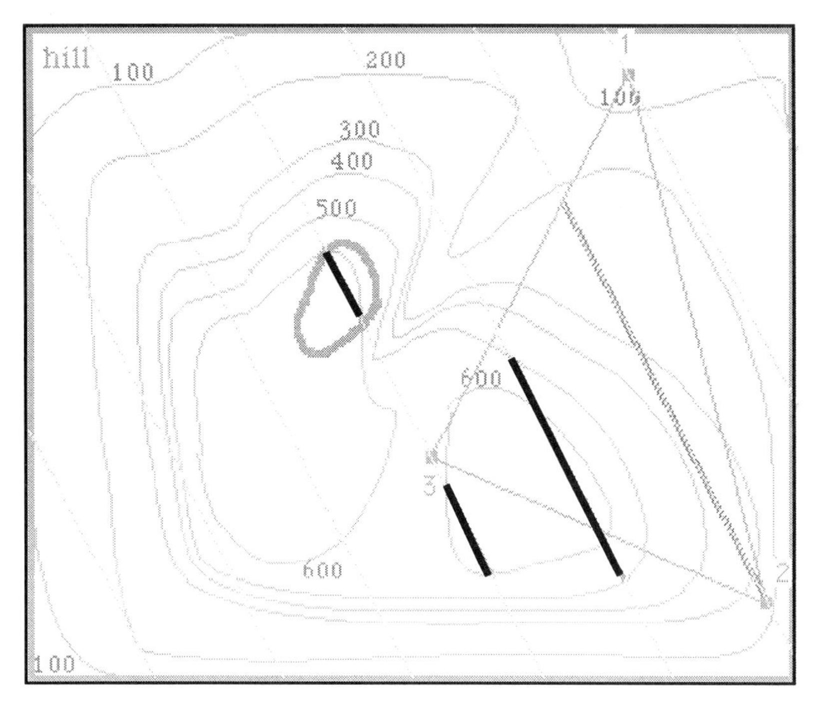

Fig. 2. The orientation and location of a planar structure are modified interactively by dragging points 1,2,3 and keying in corresponding altitudes. The thick dark loop defines an outlier if the plane is a bed or a klippe if it is a fault. Strike lines are black below ground, gray above. Thick gray line is drawn through median triangle vertex.

as outliers, inliers, tectonic klippe, and tectonic windows may be created by trial and error – an instructive exercise in itself (Fig. 2). Three point problems are solved by dragging triangle vertices to pre-assigned points marked on the map and entering vertex elevations in corresponding edit fields (not shown). There are options for solving apparent dip from true dip, strike, and apparent dip direction, and for determining true dip from two apparent dips and apparent dip directions.

In addition to learning about outcrop patterns in map view, students can learn to relate map, cross section, and stereonet information by making menu selections (Fig. 3). The computer waits for the user to click and drag the mouse along a desired line of section and then constructs the corresponding cross section, complete with the trace of the planar structure with the correct apparent dip. When a plane's orientation is changed on the map, its apparent dip in the chosen line of section also changes. Similarly, the map view may be related to a stereonet (Fig. 4).

How It Works

Apart from maintenance of the graphical user interface (90% of every programming effort), most of the code for this program is devoted to algebraic calculations

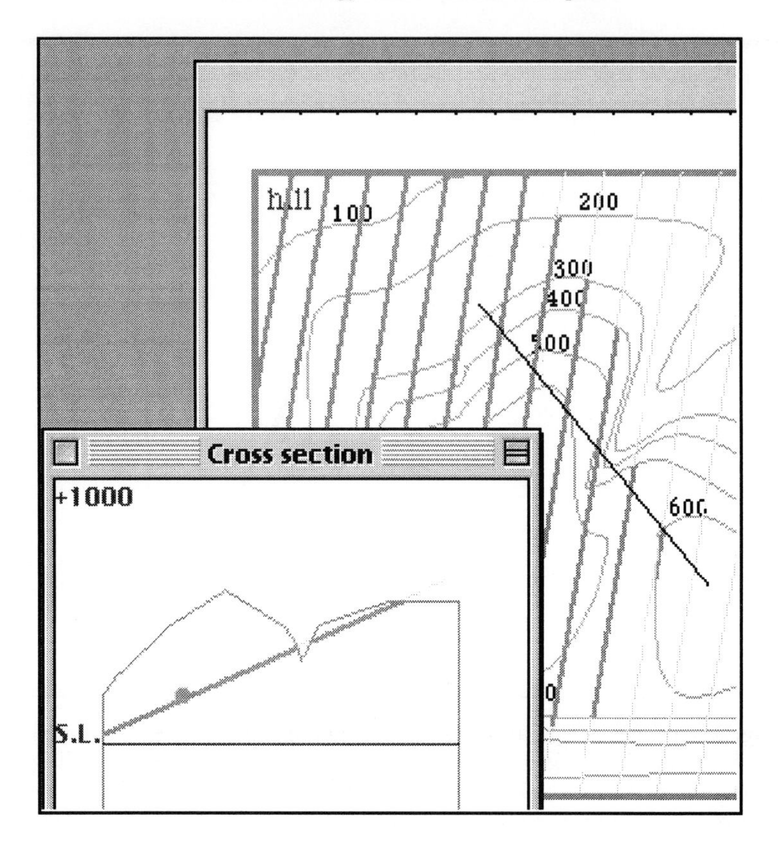

Fig. 3. Map and cross section windows. Thin line on map is line of section. Planar structure represented by dark grey line underground and by light gray line where eroded. When plane is edited on map, its cross section changes interactively.

involving either orthogonal tensor transformations or spherical trigonometry. However, the representation of strike lines in a CNTR file required an innovative approach which is described here in some detail as it may have a variety of other useful applications.

Every Macintosh file has two sections or "forks"; the data fork and the resource fork. The data fork contains sequential bytes of data which may be interpreted as text, spreadsheet data or picture drawing instructions, *etc*. The resource fork contains "resources" which are special types of organized data. Resources are used to store a user interface element (cursor, menu, window, *etc.*), a file's icon, or an application-specific data structure. In the case of a CNTR file, the problem map is stored as a PICT and each of its contours is stored as a resource of type "RGN " (the trailing space is necessary). A region is a Mac Toolbox representation of an arbitrary geometric shape which can consist of one or more closed loops. Regions are defined by toolbox calls. Syntax varies slightly with language but the format is illustrated by the following *FutureBASIC*™ code:

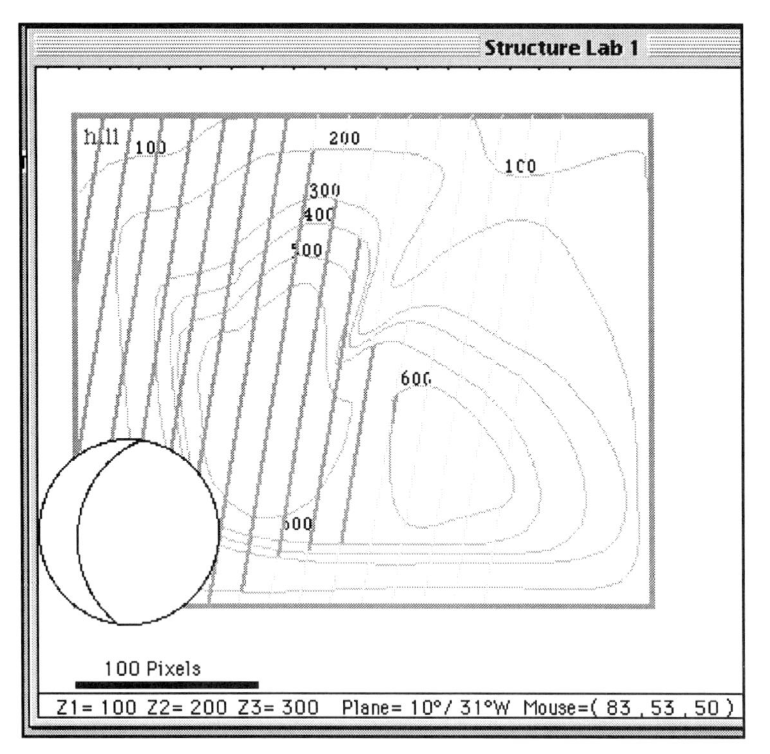

Fig. 4. Plane represented on map and stereonet. Students learn to relate great circle orientation with strike line spacing.

```
theRgn& = FN NEWRGN
LONG IF theRgn&
  CALL OPENRGN
  CALL MOVETO (X0,Y0)
  CALL LINE (0,0)
  CALL LINETO (X1,Y1)
  CALL LINETO (X2,Y2)
  '
  '
  CALL LINETO (X0,Y0)
  CALL CLOSERGN (theRgn&)
  '
  'use region here
  '
  CALL DISPOSERGN (theRgn&)
END IF
```

The first line requests a handle to a 32k memory block in which to store the region data. If the request is successful, the region is defined by a series of line drawing commands, ending in a closed polygon. After the region is defined, it may be used as an entity. Then the assigned memory is released. A range of Toolbox calls make it easy to use regions thus defined. For example, the two lines

```
CALL GETMOUSE (thePt&)
boolean = FN PTINRGN (thePt&,theRgn&)
```

may be used to check whether the mouse is currently in the defined region. To draw within a confined region, the following code segment may be used:

```
CALL GETCLIP (oldClipRgn&)
CALL SETCLIP (theRgn&)
CALL MOVETO (0,0)
CALL LINETO (1000,1000)
CALL SETCLIP (oldClipRgn&).
```

Finally, to draw outside of the region, one may use code such as this:

```
CALL SETRECT (mapRect&,0,0,WINDOW(_width),WINDOW(_height))
CALL RECTRGN (mapRgn&,mapRect&)
CALL DIFFRGN (mapRgn&,theRgn&,outerRgn&)
CALL SETCLIP (outerRgn&)
etc.
```

(Of course, each memory handle in the above code segments must be properly initialized and disposed of). In the utility program *Contour File Maker*, regions are defined for each contour level in the PICT. The PICT is displayed and the user is prompted to digitize a series of points. Regions are stored in the file's resource fork and the file is saved as a CNTR file type. When *Structure Lab 1* opens a CNTR file, the stored region resources are read into memory and the code outlined above is used to clip the drawing of strike lines so that they appear in different colors (on a color monitor, or shades of gray on a grayscale monitor) depending on whether the planar structure is present underground or has been eroded away.

Conclusion

The authors believe that it is important for students to learn to visualize geological structures in three dimensions by manual laboratory exercises and they use the program described above in addition to, not instead of, traditional pencil and tracing paper exercises. The main advantage of using the program is that students can get a feel for how things work by viewing dozens of different map, cross section, and stereonet configurations in a matter of minutes. In contrast, it is difficult to fit more than one or two manual strike line or apparent dip problems into the lab time available. Once students have 'seen the light' using the program, their progress in manual exercises is in general much faster.

References

Davis, G. H. & Reynolds, S. J. 1996. *Structural Geology of Rocks and Regions.* (2nd. ed.) J. Wiley, New York.

Marshak, S. & Mitra, G. 1988. *Basic Methods of Structural Geology*. Prentice Hall, Englewood Cliffs, New Jersey.

Using Graphics Programs to Help Students Understand Strain

Barbara J. Tewksbury
Department of Geology, Hamilton College, Clinton, NY 13323, U.S.A.
btewksbu@hamilton.edu

Abstract– Commercially-available computer graphics programs can be effectively used to help students overcome their inaccurate expectations of what happens during deformation of objects by homogeneous pure and simple shear. Many computer graphics packages allow students to draw an object on the screen and then to distort the object using commands that are the equivalents of pure shear and simple shear. As students create and deform objects, they develop a better visual understanding of 1) the behavior of lines and angles during strain, 2) the differences between pure and simple shear, and 3) the significance of the principal directions of strain. Once students have developed an accurate intuitive understanding of strain, they can go on to tackle strain analysis armed with an arsenal of accurate visual images on which to rely as they work through the quantitative aspects of strain.

Introduction

Concepts involved in strain and kinematic analysis are notoriously difficult for students in Structural Geology to grasp. Strain is a deceptive topic for students, because their intuitions about many aspects of strain are nearly invariably incorrect. If students are shown the drawing in Fig. 1, for example, most immediately leap to the conclusion that the deformation involved simple shear. As another example, most students have a difficult time understanding that amount of strain depends on the orientation of what is being measured. They have a difficult time visualizing how homogeneous deformation can change one angle by 30° and not change another angle at all. As yet another example, most students assume that lines perpendicular and parallel to the shear direction rotate and become parallel to the principal directions during simple shear. The actual behavior of principal directions during progressive simple shear is a concept that has defeated all but my best students in the past.

The job of getting students past their incorrect intuitions is a challenge, and a mathematical treatment of the subject simply doesn't solve the problem for most students. After trying a number of approaches, I have become convinced that

Fig. 1. Student intuition about strain is commonly faulty. Most students, for example, leap to the conclusion that deformation of the object in this figure must have involved simple shear.

only good firsthand, visual experience seems to carry students past the stage where their intuitions lead them astray. Once their understanding has improved to the point that their intuitions are good, students can be introduced to quantitative aspects of strain, which means more to them at that point than simple number-crunching.

Approaches that Others Have Used

The many exercises that have been published in various books and lab manuals using computer cards have gone a long way toward meeting the goal of helping students to visualize and better understand strain, but card deck models have many limitations, not the least of which is the fact that old-fashioned computer cards are impossible to get any more and fresh file cards simply don't work very well. In terms of teaching strain, the old card deck models did an adequate job with simple shear but did not allow students to explore pure shear or complex strain.

Others have also advocated using computers to teach students various aspects of strain (e.g., McEachran and Marshak, 1986), and several good products designed specifically for strain analysis are available from software authors and distributors. Programs such as these are available from Rockware (2221 East St., Suite 101, Golden, CO 80401) and Earth'nWare (148 Cadish Avenue, Hull MA 02045, U.S.A.)

The Approach in This Paper

This paper outlines a method for using generic computer graphics programs to help students develop an accurate intuitive grasp of pure and simple shear by allowing them to draw objects, deform them, and watch what happens as they deform. The aim is not to teach quantitative aspects of strain but rather to develop a visual understanding of strain that can form the basis for subsequent strain analysis. As such, this method is designed to be an introduction to strain that should be followed with coverage of additional topics in strain theory and practical strain analysis.

The technique described here is interactive and not lecture-based. It is not a self-run tutorial for students to complete independently; rather, it is the outline for a series of class meetings where students work individually at computers, and the instructor periodically brings the class together again for summary. This paper describes in detail the method for teaching these classes interactively and outlines the advantages of using the computer approach.

The strategy described here can be adopted with no modification if Macintosh computers are available with Aldus *SuperPaint*™. Other commercially-available programs could be used on either a Macintosh or an IBM PC-compatible computer, and one of the final sections of this paper ("Using a Different Graphics Program") offers suggestions of what to look for in choosing a program that would be suitable as a substitute for *SuperPaint*.

The Exercise, Part I: Establishing the Behavior of Circles, Lines, and Angles during Homogeneous Deformation

Students have little trouble visualizing that homogeneous deformation of a series of circles ought to result in a series of ellipses of similar shape and orientation. They rarely, on the other hand, intuitively grasp what homogeneous deformation means for a group of lines of various orientations or for the angles between lines. They intuitively apply their notion of the "sameness" of homogeneity to everything, expecting that all lines should behave the same and all angles between lines should behave the same during homogeneous deformation, just as all circles behave the same. Using the computer, they can see first hand that line length change and change in angle between lines depends upon the initial orientation of lines and pairs of lines. The following series of exercises allows students to deform a series of circles, lines, and crosses homogeneously by both pure and simple shear in order to see what actually happens. Part I can be completed in about two hours, including teaching students how to use *SuperPaint*, a fairly simple Mac program.

Establishing that Particular Commands Produce Homogeneous Deformation.

Students must first be convinced that they can expect homogeneous deformation of objects on the screen when they use certain computer commands. In *SuperPaint*, these commands are `stretch` and `slant`. Throughout this paper, I have used these terms to refer to the computer command equivalent to pure shear and simple shear, respectively. Programs other than *SuperPaint* may use different names for these commands.

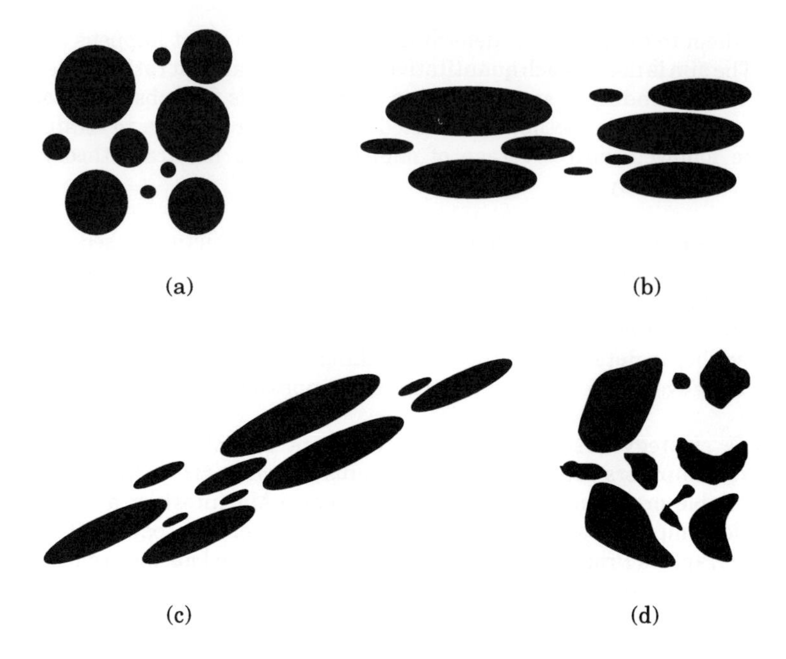

Fig. 2. a) Group of initially circular objects drawn on the computer screen. b) Same group of circles deformed by pure shear (`stretch` command in *SuperPaint*). c) Same group of circles deformed by simple shear (`slant` command in *SuperPaint*). Students should note that both b and c show homogeneous strain. d) Same group of circles deformed heterogeneously.

–*Student*: Draw a series of circles of various sizes (Fig. 2a), and `save`. Select the group of circles. `Stretch` the circles, and notice what happens as the circles deform (Fig. 2b). Try several different amounts of `stretch`, and notice the shapes and orientations of the resulting ellipses.

–*Instructor*: Ask students to discuss what aspects suggest that the `stretch` command produces deformation that is homogeneous. Students have little trouble visualizing that similar shape and orientation are the critical factors, whereas differences in sizes among ellipses are due to initial differences in sizes of circles.

–*Student*: `Undo` or `revert` to saved so that the screen image has undeformed circles. Select the group of circles, and `slant` them (Fig. 2c). Try several different amounts of `slant`, and notice the shapes and orientations of the resulting ellipses.

–*Instructor*: Ask students to discuss what aspects suggest that the `slant` command produces deformation that is homogeneous.

–*Student*: `Undo` or `revert` to saved so that the screen image has undeformed circles. Select the group of circles, and choose the `distort` command (Fig. 2d). Experiment with several different distortions, and notice the shapes and orientations of the resulting ellipses.

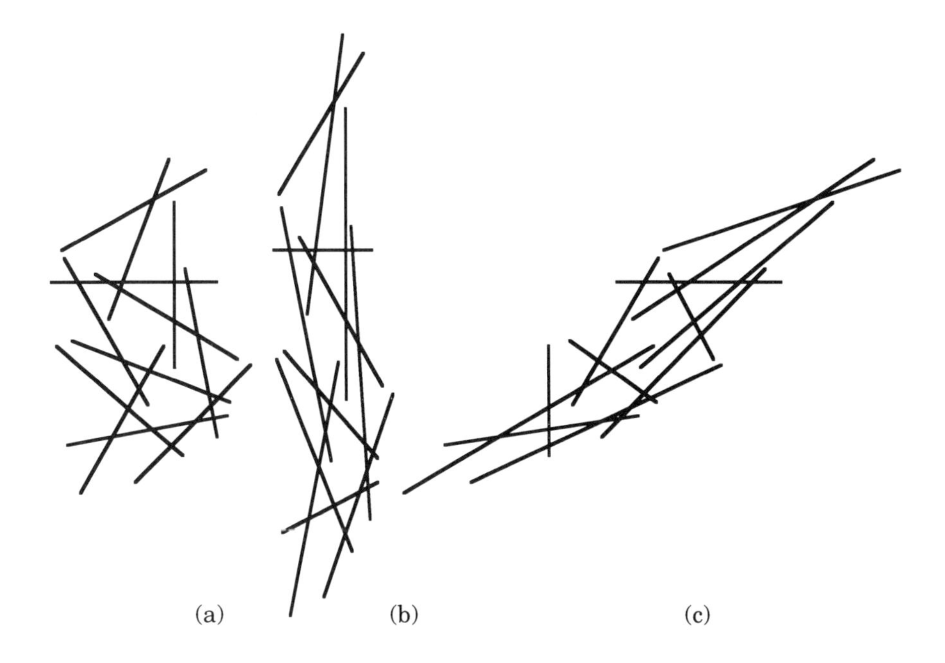

Fig. 3.a) Group of lines of same initial length drawn on the computer screen. b) Same group of lines deformed by pure shear. c) Same group of lines deformed by simple shear. Students should note that change in length of a line depends upon its orientation.

–*Instructor*: Ask students to discuss what makes this operation inhomogeneous. Compare it with `stretch` and `slant`.

Behavior of Lines during Homogeneous Deformation.

The following activity shows students that lines of different orientations behave differently even when deformation is homogeneous. Working with a program that has continuous redraw allows students to watch line length change as it happens.

–*Student*: Draw a series of lines that are all the same length but that have different orientations (Fig. 3a), and `save`. Select the group of lines. `Stretch` the lines, and notice what happens as the lines deform (Fig. 3b). Try several different amounts of `stretch`, and notice the lengths of the resulting lines.

–*Instructor*: Emphasize that they already know that the `stretch` command produces homogeneous deformation. Have students discuss why homogeneous deformation has resulted in lines of different lengths.

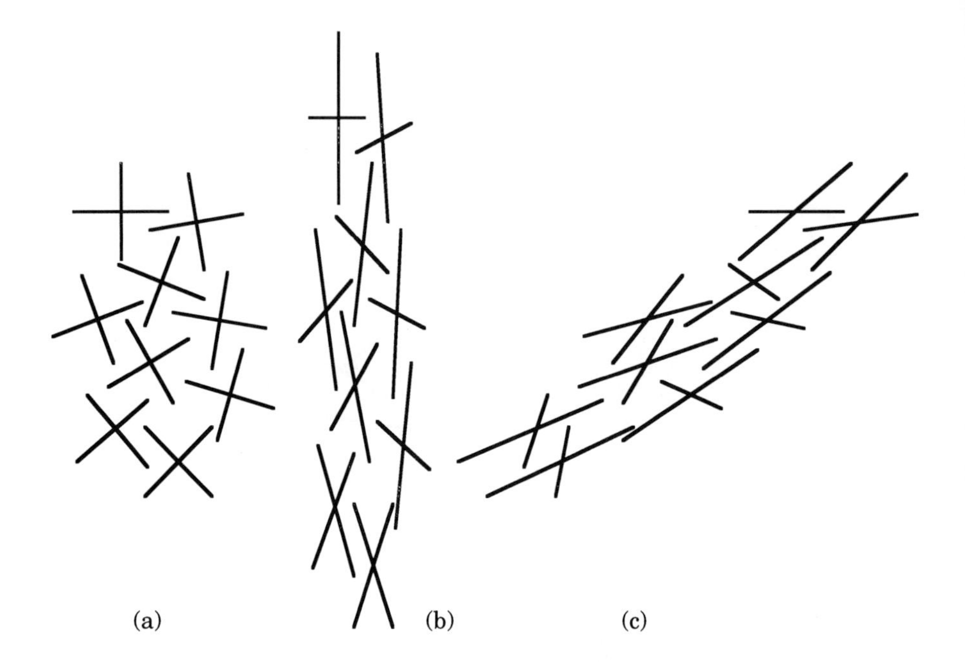

Fig. 4. a) Group of perpendicular lines of various initial orientations drawn on the computer screen. b) Same group of lines deformed by pure shear. c) Same group of lines deformed by simple shear. Students should note that change in angle depends upon the orientation of the pair of lines.

–Student: `Undo` or `revert` to saved so that the screen image has undeformed lines. Select the group of lines, and `slant` them (Fig. 3c). Try several different amounts of `slant`, and notice the lengths of the resulting lines.

–Instructor: Remind students that the `slant` command, too, produces homogeneous deformation. Ask students to summarize what happens to the lengths of lines during `stretch` and `slant`.

Behavior of Angles Between Lines During Homogeneous Deformation

The following activity shows students that pairs of perpendicular lines of different orientations behave differently even when deformation is homogeneous. Students can also see clearly that only certain pairs of lines remain perpendicular to one another. Working with a program that has continuous redraw allows students to watch change in angle between lines as it happens.

–Student: Draw a series of crosses with perpendicular arms; rotate each cross to a different orientation (Fig. 4a). `Save`. Select the group of crosses. `Stretch` the crosses, and notice what happens as the crosses deform (Fig. 4b). Try several

different amounts of `stretch`, and compare the angles between arms of individual crosses. As you `stretch` the crosses, watch the arms flex.

–*Instructor*: Remind students that they already know that the `stretch` command produces homogeneous deformation. Have students discuss what happened to the angles between arms of individual crosses. Be sure that they note the orientation of crosses that still have perpendicular arms after `stretch`.

–*Student*: `Undo` or `revert` to saved so that the screen image has undeformed crosses. Select the group of crosses, and `slant` them (Fig. 4c). Try several different amounts of `slant`, and note the angles between arms of individual crosses.

–*Instructor*: Remind students again that the `slant` command produces homogeneous deformation. Be sure that they notice that some of the crosses may have perpendicular arms after `slant` if the amount of `slant` is just right. Ask students to summarize what happens to the angles between arms of individual crosses during `stretch` and `slant`.

Behavior of Straight, Parallel Lines During Homogeneous Deformation

The following activity shows students that straight and parallel lines remain straight and parallel after homogeneous deformation, regardless of the orientation of the lines.

–*Student*: Draw a grid of several horizontal and vertical lines (Fig. 5a), and `save`. Select the group of lines. `Stretch` the grid, and notice what happens as the grid deforms (Fig. 5b). Try several different amounts of `stretch`, and notice whether the lines remain straight and parallel or not. `Undo` or `revert` to saved, and try the same thing with `slant` (Fig. 5c).

–*Instructor*: Have students discuss what they have observed about the behavior of straight and parallel lines during homogeneous deformation.

–*Student*: `Undo` or `revert` to saved so that the screen image has an undeformed grid. Select the grid, and `rotate` it about 45° (Fig. 5d). Try several different amounts of `stretch`, and notice whether the lines remain straight and parallel or not. `Undo` or `revert` to saved, and try the same thing with `slant` (Fig. 5e).

–*Instructor*: Have them discuss whether the behavior of the rotated grid made them change their generalization about the behavior of straight and parallel lines during homogeneous deformation.

Consolidating the Behavior of Circles, Lines, and Angles During Homogeneous Deformation

After spending time in front of the computer, I have found it necessary to step back and consolidate what students know before going on to define strain ellipses and to explore what makes `stretch` and `slant` different from one another.

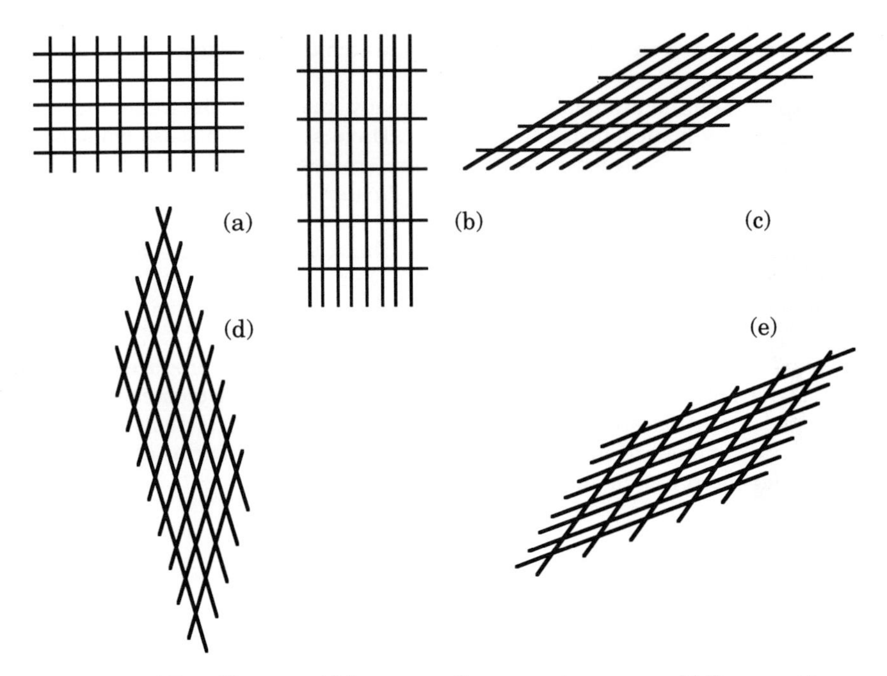

Fig. 5. a) Rectilinear grid drawn on the computer screen. b) Same grid of lines deformed by pure shear. c) Same grid of lines deformed by simple shear. d) Same grid rotated 45° and deformed by pure shear. e) Same grid rotated 45° and deformed by simple shear. Students should note that, because strain is homogeneous, all lines of the grid remain straight and parallel after strain, despite the fact that the angles have been distorted.

I typically start by having students work out a definition for homogeneous deformation, which commonly comes out initially with some flavor of "everything is deformed the same." I prod them with questions about what they mean by "everything", and they eventually work out a more precise definition reflecting the fact that circles deform homogeneously into ellipses of identical shape and orientation (which is where their sense of "everything is deformed the same" comes from) but that individual lines and angles do not all behave "the same" during homogeneous deformation, and that both change in line length and change in angle between mutually perpendicular lines depend on orientation of lines and pairs of lines. This apparent inhomogeneity resulting from different orientation is very difficult for most students to grasp without having seen it happen before their very eyes. Working it out on the computer helps enormously in this regard.

I then introduce the terms pure shear and simple shear for the computer operations of stretch and slant respectively. I typically go on to ask students if they can tell whether a set of ellipses was produced by pure or simple shear (Fig. 6a and b). Invariably, their reaction is that Fig. 6a reflects pure shear, while Fig. 6b reflects simple shear, because their sole experience with strain has

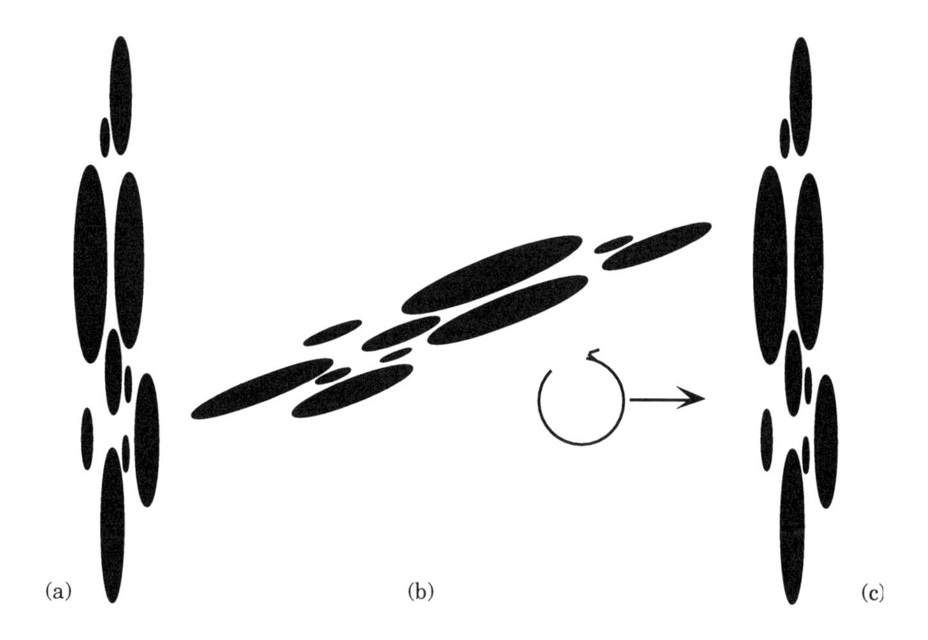

Fig. 6. (a,b,c) Students typically assume that `slanted` packages must have been produced by simple shear, when, in fact, similar geometries can be produced by either pure shear or simple shear.

conditioned them to see ellipses as remaining "upright" after pure shear, and ellipses as being "skewed" during simple shear. I then select and rotate the ellipses in Fig. 6b to show them that the ellipses look exactly the same as those in 6a. This typically amazes them.

Before going on to Part II, I introduce the notion of strain ellipses and the principal directions of strain.

The Exercise, Part II: Establishing the Differences Between Pure and Simple Shear

Most students have little difficulty visualizing the lack of rotation of principal directions during pure shear and the rotation of principal directions during simple shear, even when this is simply presented in a series of diagrams on a blackboard. What they have trouble seeing, however, is the behavior of the principal directions during progressive simple shear. Most students expect both that the same material lines will track the principal directions and that those material lines will start out parallel and perpendicular to the shear direction. Part II helps students visualize that neither of these expectations is correct. Part II can be completed easily in about an hour and a half.

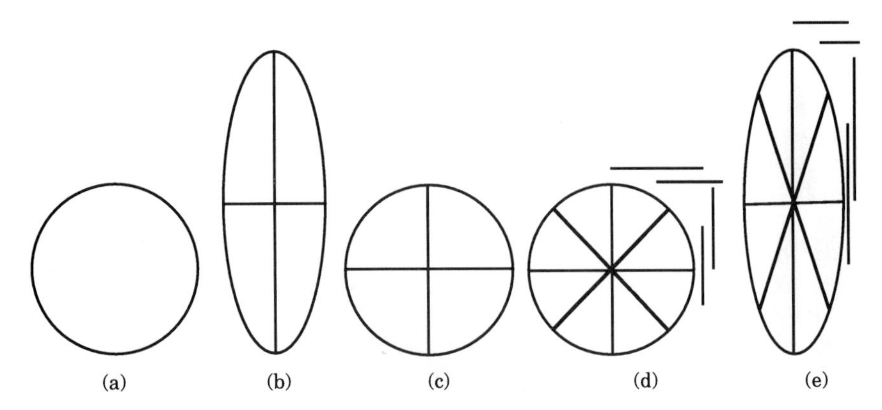

Fig. 7.a) Original circle drawn on the computer screen. b) Circle deformed by pure shear, principal directions drawn in. c) Ellipse in (b) retrodeformed to a circle. d) Lines at 45° to principal directions and lines outside the circle added. e) Circle and lines in (d) deformed by simple shear. Students should note that, in pure shear, principal directions do not rotate and that only lines parallel to the principal directions will be perpendicular both before and after deformation.

Investigating Pure Shear

The following activity shows students that the principal directions do not rotate during pure shear and that lines parallel to the principal directions are perpendicular both before and after deformation.

–*Student*: Draw a circle (Fig. 7a), and `save`. Deform the circle by pure shear into a moderately elongate ellipse, and carefully draw in the principal directions (Fig. 7b). Select the circle and axes, and retrodeform the ellipse into a circle (Fig. 7c). Deform the circle and axes again, trying several different amounts of pure shear, and note the behavior of the principal directions.

–*Instructor*: Ask students to discuss the behavior of the principal directions during pure shear.

–*Student*: Retrodeform the ellipse with principal directions into a circle. Add two more lines that are mutually perpendicular but are not parallel to the principal directions. Draw a few lines outside the circle that are parallel to the principal directions (Fig. 7d), and save. Select the circle, lines, and axes, and deform them by simple shear (Fig. 7e). Try several different amounts of pure shear, and note the behavior of the principal directions and the other lines.

–*Instructor*: Remind students that pairs of mutually perpendicular lines that are not parallel to the principal directions will not remain perpendicular after deformation (students should remember this from Part I, but it is important to reinforce the idea here). Be sure that students also see that any lines parallel to the principal directions will be perpendicular both before and after deformation. Some students have the erroneous impression that the ellipse axes themselves are unique and can be mentally divorced from the "regular" lines in an object.

Investigating Simple Shear

The following activity shows students that the principal directions rotate during simple shear, that lines parallel to the principal directions are perpendicular both before and after deformation, and that lines parallel to the principal directions in simple shear depend on the amount of shear.

–*Student*: Draw two circles (Fig. 8a), and save. Select both circles, and deform them by simple shear into moderately elongate ellipses. Carefully draw in the principal directions for the left hand ellipse only (Fig. 8b). Select the circles and axes, and deform them again by simple shear into highly elongate ellipses. Carefully draw in the principal directions for the right hand ellipse (Fig. 8c). Note what has happened to the principal directions in the left hand ellipse. Select both circles and axes, and retrodeform the ellipses into circles (Fig. 8d). Deform the circles and axes again, trying several different amounts of simple shear, and note both the behavior of the principal directions and the positions of the axes-to-be in the undeformed state.

–*Instructor*: Ask students to discuss the behavior of the principal directions during simple shear. Be sure that students notice the following: 1) that the principal directions rotate during simple shear, 2) that lines parallel to the principal directions are perpendicular both before and after deformation but not during deformation, 3) that the amount of simple shear governs which sets of material lines are perpendicular before and after deformation, and 4) that lines parallel and perpendicular to the shear direction can never become the principal directions. Finish this section by having students summarize the differences between pure and simple shear.

The Exercise, Part III: Consolidating Strain with an Example Using Brachiopods

Once students have established the general character of pure and simple shear, they need practice applying the concepts in order to consolidate what they know. Part III gives students practice and can be completed easily in about an hour.

Pure and Simple Shear

The following activity shows students that the same shape changes can be produced by both pure and simple shear. For this section, you need to provide students with a document containing an image of brachiopods all oriented the same way and interspersed with circles. The document should contain 3 identical groups of brachiopods and circles, as shown in Fig. 9a.

–*Student*: Open document 1. Select the left hand group of brachiopods and circles, and deform them by pure shear (Fig. 9b). Select the bottom group of brachiopods and deform them by simple shear (Fig. 9c). Admire the differences in appearance.

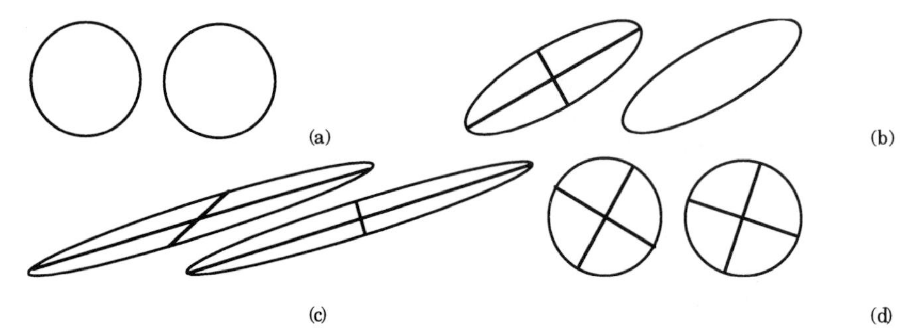

Fig. 8.a) Original pair of circles drawn on the computer screen. b) Circles deformed by simple shear into moderately elongate ellipses, principal directions drawn in on one of the ellipses. c) Ellipses in (b) deformed again by simple shear, principal directions drawn in on the second ellipse. d) Ellipses in (c) retrodeformed to circles. Students should note that principal directions rotate, that lines parallel to the principal directions are perpendicular both before and after deformation but not during deformation, that the amount of simple shear governs which sets of material lines are perpendicular before and after deformation, and that lines parallel and perpendicular to the shear direction can never become the principal directions.

–*Instructor*: Ask students whether, given a group of deformed brachiopods, they would be able to distinguish whether the brachiopods had been deformed by pure or simple shear. Some astute students will realize that this is no different from the circles-to-ellipses situation from Part I, but most will have difficulty visualizing how the "slanted" brachiopods could have been produced by pure shear.

–*Student*: Select an undeformed group of brachiopods and circles, and rotate them about 45° (Fig. 9d). Deform the group by pure shear (Fig. 9e). Compare the results with your previous pure and simple shear deformations.

–*Instructor*: This demonstration should convince even the sceptics that homogeneous pure shear and simple shear deformation cannot be distinguished from one another simply by shape and orientation of objects in the deformed state.

Locating the Principal Directions

The following activity consolidates the idea of principal directions in both pure and simple shear. For this section, you need to provide students with two documents, one containing an image of a number of brachiopods each oriented a different way and interspersed with circles, as shown in Fig. 10a, and one containing an image of an ellipse and deformed brachiopod, as shown in Fig. 11a.

–*Student*: `Open` document 2. Select the group of brachiopods and circles, and deform them by pure shear (Fig. 10b). Note the differences in shapes among brachiopods, and note the angle between hinge and median line in each brachiopod.

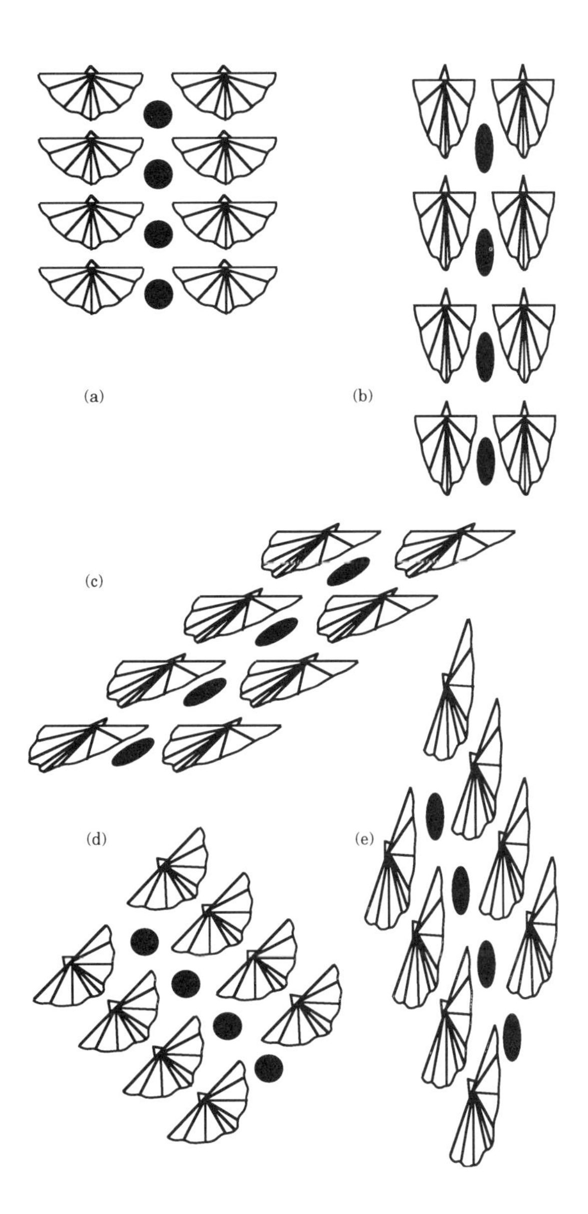

Fig. 9. a) Original group of brachiopods and circles drawn on the computer screen. b) Brachiopods and circles deformed by pure shear. c) Brachiopods and circles deformed by simple shear. d) Original set of brachiopods and circles rotated 45°. e) Brachiopods and circles in (d) deformed by pure shear. Students should note that whether a brachiopod is "skewed" during pure shear depends upon its orientation with respect to the principal directions.

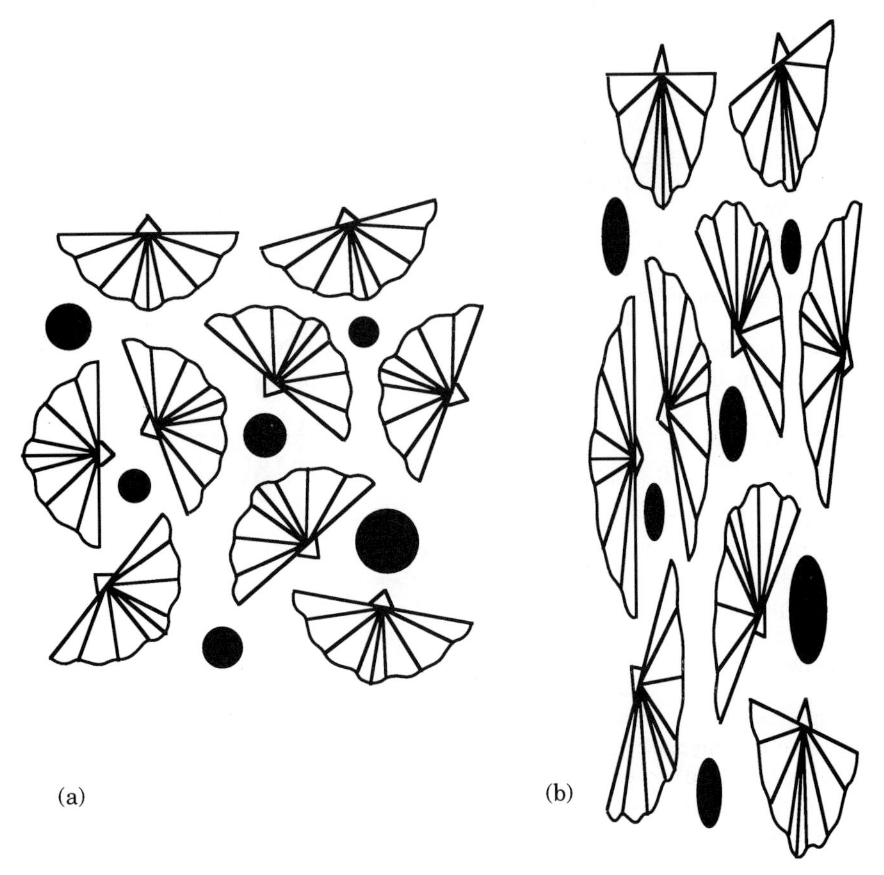

Fig. 10. a) Group of circles and randomly-oriented brachiopods drawn on the computer screen. b) Brachiopods and circles deformed by pure shear. Students should note that brachiopods oriented with hinge and median lines parallel to the principal directions will be the only brachiopods in which the hinge and median lines are still perpendicular after deformation.

–*Instructor*: Ask students whether deformation was homogeneous or inhomogeneous. Once they have reminded themselves that this must, in fact, be homogeneous deformation, ask them why each brachiopod has a different shape. This gives an opportunity to reinforce the idea that lines and angles behave differently depending upon orientation.

Ask students whether the principal directions could be located based upon the deformed brachiopods alone. Most students will have little trouble working out that, if the deformed set contains a brachiopod whose hinge and median lines are still perpendicular, the principal directions must lie parallel to the hinge and median line of that brachiopod.

Ask students whether this same strategy for locating the principal directions could be applied to brachiopods deformed by simple shear. Most students have an intuitive reaction that the strategy would be useful for pure shear, but not simple shear. The next activity helps correct the misapprehension.

–*Student*: `Open` document 3. Note that the hinge and median lines of the brachiopod are perpendicular to one another and parallel to the principal directions of the ellipse (Fig. 11a). Select the brachiopod and the ellipse, and retrodeform them by simple shear until the ellipse becomes a circle (Fig. 11b). Note that, while the brachiopod has changed shape, the hinge and median lines are still perpendicular, as are the lines that were parallel to the principal directions in the ellipse.

–*Instructor*: Reinforce the notion that a brachiopod can have perpendicular hinge and median lines after simple shear deformation even though its shape has changed. This should show students that locating the principal directions using lines that are perpendicular both before and after deformation works equally well for objects deformed by pure and simple shear.

–*Student*: Return to document 2. Select the group of brachiopods and circles (Fig. 12a), and deform them by simple shear (Fig. 12b). Note the differences in shapes among brachiopods, and note the angle between hinge and median line in each brachiopod. Decide whether there are any brachiopods that can give you the orientation of the principal directions. Then, give the brachiopods another increment of simple shear, and look for the principal directions again.

–*Instructor*: Reinforce the notion that most individuals in a group of randomly-oriented brachiopods will both change shape and be left with skewed hinge and median lines. Only brachiopods of a particular orientation will have perpendicular hinge and median lines both before and after deformation. Make sure, too, that students see that different brachiopods will retain hinge/median perpendicularity with different amounts of simple shear.

Using a Different Graphics Program

The aims of the activities described in this article can be achieved with programs for the Macintosh such as *SuperPaint*, *Canvas*™, *Freehand*™, and *Illustrator*™, despite the fact that specific commands differ from program to program. Similar programs are also available for the *DOS*™ world.

Some programs, however, are better than others for the specific aims outlined in this paper. I have found that Aldus *SuperPaint* works extremely well in this exercise. If you choose not to use *SuperPaint*, here are several important criteria for selecting a graphics package for this type of exercise:

–The program must be able to do more than change the size and orientation of an object. It must also have an operation equivalent to simple shear, an operation commonly used in graphics programs to `slant` objects or text.

–A program with a specific operation equivalent to pure shear and one equivalent to simple shear allows students to make the mental transition to pure and simple shear more easily. *SuperPaint*, for example, has a `stretch` command equivalent to pure shear and a `slant` command equivalent to simple shear.

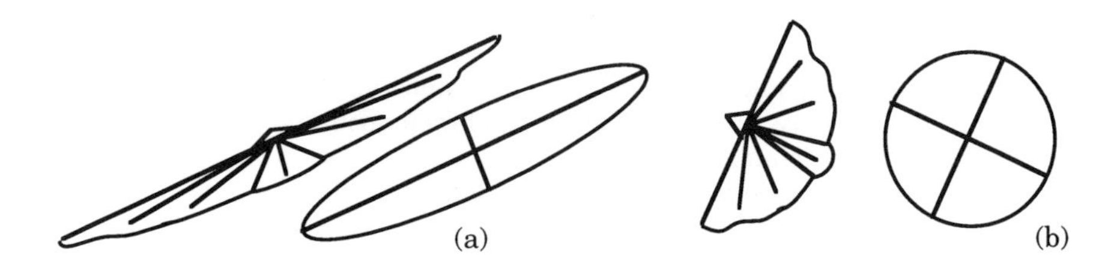

Fig. 11. a) Brachiopod deformed by simple shear. b) Brachiopod and ellipse from (a) retrodeformed by undoing the simple shear. Students should note that hinge and median lines are perpendicular after simple shear deformation if they are parallel to the principal directions.

–The program should allow continuous redraw during shape change, allowing students to watch the deformation as it happens. Many programs show only the initial and final states when an operation is performed. Choose a program that allows a student to deform an object and watch as, for example, ellipses become progressively flatter with more strain, or as lines flex out of perpendicularity, and so on. This is particularly critical when working with the section on strain ellipses, where a student must be able to "unstrain" an ellipse. Without being able to watch as the ellipse becomes progressively more circular, the student won't be able to easily tell precisely how much strain to remove. While a program without continuous redraw, such as Aldus *Freehand*, can still be used, it makes it more difficult for students to visualize what is going on.

–The program should ideally have an operation that allows inhomogeneous deformation, an operation available in some programs to distort objects non-uniformly. Many paint-type programs have this operation; many object-oriented programs, such as Aldus *Freehand*, do not.

Conclusion

Since I began this approach with my students, I have been delighted to see far fewer blunders of the "wrong-headed intuition" type that had plagued my students' work in the past. Not only do my students seem to have internalized a better grasp of the qualitative aspects of strain, they also have strong mental images with which they can draw analogies when they are asked to interpret a new situation. These students are in a far better position than those who copy strain ellipses from the chalkboard into their notes and memorize the attributes of pure and simple shear. Students who have been through the *SuperPaint* exercises have subsequently dealt more successfully with strain analysis than did my former students.

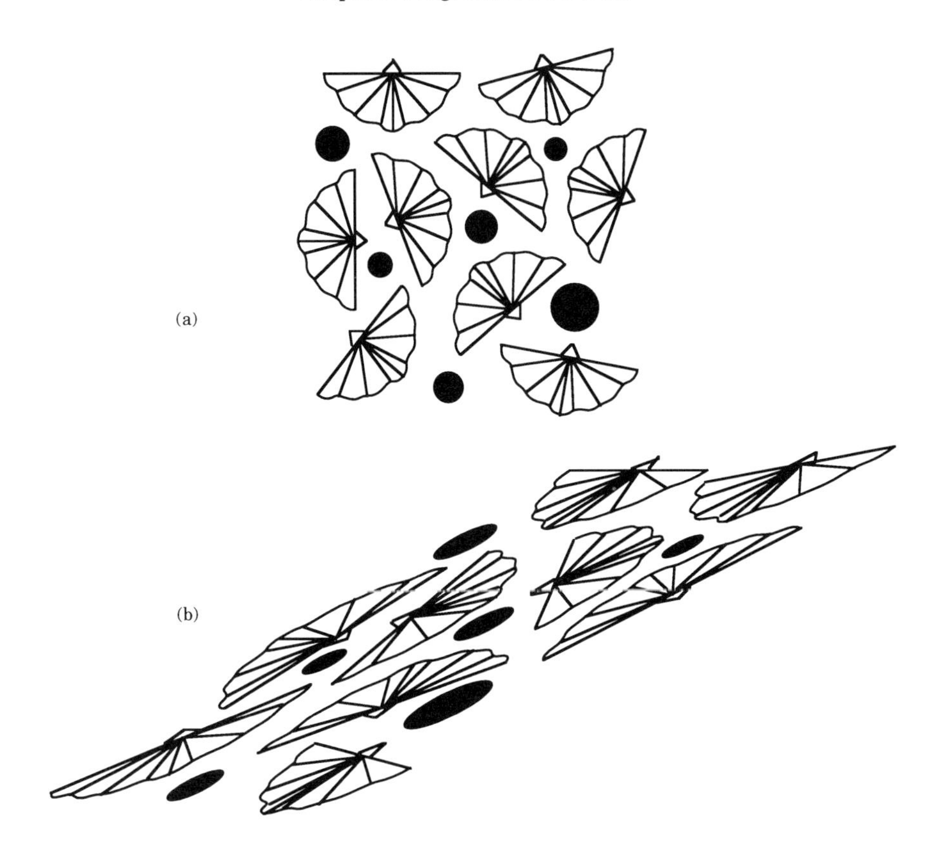

Fig. 12. a) Original group of circles and randomly-oriented brachiopods. b) Brachiopods and circles deformed by simple shear. Students should note that most brachiopods will be "skewed" unless their hinge and median lines lie parallel to the principal directions.

Acknowledgements

I'd like to thank several years of students in Structural Geology at Hamilton College for both their enthusiasm and their patience as I have blundered my way through many odd techniques for helping them visualize strain before arriving at one that works reasonably well. I am also grateful to both Steve Marshak and Declan De Paor for their careful reviews of this paper, which is much improved as a result of their suggestions.

Reference

McEachran, D. B. & Marshak, S., 1985. Teaching strain theory in structural geology using graphics programs for the Apple Macintosh computer. *J. Geol. Ed.* **34**: 191-195.

Visualization of Deformation: Computer Applications for Teaching

Basil Tikoff
Department of Geology and Geophysics, University of Minnesota, Minneapolis, MN 55455, U.S.A. teyssier@vx.cis.umn.edu

Haakon Fossen
Department of Geology, University of Bergen, N-5007 Bergen, Norway.

Abstract– In this article, we present computer programs designed to assist visualization of deformation. Instantaneous strain quantities (velocity fields, flow apophyses, instantaneous strain axes, and kinematic vorticity) and finite strain quantities (particle paths and finite strain ellipses) can be explored using the program *Strain Theory* or the simpler *Stress vs. Strain*. Finite strain and strain history are addressed through progressive deformation of a box (*Shear Box*), in which the rotation and extension or shortening of material lines can be observed. The trajectories of material points can be traced using *Flow Lines*. *General Shear* and *General Flow Lines* allows one to incorporate volume loss, isotropic or anisotropic, into finite strain calculations. Other programs examine the rotation of rigid elliptical clasts (*Rotating Clasts*) or passive planar markers (*Passive Clasts*) in a homogeneously deforming matrix, to determine the effect of clast aspect ratio, initial orientation, and kinematic vorticity on clast rotation. Because strain does not always accumulate in one episode of deformation, *Strain-o-Matic* allows multiple deformations to be superimposed on a single block. The program *3-D Strain* calculates the instantaneous and finite strain parameters associated with deformation, as well as rotation of material lines and planes, for input of any upper-triangular deformation matrix.

Introduction

Continuum mechanics provides a very useful framework from which to view kinematic analysis (*e.g.*, Malvern 1969). The basic concept is that all aspects of deformation, such as relative motion and finite strain, are interdependent. This approach has gradually been introduced into the geological literature through Biot (1965), Ramsay (1967), Ramberg (1975), and Means (1976, 1990), and has been useful in the quantitative kinematic analysis of deformed rocks (*e.g.*, Passchier 1988, 1990; Vissers 1989, Wallis 1992).

This article is an attempt to establish a framework for quantifying deformation, through computer simulation. The computer programs are intended to provide both quantification and an intuitive "feel" for the processes. Graphics concentrate on 1) how instantaneous and finite strains are related, 2) how different boundary conditions cause different deformations, 3) the complexities of heterogeneous flow, and 4) strain superposition. These programs are designed for teaching upper-level undergraduates and beginning graduate students.

Mathematical Background

This section outlines the mathematics utilized by the computer programs and introduces basic concepts of strain to the readers. However, knowledge of this section is not required to run the computer programs: one can go directly to the "Programs" section if desired.

Instantaneous Strain Quantities

Consider any material point (x,y) in a body in two-dimensional reference space before deformation. To understand how this material point will respond to deformation, we must look at the velocity field of that deformation. We will assume that the velocity field is constant over time, a condition that is called steady-state deformation. At a moment in time, the velocity of a particle is given by multiplying the position vector and the velocity gradient tensor, such that:

$$\begin{bmatrix} \dot{x} \\ \dot{y} \end{bmatrix} = \mathbf{L} \begin{bmatrix} x \\ y \end{bmatrix} \tag{1a}$$

where

$$\mathbf{L} = \begin{bmatrix} L_{11} & L_{12} \\ L_{21} & L_{22} \end{bmatrix} \tag{1b}$$

(See Fig. 1). This equation gives only an instantaneous velocity of a material particle. Once the particle moves slightly to a different position, it may have a different velocity. However, for a given velocity gradient tensor, the velocity of a particle as it passes through a particular spatial point (say $x = 1, y = 1$) is always the same. Because we can split the deformation corresponding to eqn. 1 into pure shear and simple shear components, the velocity gradient should be expressed in terms of the rates of these processes. Therefore,

$$\mathbf{L} = \begin{bmatrix} \dot{\varepsilon}_1 & \dot{\gamma} \\ 0 & \dot{\varepsilon}_2 \end{bmatrix} \tag{2}$$

where $\dot{\varepsilon}_1$ and $\dot{\varepsilon}_2$ are the instantaneous pure shear strain rates and $\dot{\gamma}$ is the simple shear strain rate. The boundary conditions of this deformation are those of sub-simple shearing (Simpson & De Paor 1993), a simultaneous combination of pure

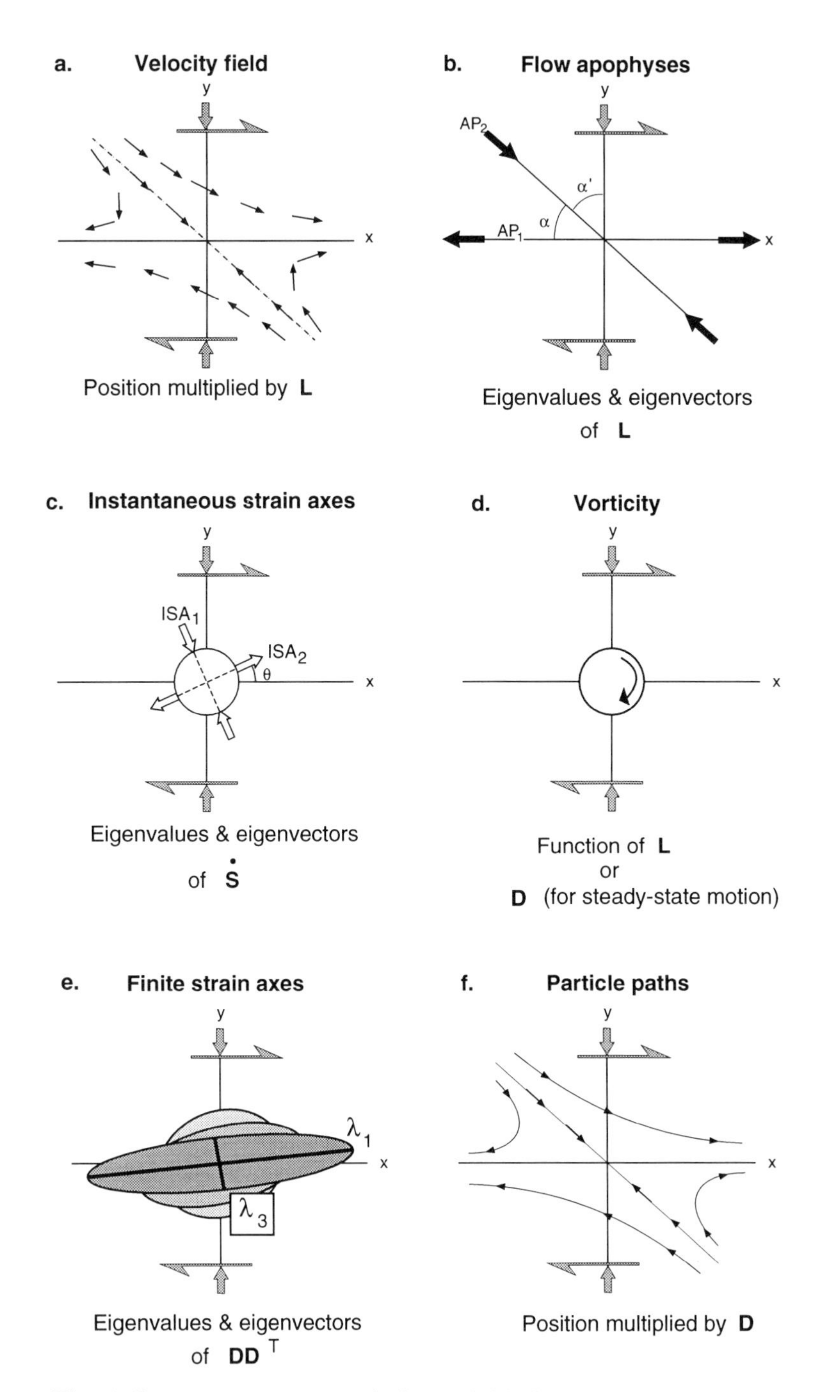

Fig. 1. Instantaneous strain (velocity field, flow apophyses, instantaneous strain axes, vorticity), finite strain parameters (particle paths, finite strain axes), and mathematical quantities they represent, for sub-simple shearing deformation.

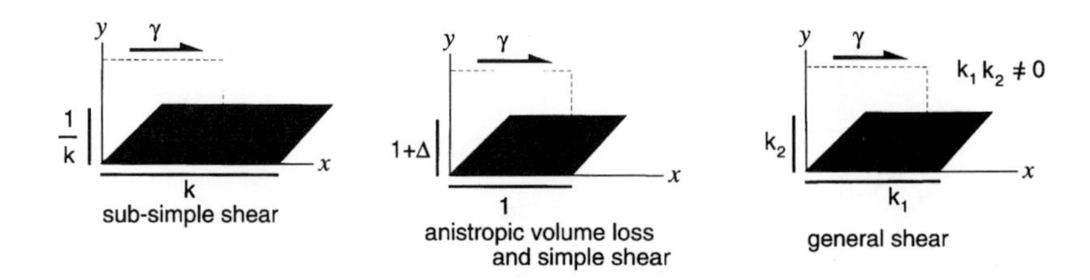

Fig. 2. Elements in an upper-triangular deformation matrix required as input for programs. These do not contain any "spin", and do not rotate coordinate system. Deformation shown are: a) sub-simple shearing, b) simultaneous anisotropic volume loss (in y-direction) and simple shearing, and c) general shearing (simultaneous pure shearing, simple shearing, and volume loss).

and simple shearing. Volume loss or gain can also be included in eqn. 2. For instance, anisotropic volume change is represented as a strain rate $\dot{\varepsilon}_1$ that is not compensated by an equal and opposite strain rate in the perpendicular direction $\dot{\varepsilon}_2$, such that $\dot{\varepsilon}_1 \neq \dot{\varepsilon}_2$ (Sanderson 1976).

Although we are addressing instantaneous strain quantities, we generally do not know the rates involved in geological deformation. However, we can put these quantities into time-independent values by making the following substitution:

$$k_1 = \exp(\dot{\varepsilon}_1 t) \quad \text{or} \quad \dot{\varepsilon}_1 = \ln(k_1 t)/t, \tag{3a}$$

$$k_2 = \exp(\dot{\varepsilon}_2 t) \quad \text{or} \quad \dot{\varepsilon}_2 = \ln(k_2 t)/t, \tag{3b}$$

$$\gamma = \dot{\gamma} t \quad \text{or} \quad \dot{\gamma} = \gamma/t. \tag{3c}$$

(Merle 1986, Tikoff & Fossen 1993). These relations assume that deformation is steady-state, which means that the kinematics are constant through deformation. For a sub-simple shearing deformation, the ratio of pure shearing and simple shearing is constant for a steady-state deformation. Throughout the rest of this section, we will assume steady-state deformation and write all quantities in terms of the finite strain parameters k and γ (Fig. 2). k (or k_1) represents the elongation of material in the x-direction caused by the pure shear component of deformation, while γ is the shear strain.

There are generally two directions (for two-dimensions, or three for three-dimensions) of the velocity field in which we are particularly interested. These directions, termed the flow apophyses (Ramberg 1975, Passchier 1988), represent maximum, intermediate, and minimum gradients of the velocity field (Tikoff & Teyssier 1994b) and are instantaneously irrotational (Bobyarchick 1986). The flow apophyses are described by the eigenvectors of the velocity gradient tensor **L** (Bobyarchick 1986). In sub-simple shear, there are two flow apophyses defined by straight particle paths asymptotic to the hyperbolic flow pattern, and material particles move either directly toward or away from the origin (Figs. 1 & 3). For

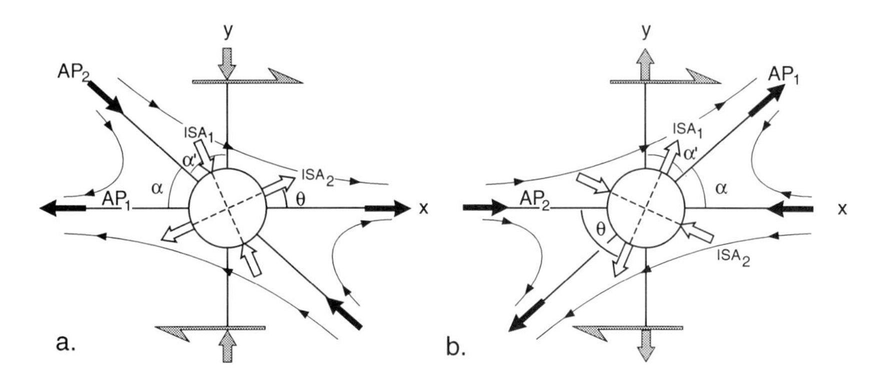

Fig. 3. Flow lines (thin lines), instantaneous strain axes (ISA; white arrrows), and flow apophyses (AP; black arrows) for two sub-simple shearing deformations. In both cases, the kinematic vorticity (*W*k) is identical. However, extension caused by the pure shear component of deformation occurs in the *x*–direction in a) and in the *y*–direction for b).

plane strain deformations, the flow apophyses are

$$\mathrm{AP}_1 = \begin{pmatrix} 1 \\ 0 \end{pmatrix}, \tag{4a}$$

$$\mathrm{AP}_2 = \begin{pmatrix} \dfrac{-\gamma}{\ln(k_1/k_2)} \\ 1 \end{pmatrix} \tag{4b}$$

(modified from Tikoff & Fossen 1993, eqn. 45). The acute angle α between the two flow apophyses, assuming steady-state deformation, is given in terms of the pure shearing, simple shearing, and volume change components as

$$\alpha = \tan^{-1}\left(\frac{\ln(k_1/k_2)}{\gamma}\right). \tag{5}$$

For sub-simple shearing, one flow apophysis is oriented along the x-axis (or, more generally, parallel to the shear direction). The other flow apophysis is oriented along the y-axis for pure shearing ($a = 90°$) and oblique to both axes ($0° < a < 90°$) for sub-simple shearing (Fig. 3). For simple shearing deformations, both flow apophyses are parallel to the x-axis ($a = 0°$).

For deformations that involve simultaneous simple shearing and anisotropic volume loss in the y-direction (Fig. 2), one must substitute 1 for k_1 and $1+\Delta$ for k_2 in the above two equations. Δ represents the volume loss component and is generally negative. For these cases, all particle paths are straight and move directly toward the origin (Fig. 4). One flow apophysis is oriented parallel to the particle movement while the other is parallel to the x-axis.

The velocity gradient tensor **L** can be decomposed into $\dot{\mathbf{S}}$, the symmetric stretching tensor, and **W**, the skew-symmetric vorticity tensor (Malvern 1969; Bobyarchick 1986), as such:

$$\mathbf{L} = \dot{\mathbf{S}} + \mathbf{W} \tag{6}$$

where

$$\dot{\mathbf{S}} = \begin{bmatrix} \dot{\varepsilon}_x & \frac{1}{2}\dot{\gamma} \\ \frac{1}{2}\dot{\gamma} & \dot{\varepsilon}_y \end{bmatrix} \tag{7}$$

and

$$\mathbf{W} = \begin{bmatrix} 0 & \frac{1}{2}\dot{\gamma} \\ -\frac{1}{2}\dot{\gamma} & 0 \end{bmatrix}. \tag{8}$$

The stretching tensor provides information about the orientation of the ISA (instantaneous strain axes of Tikoff & Teyssier (1994b) or instantaneous stretching axes of Passchier (1988)). The ISA record the directions of maximum elongation / shortening rates of material lines. The eigenvectors of the symmetric stretching tensor $\dot{\mathbf{S}}$ define the orientation of the ISA and, in terms of finite strain coefficients, are given by the equation,

$$\theta = \tan^{-1}\left(a \pm \sqrt{a^2 + 1}\right) \tag{9a}$$

where

$$a = \frac{\ln(k_2) - \ln(k_1)}{\gamma} \tag{9b}$$

(modified from Tikoff & Fossen 1995). θ defines the angle between the maximum instantaneous stretching direction $\dot{\mathbf{s}}_1$ and the shear direction (x-axis) (Fig. 1). Because $\dot{\mathbf{S}}$ is symmetric, the two ISA are necessarily orthogonal to each other. The orientation of the ISA remain fixed throughout deformation, and the relative magnitude of the axes is given by the eigenvalues of $\dot{\mathbf{S}}$.

The form of the velocity gradient tensor **L** (eqn. 2) is termed upper-triangular, because all elements below the main diagonal are zero. A more general velocity gradient tensor **L** would include terms in each element and this type of deformation was described by Ramberg (1975). However, these types of deformations often cause an external spin of shear zone boundary, the ISA, and flow apophyses with respect to an external coordinate frame (*e.g.*, Lister & Williams 1983). Since we are principally interested with the relationship of shear-induced rotation and stretching, we will use only upper-triangular, non-spinning velocity gradient and position gradient tensors.

The kinematic vorticity number *Wk* is often used to characterize different types of deformation arising in sub-simple shearing (*e.g.*, Simpson & De Paor 1993). *Wk* records the rate of rotation relative to the rate of stretching of deformation. Alternatively, *Wk* is described by a non-linear ratio between the pure and simple shearing components, given by the relationship

$$W_k = \gamma / \sqrt{2\left(\ln^2(k_1) + \ln^2(k_2)\right) + \gamma^2} \tag{10a}$$

or, in the case of no volume change ($k = k_1 = 1/k_2$),

$$W_k = \cos\left(\tan^{-1}\left(\frac{2\ln(k)}{\gamma}\right)\right) \tag{10b}$$

(Bobyarchick 1986, Tikoff & Fossen 1993). Sub-simple shearing records W_k values between 0 (pure shearing) and 1 (simple shearing). One complication is that any W_k value actually correlates to two separate types of sub-simple shearing deformation (Fig 3; Weijermars 1991). The difference between these cases is whether the pure shear component which acts in the x-direction is extensional (*e.g.*, $k_1 = 2$ and $k_2 = 0.5$) or contractional (*e.g.*, $k_1 = 0.5$ and $k_2 = 2$). As can be seen from eqn. 10, the *Wk* values are identical for these deformations. Some authors differentiate these cases by using a negative sign for deformations where $k_2 > 1$ (Simpson & De Paor 1993).

Anisotropic volume loss, when combined with simple shearing, creates a "volume" strain and has an influence on W_k estimates. The kinematic vorticity number for a two-dimensional steady-state case, with an anisotropic volume loss acting perpendicular to a simple shear deformation is

$$W_k = \frac{\gamma}{\sqrt{2\ln^2(1 + \Delta) + \gamma^2}} \tag{11}$$

(Fossen & Tikoff 1993). For example in Fig. 2, $1+\Delta = .5$ and $\Delta = -0.5$.

Figures 4a & b represent different flow lines and flow apophyses for 1) sub-simple shearing and 2) anisotropic volume loss and simple shearing, respectively. Strain histories that represent $W_k = 0, 0.25, 0.5, 0.75$, and 1.0 are shown for both deformations. The flow lines are generally curved for sub-simple shearing, while they are always straight for anisotropic volume loss and simple shearing. Additionally, for a given W_k, the acute angle between the flow apophyses is smaller for anisotropic volume loss and simple shearing than for sub-simple shearing (Note that one flow apophysis is always parallel to the x-axis in both cases). This result is caused by the inability of the anisotropic volume loss component to stretch material lines in the x-direction. Given identical orientation of flow apophyses, the relative rates of stretch are higher (W_k is larger) for sub-simple shearing deformations than for anisotropic volume loss and simple shearing.

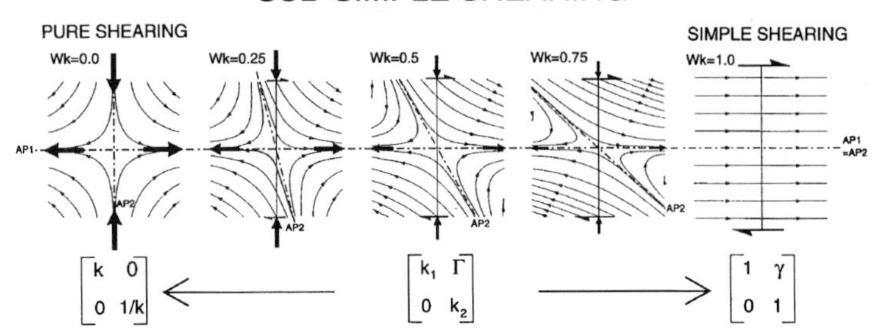

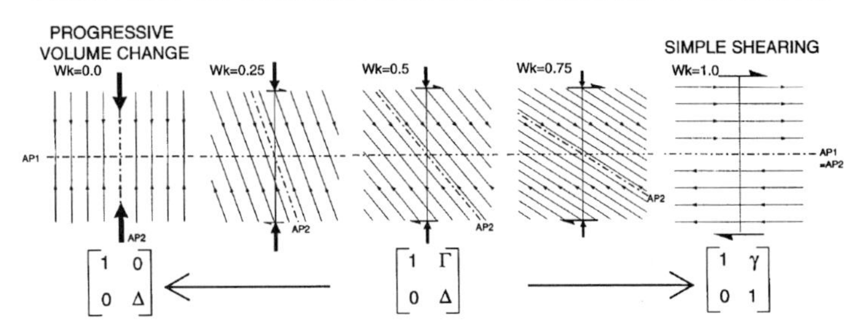

Fig. 4. Different orientations for flow lines for sub-simple shearing and anisotropic volume loss + simple shearing; *Wk* = 0, 0.25, 0.5, 0.75, 1. Flow lines are hyperbolic for sub-simple shearing and straight for anisotropic volume loss + simple shearing. Angle between flow apophyses is greater for sub-simple shearing than anisotropic volume loss + simple shearing for any *Wk*.

The velocity field, flow apophyses, ISA, and kinematic vorticity are all instantaneous quantities (Fig. 1). These parameters are placed in terms of finite strain parameters to emphasize the relation between instantaneous and finite strain parameters and to facilitate use of the computer programs. As instantaneous quantities, the flow apophyses, the ISA, and the velocity field do not rotate with respect to each other (Passchier 1990); they have a fixed orientation for a given velocity gradient tensor **L**. However, the ISA and flow apophyses are coincident only for cases of coaxial strain (*i.e.* pure shearing).

Finite Strain Quantities

The deformation (or position gradient) tensor **D** is a finite strain quantity that relates all material points before (x,y) and after (x',y') deformation, such that

$$\begin{bmatrix} x' \\ y' \end{bmatrix} = \mathbf{D} \begin{bmatrix} x \\ y \end{bmatrix}. \tag{12}$$

As suggested above, a relationship exists between instantaneous and finite strain quantities. The deformation tensor **D** (and its time derivative $\dot{\mathbf{D}}$) can be related to the velocity gradient tensor **L** through the following relationship:

$$\mathbf{L} = \dot{\mathbf{D}}\mathbf{D}^{-1}. \tag{13}$$

(Malvern 1969). Using this approach, the deformation matrix that corresponds to the velocity gradient tensor **L**, given in eqn. 1, can be expressed by the matrix **D** (Ramberg 1975; Tikoff & Fossen 1993):

$$\mathbf{D} = \begin{bmatrix} k_1 & \gamma \dfrac{(k_1 - k_2)}{\ln(k_1/k_2)} \\ 0 & k_2 \end{bmatrix} \tag{14}$$

Calculating the movement of particles during deformation is simply accomplished by multiplying the particle position by **D** (Fig. 1). These paths can also be considered as an accumulation of velocity vectors (compare Figs. 1a and 1f). These paths assume that deformations are steady-state, which means that the velocity gradient tensor **L** does not change during deformation and that the ratio of pure to simple shear components is fixed for a given deformation (cqn. 13).

The magnitudes and orientations of the principal axes of the finite strain ellipse are given by the eigenvalues *l* and the eigenvectors of the matrix $\mathbf{DD}^T$, known as the Finger tensor (Fig. 1; Flinn 1979), so that

$$\lambda = \frac{(\Gamma^2 + k_1^2 + k_2^2) \pm \sqrt{(\Gamma^2 + k_1^2 + k_2^2)^2 - 4k_1^2 k_2^2}}{2} \tag{15a}$$

where

$$\Gamma = \frac{\gamma(k_1 - k_2)}{\ln(k_1/k_2)} \tag{15b}$$

Since $\mathbf{DD}^T$ is a symmetric tensor, the axes of finite strain are orthogonal.

The relations shown in Fig. 1, although presented in two dimensions, are equally applicable in three dimensions and are used in the program *3-D Strain*. The result is that knowledge of either the finite strain axes or velocity field allows one to calculate the other, assuming steady-state deformation. The computer programs described below are based on steady-state deformation, so the user only has to input the finite strain components to calculate instantaneous strain values (velocity field, flow apophyses, instantaneous strain axes, and W_k), the flow field, and the finite strain axes.

Rotation of Elliptical and Planar Markers

The rotation of elliptical and planar markers introduces heterogeneity into the deformation field. However, if the markers do not interact (*e.g.*, Tikoff & Teyssier 1994a) and do not affect each other's flow field (*e.g.*, Ildefonse & Mancketelow 1993), their rotation can be predicted from homogeneous strain parameters. There are essentially two end-members of rigid clast rotation. One model, based upon Jeffrey (1922), assumes the clasts are rigid elliptical clasts in a ductile matrix. Ghosh & Ramberg (1976), following Jeffrey (1922), derived the equations for an elliptical clast of ratio R (= a/b or ratio of long to short axis of clast) in a two-dimensional flow in terms of strain parameters. They found that the rate of rotation $\dot{\phi}$ is given by the equation

$$\dot{\phi} = \frac{\dot{\gamma}\left(R^2\cos^2\phi + \sin^2\phi\right) + \dot{\varepsilon}_1\left(R^2 - 1\right)\sin 2\phi}{R^2 + 1} \tag{16}$$

where $\dot{\gamma}$ is the simple shear strain rate and $\dot{\varepsilon}_1$ is the pure shear strain rate in the x-direction ($\dot{\varepsilon}_1 = -\dot{\varepsilon}_2$), and ϕ is the original orientation of the long axis of the elliptical clast with respect to the x-axis (Ghosh & Ramberg 1976, eqn. 3). The first term in the eqn.16 reflects the rotation caused by the simple shear component of deformation and the second term reflects the rotation caused by the pure shear component of deformation. These terms may be added directly together since they are instantaneous quantities. Using eqn. 3, these instantaneous quantities can be put in terms of finite strain components. Therefore, a finite rotation is easily obtained for a finite strain, simply by using eqns. 14 and 15.

An alternative model of clast rotation is given by March (1932), which assumes that the marker will rotate as a material plane. This model only technically applies to planar markers, although some experimental and field evidence suggests that it may be a valid model for some instances of clast rotation. The initial orientation of a pole to a plane **p**, a unit vector consisting of direction cosines, is related to its final position **p**' by the equation:

$$\mathbf{p}'^{\mathrm{T}} = \mathbf{p}^{\mathrm{T}}\mathbf{D}^{-1} \tag{17}$$

(Flinn 1979). Since the relationship only involves the inverse of the deformation matrix **D**, a finite rotation is easily determined by a finite strain. The angle of rotation ϕ of the plane can be found by taking the dot product, from which one may derive the formula:

$$\cos\phi = \frac{\mathrm{pp'}}{|\mathrm{p}||\mathrm{p'}|} \tag{18}$$

(Tikoff & Teyssier 1994a).

Programs

Computer programs presented here were written in Microsoft *QuickBASIC*™ and *FORTRAN* for the Macintosh microcomputer and illustrate the important concepts of two-dimensional strain, using the theory described above. Some of these programs (*Shear Box, Flow Lines, Rotating Clasts*) were previously described in Tikoff *et al.* (1993) and some sample problems were given. The programs presented here require input of finite strain values (pure shear, simple shear, and volume loss components). Consequently, all programs assume steady-state deformation. Color monitors work particularly well to distinguish between color-coded deformation increments (*Flow Lines*) or parameters (*Strain Theory*). The programs are very simple to use and one can choose default options by selecting the OK button with the mouse. The size of the shear box is generally determined by the total amount of deformation, in order to keep the shear box on the screen. If a larger box is required, one should use smaller deformations. Each program is briefly described below.

Shear Box and Flow Lines Programs

The programs *Shear Box* and *Flow Lines* create deformations that result from sub-simple shearing. These programs are both designed to demonstrate the effects of finite strain. In each program, the user defines the elongation of the box in the positive x-direction (k), amount of simple shear (γ), and number of increments of deformation (Fig. 2). The computer generates a box, the size of which is determined by the employed screen and the amount of deformation, which is subsequently deformed (Fig. 5). If a simple shearing deformation is desired, the user should choose $k = 1$. For $k < 1$ the box shortens in the x-direction, while for $k > 1$, the box extends in the x-direction. Because we are limited to sub-simple shearing (*i.e.*, a volume conserving deformation), elongation of the box in the x-direction exactly compensates shortening in the y-direction, and vice versa. The top of the box moves to the right for positive shear deformations and the box is pinned at the origin.

Shear Box provides a quantitative look at progressive, sub-simple shearing deformation. The orientation and magnitude of the finite strain axes are measured at each increment of the deformation. One can also choose an arbitrarily oriented material line and follow its orientation and length through the deformation. A particularly useful exercise is to watch the rotation of material lines through either the long or short axis of the finite strain ellipse. For example, a material line oriented at 50° (counterclockwise from the positive x-axis) will rotate through the long axis of the finite strain ellipse for a deformation of $k = 1.2$ and $\gamma = 2$.

The *Flow Lines* program displays the movement path of particles during deformation, in which all material points move relative to the origin. Using smaller strains and more increments of deformation creates "smoother" particle paths. Each color records a different deformation increment so that relative velocities throughout deformation can be determined. Simple shearing deformations are unexciting since all motion is parallel to the x-axis. To slightly enhance an otherwise slumbersome experience, one should choose a small shear increment ($\gamma_{increment} < 0.25$), otherwise the particle paths will just blur into continuous lines. Pure shearing and sub-simple shearing are much more interesting. Pure shearing creates hyperbolic particle paths. Increasing the simple shear component of sub-simple shearing creates increasingly asymmetric hyperbolic paths (*e.g.*, Fig. 4).

General Shear Box and General Flow Lines Programs

The program *General Shear Box* is identical to the program *Shear Box*, except that the user can specify all three parameters of an upper triangular deformation

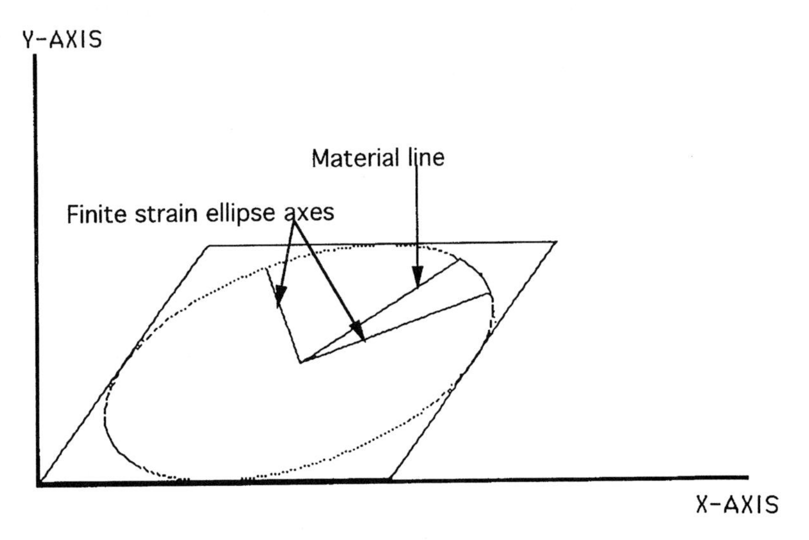

Fig.5. Sample output from *Shear Box* for a sub-simple shearing deformation matrix. The finite strain ellipse and axes are red, and the user-oriented material line is blue. All angles are measured from the positive x-axis.

matrix (eqn. 14). As a result, the user can specify sub-simple shearing, anisotropic volume loss/gain, isotropic volume loss/gain, or any combination of the above (Fig. 2c). Anisotropic volume loss in the y–direction, for example, is accomplished by setting $k_1 = 1$ and $k_2 < 1$. Isotropic volume loss occurs when both k values are equal and less than one. If sub-simple shearing deformation without volume loss is required, k_1 multiplied by k_2 must equal one.

General Flow Lines, likewise, is the same as *Flow Lines* except that volume loss can be incorporated into deformations. Anisotropic volume loss, in either the x– or y–direction, combined with simple shearing, causes flow lines that are straight lines (*e.g.*, Fig. 4). Isotropic volume loss (*e.g.*, $k_1 = k_2 = 0.5$) combined with simple shearing creates down-the-drain type flow patterns (Fig. 6), where particle paths spiral into the origin (Fig. 6b). Isotropic volume gain (*e.g.*, $k_1 = k_2 = 2$) combined with simple shearing, similarly, creates an out-of-the-drain (a frightening concept) type flow patterns (Fig. 6c). Including pure shear, simple shear, and volume loss components (*e.g.*, $k_1 = 2, k_2 = 0.2, \gamma = 2$) is also possible (Fig. 6d). In general, these types of particle paths are intermediate between the asymmetric hyperbolic paths (Fig. 6a) and the down-the-drain paths (Fig. 6b).

Strain Theory and Stress vs. Strain Programs

While *Shear Box* and *Flow Lines* address only finite strain, *Strain Theory* calculates instantaneous quantities (instantaneous strain axes, flow apophyses,

W_k, and velocity field) and finite strain quantities (flow lines and finite strain) (Fig. 7). Again, the user must input all parameters of an upper-triangular deformation matrix. The program first calculates W_k, which is constant for the entire deformation because of the assumption of steady-state deformation. The velocity field, ISA, and flow apophyses are shown graphically. Because deformation is steady-state, all instantaneous strain quantities are constant throughout deformation. Superimposed on these quantities, are the particle paths and finite strain axes, which vary with time. These quantities are color coded, so that they can be distinguished.

Strain Theory provides a graphic display of the inter-relations among strain parameters. For instance, in the first increment of deformation, the velocity field is the same as the particle paths. After the initial step, the particle paths grow and often curve, while the velocity field remains fixed and straight (Note that this distinction is not visible in Fig. 7). Another interesting relation is the orientation of the flow apophyses and ISA. In general, their orientations are parallel only for the case of coaxial deformation, which could be pure shear, anisotropic volume loss / gain, or isotropic volume loss / gain.

One of the more confusing issues in introductory structural geology is difference between stress and strain. Students, particularly engineering students, often have problems with this distinction. The program *Stress vs. Strain*, a simplification of *Strain Theory*, is an attempt to clarify this misunderstanding. The stress axes are assumed to parallel to the ISA, as would occur for an isotropic Newtonian material (see Tikoff & Fossen 1995 for discussion of the geological applicability of this assumption). If a small finite strain and many increments of deformation are chosen, the stress and finite strain axes are nearly parallel for the first increments of deformation. With additional deformation, the finite strain axes become increasingly elongate and, if a simple shear component is chosen, rotate with respect to stress axes (Weijermars 1991).

Strain-o-Matic Program

Strain-o-Matic is a program that allows the user to enter multiple episodes of deformation, and watch the evolution of finite strain. Again, the user enters the components of an upper-triangular deformation matrix and the number of increments of deformation. When the computer is finished deforming the markers, the user has the option of deforming the same box again. The next round of deformation starts with the pre-strained box caused by the first deformation. As a result, the user can input many episodes of deformation. One can superimpose pure shear on simple shear deformation, simple shear on anisotropic volume loss, *etc*. Figure 8 is an example of three superimposed deformations: pure shearing, simple shearing, and isotropic volume loss.

In this program, dextral shearing corresponds to a positive γ, while sinistral shearing corresponds to a negative γ. k_1 and k_2 can be varied from 0.1 to 10. Therefore, one can return to an unstrained box by, for example, using the default values ($\gamma = 1$, $k_1 = 2$, and $k_2 = 0.5$) for the first deformation, and then returning the box to an unstrained position by choosing the appropriate second deformation ($\gamma = -1$, $k_1 = 0.5$, and $k_2 = 2$).

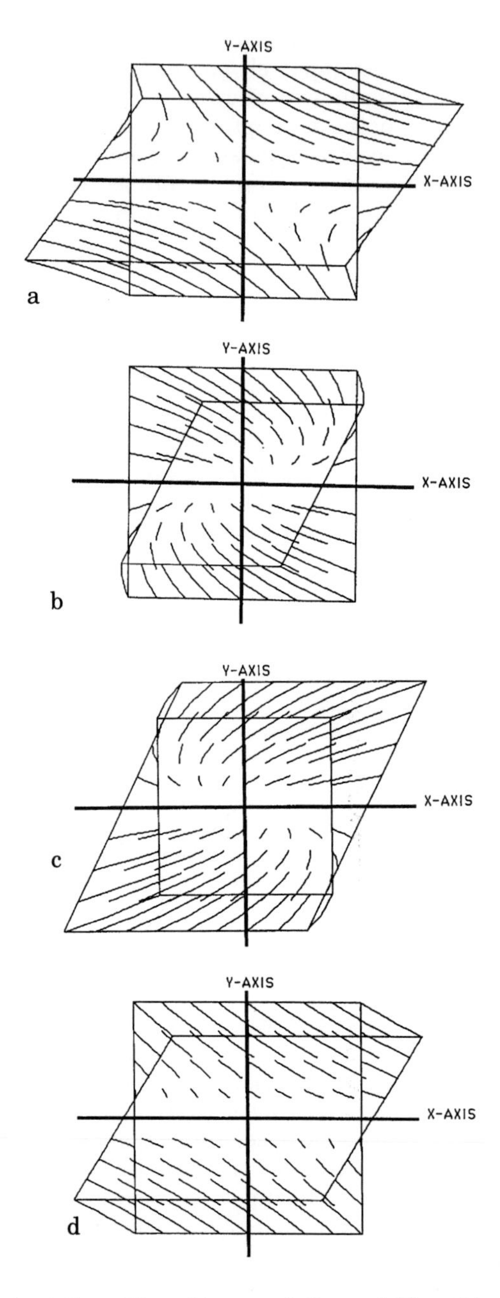

Fig. 6. Output from *Flow Lines* and *General Flow Lines* for a) sub-simple shearing, b) isotropic volume loss + simple shearing, c) isotropic volume gain + simple shearing, d) sub-simple shearing + isotropic volume loss. Sub-simple shearing creates asymmetric hyperbolic flow lines (a); isotropic volume loss/gain + simple shearing creates down-the-drain or out-of-the-drain type flow lines (b,c).

Rotating Clasts and Passive Clasts Program

The programs *Rotating Clasts* and *Passive Clasts* examine the rotation of rigid clasts. The two programs represent different models of clast rotation. *Rotating Clasts* is based upon Jeffrey (1922) and Ghosh & Ramberg (1976) and represents the rotation of rigid elliptical clasts in a ductile matrix. *Passive Clasts* is based on the mathematics of March (1932), which approximates the rotation of material planes. The Jeffrey model of rotation (*Rotating Clasts*) is generally considered a more appropriate model for clasts of relatively low ellipticity (clast elliptical ratio $R_{ell} < 5$). However, recent studies have shown that the March model may be a better description of clast rotation in the laboratory (Ildefonse & Mancketelow 1993) as well as in the field (Tikoff & Teyssier 1994a), even for low clast ellipticity. The programs allow the user to compare and contrast the rotational dynamics of these two end-member models.

The size and orientation of four clasts can be chosen by the user — default values are available — as well as the number of increments of deformation. Again, the user can choose any sub-simple shearing deformation. After each increment, the program draws the clasts and gives the orientation of each clast and the amount of rotation achieved during the last increment of deformation. There is also a "movie" option, where the user can watch the deformation without calculation of rotation angles.

Notice that the clasts should not only rotate, but translate in space. However, each clast is pinned at its center and only allowed to rotate. The clasts are also treated as rigid and do not deform. For *Rotating Clasts*, rotation rate depends on the shape of the clast, its orientation, and the ratio of pure to simple shearing. For *Passive Clasts*, because clasts are treated as rigid, planar markers, rotation rate only depends on orientation and the ratio of pure to simple shearing.

Several simple experiments are useful to explore clast rotation, some of which are shown in Fig. 9. (1) Change the orientation of the clasts and watch the result on the speed of clast rotation. (2) Change the aspect ratio of the clasts. Generally, this will have a strong effect on rotation in the *Rotating Clasts* and no effect in the *Passive Clasts* program. As clasts are made increasingly elongate ($R_{ell} > 5$) in the *Rotating Clasts* program, they will approximate clast rotation in the *Passive Clasts* program (Fig. 6). (3) Observe the effect of both initial clast orientation and strain history (relative amounts of pure and simple shear) on sense of rotation. Clasts in both programs are affected by both parameters, and both clockwise and counterclockwise rotation is possible in either model. (4) Observe the stable orientations of clasts, based on clast aspect ratio and strain history. For this example, the programs show very different behaviors. *Passive Clasts* causes all clasts to rotate into the shear plane (x-direction). The behavior in *Rotating Clasts* is much more complex. All clasts rotate past the shear plane for simple shearing deformations, while no clasts rotate past the x-direction during pure shearing. For combinations of pure and simple shear, clasts of a minimum aspect ratio become stuck depending on the relative proportion of pure and simple shearing (Ghosh & Ramberg 1976, Passchier 1987).

These concepts of clast rotation are complex and readers are referred to Ghosh & Ramberg (1976) and Passchier (1987) for a more rigorous explanation.

FLOW APOPHYSES
FINITE STRAIN ELLIPSE RATIO IS 2.75 ORIENTED AT 9.9 °.
THIS IS THE VELOCITY FIELD - CLICK MOUSE TO CONTINUE
THE LINES ARE THE DISPLACMENT FIELD - CLICK MOUSE TO CONTINUE

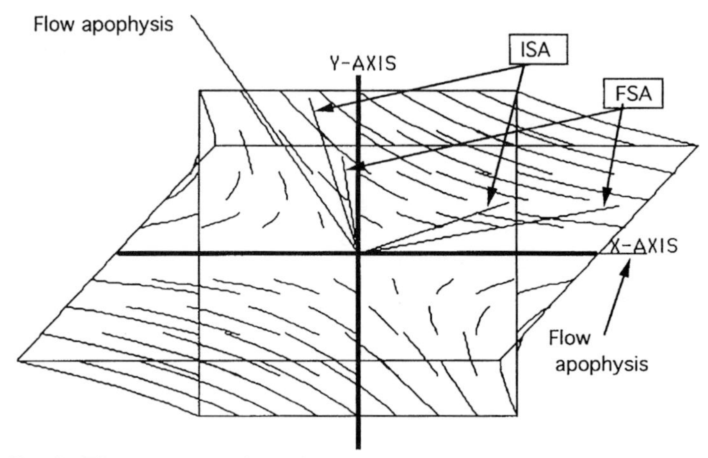

Fig. 7. *Strain Theory* output for sub-simple shearing. On color screen, velocity field is black, instantaneous strain axes are green, flow apophyses are blue, and finite strain axes are red. Continuous lines represent movement of material points during deformation (flow lines); deformation increments are recorded in different colors.

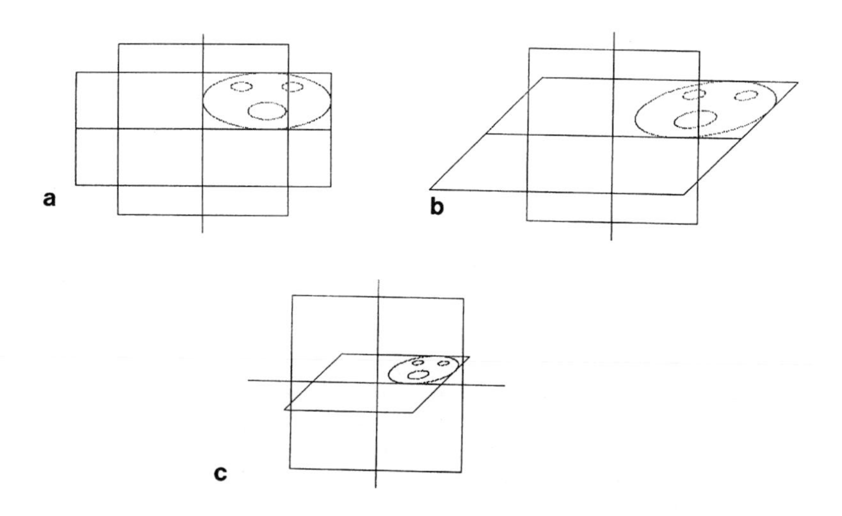

Fig. 8. Three deformations superimposed with *Strain-o-Matic*: a) pure shearing, followed by b) simple shearing, followed by c) isotropic volume loss. This program shows that deformation is path dependent; simple shearing followed by pure shearing does not have the same outcome as vice versa.

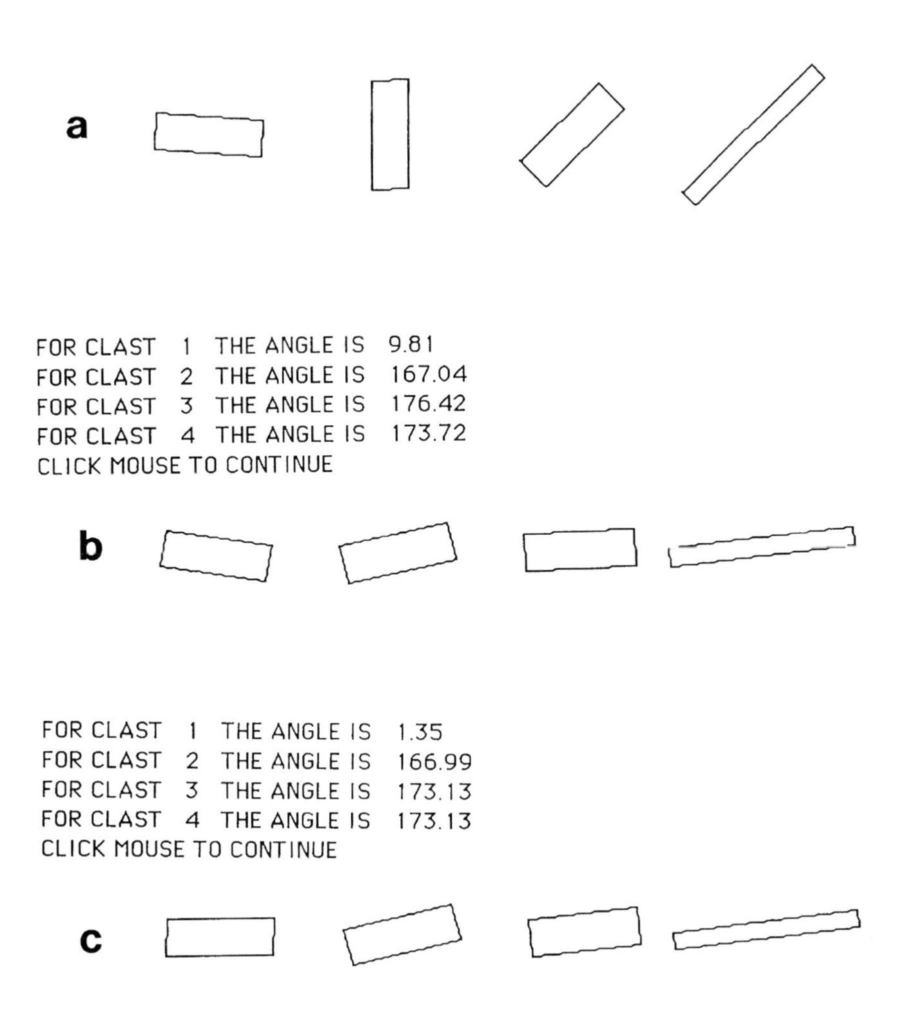

Fig. 9. Clast rotation starting with clasts initially oriented (a) at different positions (1-3) or at same position with different elliptical ratios (3-4). Rectangles used for computational speed. Clasts rotate differently in *Rotating Clasts* (b; Jeffrey 1922) and *Passive Clasts* (c; March 1932) programs. Deformation is dextral sub-simple shearing ($k = 2, \gamma = 5$). Clasts can rotate counter-clockwise in both models. All clasts rotate clockwise in (b) while clast 1 rotates counter-clockwise in (c). Rotation rate depends on initial clast position in both models. Axial ratios do not affect rotation rate in (b). (b) approximates March model rotation – compare final angle to (c).

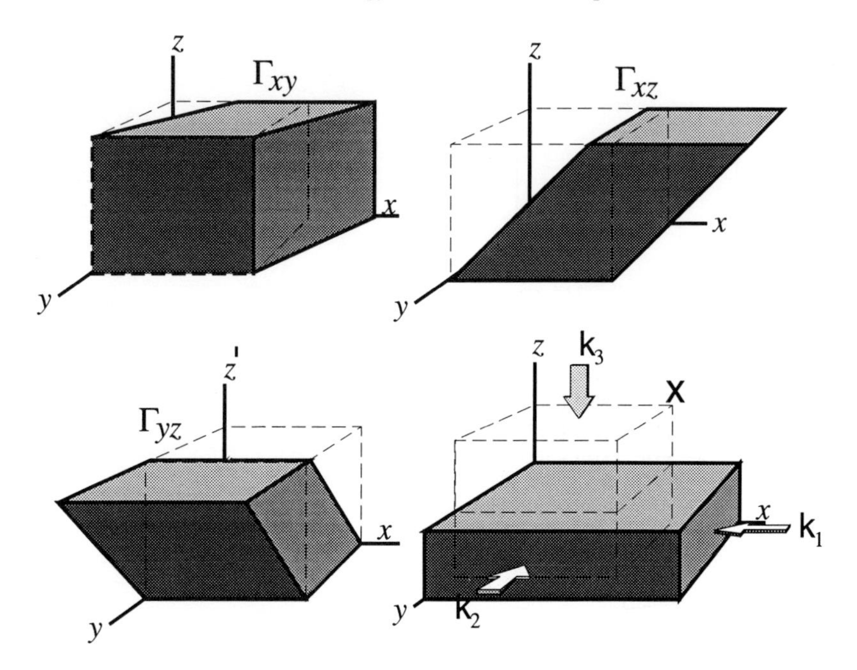

Fig. 10. Types of deformation recorded by three dimensional upper-triangular deformation matrix.

One caveat about the clast rotation programs is that the amount of rotation actually depends on the number of deformation increments. This problem arises because equation eqn. 16 is strictly true only for instantaneous rotations and large deformation increments violate this assumption. Therefore, identical finite strain values (*k* and *γ*) with different numbers of increments will lead to very small differences in rotation angles, with the smaller increments giving more accurate results.

3D-Strain Program and Three-Dimensional Deformations

While the analytical expressions for both instantaneous and finite strain values are given for two dimensional deformations, it is very difficult and tedious to do the same for three-dimensional deformations. In these cases, it is easier to utilize the tensor relations outlined for two dimensions and study the deformation analytically – see Tikoff & Fossen (1993) and Tikoff & Fossen (1995), for more details of three-dimensional deformations. As with the case for two-dimensional deformations, we only consider an upper-triangular deformation matrix, because we do not wish to incorporate "spin" or external rotation of the coordinate axes. The deformation matrix is given by

$$\mathbf{D} = \begin{bmatrix} k_1 & \Gamma_{xy} & \Gamma_{xz} \\ 0 & k_2 & \Gamma_{yz} \\ 0 & 0 & k_3 \end{bmatrix} \qquad (19)$$

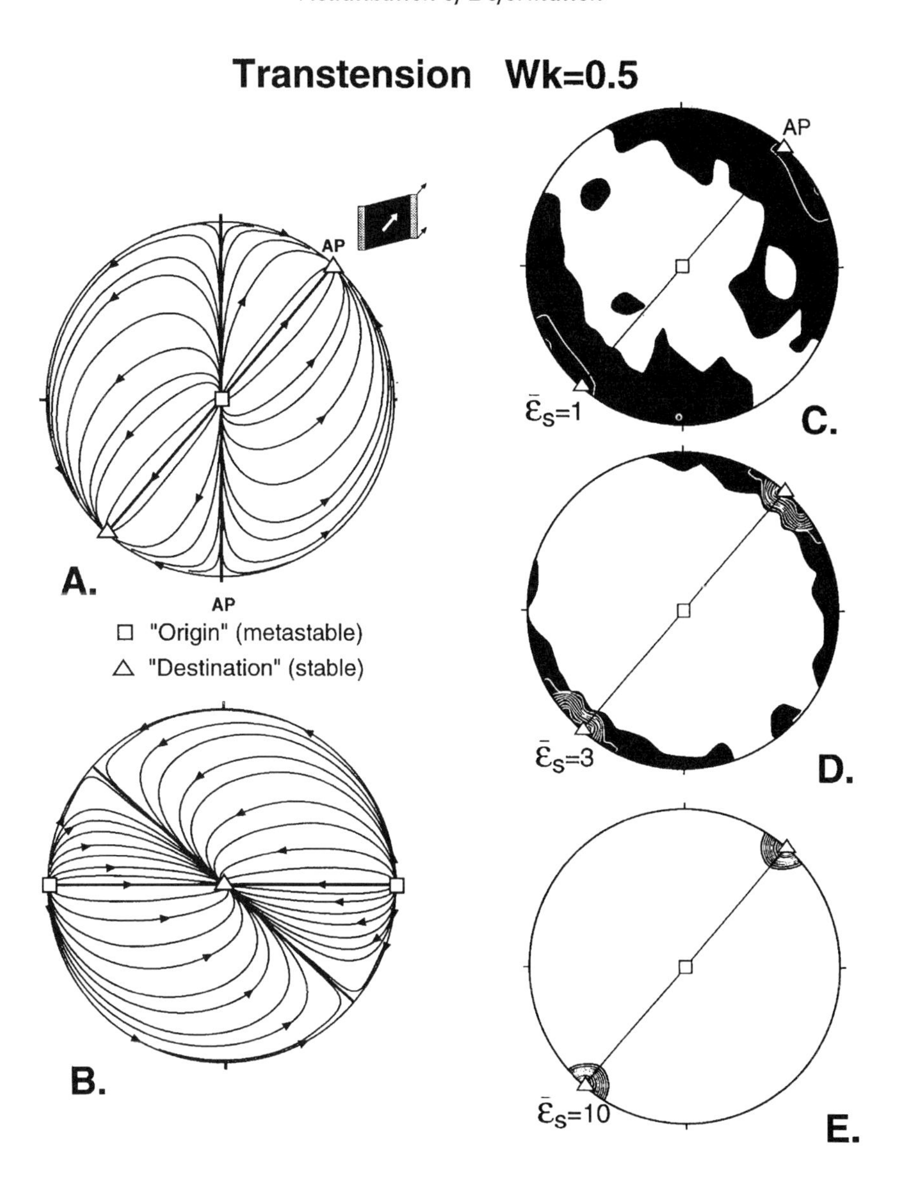

Fig. 11. Results from *3-D Strain*: sinistral transtensional deformation, lower hemisphere, equal area projection (from Fossen *et al.* 1994). Rotation paths for (a) passively deforming lines, (b) poles to planes. Lines rotate toward extensional movement direction (one flow apophysis). Planes rotate to horizontal, with poles vertical. (c)-(e) Progressive reorientation of passively rotating lines with increasing octahedral shear strain ε_s (a measure of three-dimensional strain) for same deformation. Contours by Kamb method: 1, 10, 30, 50, & 70% intervals. With increasing deformation, randomly oriented lines are progressively oriented parallel to extensional flow apophysis.

where the Γs are functions of γs and ks. The types of deformation described by eqn. 16 are shown in Fig. 10. Only steady-state deformations are considered in these programs.

The program *3D-Strain* is a purely analytical program. Here, the user inputs the coaxial components (k_{1-3}) and noncoaxial components (γ_{12}, γ_{23}, γ_{13}) of deformation. The program then calculates the deformation matrix, finite strain axes, octahedral shear stress, and Flinn values. Instantaneous strain quantities, such as W_k, are also calculated. The program also deforms lines and planes for a given initial orientation. The result, the new orientation and extension of line is shown on the screen and output to a file. It is also possible to deform a population of lines or planes for a given finite strain. Figure 11 is an example of deformation of lines and planes during a transtensional deformation. In this case, 500 randomly oriented lines were used to construct the diagrams (*c.f.* Fossen *et al.* 1994).

Although *3D-Strain* may be ultimately more useful as a research tool, it also has potential teaching applications. The basics of three-dimensional strain could also constitute an undergraduate lab. The rotation of lines and planes in three-dimensions may be a useful class exercise.

Availability of Programs

These programs are available on the worldwide web, by using a FTP to www.geo.umn.edu. Your e-mail address will act as your password and the programs can be found under /pub/structure. These programs are available from the authors. Interested readers should send an unformatted 3.5" disk and a self-addressed, stamped package (colleagues not from the U.S. should not worry about postage). We are willing to modify the programs to avoid specific problems, so please let us know if you encounter any. The RAM Cache needs to be turned off when using 040 machines and a FPU simulator is necessary for PowerMacs. Any ideas for improvements are also welcome.

Acknowledgements

The authors wish to thank Reidun Helming for mathematical and computer consultation and Christian Teyssier for useful ideas concerning the programs. Peter Hudleston, Carol Ormand, Labao Lan, Declan De Paor, and two anonymous reviewers are acknowledged for helpful reviews of this paper. The work was supported by a grant from Apple Computer Inc., through the Minnemac Program, University of Minnesota. Mr. Bill was the primary source of inspiration for *Strain-o-Matic*.

References

Biot, M. A. 1965. *Mechanics of Incremental Deformation*. Wiley, New York.

Bobyarchick, A. R. 1986. The eigenvalues of steady flow in Mohr space. *Tectonophysics* **122**: 35-51.

Flinn, D. 1979. The deformation matrix and the deformation ellipsoid. *J. Struct. Geol.* **1**: 299-307.

Fossen, H. & Tikoff, B. 1993. The deformation matrix for simultaneous simple shearing, pure shearing, and volume change, and its application to transpression/transtension tectonics. *J. Struct. Geol.* **15**: 413-422.

Fossen, H., Tikoff, B., & Teyssier, C. 1994. Strain modeling of transpressional and transtensional deformation. *Norsk Geol. Tidskrift* **74**: 134-145.

Ghosh, S. K. & Ramberg, H. 1976. Reorientation of inclusions by combination of pure and simple shear. *Tectonophysics* **34**: 1-70.

Ildenfonse, B. & Mancktelow, N.S. 1993. Deformation around rigid particles: the influence of slip at the particle / matrix interface. *Tectonophysics* **221**: 345-359.

Jeffery, J. B. 1922. The motion of ellipsoidal particles immersed in viscous fluid. *Proc. R. Soc. Lond.* **A102**: 161-179.

Lister, G. S. & Williams, P. F. 1983. The partitioning of deformation in flowing rock masses. *Tectonophysics* **92**: 1-33.

Malvern, L. E. 1969. *Introduction to the Mechanics of a Continuous Medium*. Prentice-Hall, New Jersey.

March, A., 1932, Mathematische Theorie der Regelung nach der Korngestalt bei affiner Deformation. *Zeitschr. f. Kristallogr.* **81**: 285-297.

Means, W. 1976. *Stress and Strain*. Springer-Verlag, Berlin.

Means, W. D. 1990. Kinematics, stress, deformation and material behaviour. *J. Struct. Geol.* **12**: 953-971.

Merle, O. 1986. Patterns of stretch trajectories and strain rates within spreading-gliding nappes. *Tectonophysics* **124**: 211-222.

Passchier, C. W. 1987. Stable positions of rigid objects in non-coaxial flow – a study in vorticity analysis. *J. Struct. Geol.* **9**: 679-690.

Passchier, C. W. 1988. Analysis of deformation paths in shear zones. *Geol. Runds.* **77**: 309-318.

Passchier, C. W. 1990. Reconstruction of deformation and flow parameters from deformed vein sets. *Tectonophysics* **180**: 185-199.

Ramberg, H. 1975. Particle paths, displacement and progressive strain applicable to rocks. *Tectonophysics* **28**:1-37.

Ramsay, J. G. 1967. *Folding and Fracturing of Rocks*. McGraw-Hill, New York.

Sanderson, D. J. 1976. The superposition of compaction and plane strain. *Tectonophysics* **30**: 35-54.

Simpson C. & De Paor, D. G. 1993. Strain and kinematic analysis in general shear zones. *J. Struct. Geol.* **15**: 1-20.

Tikoff, B. & Fossen, H. 1993. Simultaneous pure and simple shear: the unified deformation matrix. *Tectonophysics* **217**: 267-283.

Tikoff, B., Fossen, H., & Teyssier, C. 1993. Computer applications for teaching general two-dimensional deformation. *J. Geol. Ed.* **41**: 425-432.

Tikoff, B. & Teyssier, C. 1994a. Strain and fabric analyses based on porphyroclast interaction. *J. Struct. Geol.* **16**: 477-491.

Tikoff, B. & Teyssier, C. 1994b. Strain modeling of displacement field partitioning in transpressional orogens. *J. Struct. Geol.* **16**: 1575-1588.

Tikoff, B. & Fossen, H. 1995. The limitations of three-dimensional kinematic vorticity analysis. *J. Struct. Geol.* **17:** 1771-1784.

Vissers, R. L. M. 1989. Asymmetric quartz *c*-axis fabrics and flow vorticity: a study using rotated garnets. *J. Struct. Geol.* **11**: 231-244.

Wallis, S. R. 1992. Vorticity analysis in metachert from the Sanbagawa Belt, SW Japan. *J. Struct. Geol.* **14**: 271-280.

Weijermars, R. 1991. The role of stress in ductile deformation. *J. Struct. Geol.* **13**: 1061-1078.

Computer-aided Understanding of Deformation Microstructures

Carol Simpson
Department of Earth Sciences,
Boston University, 675 Commonwealth Ave.,
Boston MA 02215, U.S.A. csimpson@bu.edu

Declan G. De Paor
Department of Earth and Planetary Sciences,
Harvard University, 20 Oxford Street,
Cambridge MA 02138, U.S.A. depaor@eps.harvard.edu

Abstract– Using the Kodak PhotoCD™ file format, which can be viewed on either IBM PC or Macintosh computers, along with a set of stand-alone Macintosh appplications, we have developed an aid for teaching microstructural analysis in introductory through advanced structural geology courses. The PhotoCD stores digital images of 100 photomicrographs to serve as a reference atlas for research workers in the field. Each image is accompanied by a brief description of the illustrated microstructure. The collection is aimed at students and professional researchers who wish to evaluate their own microscopic samples of kinematic indicators, deformation fabrics, *etc*., by comparing and contrasting features. The images also serve as a useful resource for instructors who do not have access to a comprehensive teaching collection.

Introduction

Teachers who wish to incorporate the study of microscopic deformation structures in their structural geology courses face a number of hurdles. Because of the decrease in class time devoted to optical mineralogy and microscopy in recent years, students today are generally ill-prepared for such analytical work. They may waste a lot of lab time because they have difficulty operating their microscopes; indeed, at introductory level, there may not be enough microscopes to go around in many departments and those that exist may be poorly maintained. Students may have difficulty in locating good examples of structures in their thin sections or in relocating such structures after the slide has been moved, and they may be confused by birefringence changes accompanying stage rotation. Professors whose research interests are on a regional or global tectonic scale may not have extensive collections of thin sections or photomicrographs suitable for

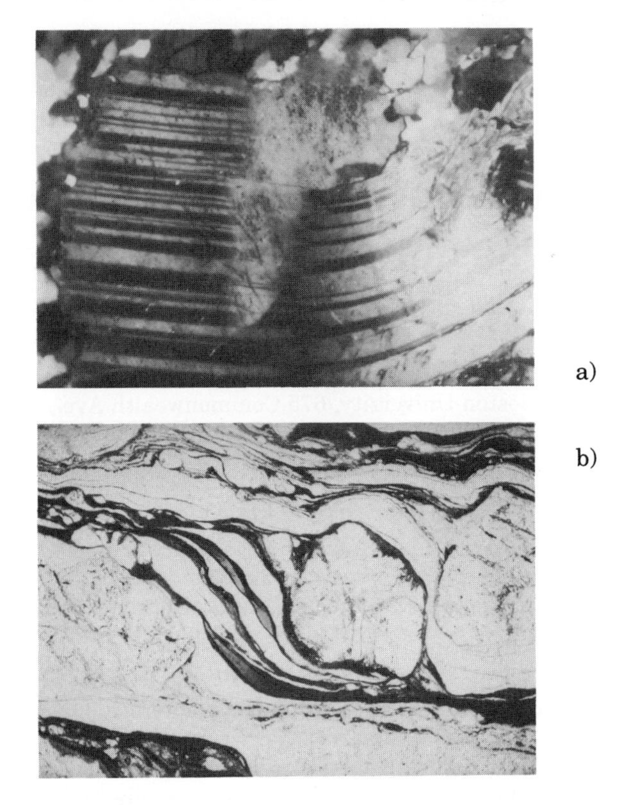

Fig. 1. Images from the Microstructures PhotoCD ROM. a) Strain induced removal of twin planes in a bent plagioclase porphyroclast deformed at upper greenschist facies. Borrego Springs mylonite zone. b) σ-type porphyroclast in *S-C* mylonite showing sigma grain with recrystallized feldspar tails on upper left and lower right, consistent with left lateral shear.

teaching large classes. All teaching collections are continuously eroded by careless students who drop and smash glass-mounted thin sections and even by careful, dedicated students who get excited by the discovery of a particularly good specimen of a microstructure and ram it with their high-power objective.

In addition to the above difficulties associated with the viewing of natural examples of deformation microstructures, students tend to have difficulty visualizing submicroscopic scale processes such as dislocation glide, sub-grain formation, or the generation of crystal defects at a Frank Read source. Static illustrations in text books and research papers do little to reveal the nature of these dynamic processes.

In order to help teachers overcome the above barriers to learning and to incorporate deformation microstructures both in their undergraduate and graduate level courses, we have assembled a group of computer-based teaching aids. These are based upon our own teaching collections and lecture notes but they draw upon many excellent published works of theory and field observation (*e.g.*, Frost & Ashby1982, Hanmer & Passchier 1991, Hull & Bacon 1984, Nicolas & Poirier 1976, Passchier & Trouw1996, Wenk 1985)

The Microstructures CD ROM

The Microstructures CD ROM evolved out of trivial origins. Many years ago, one of us (C.S.) made poster-sized photographic enlargements of her best photomicrographs to hang as decorations in her structure laboratory. They attracted a lot of attention over the years and, with the advent of color computer scanners, we decided to make a collection of a half dozen similar photomicrographs to use as a slide set with the *After Dark*™ screen saver. These images were so eye-catching and attracted so much attention from colleagues and students alike that we decided to assemble a larger collection to serve as a teaching aid and research reference set, or "electronic atlas". We chose the Kodak PhotoCD format (Fig. 1) because it was easy (relatively speaking) and cheap (US$120) to produce an initial disc with the help of our local Kodak processing agent and because the format could be accessed using either IBM or Macintosh platforms.

In order to manufacture replicas of the master CD ROM, we contacted a disc manufacturer licensed by Kodak (we were surprised at how few license holders there are in this rapidly growing market). Readers interested in replicating their own CDs should note that a minimum of 1000 discs had to be manufactured for a much smaller potential usership so that a sizable remainder had to be calculated into the project budget. Nowadays, many universities have CD authoring systems but these are not generally licensed to replicate Kodak format discs. Kodak also distributes more powerful multimedia authoring software called Kodak *Portfolio*™ which permits users to create buttons and text fields linked to images. Our CD is strictly a slide set!

Viewing Images

Initially, users had to purchase an application such as Kodak *Access*™ to read the Kodak format but now such capability is built into almost all personal computers and graphics editing programs and it is rumored that it may be incorporated on the PhotoCD medium itself in the future.

An uncropped standard size image displayed in millions of colors requires 1 MB of memory. The memory required can be reduced by setting the monitor to display fewer colors but images will not appear natural in fewer than 256 colors.

We strongly recommend that users view the microstructures imagery with Adobe *Photoshop*™ software. This application permits several levels of zoom analogous to changing lens on a traditional microscope. When a student zooms in, she or he sees fine details, such as tiny rutile needles in quartz grains, that were not visible at the lower magnification. Instructors can retouch and annotate copies of individual images and incorporate them in their own multimedia presentations, lecture notes, or overhead projection transparencies.

The CD images can also serve as reference collections of natural structures; for example, if students think they see a particular structure or texture in their research materials, they can compare the samples with type specimens on the PhotoCD. Figure 1 shows representative (though greatly reduced) samples of the images.

Simulations of Deformation Mechanisms

One of us (D.P.) has developed a set of microstructure simulations for the Macintosh computer. They work on any model from Mac Plus through PowerMac and PowerWave clones and require between 1 and 2 MB of free RAM depending on screen size and monitor color depth. In order to minimize the learning curve, each deformation mechanism is illustrated in a separate stand-alone application with a minimum number of menu options and dialog boxes. Thus students can concentrate on the process they are supposed to be studying rather than the details of the computer's operation.

Glide

The first application in the group, called *Glide* version 7.0, demonstrates the processes of dislocation glide and climb and the migration of screw dislocations (Fig. 2). It is impossible to convey the full operation of this program with still illustrations. For example, all of the atoms in Fig. 2 undergo rapid random oscillations about their mean positions as they would in a crystal structure at a finite temperature (to get an idea of what this looks like, hold Fig. 2 close to your face and shake it violently). When a sinistral simple shear strain is applied to the crystal lattice, the lower atoms move to the right, stretching crystal bonds. At a critical point, the stretched bonds break sequentially and new, shorter bonds form, effectively causing a half plane to migrate through the crystal lattice. Optionally, students may display a 'T' symbol representing the edge of the half plane and a Burger's vector (Fig. 2a).

Students have particular difficulty visualizing the migration of a screw dislocation. This is simulated in Fig. 2b using a three dimensional view of the crystal structure. The process of dislocation climb also causes students a lot of trouble. It is illustrated upon selection of a menu option. An impurity atom is added to the crystal structure and the bonds between it and ordinary atoms are highlighted. If these bonds are stronger than normal, then the passage of a dislocation through the crystal may be impeded (Fig. 2c). Students must wait and watch for a vacancy which migrates along a random path through the structure. When the vacancy reaches the tip of the half plane (Fig. 2d), the dislocation is said to climb (it migrates down the screen in this example, reinforcing the point that "climb" can occur in any direction, not just upwards). After the climb event, the dislocation continues to migrate.

Porphyroblast

A second application, called *Porphyroblast* illustrates the concepts of pre-, syn-, and post-tectonic crystallization (Fig. 3). The program displays a rock fabric in its

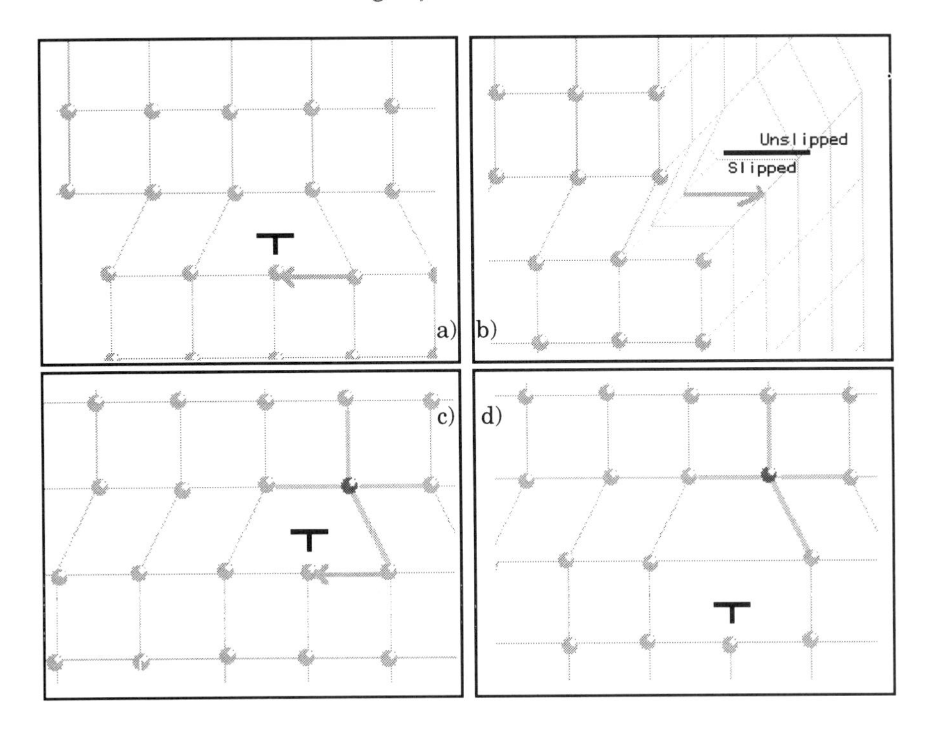

Fig. 2. Screen shot of the *Glide* program. a) When running, the location of the half plane sweeps through the atomic array as bonds break in front of it and are reformed in its wake. The T-bar and Burger's vector are optional features. b) A three dimensional simulation demonstrates the glide of a screw dislocation. c, d) Demonstration of dislocation climb. Dark atom with thick bonds represents impurity. c) Dislocation glide halted by strong bonds. d) Vacancy migration reduces height of half plane, releasing dislocation.

initial state and lets the student select the crystallization history desired. Upon pressing the Go button, the student watches as crystals grow and as the fabric becomes deformed. Figure 3 illustrates the case of pre-tectonic crystallization. The fabric is enclosed in the growing crystal and the enclosed portion is protected from further deformation. For simplicity, the imposed deformation in this case is a horizontal pure shear, as illustrated by the straight horizontal fibers that grow progressively from the ends of the crystals.

The program works with the aid of two off-screen bitmaps. One holds the initial image and the second is a scratchpad for assembling images before transfer onto the screen. To minimize memory requirements, a small image of the initial fabric is tiled onto the screen. If you look closely, you will detect a wallpaper-like repetition of the pattern. The growth of crystals is controlled by mathematical functions and the appropriate sizes and locations of circles are marked on the scratchpad at each increment. The fabric inside each crystal is copied from the appropriate part of the initial image without any distortion whereas the fabric outside the crystals is stretched using a function that maps a bitmap from a source

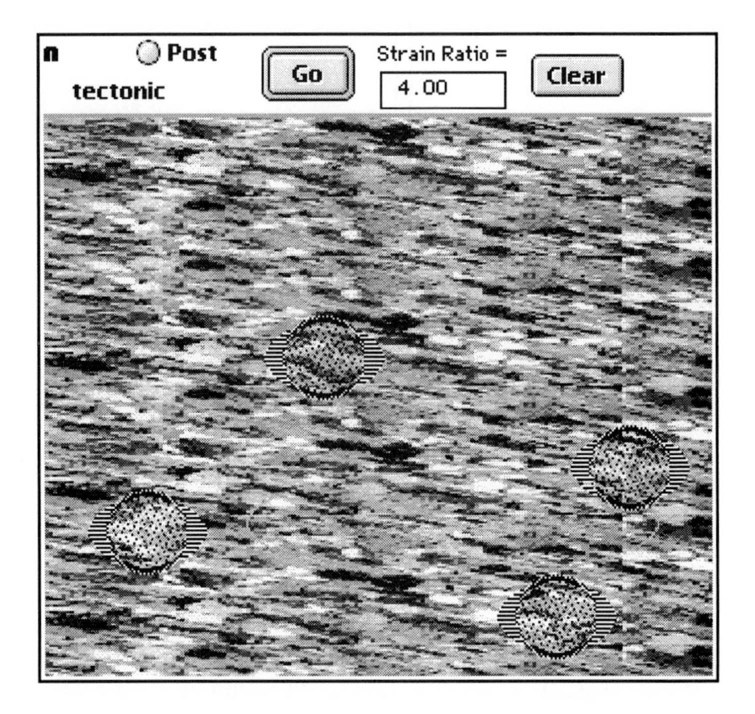

Fig. 3. Screen shot from the application *Porphyroblast* for the case of pre-tectonic crystallization.

rectangle to a destination rectangle. The assembled scratchpad image is transferred to the screen very quickly and no intermediate steps are visible.

Other Applications

Glide and *Porphyroblast* are two examples of a set of teaching mini-applications that we use in microstructure laboratory classes. Others include simulations of the generation of dislocations at a Frank Read source, the formation of sigma and delta type porphyoclast systems, grain boundary migration recrystallization, grain boundary sliding, basal <a> and prism <a> slip systems, syntaxial *vs.* antitaxial fiber growth, the formation of tiltwalls, the domino effect in fractured feldspars, vacancy migration, and stylolite formation

Discussion

A PhotoCD is like a carefully indexed set of photographic negatives (when did you last file your slides or negatives?). Unlike negatives or transparencies, individual images on CD won't get lost or spoiled. Unlike glass microscope slides, CDs don't deteriorate with age and they don't easily break.

The Microstructures CD ROM is intended to make available to students on remote or under-financed campuses the benefits of a thin section and photomicrograph collection that might not otherwise be accessible to them. We believe that this technology has great potential for democratization of educational access. Without large financial expenditure, many readers could create their own archives of teaching and research materials on Kodak PhotoCD ROM. The accompanying applications simulating deformation microstructures seem to aid visualization, especially among students with weak backgrounds in optical or sub-optical mineralogy and atomic physics.

References

Frost, H. J. & Ashby, M. F. 1982. *Deformation-mechanism Maps. The Plasticity and Creep of Metals and Ceramics.* Pergamon Press, Oxford.

Hanmer, S. & Passchier, C. W. 1991. Shear sense indicators: A review. *Geol. Surv. Canada Pap.* **90**: 1-17.

Hull, D. & Bacon, D. J. 1984. *Introduction to Dislocations* (3rd. ed.) Pergamon Press, Oxford.

Nicolas, A. & Poirier, J. P. 1976. *Crystalline Plasticity and Solid State Flow in Metamorphic Rocks.* J. Wiley & Sons, New York.

Passchier, C. W. & Trouw, R. A. J. 1996. *Microtectonics.* Springer- Verlag, New York.

Wenk, H-R. (ed.) 1985. *Preferred Orientation in Deformed Metals and Rocks: An Introduction to Modern Texture Analysis.* Academic Press, New York.

Understanding information [illegible]

The Microcomputer CD-ROM [illegible] on campus of under-financed campuses [illegible] microfilm collections [illegible] that the technology [illegible]. Without large financial [illegible] collection of teaching and research [illegible] [illegible]

Notes

[illegible]

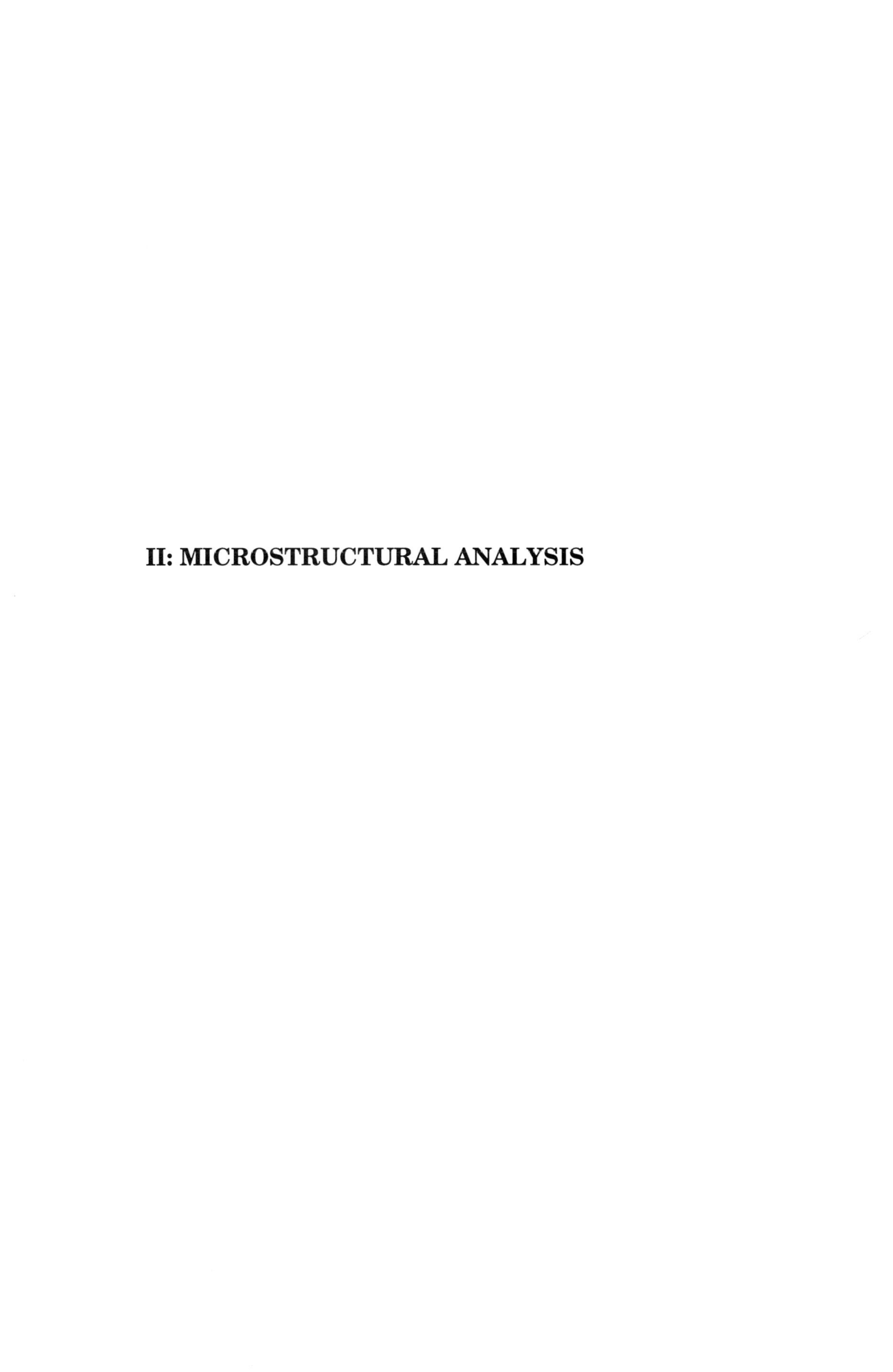

II: MICROSTRUCTURAL ANALYSIS

Image Analysis in Structural Geology Using *NIH Image*

Marcia G. Bjørnerud
Geology Department, Lawrence University, Appleton WI 54912, U.S.A.
Marcia.G.Bjornerud@Lawrence.edu

Brian Boyer
Geology Department, Miami University, Oxford Ohio 45056, U.S.A.

Abstract— *Image* is a versatile, public-domain Macintosh image analysis program developed at the US National Institutes of Health for quantitative characterization of biological materials. Available at no cost over the Internet, the program has tremendous potential for use in the geosciences. *Image* can import and analyze color and gray-scale PICT, TIFF, PICS and *MacPaint*-type files from all standard graphics programs, and it includes many of the built-in filters and enhancement functions of commercial image processing applications. Its analysis functions include determination of object areas, perimeters or lengths, color/grayscale values and, for elliptical particles, axial lengths, orientations, and *x-y* centers. Used creatively, these functions make it possible to automate tasks including point-counting, strain analysis, areal estimation, and assessment of lattice and grain-shape preferred orientation. Results can be exported to any standard spreadsheet. The program can also create animated sequences from successive images and generate three-dimensional visualizations from serial sections. A spin-off program, *ImageFractal*, can perform both Richardson and box-counting fractal analysis of objects in binary images. The potential of *Image* and *ImageFractal* for use in structural geology is illustrated in several examples.

Introduction

Geometric characterization of deformational features remains central to most studies in structural geology. Spatial patterns in maps, satellite images, cross sections, outcrop photographs, hand specimens and photomicrographs are analyzed in order to assess the extent, nature and conditions of rock deformation. Traditionally, such analyses have been either subjective and descriptive or quantitative but highly labor-intensive. The emergence of sophisticated image processing software for personal computers makes it possible to expedite or even automate many of the procedures previously carried out at the drafting table or microscope. Although excellent commercial image analysis software is available, one of the most versatile programs of this type is in the public domain and accessible via the Internet. *Image*, developed by Wayne Rasband at the US National Institutes of Health (Rasband 1994a), is a Macintosh program that can import, process, and analyze PICT, TIFF, PICS (animation), and *MacPaint*-type graphics files. After an image is appropriately filtered and enhanced, the program can perform a variety of measurements on specified objects or regions within the image and save the results in spreadsheet-compatible format. By 'stacking' a series of related two-dimensional images, the program can also render three-dimensional images and create animated sequences. Although it was originally written for the biological sciences, *Image* is flexible enough to find uses in any field that involves quantitative analysis of geometric features (Lennard 1990). *Image* has also inspired a number of spin-off programs, including *Image Fractal*, which further extend its analytical capabilities.

There are two principal modes of image processing and analysis:

1) enhancement of features to delineate forms more clearly, for graphical presentation or quantitative analysis; and

2) use of color/tonal characteristics in order to create "topographic" maps of variables and/or extract quantitative information from images.

A third mode of analysis is to "stack" images in spatial or temporal series, merging information from them to generate new portrayals (cross sections, projections, animated sequences). Table 1 lists examples of ways in which these distinct modes of image analysis can be applied to structural studies. Detailed illustrations of several of these applications appear later in the paper.

Using *NIH Image*

Access to the Program

The program can be obtained at no charge: 1) by an anonymous file transfer (ftp) from 'zippy.nimh.nih.gov' at the /pub/nih-image directory; or 2) at the *NIH-Image* web site: http://rsb.info.nih.gov/nih-image/ . This site contains the *Image* application program as a *BinHex*™-encoded self-extracting archive, in versions for Macintosh computers with and without floating-point coprocessors (fpu's). The program can be downloaded using standard decompression programs (several of which are included in the 'Programs' directory at the ftp site). The site also includes a variety of other files and applications related to *Image*. Most useful is the 90-page Users Guide (Rasband 1994b), which provides comprehensive descriptions of all of the program's features. Other directories contain example images, update notices, bug reports, macros, spin-off programs, contributions from *Image* users, and 'plug-ins' for image acquisition from various types of digitizing hardware.

The present version of *Image* (v. 1.60) requires at least 8 MB of RAM, but more is needed for very large graphics files or if the `stacking` option is used for animation or creation of 3-D graphics. The default RAM allocation for the program is 4000K, though this is readily changed in the usual way, by highlighting the program icon and then selecting `Get Info...` from the `File` menu of the Macintosh Finder.

Importing Images

The first step in using *Image* is to convert the original graphic to be analyzed (map, cross-section, field photograph, photomicrograph, video frame) to digital format using a scanner, frame grabber or other device. (Digitized images can also be imported from CDs or downloaded from the Internet). Many image acquisition peripherals come with Adobe *Photoshop*™-compatible plug-in modules, some of which are also compatible with *Image*. Plug-ins known to work with *Image* are listed in the User Guide, and several of these are available on-line at the *NIH-Image* FTP site. These plug-ins must be stored either in the Macintosh system folder or in the same folder as the program itself. Even if image acquisition must occur outside *Image*, most acquisition software is capable of saving graphics in formats that *Image* supports. Any Macintosh 8-bit TIFF, PICT, PICS or *MacPaint* file can be opened directly in *Image*, by selecting `Open` from the `File` menu. Sixteen-bit TIFF files and TIFF files created on PCs can also be read by the program but must be opened using `Import` from the `File` menu. Twenty-four bit RGB-color images can be imported and converted to three 8-bit 'slices' (see discussion of `Stacks` menu below).

Main Windows and Menus

Screen Layout

When a graphics file is opened in *Image*, it appears in a large central window framed by the main menu bar and four smaller windows. The image window can be resized using the zoom box (upper right; makes image as large as possible) or grow box (lower right; allows user to vary image size when `Scale to Fit` is selected in `Options` menu). The smaller windows include 1) the video look-up table (LUT), which displays the screen color palette (the range of colors corresponding to the 256 possible pixel values in an 8-bit file); 2) the `Map` window, which permits adjustment of image brightness and contrast and modifies the LUT; 3) the `Info` window, whose contents depend on which window the cursor occupies (*e.g.*, the *x-y* coordinate position of the cursor within the main window); and 4) the `Tool` window, which contains not only standard graphics tools (*e.g.*, pencil, paint brush, eraser, text tool, line-drawing tool, selection tools) but also several for image enhancement and analysis. The graphics tools can be used to annotate images for presentation purposes; for example, to label and demarcate particular beds in a field photograph, or to identify specific grains in a photomicrograph. Simple measurements can also be made using implements from the `Tool` window. As a line is drawn with the `line` tool, its length appears in the `Info` window. Similarly, angular measurements can be made interactively with an `angle` tool. Manual point counting can be done with the `cross-hair` tool, which labels particles and keeps track of their *x-y* coordinates. More sophisticated measurement techniques require use of the Analysis menu, discussed in the next section.

All of the windows can be moved around the screen by their header bars. When Print... is selected from the File menu, only the contents of the active window are printed. Several other optional windows, described below, can be displayed at the user's discretion. The functions of the various windows are discussed in greater detail in the context of corresponding menu options.

Summary of Main Menu

The File and Edit menus include the standard Macintosh options and only a few require further explanation. Using the Save As... and Export options from the File menu, the contents of the LUT and any current measurements can be saved separately from the image itself. In addition to Undo, Cut, Copy, Paste, Clear and Select All, the Edit menu includes a variety of options (Rotate, Flip, Scale, Fill, Invert [colors]) for changing the orientations, sizes and tones of selected areas or objects in the image. Selections are made using any of eight tools in the Tool window: four for area selection (rectangle, ellipse, polygon, freehand closed area); three for line selection (straight, segmented, freehand); and one (the wand) for automatic outlining of discrete objects — *i.e.*, groups of contiguous pixels — in binary or thresholded images (see following discussion of Options menu).

The principal function of the Options menu is modification of the look-up table (LUT). Adjusting the LUT is usually the first step in image enhancement; a well designed LUT will highlight features of interest and subdue background noise. Another reason for adjusting the LUT is to 'tune' it to the spectrum of colors or tones in an image where colors have quantitative meaning (*e.g.*, birefringence colors in a photomicrograph, or shades on a color-indexed topographic map). Cross sectional profiles and "topographic" maps of the color-encoded variable can then be generated (see discussion of the Analyze menu and example applications below).

The Options menu offers a number of built-in LUTs, including the standard Grayscale table (256 gray values, from 0 = white to 255 = black) and various Color tables (*e.g.*, 20-, 32- or 256-color spectra). Customized LUTs can be created using the dialog box under the LUT Options submenu. This allows the user to set the number of distinct color or gray values, to designate 'reserved' colors that remain unchanged when LUT editing is done, or to invert the LUT to create a negative of the original image. [The LUT can also be modified directly using tools from the Tool window. The palette tool, a double-headed arrow, can stretch or rotate the LUT spectrum, and the eyedropper tool can be used to modify particular colors in the LUT].

Threshold and Density Slice under the Options menu provide other means of manipulating the LUT and are prerequisites for most quantitative geometric analyses. Thresholding changes all pixels with LUT values greater than a threshold value to black and those below the threshold to white. The 'height' of the threshold is adjusted directly within the LUT window, using the palette tool (which automatically replaces the cursor when Threshold is selected). The numerical value of the threshold is given in the Info window. Density slicing permits the user to highlight pixels with values within a certain range (interactively displayed in the Info window), making it possible to isolate objects with distinctive colors — *e.g.*, grains with particular birefringence values. The width of the density slice is again set manually in the LUT window by using the palette tool. Once the density slice is set at the appropriate level, a binary image showing only pixels with values in the specified range can be created using the Binary command within the Enhance menu (see below). Most of the geometric analysis functions offered by the program can be performed only on thresholded or binary images. An alternative to density slicing, for cases where the objects of interest have pixel values ≤1 or ≥254, is to select Highlight saturated pixels from the Preferences dialog box in the Options menu.

A particular LUT can be applied simultaneously to all open images using the Propagate command of the Options menu. This is useful in cases where a series of images with similar color, brightness or contrast characteristics are to be analyzed (e.g. for a series of photomicrographs from the same thin section, or when an entire roll of film is under- or over-exposed). Alternatively, an LUT created for a particular image can be saved as a separate file, then applied to another image later simply by opening the LUT file.

A carefully specified LUT may not be sufficient to enable the program to recognize the objects you consider to be most important. Filtering functions in the Enhance menu allow additional fine-tuning of entire images or selected parts. The program includes standard grayscale filters such as Smooth, Sharpen, Shadow, Reduce noise, Dither, and Trace edges. Custom filters can also be designed or imported from other image processing programs. (*Photoshop*-type plug-in filters must be stored either in the Macintosh system folder or in the folder containing *Image*). Enhance contrast and the converse operation, Equalize, are built-in filters that generate new LUTs, which can then be applied to other images. As mentioned above, the Binary command creates a black and white image from a thresholded or density-sliced file. Once a binary image has been created, special filters including Erode, Dilate, Outline and Skeletonize (within the Binary submenu) can be applied until objects have the desired form. Of course, these filters must be used judiciously, so that the image processing does not spuriously alter the features to be analyzed.

The Arithmetic option under the Enhance menu allows the user to modify pixel values by specifying constants for addition, subtraction, multiplication or division. With the Image Math submenu, arithmetic operations can also be performed with the pixel values of two images to yield a third, resultant image. For example, by finding the difference or ratio of pixel values for two coincident images of a thin section taken under plane and cross-polarized light, it may be possible to create an image in which grain boundaries are more sharply defined than in either of the original images. Subtract background makes it possible to edit out an unwanted background (*e.g.*, a fine-grained groundmass in a photomicrograph), as long as there are not abrupt discontinuities in the pixel values.

The Analyze menu contains most of the program's quantitative measurement functions. Measure is the command used for analysis of a single selection.

If no selection is specified, the entire image is analyzed. Selections can be made using the area or line selection tools, but measurements of irregularly shaped objects will be more accurate if the auto-outlining `wand` tool is used, and this requires that the image be thresholded or density sliced. Parameters to be measured are specified through the `Options` submenu. Choices include area; length or perimeter; *x-y* centers, axial lengths, and orientations of best fit ellipses; and density (grayscale) information (*e.g.*, minimum, maximum, and modal values, standard deviation). Real-world scale and density calibrations can be made by selecting `Set Scale` and `Calibrate`. Automated measurements of distinct (non-contiguous) objects are possible for thresholded or density-sliced images using the `Analyze particles` command. This is the most efficient way to gather information about a large number of objects — for example, the axial lengths and orientations of strain markers (see example applications below). Careful image enhancement is necessary, however, for automated measurements, and it is important to confirm that the program is analyzing the intended objects. This can be checked by selecting `Label particles` and/or `Outline particles` within the `Analyze particles` dialog box. This box also allows the user to specify whether or not the measurements should include particles at the edge of the image or interior holes within particles.

Whether measurements are done one at a time using `Measure` or collectively using `Analyze particles`, they are saved temporarily in a window that can be viewed by selecting `Show Results`. Individual measurements can be repeated or omitted using `Redo` or `Delete`. Results can be saved in tab-delimited (spreadsheet-compatible) format by selecting `Export` from the `File` menu.

The `Analyze` menu also includes several options for analysis of color or grayscale values in an image. These options could be useful, for example, in characterizing the degree of lattice-preferred orientation in a thin section on the basis of birefringence (see example applications below). `Show Histogram` displays a plot showing the number of pixels with particular values. `Plot Profile` creates a density profile along a specified line or rectangular strip. `Surface Plot` shows the variations in pixel values as three-dimensional 'topography'. By inverting the LUT, valleys can be made into mountains and vice versa. `Surface Plot` makes it possible to generate dramatic three-dimensional views of landscapes from maps with color-coded topography (Gordon 1994).

The `Special` menu is the interface for capturing video images using a frame grabber card. Users can specify the number of frames to be captured and whether the frames should be averaged or summed. Brightness, contrast, and other characteristics can also be controlled interactively during image capture. The second major function of the `Special` menu is to load macros designed to expedite repetitive tasks. The macros must be written in a Pascal-like programming language that is specific to *Image*. Several useful macros — for example, one that can convert analog line plots to digital *x-y* data — are included at the *Image* ftp site. Information on writing custom macros appears in an appendix to the user manual (Rasband, 1994b).

The `Stacks` menu makes it possible to link a series of images for purposes of three-dimensional analysis, creation of animated sequences, or preservation of RGB color information. Creating and working with `stacks` is memory intensive. To use the full capabilities of the Stacks menu, it is usually necessary to have at least 16 MB RAM and to increase significantly the program's memory allocation. If memory permits, as many as 1000 images can be grouped into a stack. A stack is created by opening in sequence all images (slices) to be included, then selecting Windows to stack. A stack is saved as a single file, though most of the menu commands operate only on the active slice. Stacks can be dismantled by choosing Stack to windows, which converts each slice back to an independent file. Stacks can be viewed and edited using Next Slice (or the '>' key), Last Slice (or '<'), Add slice and Delete slice. Animate displays the slices in sequence at a rate that can be controlled by pressing keys '1' (slowest) through '9'. New composite images can be created using Average, which shows the arithmetic mean of all slices in the stack, and Make montage, which displays some or all of the slices in row and column format in a single screen image. (Figure 1 was generated in this way).

Three dimensional analysis of a sequence of slices requires that the program recognize corresponding points on successive images. If the spatial coordinates of a series of slices do not coincide after images are scanned or imported, the Register submenu can be used to align them. Registration can be achieved either by 1) clicking on two or more 'fiducial' points in successive images; or 2) creating a simple text file containing alignment information. The first method is the simplest and most direct, but the second makes it possible to register slices in which fiducial marks (*e.g.*, those in an SEM micrograph) are beyond the borders of the screen image. Once registration in established, cross sections in new orientations can be generated using the Reslice submenu. Users manually specify the position and orientation of the new section using the line selection tool, then are prompted for information on slice spacing in pixels. The larger the number of slices and the closer the slice spacing, the more detailed the resultant section (and the greater the RAM requirements). The Project submenu creates planar projections of the three-dimensional data set from specified directions. If memory is sufficient, an animated sequence is generated showing how the projected image changes with viewing angle. A dialog box allows the user to specify the axis of rotation and includes options for highlighting or eliminating interior or exterior points in the projections. The Project option can be used even for a one-slice stack and is useful for generating down-plunge views of folds. Other options within the Stacks menu permit conversions between 24-bit RGB and 8 bit formats and creation of stacks from video frames.

Spin-off Program ImageFractal

ImageFractal, a modified version of *Image* developed by William Sheriff at NIH (Sheriff 1992), can perform Richardson (caliper) and box-counting fractal analyses of objects in binary or thresholded images. It shares the menu structure of *Image* except that it lacks Stacks (analyses are restricted to a single plane) and has a Fractal menu that appears when the Fractal Analysis option is chosen from the Analyze menu. Two modes of analysis are possible through the Fractal menu. In the Grid Screen mode (the default), the program carries out a box-counting analysis of all pixels in a binary image. In the Get Outline mode, the program analyzes individual objects selected by the user. When Get Outline is chosen, the thresholding/density slicing is automatically enabled, and

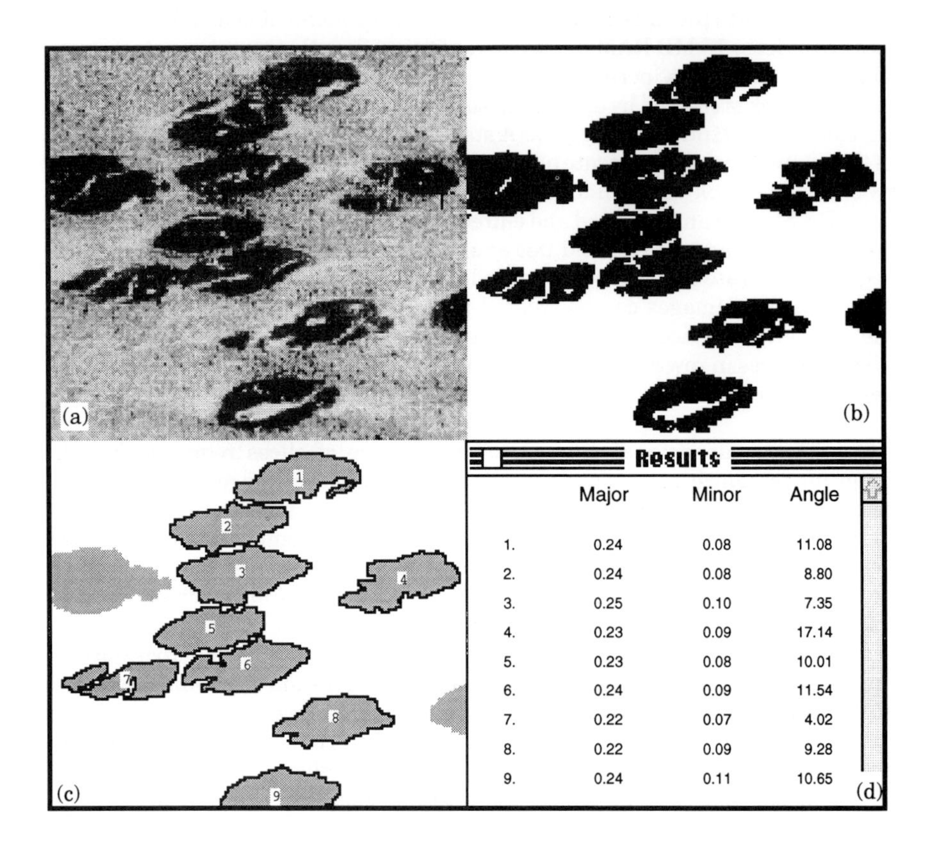

	Major	Minor	Angle
1.	0.24	0.08	11.08
2.	0.24	0.08	8.80
3.	0.25	0.10	7.35
4.	0.23	0.09	17.14
5.	0.23	0.08	10.01
6.	0.24	0.09	11.54
7.	0.22	0.07	4.02
8.	0.22	0.09	9.28
9.	0.24	0.11	10.65

Fig.1. Strain analysis using *Image*. a) Deformed amygdules in metabasalt. Grayscale image from plane-light photomicrograph. Width is approximately 1 cm. b) Binary image created after density slicing to highlight amygdules. c) Amygdules outlined and numbered during measurement. d) Axial length (pixels) and orientations (relative to the x-axis) of best-fit ellipses.

the `palette` tool is used to modify the LUT so that the object(s) to be analyzed is (are) highlighted. Next, the `wand` tool is used to select the object and either `Caliper` (Richardson method) or `GridXY` (box-counting method) are specified in the `Fractal` menu. The x-y coordinates of the selected object may be saved as a text file if desired. In both the `Grid Screen` and `Get Outline` modes, the results (ruler/tile size and number of rulers/tiles) can be displayed by choosing `Show Results` from the `Analyze` menu. To save the results as a tab-delimited file, the user must click the `Export` button at the bottom of the `Results` window. NOTE: Some users of *ImageFractal* report that the program may yield apparently spurious results at the longest and shortest ruler lengths. Each user should check to confirm that the data generated are meaningful.

Example Applications of Image

The versatility of *Image* makes it impossible to catalog all of its possible uses in the geosciences; each user can discover new applications and define corresponding procedures. The following examples illustrate how *Image* might be used for several common tasks in structural geology: 1) performing strain analyses and point counts; 2) assessing the extent of recovery and lattice-preferred orientation; 3) conducting fractal analyses; and 4) visualizing three-dimensional structures from two-dimensional sections. In all cases, the goal is to enhance an original image in a manner that enables the program to recognize geologically significant features. This is facilitated by judicious choice of starting material. If thin sections are to be analyzed, for example, image processing should be borne in mind at the time that photomicrographs are taken.

Strain Analysis / Point Counting

The capacity of *Image* to calculate best-fit ellipses for distinct particles makes it an ideal tool for strain analysis. A plane-light photomicrograph of deformed amygdules in a greenschist-facies metabasalt (Fig. 1a) was digitized and saved as an 8-bit TIFF file, then opened in *Image* and converted to grayscale. It was density sliced (`Options` menu) so that only the amygdules, which were uniformly dark in the original image, were highlighted. A binary image was then created (`Enhance` menu), leaving just the amygdule silhouettes (Fig. 1b). From the `Options` dialog box of the `Analyze` menu, `Ellipse major axis`, `Ellipse minor axis`, and `Angle` were selected as parameters to be measured. From the `Analyze particles` box (also in the `Analyze` menu), `Label particles` and `Outline particles` were chosen in order to keep track of the objects the program was measuring (Fig. 1c). The option `Ignore particles touching edges` was also selected in order to avoid erroneous results from incomplete strain markers. The measurements were displayed (Fig. 1d) by selecting `Show Results` (`Analyze` menu), then saved in tab-delimited format for later analysis in a spreadsheet or R_f/ϕ analysis program.

A similar procedure could be used for almost any point-counting task, provided that the objects to be tallied had properties that allowed them to be distinguished from their neighbors (Table 1). By selecting `Area` as one of the variables to be measured (`Analyze / Options`), it is possible, for example, to determine the areal percentages of particular minerals or grains in a thin section (e.g. porphyroclasts in an ultramylonite; cumulate phases in an igneous rock; clasts vs. matrix in a sediment).

Assessing the Extent of Recovery and Lattice-preferred Orientation

In microstructural analysis, many inferences about deformational history are based upon spatial variations in the birefringence of grains. Statements about the extent of intracrystalline strain, subgrain development, or lattice preferred orientation in a thin section reflect the viewer's perception of extinction characteristics within the section. In general, these statements are qualitative (*e.g.*, "strong undulose extinction"; "partial recovery", *etc.*). Image analysis offers the

Table 1. EXAMPLE APPLICATIONS OF IMAGE ANALYSIS IN STRUCTURAL GEOLOGY

Mode 1: Enhancement of features to delineate forms more clearly, for graphical presentation or quantitative analysis

Point counting
Grain size analysis
Grain shape analysis (via perimeter/area relationships)
Strain analysis
Assessing degree of grain-shape preferred orientation
Determining areal percentages (clast/matrix; cumulate/melt, etc.)
Delineating porphyroblast inclusion trails

Mode 2: Using color / tonal characteristics to create "topographic" maps and / or extract quantitative information from images.

Quantifying undulatory extinction
Assessing extent of subgrain development
Characterizing deformation lamellae and twins
Determining degree of lattice-preferred orientation
Mapping zoned crystals
Creating 3-D representations from color-indexed maps

Mode 3: Using image `stacking` to create tomographic portrayals or animations from images in spatial and temporal sequences.

Generating oblique cross sections from across-strike sections
Creating "down-plunge" views of structures
Viewing porphyroblast inclusion trails in different sections
Making movies to illustrate progressive deformation

possibility of quantifying previously qualitative descriptions of microstructural characteristics, by converting colors and tones in the original image to numerical data (Table 1).

To illustrate the potential of this type of analysis, three microstructural images (Figs. 2 and 3) were imported into *Image* via Adobe *Photoshop* from a photo CD by Simpson & De Paor (1994). To be opened in *Image*, the files must be saved in 8-bit TIFF or PICT format. The principal challenge in extracting quantitative data from colors and tones in photomicrographs is to preserve original color/tone information as faithfully as possible. Because *Image* is limited to 8-bit files, the colors in the original must be mapped into 256 tones. For low-birefringence minerals like quartz and feldspar, the best approach is to convert to the image to grayscale, which can usually be done without significant loss of information. Analyzing microstructural images with higher-order interference colors, however, requires careful tuning of the look-up table to the appropriate section of the Michel-Levy chart. This is most effectively done in *Photoshop* , before the image is imported into *Image*, by saving the file in `indexed color` mode. An exact match can then be achieved by loading the color table from a 256-color scan of the relevant color cycle on the Michel-Levy chart. If it is not possible to scan an interference color chart, a reasonable approximation can be made by measuring the widths of color bands on the chart and creating color gradients over proportional distances in the 256-entry custom table, following the instructions in the *Photoshop* manual. (In *Photoshop*, saving an image in the `Indexed color` mode creates a 256-color file that looks much like the original but has a look-up table in which there may be a non-systematic relationship between color and associated pixel values. Building a custom LUT in *Photoshop* allows the user to control this relationship and to create smoothly varying color gradients). Less delicate manipulation of the LUT is also possible within *Image*. It is generally easiest to begin by applying the built-in `Spectrum` color table (`Options` menu/ `Color tables`), then to use the palette tool to stretch and/or rotate the spectrum until it spans the appropriate range of interference colors. Building a custom LUT is somewhat laborious in both *Photoshop* and *Image*, but once created, a given table can be applied to other microstructural images with the same range of birefringence colors. Users may want to develop a library of LUTs for different minerals.

Once an appropriate LUT has been established, quantitative microstructural analysis is remarkably easy. Drawing a line with the profile tool across any part of an image creates a plot showing how birefringence (pixel values) vary along that traverse (Fig. 2). The data in such a plot can be exported to a spreadsheet for statistical analysis. The differences between undulose extinction (Fig. 2, traverse A), deformation bands (traverse B), and subgrains in quartz (traverses C and D) are clearly visible in these profiles and could be described quantitatively by developing numerical indices to characterize the 'roughness' of the birefringence 'topography'. For a group of specimens from a particular area, it would then be possible to make quantitative comparisons of the degree of recovery, *etc*.

The extent of lattice preferred orientation in a monomineralic thin section can be dramatically conveyed using the `Analyze / Surface plot` command. The photo-micrograph in Fig. 3a shows the same field of view as the lower image in Fig. 2, but was taken with the gypsum plate inserted to emphasize areas with like crystallographic orientation (Simpson & De Paor 1994). A custom color table was created for the image in *Photoshop*, to match the heightened birefringence colors to pixel values. (Figure 3 is a grayscale printed version of the image). The surface plot of the image (Fig. 3b) highlights areas where crystallographic orientations deviate from the average. A histogram showing the frequency of occurrence of pixel values (relative lattice orientations) for all or selected parts of the image can be generated by selecting `Analyze / Show histogram`. The histogram data can be exported to a spreadsheet. The standard deviation of the pixel values would reflect the extent of lattice preferred orientation in a thin section.

Fractal Analysis

"Small scale structures mimic large" has long been the mantra of structural geologists but the principle was difficult to test quantitatively before the advent of fractal mathematics. Although fractal analysis of some structural features shows that they are indeed self-similar over a wide range of scales (*e.g.*, Turcotte

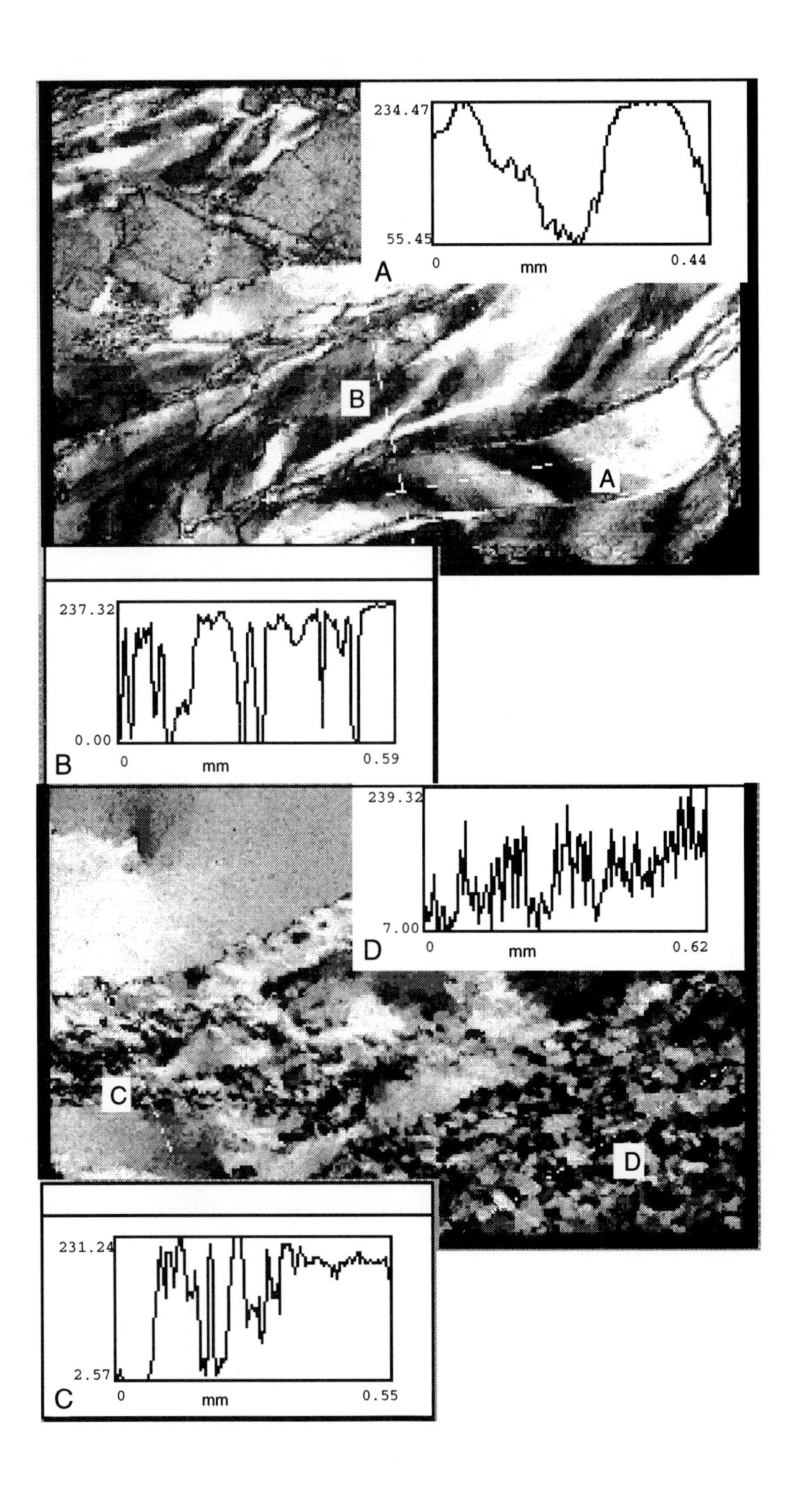
234.47
55.45
0
mm
0.44
A
B
A
237.32
0.00
0
mm
0.59
B
239.32
7.00
0
mm
0.62
D
C
D
231.24
2.57
0
mm
0.55
C

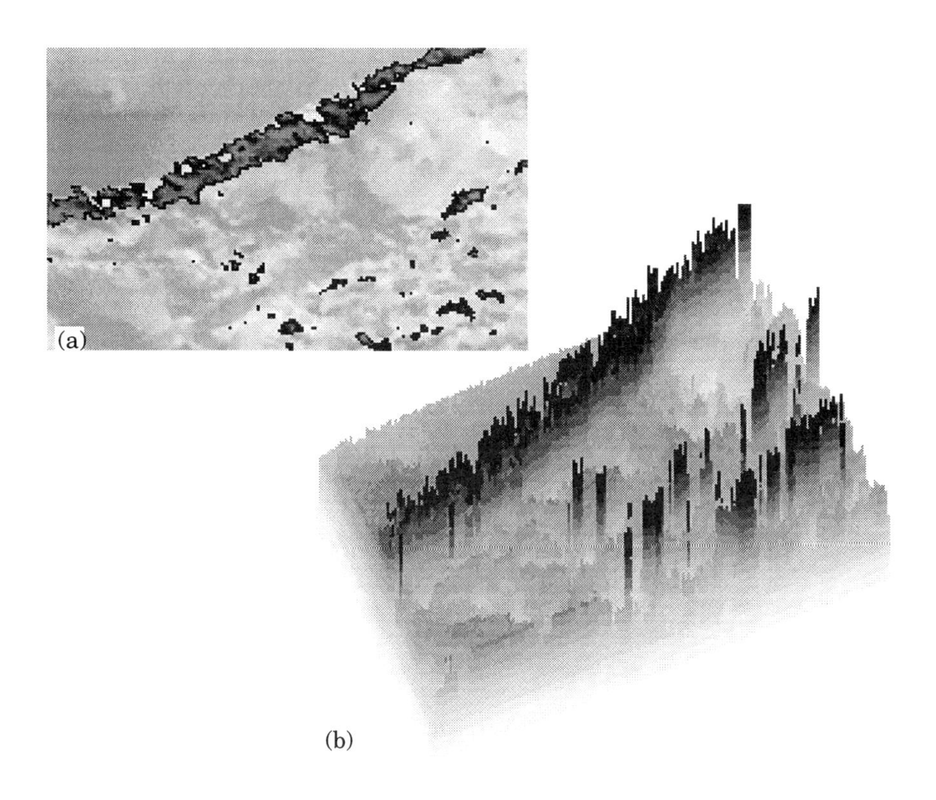

Fig. 2 (Opposite). Characterizing extent of recovery in thin sections. Upper image (Simpson & DePaor 1994, #18) shows undulose extinction in quartz ribbons within protomylonitic greenschist-facies granodiorite (field of view 1.5 mm, cross-polarized light). Profile for traverse a) parallel to length of a ribbon shows smoothly varying pixel values. In contrast, plot for traverse b), across ribbons, shows abrupt changes in birefringence associated with deformation bands within grains. Lower image (Simpson & De Paor 1994, #22) is from a quartz vein deformed under lower greenschist facies conditions (field of view 1.5 mm, cross-polarized light). Plot for traverse c) shows distinction between relatively flat 'topography' of surviving host grains and fine-scale 'roughness' where subgrains have formed. Traverse d) lies entirely within subgrain terrain.

Fig. 3 (above). Assessing extent of lattice preferred orientation. a) Same image as at bottom of Fig. 2, but taken with gypsum plate inserted to emphasize areas with like crystallographic orientation (Simpson & De Paor 1994, #23). (This is a grayscale printed version). b) Surface plot of image highlights areas where lattice orientations deviate from average.

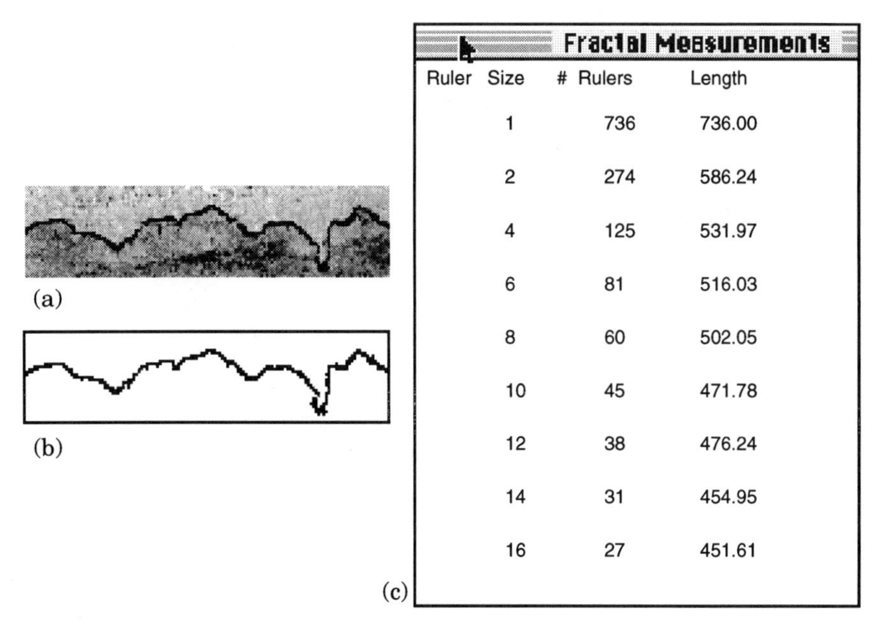

Ruler Size	# Rulers	Length
1	736	736.00
2	274	586.24
4	125	531.97
6	81	516.03
8	60	502.05
10	45	471.78
12	38	476.24
14	31	454.95
16	27	451.61

Fig. 4. Fractal analysis using *Image*. a) Grayscale image of stylolite in limestone slab. Width is approximately 3 cm. b) Binary image created after density slicing. c) Results of Richardson analysis, with lengths given in pixels.

1992), perhaps the more interesting cases are those in which self-similarity breaks down at some scale (as in the Gutenberg-Richter relationship, for example). Changes in scaling laws may reveal significant transitions in the importance of certain deformational processes.

As part of a study to define the fractal character of ductile deformational structures and mechanisms (Boyer & Bjørnerud, *in prep.*), we have analyzed stylolites in order to identify the range of scales over which pressure dissolution operates. The slab in Fig. 4a was scanned directly on a flat bed scanner, saved as a TIFF file using the scanner software and opened in *ImageFractal*. After being converted to grayscale, the contrast was enhanced. `Analyze / Fractal Analysis` and `Fractal / Get outline` were selected in sequence. Density slicing allowed the dark stylolitic seam to be distinguished from the surrounding matrix, and the image was converted to binary format (Fig. 4b). The `wand` tool was used to select the stylolite, and `Caliper` was chosen from the `Fractal` menu. In *ImageFractal*, results (Fig. 4c) are always given in pixels, but a conversion factor for real distances can be found by using the `line` tool to indicate a known distance on the image and entering this value in the `Set Scale` dialog box (`Analyze` menu). The fractal dimension D can be found by finding the slope S of a plot of log [total length] *vs.* log [ruler length], and solving for D from the relationship $D = 1-S$. (Note that because total length decreases with increasing ruler length, S will be ≤ 0 and the fractal dimension will be ≥ 1). Typical fractal dimensions for compaction-related stylolites in micritic limestones are 1.15 to 1.30.

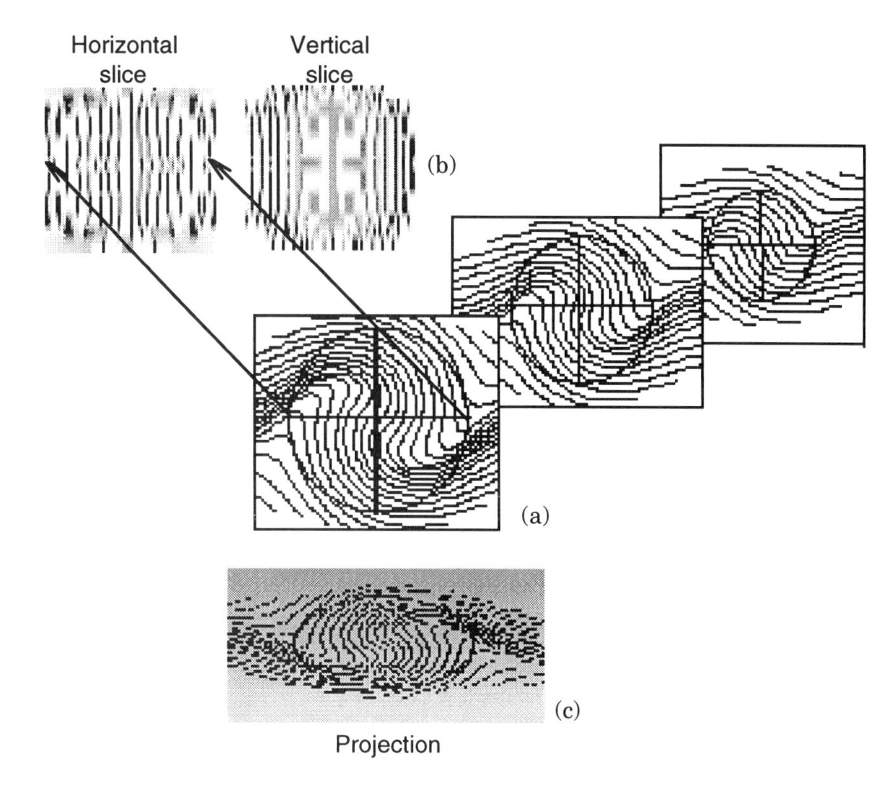

Fig. 5. 3-D visualizations using *Image*. a) Representative slices from a 21-image stack depicting snowball inclusion trails in a porphyroblast (Bjornerud & Zhang, 1994). b) Horizontal and vertical reslices through stack (lines of section shown in (a). c) Nearest-point projection through porphyroblast center, from an angle of 60° above x-axis.

Three-dimensional Visualizations from Cross Sections

The `Reslice` and `Project` options in the `Stacks` menu make it possible to extract three dimensional information from structural cross sections. In order to illustrate the capabilities of these commands, a 21-slice stack depicting 'snowball' inclusion trails in a porphyroblast was created using images generated in a simulation program by Bjørnerud & Zhang (1994). The original images show inclusion trail geometries in sections perpendicular to the porphyroblast's rotation axis (three representative slices are shown in Fig. 5a). The twenty one sections were opened in sequence, then converted to a stack (`Stacks / Windows to stack`). Next, the spacing between slices was specified (`Stacks / Options`). Because the slice spacing must be given in pixels, it is usually necessary to use `Analyze / Set Scale` (and the `line selection` tool) on one of the slices to determine the appropriate unit conversion factor. Large stacks with closely spaced slices will generate higher resolution reslices and projections, but quickly consume RAM. In the present example, it was necessary to set the memory allocation at 12 MB and to increase the `Undo` (clipboard) buffer to 2500 KB.

To generate a new cross section through the stack, the user marks the intended line of section on any slice using the `straight-line` selection tool, then selects `Reslice`. The new section is displayed in a separate window that may be saved and printed. Horizontal and vertical section through the simulated porphyroblast are shown in Fig. 5b. The horizontal slice shows the 'ɔc' inclusion pattern described by Powell & Treagus (1970) in a study of cut effects in thin sections of natural synkinematic porphyroblasts.

Projections generated through the `Stacks / Project` command can be thought of as shadows created when incident light shines at specified angles on the object depicted in a stack. Three types of projections are possible: `Nearest point`, `Brightest point` and `Mean value`. Nearest-point projections, like that shown in Fig. 5c, are the 'shadows' cast by the first points encountered by 'light rays' passing through the stack. The shadows have the same value as the pixels in the original slices. This option can be used to find the true profile of a plunging fold by setting the projection angle at the value of the plunge. Brightest-point projections depict only the lightest pixels encountered along a given ray. Mean-value projections sum the values of all points on a raypath and display the averages on the projection plane. Projections are stored as slices in a new stack. Displayed in sequence, they allow the viewer to 'walk' around, over or through the object represented.

Summary

NIH Image is a potentially powerful tool for research and teaching in structural geology and other geoscience disciplines. Comparable to commercial image analysis programs in its versatility, *Image* is available at no cost and has the additional advantages of frequent upgrades and an extensive community of users connected by the Internet. We encourage geologists to sample from and contribute to the growing directory of macros, examples, and tutorials available at the *NIH Image* ftp site.

References

Bjørnerud, M. & Zhang, H. 1994. Rotation of porphyroblasts in non-coaxial deformation. Insights from computer simulations. *J. Metamorphic Geology,* **12**: 135-139.

Boyer, B. & Bjørnerud, M. *(in prep.).* Fractal characterization of ductile deformational structures.

Gordon, J. 1994. Bringing computerized tutorials and image processing into the classroom on a shoestring budget using "NIH Image" for Macintosh. *Geological Society of America Abstract with Programs*, **26**: A44.

Lennard, P. 1990. Image analysis for all. *Nature*, **347**: 103-104.

Powell, D. & Treagus, J. 1970. Rotational fabrics in metamorphic minerals. *Mineralogical Magazine*, **37**: 801-813.

Rasband, W. 1994. *NIH Image* v. 1.55. The National Institutes of Health, Washington D. C. (Public domain software).

Rasband, W. 1994. *NIH Image User Guide*. The National Institutes of Health, Washington D. C. (Unpublished).

Simpson, C. & De Paor, D. G. 1994. *One hundred photomicrographs of microstructures* (PhotoCD ROM). Earth'nWare Inc., Hull, Massachusetts.

Sheriff, W. 1992. *ImageFractal: Public domain fractal analysis software*. The National Institutes of Health, Washington D. C. (Unpublished).

Turcotte, D. 1992. *Fractals and Chaos in Geology and Geophysics*. Cambridge University Press, Cambridge.

Synkinematic Microscopic Analysis Using *NIH Image*

Youngdo Park
Department of Geological Sciences,
University at Albany, Albany NY 12222, U.S.A.

Abstract– The procedures of microstructural analysis of deforming material during synkinematic microscopy are greatly improved using a Macintosh based imaging system. The system consists of a microscope-video camera for obtaining images, a frame grabber for converting the analog signal into a digital image, and the image-processing program *NIH Image*. Two separate macro programs for *NIH Image* have been developed for extracting information from the digital images. The first macro is used to digitize the positions of marker particles, which aid in the analysis and visualization of deformation. The second macro is a modified version of the Panozzo Heilbronner & Pauli (1993) technique for constructing A.V.A. (Achsenverteilungsanalyse) diagrams, and measuring *c*-axis orientations for uniaxial minerals.

Introduction

Synkinematic microscopy is a technique in which a sample material is deformed on a microscope stage so that microstructural changes can be observed during deformation. Since the observed microstructural changes are related to the active deformation processes, the technique has been used for process studies during deformation (*e.g.*, Means 1977, Urai *et al.* 1980). One of the commonly used techniques for the analysis of deformation is to study the relative motion of marker particles (usually small inert particles such as silicon carbide grains) embedded in the sample material. Maps of deformation fields such as displacement maps and strain maps can be constructed based on the changes in positions of markers at different stages. The first task in analyzing the experimental deformation using these markers is to digitize the *XY* position of markers. A new, on-screen digitization technique has been developed for rapid acquisition of the positions of markers.

Another technique to be discussed is the construction of A.V.A. (Achsenverteilungsanalyse) diagrams which are maps of grain boundaries and the lattice orientations of grains (*e.g.*, Sander 1970). They can be used to correlate

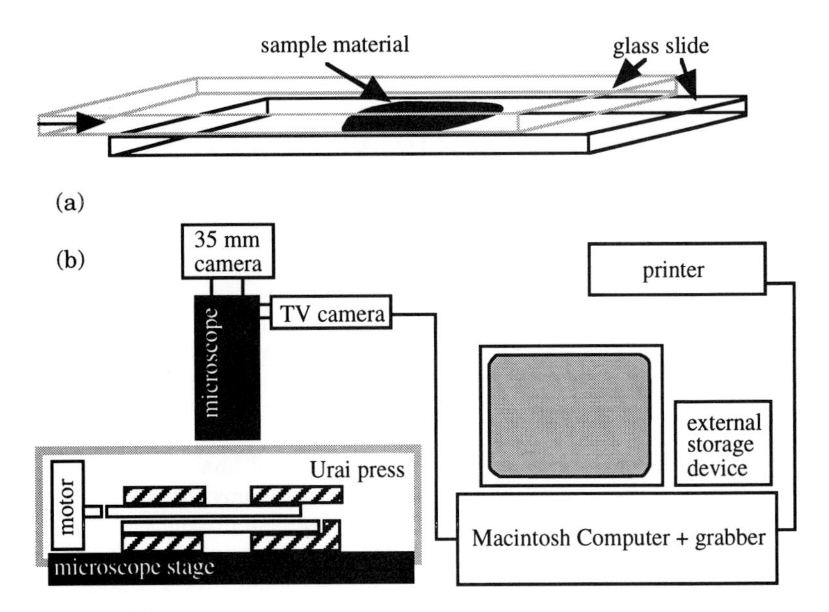

Fig. 1. Experimental sample assembly (a) and equipment for synkinematic microscopy (b).

lattice orientation domains with the domains of other microstructural parameters such as grain size or the domains of different deformation processes. In synkinematic microscopy, the combination of A.V.A. diagrams and displacement of markers in a grain can be used to determine the slip systems in a deforming grain. Since the deformation apparatus (Urai-type press) is fixed to the stage, a universal stage can not be used to measure *c*-axis orientation with this setup. The computer-aided photometry technique developed by Panozzo Heilbronner & Pauli (1993), on the other hand, can be used in order to measure c-axis orientation without a universal stage. However, a sample must be slightly tilted for this technique, and this is not possible in a stage-fixed Urai-type press. A modification of the Panozzo Heilbronner & Pauli (1993) technique has been made so that the technique can be used for measurement of *c*-axis orientation during experiments.

Experimental Apparatus and Image System

Figure 1 illustrates the experimental apparatus for synkinematic microscopy utilizing a Macintosh computer based imaging system. The sample assembly consists of a thin-film of experimental material between two glass slides (Fig. la). Soft analog materials such as polycrystalline octachloropropane, norcamphor, and ammonium based crystal-melt mixtures, are used to study high temperature deformation processes (*e.g.*, Jessell 1986, Bons 1993, Means & Park 1994). The sample-facing sides of glass slides are frosted, scratched, or stepped so that the displacement of the upper glass slide relative to the fixed lower slide causes deformation in the sample material. The sample assembly is placed into a miniature press (designed by J. L. Urai) mounted on microscope stage (Fig. lb). A video camera connected to the computer via a frame grabber, is used for imaging. Since the file size of a digital image is fairly large (typically 1MB for a 24 bit image) and there can be many images from a single experiment, a large hard disk is necessary. External storage devices such as tape drives and optical disk drives can be used for permanent storage of the images obtained during experiments. Color printers or laser printers of high resolution are suitable for printing the digital images.

Time-lapse "movies" can be assembled with the images taken during experiments, using the public domain program *NIH Image* (Rasband, 1994; *NIH Image* can be obtained at the anonymous ftp site – zippy.nimh.nih.gov). Since the program stores images in RAM, a computer system with a large amount of RAM is necessary. For example, a grayscale movie with 80 frames (640×480 pixels) can be stored in a 20 MB RAM system. For movies that exceed the RAM capacity of the system, a macro program that repeats the same function of image-resizing for each frame can be used to reduce the size.

Technique

Time-lapse movies of digital images in *NIH Image* are referred to as "stacks" which are series of images. Stacks make the procedure of comparing images simpler in that subtle changes in the positions of microstructures or particles are easier to see when played fast (up to about 10 frames per second). Stacks can also be used to obtain the grayscale average of pixels in multiple images. Two examples of superposing images are shown in Fig. 2. When interface motion occurs between two phases, changes in the position of phase boundaries can be detected by superposing two images, although the direction of the phase boundary migration is not shown (Fig. 2 a and f). The direction of migrating boundaries, however, can be seen when inverted and normal images are superimposed (Fig. 2b); the darker and lighter areas of the crystalline part (indicated by arrows) represent dissolution and growth, respectively.

Stacks or individual images can be processed manually by the provided functions or automatically by macro programs in *NIH Image*. Macro programs in *NIH Image*, which are similar to Pascal in syntax, are scripts for repeating the same procedures many times. Macro programs provide a powerful method for performing image operations and collecting information from images. Two examples are discussed below.

On Screen Digitization

When constructing maps of deformation fields using the changes in marker positions, the first step is to digitize *XY* positions of markers. Conventionally, a digitizing tablet has been used with CAD programs to digitize marker positions (*e.g.*, Ree 1994).

The usual procedures for digitization are first to label all the particles of interest on the printed pictures and second to digitize marker positions (Fig. 3).

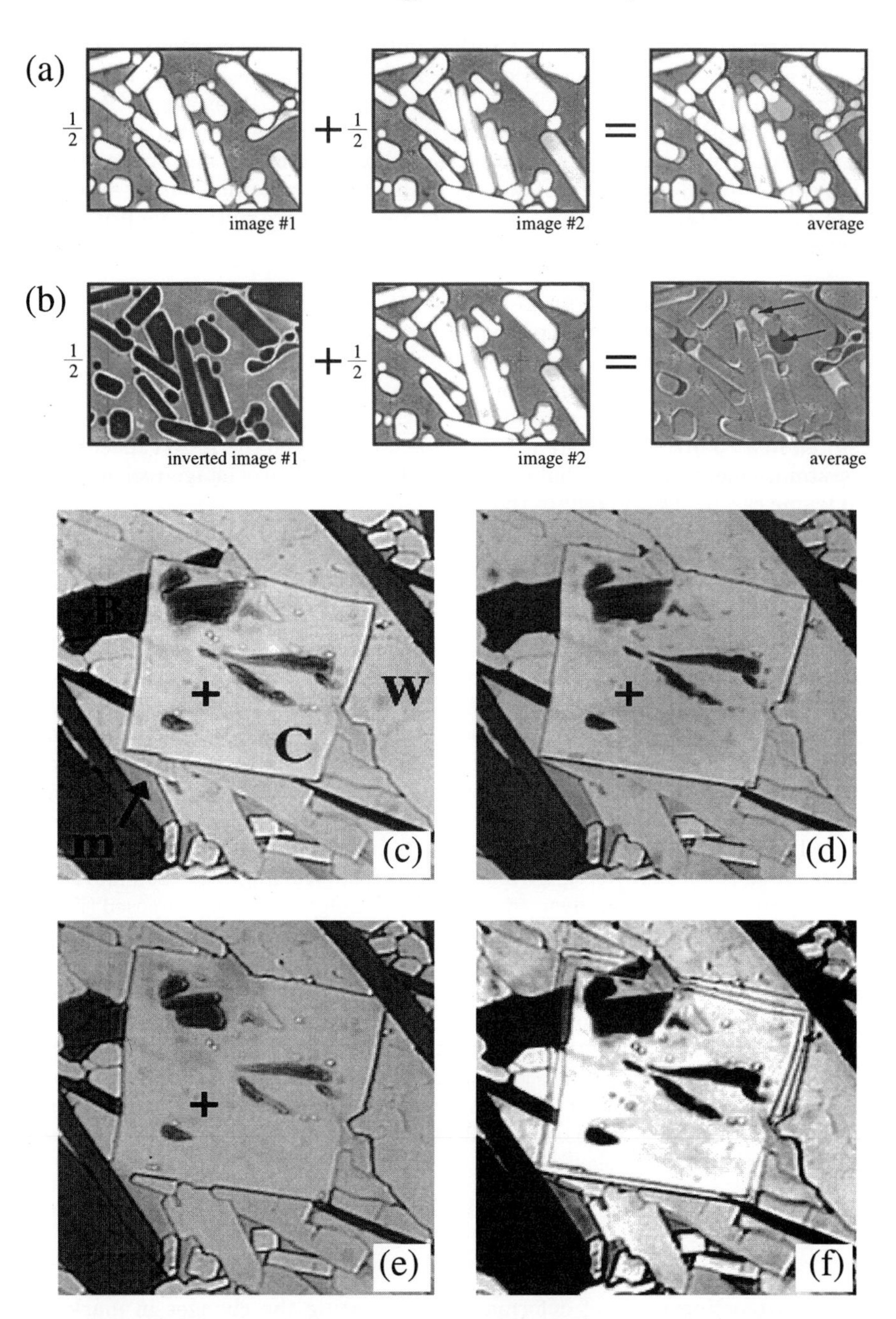

Fig. 2. Superposition of gray-scale images. a,b) Isothermal growth and dissolution of ammonium thiocyanate and ammonium chloride crystals (white) in melt (gray). Elapsed time: 10 hours. Field width: 0.5mm. c,d,e,f) Isothermal growth of porphyroblast **C** in partial melt **m**. Other crystalline phases labeled **B**, **W**. Elapsed time: d) 20 days, e) 68 days, with respect to (c). Field width: 0.3mm.

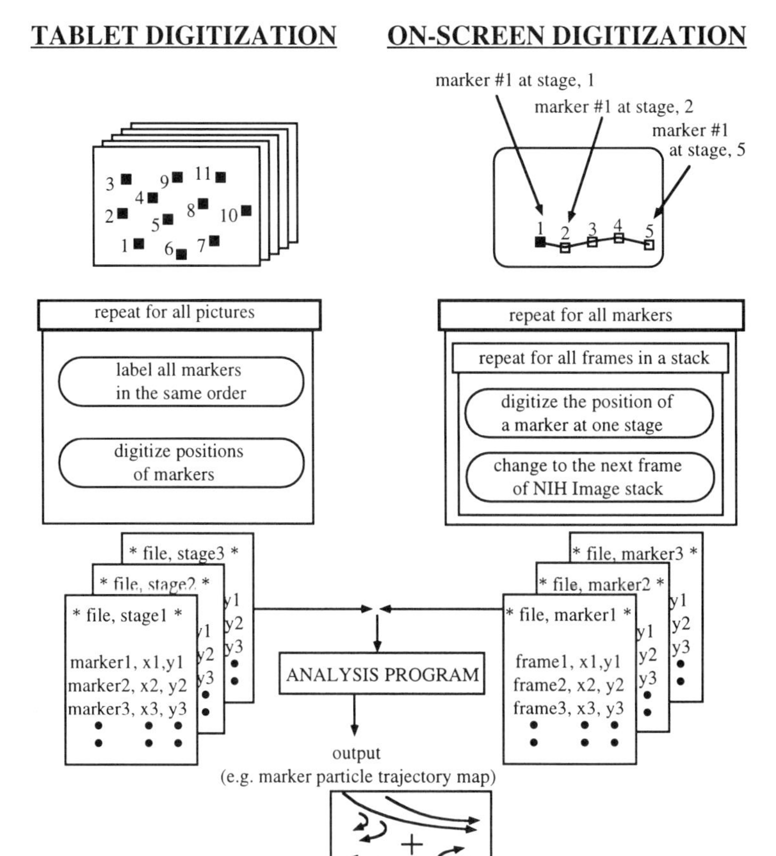

Fig. 3. Tablet digitization and on-screen digitization procedures.

Since a CAD program does not know which particle is which, particles in the pictures of different stages are usually digitized in the same sequence after labeling. The marker particles in every stage have to be identified in order to make identical sequence. Therefore, labeling takes most of the time during tablet digitization.

The time-consuming step of labeling is no longer necessary by digitizing the positions of markers on-screen. During on-screen digitization, the same marker is digitized for different stages, as opposed to digitizing different markers for the same stage (Fig. 3). When the time interval between two images taken during experiments are close enough, the changes in the position of a marker particle relative to others or relative to the edge of the computer screen is small, and identification of the same marker in the next stages is a relatively simple task. The actual steps of digitization, including the preparation steps, are; (1) assembling a time-lapse movie, or stack, in *NIH Image*, (2) loading a macro program for digitization in *NIH Image*, (3) digitization of the positions of a marker particle for

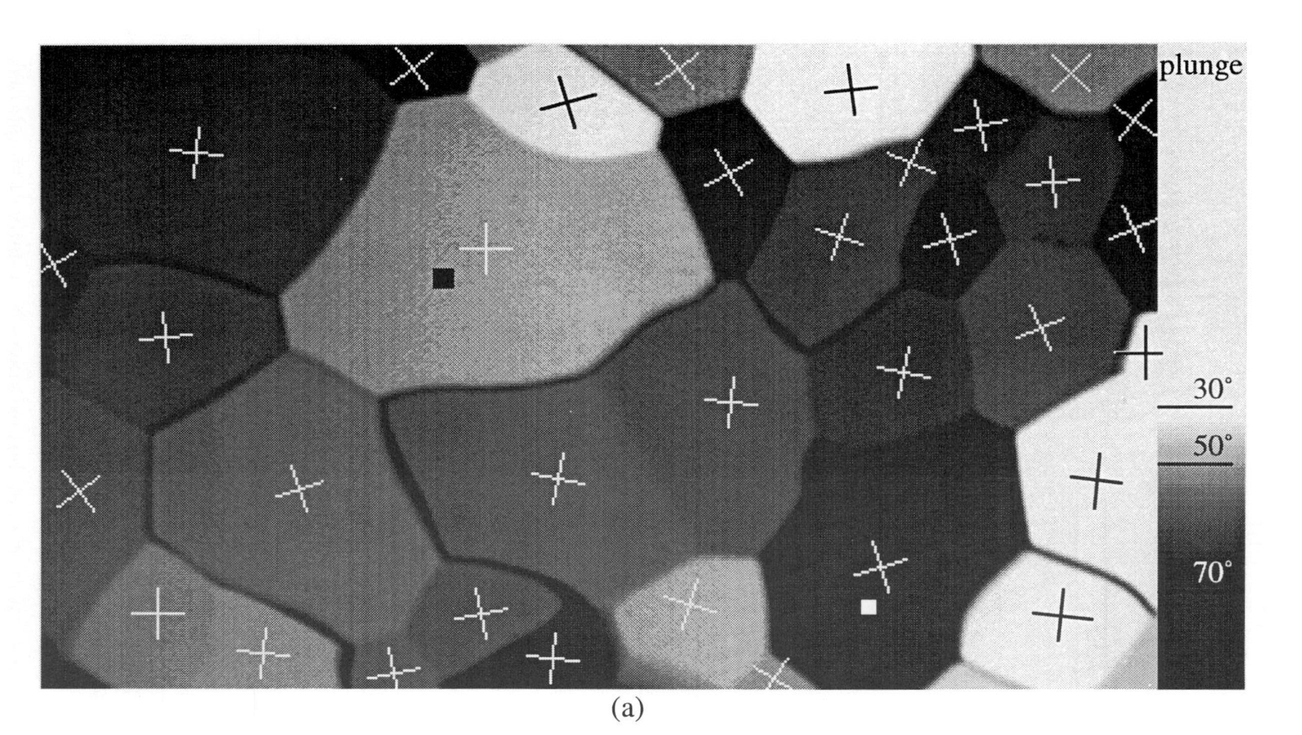

(a)

Fig. 4. A.V.A. construction and *c*-axis measurement for polycrystalline octachloropropane. (a) Simplified A.V.A. Grayscale of grains indicates plunge of *c*-axis as shown in grayscale bar. Numbers represent plunge angles of *c*-axes measured from universal stage. Crosses indicate two directions of extinction, parallel and normal to *c*-axis trend. Field width: 1 mm.

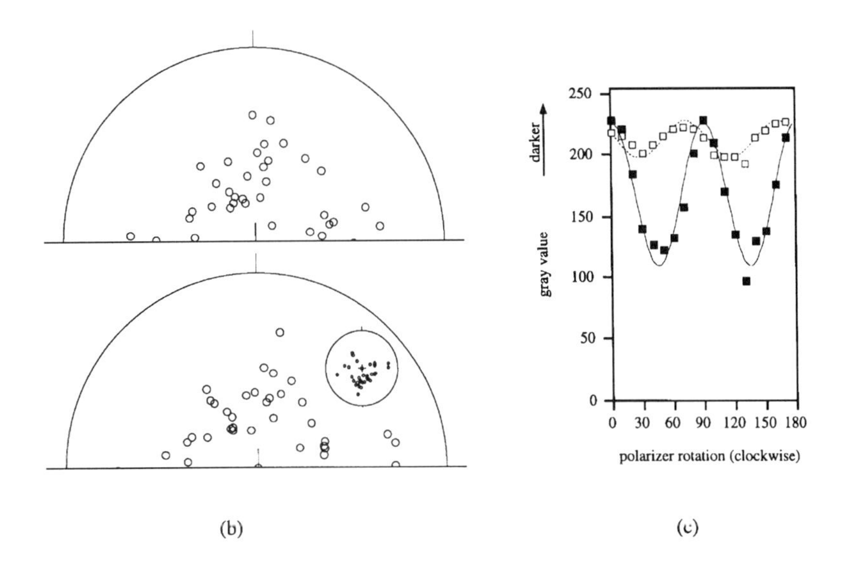

Fig. 4. (b) Half stereographic projection of *c*-axes. Top: projection using calculated results. Bottom: projection using universal stage measurements. Additional information on plunge direction is shown as full projection (inset in the lower diagram). See text for detail. (c) Illustration of steps involved during construction of (a) & (b). Symbols in the graph corresponds to the grains with the same symbols in (a).

different stages (this step is repeated for the rest of marker particles), and (4) merging the files of a particle position into a file representing one stage. This step requires a separate program which merges data files and translates the positions of marker particles relative to a marker which has been chosen as the origin. Files obtained after step (4) can be used to construct deformation field maps using other programs (The best program for doing this, at present, is *Marker* developed by Bons & Jessell, 1993). It has been found that on-screen digitization is faster than the conventional tablet digitization by at least a factor of 5.

Construction of simplified A.V.A.

The method for A.V.A. construction (Fig. 4) is based on the Panozzo Heilbronner & Pauli (1993) method. Interested readers are referred to their paper for details on principles and imaging equipment.

The major difference between the modified method and the Panozzo Heilbronner & Pauli (1993) method is the type of starting images that are used. In the modified method, grayscale images taken in crossed polarized light are used, rather than the images taken in crossed polarized light with a lambda plate. The three steps involved in the construction of A.V.A. are to; (1) obtain images, (2) collect data from the images, and (3) analyze the collected data. Eighteen images of the same area are obtained after rotating the upper and lower polarizers at 10° intervals while keeping the stage and sample fixed (accuracy can be improved by taking 36 images at 5° intervals). The images are read to the *NIH Image* program with the option 256 grayscale LUT (Look-Up Table) selected. A stack is created using these 18 grayscale images. A macro program is then used to calculate the grayscale value of a user-specified area in each of the images of the stack. This procedure is repeated for all grains of interest. After this stage, a file containing grayscale *vs.* polarizer rotation (as represented in Fig. 4c) is produced. The data points from the file of grayscale *vs.* polarizer rotation are fitted to a sine curve by the least squares method (Fig. 4c). Two peaks in the sine curve are expected, and the positions of the peaks will correspond to the two extinction orientations. Depending on the plunge of the *c*-axis of a grain, the amplitude and the average value of the sine curve will change. For example, the sine curve defined by filled squares in Fig. 4c is different from the one defined by open squares, since the orientations of *c*-axes for these two grains are different.

The extinction orientations of each grain can be superposed onto an image which is prepared by averaging the grayscale of pixels from a stack (Fig. 4a). Figure 4a conveys information on the plunges of *c*-axes (represented by average grayscale) and the trends of *c*-axes (one of the two extinction directions) as well as grain boundaries, and therefore is equivalent to A.V.A. The relationship between the grayscale value and the plunge angle must be known in order to interpret the observed, average grayscale from grains. Calibrations can be made by comparing grayscale value with the universal stage measurement of plunge. The true *c*-axis trend, which is one of the two extinction directions (crosses in Fig. 4a), can be determined using the images taken with a gypsum plate. However, the plunge direction of the *c*-axis can not be determined by this technique because the technique does not involve sample tilting. Because of the uncertainty of the plunge direction, it is impossible to present the results with the standard stereographic projection method. The true plunge direction can be either the measured trend or (180° + measured trend), therefore representation of plunge direction should include these two possibilities. This can be done by plotting both directions on a stereonet, and Fig. 4b represents half of the stereographic projection made in this way. Universal stage measurements (inset in Fig. 4b lower diagram) were treated in the same way in order to compare the results with the results of the computer measurements. The computer results are in good agreement with the universal stage measurement.

Although the information on plunge direction can not be obtained with this method, the results are still useful provided there is symmetry of 180° rotation in the *c*-axis fabric. For example, *c*-axis distributions resulting from coaxial and non-coaxial flow, will be different on half of a stereonet, just as they are on a full stereonet. Although the modified method does not give full orientation of *c*-axis like the Panozzo Heilbronner & Pauli method, it may provide most of the information necessary for microstructural analysis. Additionally, relatively rapid analyses can be performed with less sophisticated equipment. For example, it took approximately 20 minutes to generate the results (shown in Fig. 4) without any special optical equipment.

Concluding Remarks

Digital images of microstructures obtained during synkinematic microscopy can be used with the powerful image program *NIH Image* to bring out microstructural changes more clearly. Digitization of structural elements, such as grain boundaries, is also possible as shown in the example of digitizing points. The program *NIH Image*, with the macro programs described in this paper, may also be used for the analysis of natural microstructures. For example, intensity and orientation of foliation in rocks may be measured after suitable modification of digital images using the image filters, and simplified A.V.A. can be constructed for naturally deformed quartzite. Techniques such as on-screen digitization can be applied for strain analysis (*e.g.,* the center to center method by Fry, 1979) after slight modification of the macro source code for digitization.

The programs described here and other utility programs for microstructural analysis are available upon request.

Acknowledgement

I wish to thank Paul Bons and an anonymous reviewer for improvement of the manuscript, Professor W. D. Means for helpful discussions, Ben Hanson for improving the English, and Marco Herwegh for encouragement and thorough testing of the on-screen digitization program. The techniques described in this paper were developed during the studies on textural development in crystal-melt mixtures supported by NSF grants EAR-92704781 and EAR-9017478 to W. D. Means.

References

Bons, P. D. 1993. Experimental deformation of polyphase rock analogues. *Geologica Ultrajectina* (Ph.D. Thesis, Utrecht University). 110pp.

Bons, P. D. & Jessell, M. W. 1993. *Marker.* Monash University.

Fry, N. 1979. Random point distributions and strain measurement in rocks. *Tectonophysics* **60**: 89-105.

Jessell, M. W. 1986. Grain boundary migration and fabric development in experimentally deformed octachloropropane. *J. Struct, Geol.* **8**: 527-542.

Means, W. D. 1977. A deformation experiment in transmitted light. *Earth Planet. Sci. Lett.* **35:** 169-179.

Means, W. D. & Park., Y. 1994. New experimental approach to understanding igneous texture. *Geology* **22:** 323-326.

Panozzo Heilbronner, R. & Pauli, C. 1993. Integrated spatial and orientation analysis of quartz *c*-axes by computer-aided microscopy. *J. Struct. Geol.* **15:** 369-382.

Rasband, W. 1994. *NIH Image*. National Institutes of Health, U.S.A.

Ree, J. -H. 1994. Grain boundary sliding and development of grain boundary opening in experimentally deformed octachloropropane. *J. Struct. Geol.* **16:** 403-418.

Sander, B. 1970. *An Introduction to the Study of Fabrics of Geological Bodies.* (English Edition, translated by Phillips, F.C. & Windsor, G.). Pergamon Press, Oxford.

Urai, J. L., Humphreys, F. J., & Burrows, S. E. 1980. *In situ* studies of the deformation and dynamic recrystallization of rhombohedral camphor. *J. Mater. Sci.* **15:** 1231-1240.

Appendix A

```
Macro 'ResizeStack[1]';
VAR
  i: integer;
  NumFrame: integer;
  ZoomFactor: real;
  gNewStackCounter, hundred, rest, ten, one: integer;
  ChHundred, ChTen, ChOne, StrForWindow,Prefix: STRINC;
BEGIN
  gNewStackCounter := 0;
  NumFrame := trunc(GetNumber('Number ofImages :', 10));
  ZoomFactor := GetNumber('ZoomFactor :', 0.7);
  Prefix:= GetString('Filename Prefix?','a');
  FOR i := 1 TO NumFrame DO BEGIN
    gNewStackCounter := i;
    hundred := gNewStackCounter DIV 100;
    rest := gNewStackCounter MOD 100;
    ten := rest DIV 10;
    one := rest MOD 10;
    ChHundred := chr(ord('O') + hundred);
    ChTen := chr(ord('O') + ten);
    ChOne := chr(ord('O')+ one);
    StrForWindow := concat(ChHundred, ChTen);
    StrForWindow := concat(StrForWindow, ChOne);
    StrForWindow := concat(Prefix,StrForWindow) ;
    SelectSlice(i);
    SetScaling('BilinearNew Window');
    ScaleAndRotate(ZoomFactor, ZoomFactor, 0);
    SetPicName(StrForWindow) ;
    SaveAs;
    Close;
  END;
END;
```

Appendix B

```
{** On Screen Digitizer for NIH Macro **}

Var
  StartFrame, EndFrame, CurrentGrainNo: integer;
  Macro 'SetUp';
Begin
  StartFrame := trunc(GetNumber('Starting Frame :',1));
  EndFrame := trunc(GetNumber('Ending Frame :', 50));
  CurrentGrainNo := trunc(GetNumber('Current Grain No. :', 1));
End;
Macro 'PointDigitizer [*]';
Var
  i, size, SaveStack: integer;
  x, y, x1, yl, x2, y2: integer;
  GrainID: String; rest, ten, one: integer;
  ChTen, ChOne, StrForWindow: String;
Begin
  size := 3;
  SetFont('Monaco');
  SetFontSize(9);
  SetText('Bold');
  SetLineWidth(2);
  For i := StartFrame To EndFrame Do Begin
    SelectSlice(i) ;
    SetCursor('cross');
    Repeat
      CetMouse(x, y)l
    Until button;
    wait(O.5);
    beep;
    x1 := x - size;
    yl := y - size;
    x2 := x + size;
    y2 := y + size;
    MoveTo(xl,yl);
    LineTo(x2,y2);
    x1 := x - size;
    y1 := y + size;
    x2 := x + size;
    y2 := y - size;
    MoveTo(x 1, y1);
    LineTo(x2, y2);
    rUserl[i] := x;
    rUser2[i] := y;
    ShowMessage('Point No.:', i : 3,'\');
    ShowMessage('x:', x:3,', y:', y:3,'\');
  End;
  Begin
    ten:= CurrentGrainNo Div 10;
    one := CurrentGrainNo Mod 10;
    ChTen := chr(ord('O') + ten);
    ChOne := chr(ord('O') + one);
    StrForWindow := concat(ChTen, ChOne);
    StrForWindow := concat('#', StrForWindow);
  End;
```

```
  GrainID := GetString('Grain ID Number?', StrForWindow);
CurrentGrainNo := CurrentGrainNo + 1;
  NewTextWindow (GrainID, 200, 400);
  SelectWindow (GrainID);
  writeln('Grain No.:', GrainID : 5);
  writeln('Starting Frame #');
  writeln(StartFrame);
  writeln('Ending Frame #');
  writeln(EndFrame);
  writeln('Frame No.', chr(9),'x', chr(9),'y');
  For i := StartFrame To EndFrame Do Begin
    writeln(i, chr(9), rUserl[i], chr(9), rUser2[i]);
  End;
  Save;
  Close;
  SelectSlice(StartFrame);
  SetCursor('arrow');
  SaveStack := GetNumber('SaveMovie? (l:yes)', 1):
  If SaveStack Then
  Save;
  PutMessage('Digitization Complete!')
End;
```

Image Analysis of Microstructures in Natural and Experimental Samples

Paul Bons and Mark W. Jessell
Victorian Institute of Earth & Planetary Sciences,
Monash University, Clayton, Victoria 3168, Australia
paul@artemis.earth.monash.edu.au mark@artemis.earth.monash.edu.au

Abstract– Image analysis is a useful tool that allows geologists to perform some microstructural analyses in a reproducible, unbiased, and quantitative manner, often with considerable time-saving compared to manual analysis. It can perform tasks such as detailed crystallographic orientation mapping and strain analysis with pattern matching that are virtually impossible (or at least impractical) to perform manually. In other cases, such as detection of grain boundaries, manual analysis can be just as good as, or even better than, digital image analysis, although it may be more laborious.

Introduction

Why Image Analysis?

Structural geology is probably the discipline in geology that relies most on visual information. This is because the geometry of deformed rocks provides a fundamental source of data for the interpretation of kinematic and dynamic histories. This information varies in scale from remote sensing data, through aerial photographs, the appearance of an outcrop in the field, and the microscope view of a thin-section, down to TEM-images showing dislocation tangles or diffraction patterns.

Structural geologists are trained to analyse these images and to recognise and select features of interest, such as fault traces, cleavage planes, lineations, *etc*. Tasks such as the estimation of average grain size or the recognition and measurement of *S* and *C*-planes in an *S-C* mylonite can usually be done without the aid of sophisticated instruments. Some tasks are, however, more difficult or cumbersome. For instance, estimation of grain size can be performed with a micrograph and a ruler, but is a lot of work, and while recognising a Shape Preferred Orientation (SPO) is easy, it is much harder to visually determine the intensity of the SPO. Such cumbersome and tedious jobs can often be performed much faster by computers.

The human eye and brain are extremely adept at recognising patterns and other features in an image. One reason for this is that our intelligence enables us to use available information selectively to quickly do the analysis. This includes not only information that is contained in the image, but also information learned from previous experience. This means that human observations are invariably biased to some extent, which can be both useful, when it allows us to quickly recognise a microstructure seen previously, or unhelpful, when for instance more attention is paid to large grains than small ones, or when a bias resulting from training makes us unknowingly look for certain features and neglect others.

Human analysis is not only biased, it is also limited in its reproducibility. Observations become less accurate due to fatigue or boredom and may depend on mood. Observations may also depend on the orientation of a micrograph, for instance, or on lighting conditions (although this is true for digitally acquired images also). Certain measurements and analyses are therefore often better performed by computer. The main advantages are that difficult, cumbersome, or tedious jobs can be done quickly (at least not using 'human' time) and quantitatively and the results are reproducible. Image processing and analysis encompasses numerous tasks, many of which can be, and have been, done manually but are handled better and faster by computer. Image analysis also allows the analysis of more complex spatial and orientation relationships to be performed, yielding insights that may not be apparent using manual techniques.

In this paper, we describe how microstructures can be analysed using digital computer image analysis. This means that we will deal mainly with images derived from thin-sections or polished surfaces of deformed rocks and neglect image analysis routines specifically used for remote sensing or SEM/TEM images.

Digital Images

Computerised image analysis is performed on a digital image, which is a two dimensional array of brightness or colour values derived from the brightness or colour of the real world image. In most cases, one will deal with grayscale images where the brightness for each picture element (pixel) in the array is represented by an integer value typically ranging from 0 to 255 (8-bit). Colour images are usually represented by 3 8-bit values per pixel, where the values define the colour of the pixel in some standard representation such as red-green-blue or hue-saturation-intensity. The binary image is a special type of image; each pixel can only have one of two values (0 or 1, usually interpreted as black and white).

Image Acquisition

There are normally two ways of obtaining a digital image of a microstructure, (i) directly with a digital camera, or (ii) by scanning an existing analogue image such as a photograph (Fig. 1). To get digital images of thin sections one can use both methods. Video cameras are available that can be mounted on a microscope. The quality and resolution of these cameras is steadily improving while prices are

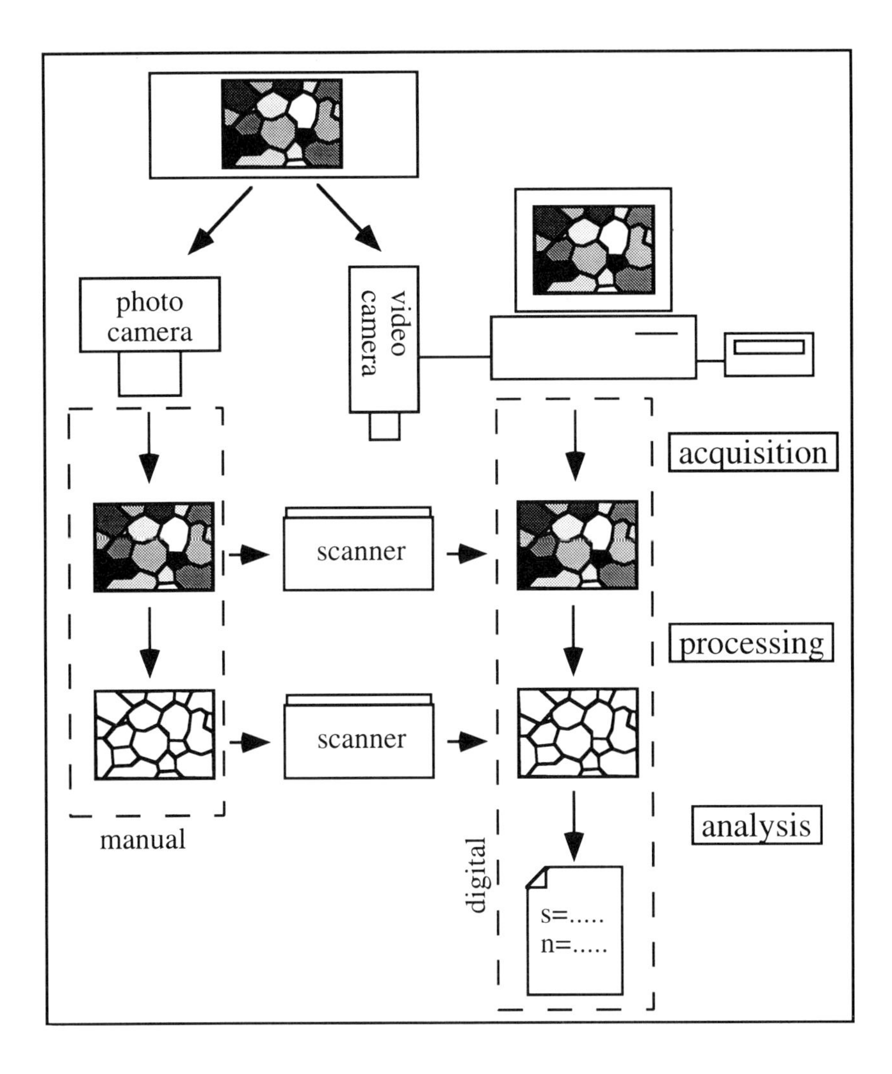

Fig. 1. Steps involved in image processing and image analysis. See text.

decreasing. Monochrome video cameras with the minimum practical resolution of about 700×500 pixels presently cost about US$500. The video camera is attached to a computer which is equipped with a board, known as a frame grabber, which converts the video input into a digital image. The video input is usually displayed on the computer display to allow 'live' selection of the frame, focussing, and adjustment of light intensity, *etc*. Once the displayed image is satisfactory, it can be saved, usually on a hard disk. Any image processing set-up should be complemented with suitable archival data-storage media such as optical disks or tape because of the large file sizes involved (350k for an 8-bit 700 × 500 pixel image).

The resolution of current video camera/frame grabber systems, determined by the resolution of the video Charge Coupled Device (CCD) chip and the frame grabber's clock speed, is inferior to the resolution of photographic film as determined by the size of the photo-sensitive grains. It can therefore be desirable to take photographs with a "classical" photo-camera and digitise prints later with a scanner. As a result, however, the immediacy of the response is lost, so that optimal focus and lighting may be not be achieved.

Whilst we are mainly concerned with microscopic images of thin sections of deformed rocks, some attention will be given to the use of image analysis for deformation experiments where a time-sequence of images of a deforming rock is available. In these experiments it is crucial to obtain the best quality images during the experiment, since the sample continues to deform and one cannot remake an image later, as one can do with a thin-section. It is then often best to make photographs or slides with a photo-camera and later digitise the prints.

Image Processing and Image Measurement

The computerised treatment of images, or image analysis, can be divided into two parts: (i) image processing and (ii) image measurement. Image processing encompasses operations on an image which produce another image. These operations are usually performed to enhance certain features within the image and to prepare the image for the next step, namely image measurement. Image measurement may involve the determination of grain size, the analysis of strain, or the calculation of *c*-axes orientations. This step can produce one or more numbers (*e.g.,* average grain size, grain size distribution) or a new image (*e.g.*, a map of *c*-axes orientations). We first briefly discuss the most common operations of image processing. Many clear textbooks cover this topic in much more detail and the interested reader is referred to Russ (1990, 1994), Gonzalez & Wintz (1987) and Dougherty (1994) on which this section is based, and to Fabbri (1984) which provides further examples of geological applications. Next we treat certain routines for image analysis that are particular to, or most useful for, microstructural analysis.

Software

Most digital video-cameras and scanners require specialised software, which is supplied with the hardware. Once the images are obtained one can perform further image analysis with certain more general software packages such as *NIH-Image* (Mac public domain package, written by Wayne Rasband, National Institutes of Health), *ImageTool* (PC shareware, Univ. Texas Health Science Center), or *IMGEO* (*UNIX* software – see Shtuka & Royer 1994). Most of these packages contain general image processing routines. For specific geological operations, one often has to rely on software made available by geologists or write one's own software (although *IMGEO* is a geologically oriented system).

Image Processing

Point Operations

The simplest operations that can be performed on an image are the so-called "point operations" or "point transformations" (Fig. 2). In these operations, the new value of each pixel only depends on the original value of the pixel itself. In general terms this can be described by

$$g'(i,j) = f(g(i,j)) \tag{1}$$

where the original value of the pixel at coordinates (i,j) is $g(i,j)$ and the new value is $g'(i,j)$. In modern systems, the pixel aspect ratio is nominally 1:1, giving square pixels, and any pixel can be referenced by its column and row numbers (i,j). The most common operation is contrast enhancement which include the application of a linear stretch (Fig. 2c) or equalisation (Fig. 2d). Contrast can be thought of as the range of brightness in an image in adjacent areas. The human eye is optimised for picking up high contrast features in a scene, so improving the contrast of particular features improves our ability to pick them out from background information. Often the image itself does not have to be changed, but one changes the Look-Up Table (LUT), which is a mapping function that describes the way a certain gray-value is actually displayed. The advantages of using a LUT are first that for an 8-bit image one only has to alter 2^8 (or 256) values instead of 700x500 (or 350,000) values to change the appearance of the image, and secondly that the underlying data are preserved.

A special point operation is called thresholding by brightness. Here a binary image is produced by replacing the original value of a pixel by 0 if its value is below a certain threshold value and by 1 if it is above the threshold. Once a binary image has been produced, area and shape measurements can be easily made. Thresholding itself is a very simple operation, but the determination of a suitable threshold can be difficult. A poor choice of the threshold can lead to large errors in subsequent measurements (Fig. 3).

Filters or Kernels

Filters or kernels are neighbourhood operators that transform one pixel using information about that pixel and its neighbouring pixels. The underlying principles of neighbourhood filters are identical to filters applied to one dimensional data, such as a moving average filter. Filters can be classed according to the type of operation that is applied to the data and to the range of spatial frequencies that are affected by the filter. In order to conceptualise the idea of spatial frequency in an image, consider a profile line drawn across the image showing one dimensional brightness variations as a function of position and treat the frequencies in the profile as you would a normal 1D signal. The concept of 2D frequencies is the same, except that a direction as well as a frequency must be considered.

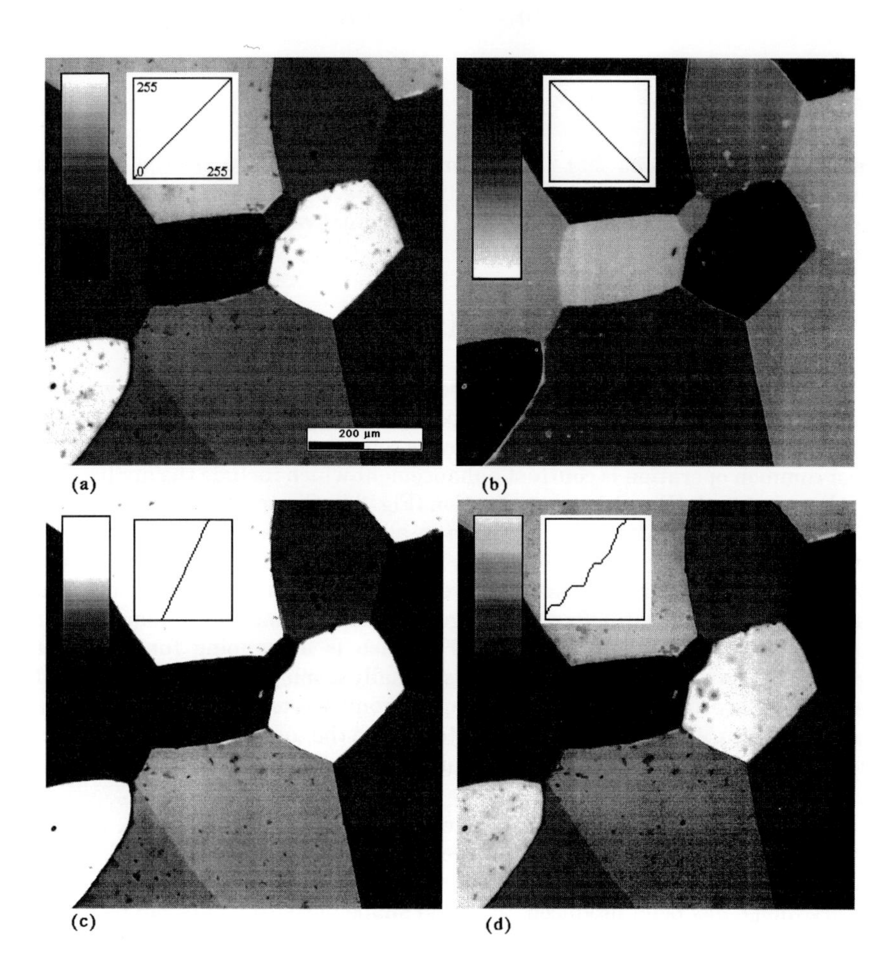

Fig. 2. Example of the effect of different Look Up Tables (LUT's). 512×512 pixel image of octachloropropane (OCP) grains viewed with a microscope with cross-polarised light. Different grains show as different shades of grey. Small dots and specks are small dust and dirt particles. *NIH Image* is used for many subsequent figures to show effect of various image processing routines. Graph inset shows relation between pixel value (ordinate) and brightness value (abscissa). Column-inset shows brightness range 0 at bottom to 255 at top. a) Normal LUT: Pixel value = brightness value. b) Inverted LUT: brightness = -pixel value, resulting in negative image. c) Contrast enhancing LUT: part of pixel value range used for whole brightness value range. Pixel value *vs*. brightness in this range still linear. d) Equalisation: Each brightness value occupies equal area of pixels in image. Graph is cumulative frequency of brightness values in whole image.

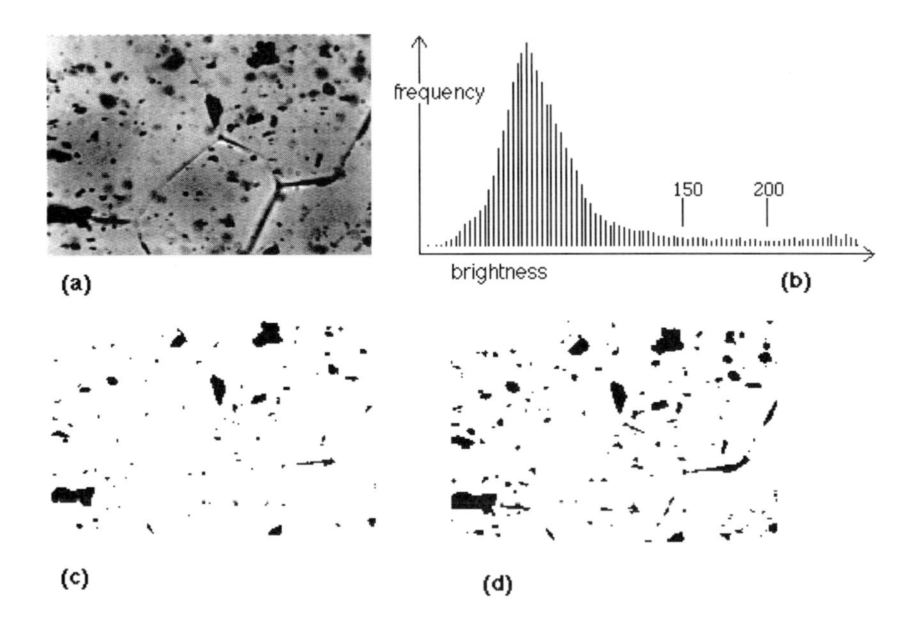

Fig. 3. Example of thresholding. Image is of OCP mixed with fine graphite powder. a) original image, b) histogram of brightness values. Graphite particles stand out in image but histogram doesn't show two clear phases. c) Binary image to show graphite particles produced by thresholding original image at brightness = 200. Only darkest graphite particles are recognised. d) Same, thresholded at 150. Most graphite particles are recognised, but dark grain boundaries are also above threshold and appear black.

Low-Pass Filters

Low-pass filters allow low frequency (long spatial wavelength) data to remain relatively unaffected but remove or reduce the impact of high frequency (short wavelength) information. Some filters can be expressed by a matrix operation, and are known as convolution filters, such as the one for simple smoothing:

$$\begin{pmatrix} 1 & 1 & 1 \\ 1 & 1 & 1 \\ 1 & 1 & 1 \end{pmatrix} \text{ or } \begin{pmatrix} 1 & 1 & 1 & 1 & 1 \\ 1 & 1 & 1 & 1 & 1 \\ 1 & 1 & 1 & 1 & 1 \\ 1 & 1 & 1 & 1 & 1 \\ 1 & 1 & 1 & 1 & 1 \end{pmatrix}$$

With this smoothing filter the new value for each pixel is the average of the original value of the pixel and the 8 (or 24) neighbours for a 3×3 (or 5×5) filter. It is calculated by multiplying each filter coefficient with the image pixel value it overlays, and then summing the result and dividing by the number of coefficients

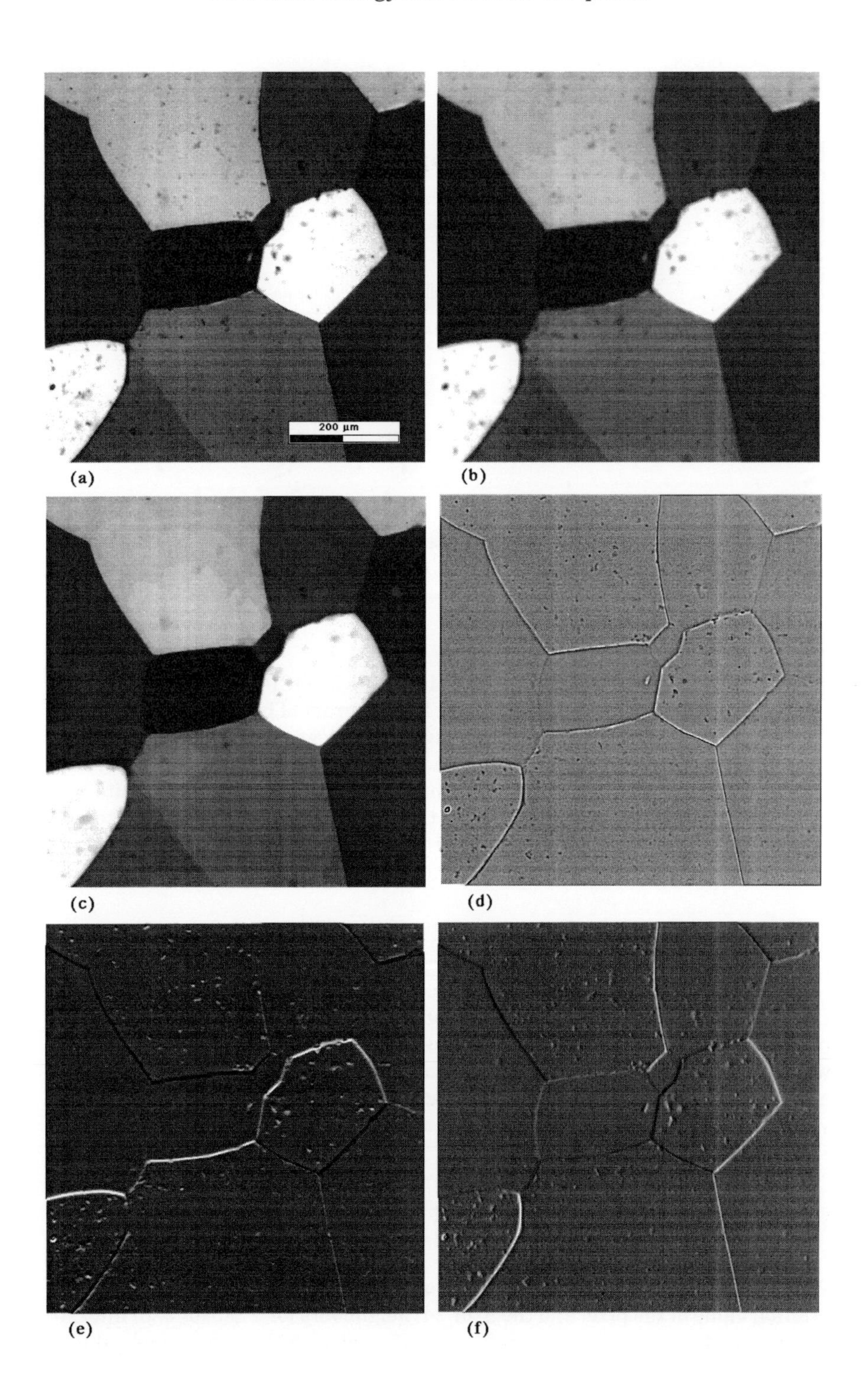
200 µm
(a)
(b)
(c)
(d)
(e)
(f)

in the filter – 9 for a 3×3 filter. The filter can be used to reduce noise (high frequency random or structured patterns unrelated to the actual image data), but at the same time it reduces the spatial resolution, because non-related pixels are also averaged. For instance, a sharp one pixel wide boundary becomes less distinct and spread over 3 to 5 pixels (Fig. 4.b). A better filter for noise reduction is the median filter (a class of rank filter) where the value of the middle pixel is replaced by the median value of the pixels under the filter (Fig. 4.c).

High-Pass Filters

High-pass filters are those that allow high frequency data to remain relatively unaffected but remove or reduce the impact of low frequency information. Detection of sharp gradients in image data can be achieved with the Laplacian convolution filter, of which 3×3 and 5×5 examples are:

$$\begin{pmatrix} 1 & 1 & 1 \\ 1 & 8 & 1 \\ 1 & 1 & 1 \end{pmatrix} \text{ and } \begin{pmatrix} 1 & 1 & 1 & 1 & 1 \\ 1 & 1 & 1 & 1 & 1 \\ 1 & 1 & 24 & 1 & 1 \\ 1 & 1 & 1 & 1 & 1 \\ 1 & 1 & 1 & 1 & 1 \end{pmatrix}$$

This filter produces zero values in uniform regions and a large positive or negative value in non uniform regions, such as grain edges, lines, and noisy regions (Fig. 4.d).

A classic image analysis problem is the measurement of grain boundaries which necessitates the selection of linear features. This can be done with directional operators:

$$\begin{pmatrix} -1 & 0 & 1 \\ -2 & 0 & 2 \\ -1 & 0 & 1 \end{pmatrix} \text{(N -S)}, \quad \begin{pmatrix} -2 & -1 & 0 \\ -1 & 0 & 1 \\ 0 & 1 & 2 \end{pmatrix} \text{(NE-SW)}, \quad \begin{pmatrix} -1 & -2 & -1 \\ 0 & 0 & 0 \\ 1 & 2 & 1 \end{pmatrix} \text{(E-W)}, \textit{etc.}$$

These filters each produce non-zero values along certain orientations of lines or boundaries between homogeneous regions of different gray-value, with their names reflecting the geographic orientation of features that they will enhance. Each of these filters gives the gradient in brightness in a certain direction (Fig. 4e, f). These filters are used by Oltra (1988) as the first part of a routine to define orientations of grain boundary segments. The edge-detection filtering is followed by thresholding to produce a binary image where 1 = boundary and 0 = grain

Fig. 4 (opposite). Enhancement by filtering. a) Octachloropropane image. b) 5×5 Mean filter smooths somewhat. Dust suppressed but grain boundaries also blurred. c) 5×5 Median filter suppresses dust more strongly, grain boundaries remain sharp. d) 5×5 Laplace filter highlights dust and grain boundaries. e), f) 3×3 directional filters highlight high frequencies in one direction: e) north-south, f) east-west.

interior. The orientation α of the boundary at each boundary pixel is then determined with two other 5×5 edge detection filters:

$$\alpha = \tan^{-1}\left(\frac{F_1}{F_2}\right) \tag{2}$$

with

$$F_1 = \begin{pmatrix} 1 & 1 & 0 & -1 & -1 \\ 2 & 3 & 0 & -3 & -2 \\ 3 & 4 & 0 & -4 & -3 \\ 2 & 3 & 0 & -3 & -2 \\ 1 & 1 & 0 & -1 & -1 \end{pmatrix} \text{ and } F_2 = \begin{pmatrix} 1 & 2 & 3 & 2 & 1 \\ 2 & 3 & 4 & 3 & 1 \\ 0 & 0 & 0 & 0 & 0 \\ -1 & -3 & -4 & -3 & -1 \\ -1 & -2 & -3 & -2 & -1 \end{pmatrix}.$$

Segmentation

Image processing is often performed by a geologist to recognise or delineate grains. Grains can be defined (i) by regions or groups of pixels in an image or (ii) by their boundaries, where certain pixels represent the grain boundaries and the grains are interpreted as occupying the space within boundaries. The character of the original image and the requirements for further processing will determine which representation is preferable in specific cases. In some cases each grain or each grain type has a distinct gray-value in the image. One can then use the gray-value of a grain itself to define the region occupied by a grain. This can be done by multiple thresholding if two or more distinct phases are present, for instance opaque minerals in a transparent matrix (Fig. 3, 5a,b). Grains of a birefringent material, such as quartz or octachloropropane (OCP), can be distinguished by grouping all pixels with similar gray-value in a connected region. This can be done in a top-down split and merge routine (Fig. 5c, d) where one starts with a large square region and splits the region in four quarters if the variation in gray level within the region is above a set threshold. Regions of similar variation in brightness are merged. One then repeats the procedure until the smallest possible regions – one pixel – are reached. The advantage of this approach is that some information on each grain is retained (*e.g.*, opacity versus transparency). The method works well for cases where each grain has a homogeneous brightness, but not where the brightness varies gradually within a grain, such as is the case with undulose extinction.

In most cases the only way, or the most desirable way, to define grains is by their boundaries. The simplest and often best (but also most time consuming) method is manual tracing of grain boundaries. The trained eye of the structural geologist can readily distinguish real grain boundaries from possible artifacts and other features. Grain boundary tracing can be done on screen or on a transparent sheet over a photographic print. The tracing is scanned to produce a digital image of the grain boundaries for further digital analysis. In the case of an image in which each grain has a different, but relatively homogeneous brightness, one can use an edge-detection routine, followed by thresholding to distinguish between grain edges and interior pixels (Fig. 6; see Oltra 1988,

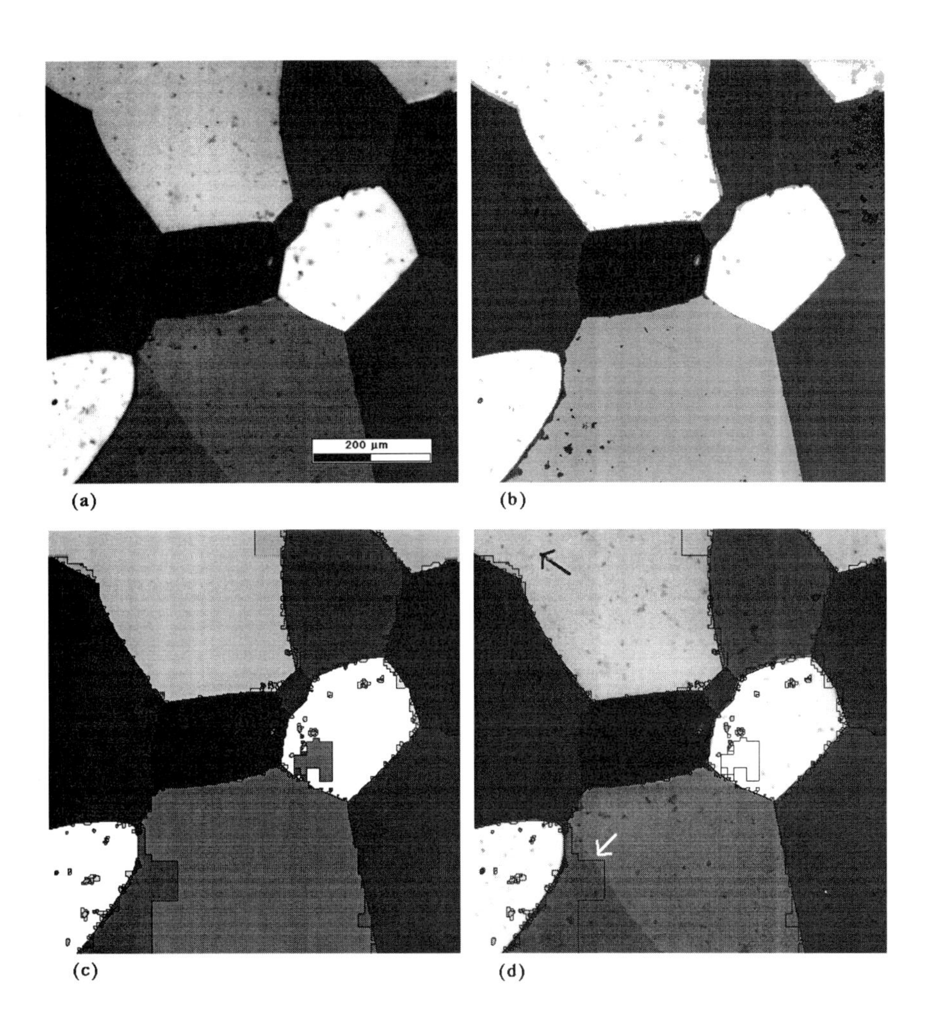

Fig. 5. Segmentation. a) Original image of OCP grains. b) Segmentation into four categories by multiple thresholding at brightness levels of 64, 128 & 192. Grains with similar brightness level grouped together. c) Segmentation by top-down split & merge routine; regions are split in four when average variation from mean in region exceeds 5. Adjacent regions are merged when their difference in mean does not exceed 10. d) Segment boundaries drawn over original image. Most grains are recognised by the routine but many small 'grains' are added at grain boundaries and around dust specks in light grains. Small differences in brightness across grain boundaries are not picked up (black arrow) or are misplaced (white arrow).

Launeau *et al.* 1990). In other cases the grain boundaries themselves may be distinct in the original picture, in which case thresholding may be applied directly on the original image. A problem with edge detection routines is that they often also pick up noise, small second phase particles, fluid inclusions, *etc*.

The method for defining grains depends on the original image. Automatic segmentation should always be checked carefully and should be followed by manual editing to correct errors.

Morphological Filtering: Dilution, Shrinking, and Skeletonising

Morphological filters are another important class of filters which, as their name implies, alter the image according to its local shape. The underlying image can be a grayscale image but most research has been performed in the field of binary (black and white) images. The morphological filter is a binary mask that is compared with the underlying image. If the mask matches the image a black pixel results, otherwise the pixel is left white. By combining a number of different masks into one filtering step, it is possible, for example, to remove isolated black pixels from an image and also to determine the connectivity of groups of pixels. An important processing step is skeletonisation to reduce a wide linear feature such as a grain boundary to a one pixel wide line (Fig. 6b, c). Gaps in the grain boundary may have to be filled using a dilation step, which increases the size of features and isolated pixels erroneously classified as grain boundary may have to be removed using an erosion step which shrinks features. Several routines exist for this processing but manual editing may be necessary as well.

Fourier Transforms

In normal images, the intensity of the pixel refers to the intensity of an (x,y) location in space. These are known as spatial domain images. In them, the individual fabric defining elements need to be selected and measured. Often it is the overall fabric that is of interest, and in this case a frequency domain description of an image can be much more informative. In the frequency domain, each pixel refers to the strength of periodicity for a particular wavelength and orientation. Images can be transformed to the frequency domain by transforms such as the Fourier transform, and filtering of images in the frequency domain can allow much more precise selection of features with a particular characteristic wavelength or orientation. The precision of the operations available in the frequency domain can be considered by looking at the possible range of oriented high-pass filters described in the section above. Only eight orientations may be easily targeted by a 3×3 convolution filter. In contrast, in a frequency domain image the division of the orientation of the filtering is on the order of one half of the number of pixels in one side of the image, which may be several hundreds or thousands.

The Fourier transform for a continuous function is defined by:

$$G_{(u,v)} = \int\int g_{(x,y)} e^{-2\mu i(ux+vy)} dxdy \qquad (3)$$

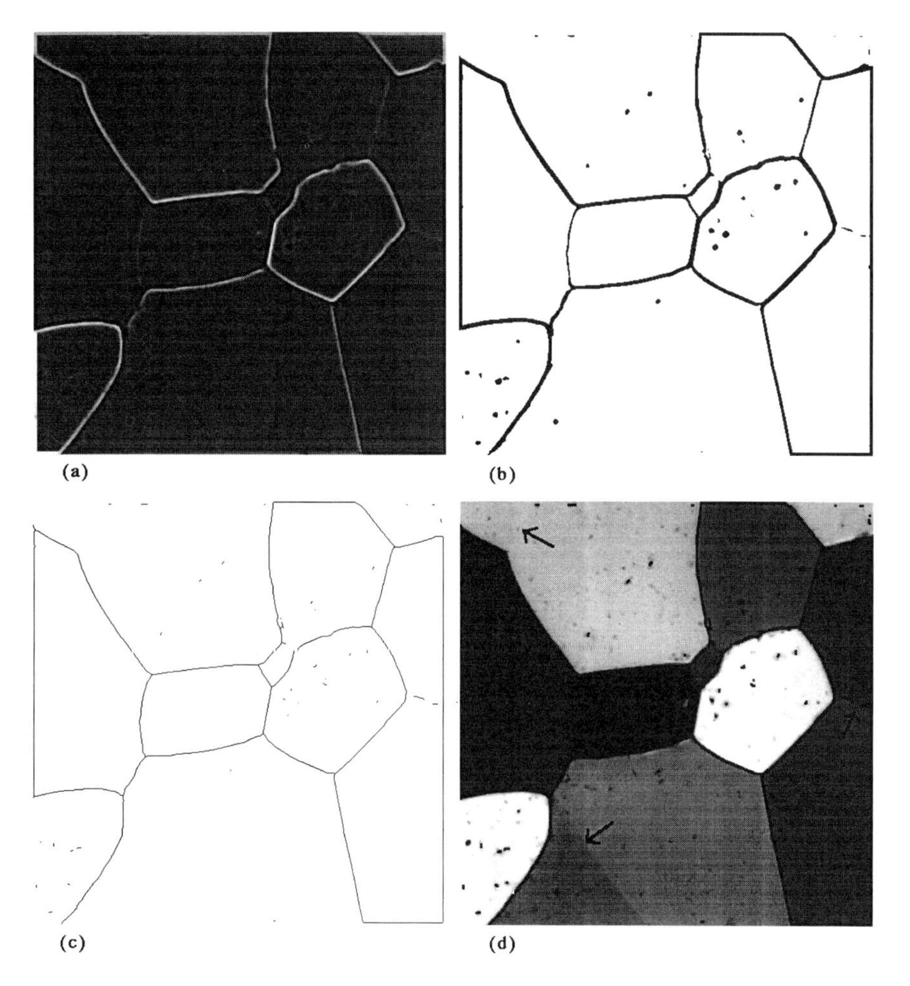

Fig. 6. Example of segmentation through edge detection of OCP image from Fig. 5a. a) Grain boundaries show up with 9×9 Laplace filter applied to 5×5 median-filtered original image (Fig 4.c). Median filtering step was inserted to suppress dust. b) Laplace filtered image is thresholded to make boundaries black and grain interiors white. Some dirt still visible, especially inside two light grains. c) Skeletonising reduces grain boundaries to 1-pixel width. d) Grain boundaries (dilated to constant 3-pixel width) drawn over original image. All distinct grain boundaries are picked up correctly, but other grain boundaries are not or are only partially recognised (arrows).

The integration can be replaced by a summation over discrete intervals (pixels), since a pixel image is not a continuous function of brightness values, giving:

$$G_{(u,v)} = \frac{1}{n^2} \sum \sum g_{(x,y)} e^{-2\pi i (ux+vy)/N} \quad (4)$$

where u, v are spatial wave numbers, the inverse of the wavelength in a particular direction.

A fast routine to do this is known as the Fast Fourier Transform or FFT. For two dimensional images the process can be simplified by first transforming the image by rows and then transforming this result by columns. This way the summation for the whole image does not have to be done again for each pixel:

$$G_{(u,v)} = \sum g_{(x,y)} e^{\frac{-2\pi i x u}{N}} \tag{5a}$$

$$G_{(u,v)} = \frac{1}{n^2} \sum G_{(x,y)} e^{\frac{-2\pi i v y}{N}} \tag{5b}$$

The horizontal and vertical dimensions of the image do not necessarily have to be the same; one can, for instance, use an image of 512 by 1024 pixels. However, a draw-back of the FFT is that the dimensions have to be powers of 2 for many of the available routines. This problem can be partially circumvented by "padding" the image on the sides with zeros or the average brightness value of the image to reach the desired image size.

The outcome of an FFT is an array of complex numbers of the same dimensions as the original image. The transform is usually displayed in the form of the power spectrum, where each pixel represents a certain frequency and the brightness represents the power p of that frequency, with:

$$p = \sqrt{(\text{real part})^2 + (\text{imaginary part})^2} \tag{6}$$

As the power often varies over several orders of magnitude it is useful to set the brightness to the logarithm of the power, or use a logarithmic LUT.

Filtering can be done in the spatial domain as we have seen before, but also in the frequency domain (Fig. 7). A convolution filter operation in the frequency domain can be described mathematically as:

$$G'(u,v) = F(u,v) \cdot G(u,v) \tag{7}$$

where $G'(u,v)$ is the filtered image in the frequency domain and $F(u,v)$ and $G(u,v)$ are the frequency transforms of the filter and the image. The filtered image in the spatial domain can be obtained from G' by the inverse of the FFT. The equivalence between convolution in the spatial domain and multiplication in the Fourier domains allows the rapid calculation of several image processing procedures, such as autocorrelation and wavelet transforms (see below).

The frequency domain is ideally suited for recognising and, if necessary correcting for, features related with a certain frequency or periodicity. For example:

(i) The power spectrum of a fabric can be used to characterise both the orientation and scale of the grains.

(ii) Noise can be periodic (for example as a result of the interference of the video cables and AC power lines). In the frequency domain, this periodic noise appears as high power (bright) spots at certain frequencies. Masking these frequencies removes the periodic noise without significantly affecting features with other frequencies or periodicities.

(iii) Random noise is usually a high frequency feature in an image. Masking the high frequency of the image in the frequency domain removes this noise. Masking can be done with a simple 'hat-filter' where the power of all frequencies higher than a set threshold are set to zero or by for instance a Butterworth filter (Fig. 7c), where the power is more gradually reduced to zero with increasing frequency:

$$p'(R) = \frac{1}{1 + c\left(\frac{R}{R_0}\right)^{2n}} \cdot p(R) \tag{8}$$

where R is the distance to the centre of the filter (usually zero frequency) and R_0 is the distance at which p' is 50% of p if the constant c is set to 1. n is the order of the filter, usually set to 1. This sort of filtering represents a second example of the precision of frequency domain filtering since the definition of the filter is precisely in terms of the wavelength of the features to be affected.

Correlation

Different images or different parts of the same region can be compared and tested for similarity or *correlation*. Such a routine is regularly performed by the human brain when the images of the left and right eye are compared and matched to produce our three dimensional vision. Correlation also has some geological applications, as we will see later. It can be used to (i) locate certain features with a known pattern in an image, (ii) to match points in a stereo-pair of images (*e.g.*, aerial photographs) to determine elevation, and (iii) to determine displacements in a time sequence of images of a deforming sample. Here we will only briefly describe the principles of and some routines for correlation.

The correlation between two regions is a measure of the similarity of the two regions. Correlation can be determined by superimposing the two regions and comparing the similarity of each superimposed pair of pixels. When comparing a binary image one can count the number of equal pixels (black on black or white on white) and use this as the correlation (Fig. 8). The correlation can then vary from 0 (completely anti-correlated) to the number of pixels in the region (complete correlation). In case of grayscale images the pixels will rarely exactly match but may vary in difference. A simple routine to determine correlation is to sum the absolute differences in the region. A more sophisticated routine is the cross correlation function (Gonzalez & Wintz 1987, p. 426):

$$r(m,n) = \frac{\sum_x \sum_y \left[f(x,y) - \bar{f}(x,y)\right]\left[w(x-m,y-n) - \bar{w}\right]}{\left[\sum_x \sum_y \left[f(x,y) - \bar{f}(x,y)\right]^2 \sum_x \sum_y \left[w(x-m,y-n) - \bar{w}\right]^2\right]^{\frac{1}{2}}} \tag{9}$$

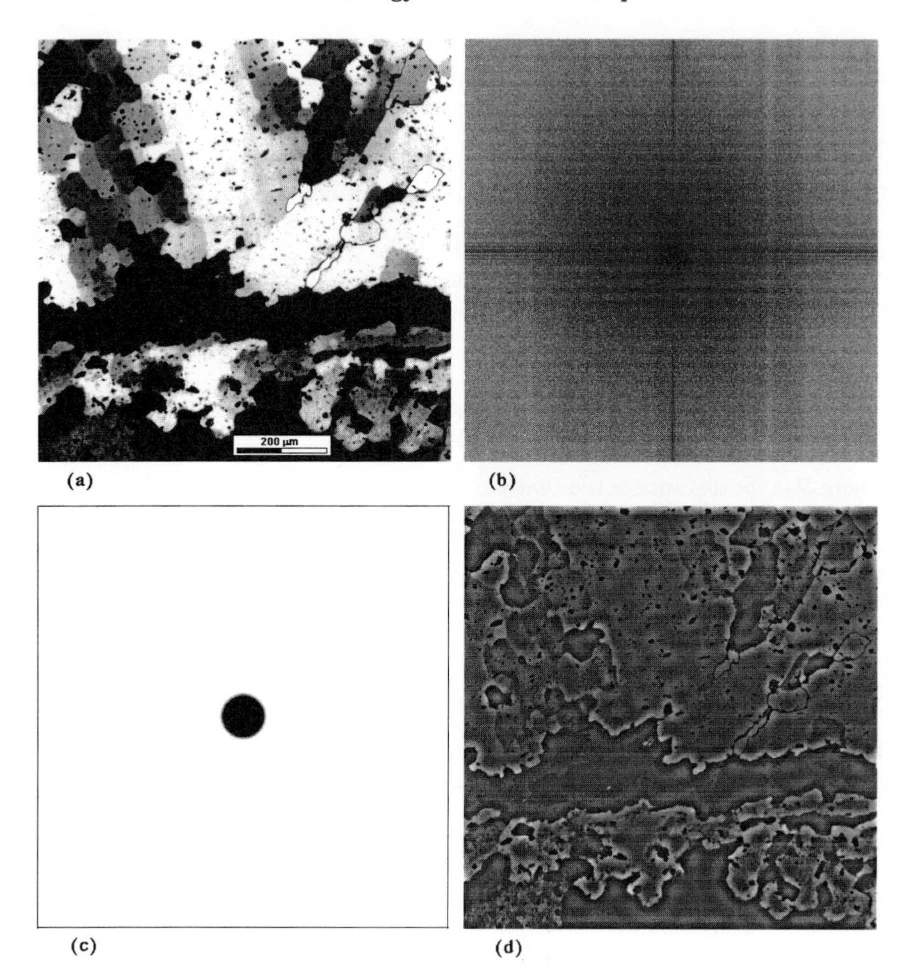

Fig. 7. Example of Fourier transforms. a) Original image of OCP grains with small black marker particles deforming in a transparent deformation cell. Cross-polarised light. b) Power-spectrum of FFT. c) 10% high pass filter – only frequencies under black dot are suppressed. d) Image resulting from application of filter to FFT image, followed by inverse FFT transformation to restore resulting image to spatial domain. All low frequency features are removed (dark and light grains), but high frequency features such as grain boundaries and marker particles show up clearly.

One region of an image, the template, can be shifted over the other image to find the best match. The correlation of the template with the region around a pixel in the second image can be translated to a brightness value for that pixel. If one repeats this procedure for each pixel in the image, one obtains an image where the best correlation is represented by the brightest pixel.

Wavelet Transforms

The application of wavelet theory to image processing was demonstrated in Antoine *et al.* (1993), and has recently been applied to the study of magmatic flow fabrics by Gaillot (1995).

The information derived from a wavelet transformation can be regarded as being intermediate between a correlation transform and a Fourier transformation. In a typical correlation, there is one template which is compared with the image at any position (x,y). The quality of the match is translated to a brightness value for (x,y) in the transformed image. In a Fourier transform, one correlates a certain sinusoidal wave with a certain direction (defined by real and imaginary parts) to the whole image once. This is repeated for a whole range of wavelengths and directions. The results of the transformation are two maps of magnitudes of the real and imaginary parts of the matches of each wave. Visualisation in the form of a power spectrum shows the match of the wave of wavelength u and orientation v as a brightness at pixel (u,v) in the power spectrum image.

A wavelet is typically a single sinusoidal waveform. Unlike a Fourier transform, the wave is defined for a certain point (x,y), where it has its maximum amplitude. The amplitude diminishes to reach zero at some distance away from (x,y). The rate at which the amplitude diminishes defines the size, S, of the wavelet. Each wavelet of scale S can be used as a template for correlation with the original image. For each point in the image one can thus determine the quality of the match for a range of values of S, with one image for each of N values of S. Each image shows how well the local pattern in the original image correlates with a wavelet of size S.

If, for each pixel, one assigns the value S (as a brightness value) of the best correlation of all N images, one can combine all of the images and produce a new image. This image shows the frequency of brightness variations in the original image. For instance, areas in the original image with rapid brightness variations (high frequency variations) show up light and areas with only low frequency variations show up as dark in the transformed image. The original (untransformed) image shows brightness variations in the spatial domain. One sees exactly what the brightness is at any point but one cannot tell the frequency and orientation of the brightness variations. The Fourier transform shows brightness variations in the frequency domain. Here one can see the intensity of brightness variations for any frequency and orientation but one cannot tell from the image where this brightness variation occurs. The advantage of wavelet transforms is that one obtains a bit of both worlds – some spatial information and some frequency information.

Wavelet transforms can be applied with other variations. One can choose many mathematical equations for the wavelet (*e.g.*, Gaussian, sinusoidal, or cosinusoidal functions). One can also use directional wavelets. Gaillot (1995) shows an interesting example of how directional wavelets can be applied to magmatic fabric-analysis (Fig. 9a). In that study, the wavelet is anisotropic in shape. Rather than varying the size of the wavelet, the direction of its long axis a is varied. For a particular size of the wavelet, this results in M matches for different a for each pixel in the original image. Out of these M matches the best

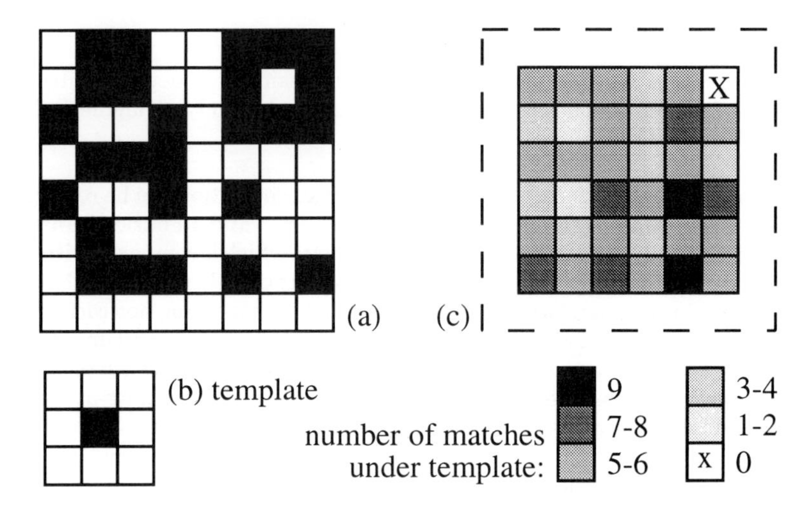

Fig. 8. Example of principle of pattern matching on binary image a) Original image. b) 3×3 template. c) Template is moved over original image and number of correct matches (0 to 9) are counted. Each match is converted to brightness value for resulting image. c) Complete match shows as black and complete anti-correlation as white (X). Image shows that template pattern found two exact matches in original image.

one is selected and converted to a brightness value in the transformed image. By repeating this exercise for different scales of wavelets, the relationship between the orientation of features and their scale can be investigated. Figure 9a shows a synthetic image with a bimodal particle size distribution, where the larger particles have a different orientation distribution compared to the smaller ones. A 2 pixel wavelet applied to this image produces the image seen in Fig. 9b, and a rose diagram of orientations in this summary image (Fig 9c) shows a strong preferred orientation at 10° anti-clockwise from the vertical, whereas a 6-pixel wavelet (Fig 9d) better matches the larger particles at an angle of -18° (Fig. 9e).

Geological Applications

Area Estimations

Measurement of features of individual grains can be performed once the original image is converted to a binary image, where for instance pixels inside grains have the value 1 and boundary pixels have the value 0. General image processing software, such as *NIH-Image* can quickly measure parameters such as size, position of centre, perimeter, aspect ratio, and orientation for each grain from such a binary image. Measurement of some global parameters can also be done easily from a binary image. For example, if solid material is represented by 0 and porosity by 1 in the binary image, then a histogram of the whole image directly provides the porosity value.

Strain Analysis Based on Digital Images

A preferred orientation of objects, grains, or grain boundaries can be used as a measure of strain assuming, of course, that the relationship between strain and microstructure is understood. The measurement of a Shape Preferred Orientation (SPO) is one of the operations in structural geology that can be easily carried out by image analysis. The contribution of image analysis in SPO-measurement can be one of four types:

(i) Image processing to delineate grain boundaries or grains: The result is stored in a non-image format (*e.g.*, a list of grain boundary segment orientations and lengths) for actual SPO-measurement (Panozzo 1983, 1984, Oltra 1988, Alliers *et al.* 1995). This list of grain boundary segments can also be used to characterise preferred grain boundary alignments.

(ii) Image processing to delineate grain boundaries or grains, followed by SPO measurement on the resulting (binary) image (Figs. 10, 11): The linear intercept method described by Launeau *et al.* (1990) falls in this group. First, a binary image is made to separate objects and matrix. Then the number of intercepts (crossings of a matrix object boundary) as a function of direction are measured (Fig. 10). The anisotropy in number of intercepts is a measure of the SPO.

(iii) SPO-measurement: This can be done directly on the original grayscale image with the autocorrelation function (Heilbronner 1992, Russ 1990). Here the correlation is calculated between an image and a copy of itself, with translation offset applied. The autocorrelation coefficient for zero offset is defined as 1 and the correlation coefficient is calculated for all offset values in which the two copies of the image still overlap. The autocorrelation function can be displayed in the form of an image with the autocorrelation coefficient translated to a brightness. This typically produces a bright spot in the middle of a dark field. The shape and orientation of the central spot is a measure of the SPO in the original image (Fig. 12). The size of the spot is related to the size of the features (*e.g.*, grains) in the original image.

(iv) Correlation: A correlation can been made between grain size and the frequency content of the image, and this in turn can be compared with strain measured with more conventional techniques (Dutruge *et al.* 1993).

Crystallographic Orientations

A number of techniques have been developed for the automated measurement of crystallographic preferred orientations, some of which fall under the heading of

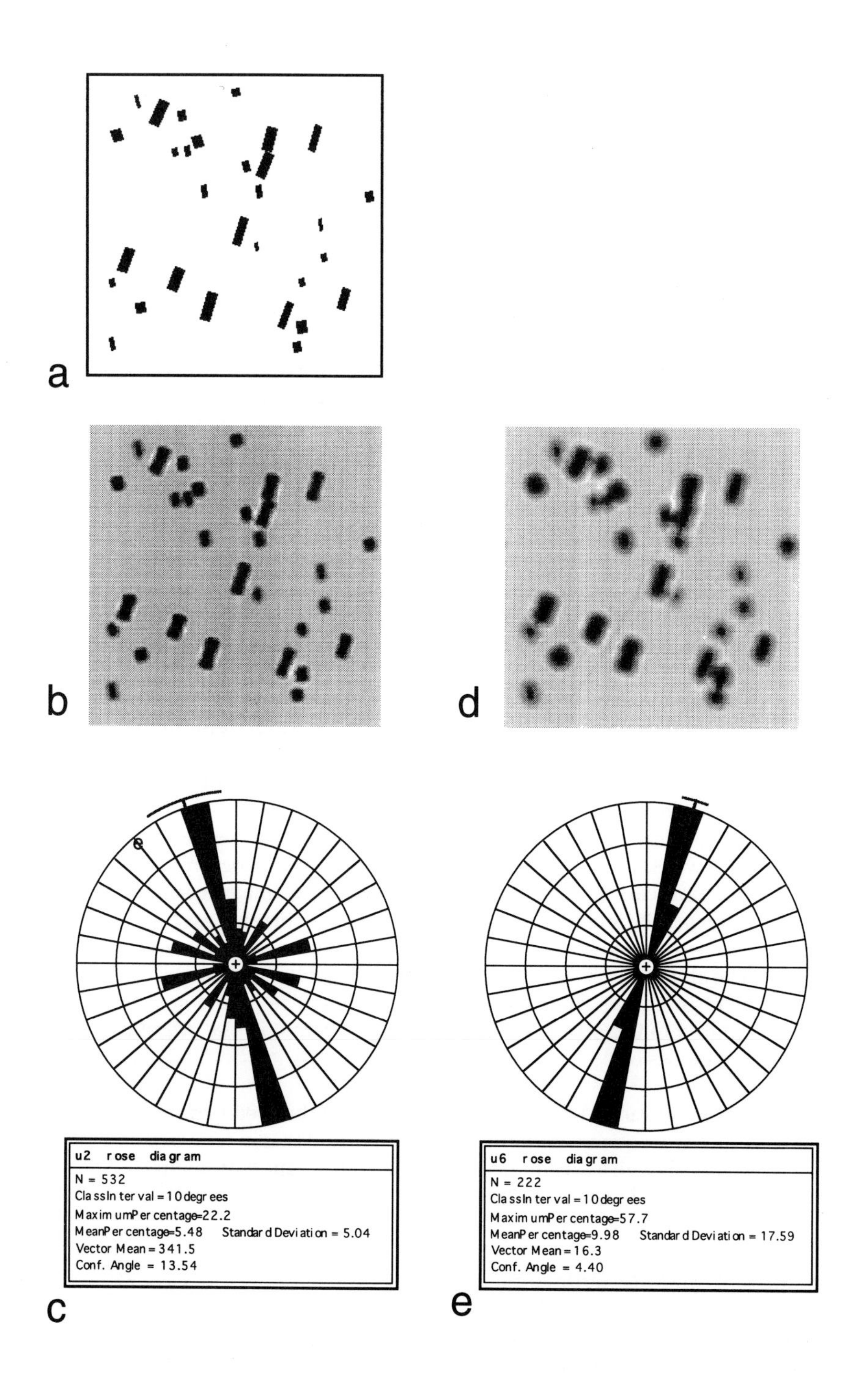
a
b
d
u2 rose diagram
N = 532
ClassInterval = 10 degrees
MaximumPercentage=22.2
MeanPercentage=5.48 Standard Deviation = 5.04
Vector Mean = 341.5
Conf. Angle = 13.54
c
u6 rose diagram
N = 222
ClassInterval = 10 degrees
MaximumPercentage=57.7
MeanPercentage=9.98 Standard Deviation = 17.59
Vector Mean = 16.3
Conf. Angle = 4.40
e

image processing. The problem itself is a classic target for automation, since the process of measuring crystal axes is infamously tedious. There are essentially two types of automated measurement systems that have been created, those which produce partial or complete orientation distribution functions for a given thin section using *X*-ray goniometry (Baker *et al.* 1969, Schmid & Casey 1986) or photometric techniques (Price 1973), and those that calculate the crystallographic orientation of a single grain or sub-grain using a universal stage or SEM for electron channelling (Lloyd & Ferguson 1986). The former inherently include a grain size weighting for the fabrics produced, whereas the latter do not.

While the techniques listed above produce accurate and detailed information, with the exception of the universal stage they are not readily accessible to the average geologist. Image processing based techniques developed in recent years (Beyna *et al.* 1990, Zhu 1992, Heilbronner & Pauli 1993) promise a relatively low cost solution, at least for certain minerals. These techniques are all essentially modifications of Price's photometric technique which is based on the knowledge that at certain wavelengths of light, there is, apart from a mirror symmetry that can be easily resolved, a unique relationship between the transmission properties as a function of stage orientation of a uniaxial mineral and the orientation of its *c*-axis. In the original Price technique, the transmission properties of the entire thin section were measured, and systematic tilting of the stage was performed to allow the calculation of the complete *c*-axis fabric. In Heilbronner & Pauli's Computer Integrated Polarization technique, the transmission profiles are calculated for each pixel in the image, and thus orientation maps can be calculated at pixel scale resolution. The potential for calculation of not only AVA diagrams (maps of crystallographic orientations, Fig. 13), but also truly accurate grain boundary maps, is demonstrated in Heilbronner & Pauli (1993).

Analysis & Tomography: 3-D Analysis

A fundamental problem in structural geology at all scales is the construction of three dimensional models from restricted information. In microstructural studies this is traditionally accomplished by examining three orthogonal thin sections, usually cut aligned to a macroscopically obvious fabric. Three dimensional strain estimates can then be derived from the analysis of these three sections (De Paor 1988, 1990, Burkhard 1990). In this section we review a number of image processing techniques which enable the full three dimensional reconstruction of the 3D geometry of a small volume of rock. This reconstruction may result in a

Fig. 9 (opposite). a) Binary synthetic image of particles possessing correlated grain sizes and orientations. b) 2-pixel size Gaussian wavelet transform applied to image 9a. Grayscale values are function of orientation of wavelet with highest local correlation coefficient. c) Rose diagram of all orientations found in 2-pixel wavelet transform image, showing strong preferred orientation at 10° anticlockwise. d) 6-pixel size Gaussian wavelet transform applied to image 9a. e) Rose diagram of all orientations found in 6-pixel wavelet transform image, showing strong preferred orientation at -18°. After Gaillot (1995).

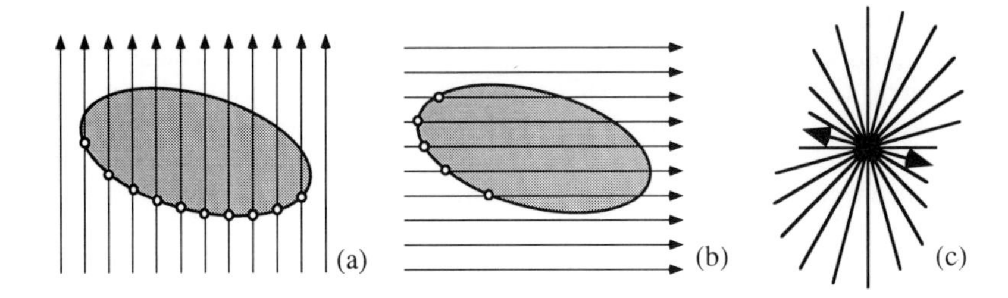

Fig. 10. Intercept method for SPO analysis. Number of object boundary entrant intercepts is counted as function of orientation of scan lines: 10 in a), 5 in b). c) Results represented in rose-diagram. Orientation of minimum score is maximum elongation of objects (15°, double arrow) and ratio of maximum : minimum scores is a measure of average object aspect ratio (= 2:1). After Launeau *et al.* (1988).

map of linear features, such as triple junction networks, curved surfaces, such as surface roughness measurements, or volumetric information such as grain shapes.

Analysis & Tomography: Reconstruction by Serial Sectioning

The use of serial sectioning for the observation of 3-D structure is important because it requires only the availability of normal thin section or polished section preparation equipment. The quality of the reconstruction depends mainly on the spacing at which the slices can be made.

Some thought needs to be given prior to sample preparation as to how the images are going to be aligned once they have been made. Collecting images haphazardly can result in images that are not uniform in scale, orientation, position or spacing, which greatly reduces the utility of the final data. With polished sections the best technique is to make a fixed mount for the section, so that position, orientation and scale are fixed, and to use the thickness of the remaining slab as a measure of image separation. Thin sections require a little more thought, and one simple solution is to physically score the sides of the rock slab prior to sectioning, so that these indents can be used for alignment later on.

Once a series of images has been properly aligned arbitrary sections through the data may be made with packages such as *NIH Image, Dicer*™ (Mac), *Slicer*™ (PC), or *XDataSlice*™ (*UNIX*). Thresholding the full volume of pixels (known as voxels in 3-D), can produce images of the interconnectivity of pore volumes (Ohashi 1992).

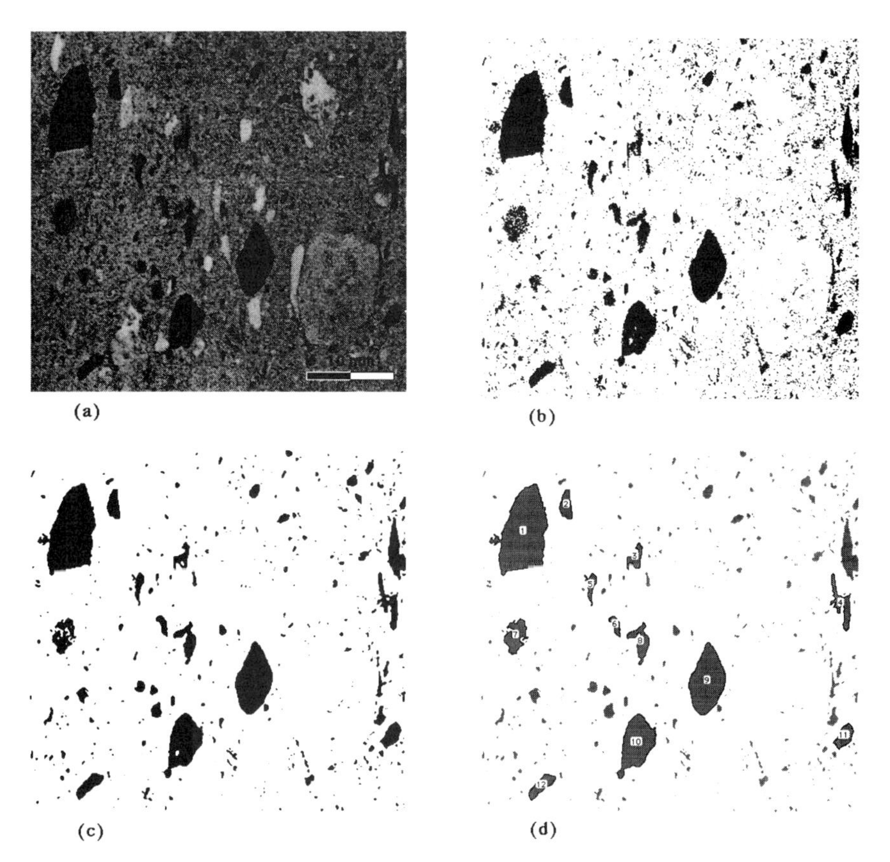

Fig. 11. Image analysis to measure shape and orientation of pebbles in polished conglomerate section. a) Original image of conglomerate with dark shale and light quartz and schist pebbles. b) Thresholded (at 200) binary image highlighting shale pebbles and small dark specks. c) One step of erosion and dilation (binary morphology operators) removes most smaller dark specks, but leaves sizes and shapes of larger shale pebbles unaltered. d) Automatic particle measurement (with *NIH Image*) has selected 12 particles with an area >200 pixels not touching edge of image for measurement.

Surface Characterisation & Reconstruction

The roughness of fracture surfaces plays an import role in controlling sliding behaviour and fluid transport rates. Using a profilometer (with a linear variable differential transformer displacement sensor) it is possible to measure the topography of a surface with a precision of $\pm 0.1 \mu m$, and at present these accuracies cannot be achieved using stereo photogrammetric techniques. However, with standard imaging equipment an accurate surface topography can be calculated (Jessell *et al.* 1995) which can be applied to rough surfaces of all scales.

The aim of all stereo photogrammetric techniques is to solve the stereo correspondence problem, which involves the matching of the same location in both

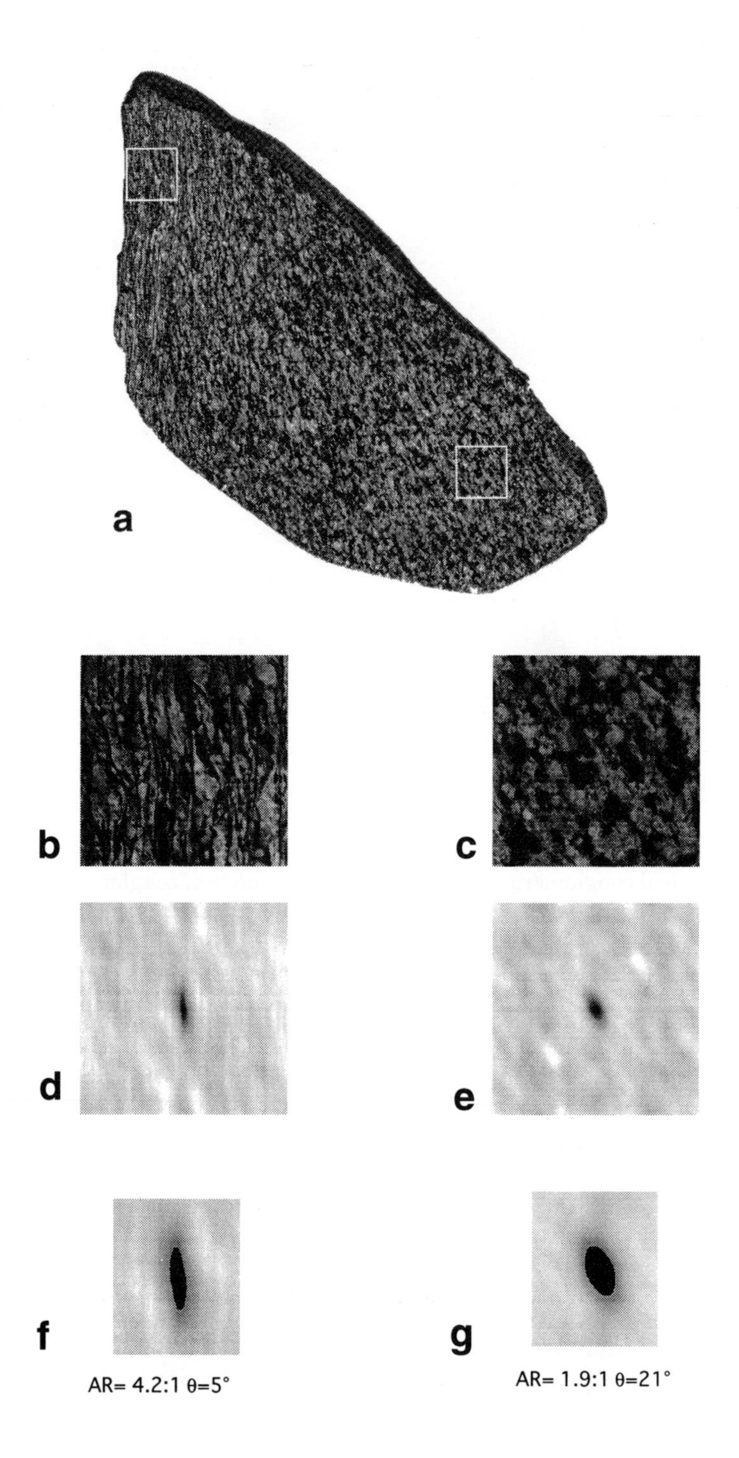
a
b
c
d
e
f
g
AR= 4.2:1 θ=5°
AR= 1.9:1 θ=21°

images of a stereo pair. Heights are calculated by comparing apparent displacements (known as disparities) between matched points in the two images. These disparities are proportional to the heights of individual points (relative to some datum) (Fig. 14), and by systematically matching equivalent features in each image a map of disparities, and hence heights, can be built up for the overlapping parts of the images at whatever resolution is required up to the pixel resolution of the image.

Tomography, Fracture, & Porosity Analysis

The tomographic reconstruction of 3D volumes has mainly been developed in a medical context, and given the price of Computer Assisted Tomography scanners, this is unlikely to change. However, if access to such a machine is available the potential for three dimensional non-destructive analysis is considerable (Mazerolle *et al.* 1989, Colletta *et al.* 1991, Carlson & Denison 1992) (Fig. 15).

The fundamental principal of any tomographic imaging technique is to collect a number of projected images (including x-ray, isotope emission, sound waves), taken at different orientations with respect to the sample, and to apply an inversion technique to reconstruct the three dimensional structure of the interior. Reconstructions can be made in both the spatial and frequency domains (Russ 1994) and normal image filtering is often applied to the raw images prior to and following reconstruction to improve the resolution of features of interest.

The major issues that need to be addressed are the spatial resolution of the technique (about 1mm for medical systems down to 10-100μm for specialised *X*-ray systems), the contrast in rock properties (such as x-ray density) which allow a particular tomographic technique to work, and the cost of the technique compared to serial sectioning.

Image Processing of Experimental Samples

Several types of experimental deformation allow the observation and recording of sample microstructure at different stages during progressive deformation. Examples are "clay-box" experiments with clay, wax, or putties (*e.g.,* Mancktelow 1991, Treagus & Sokoutis 1992) and experiments in transparent deformation cells (*e.g.,* Wilson 1986, Means 1989). In these experiments, strain can be analysed by comparing the structure at successive stages and determining the

Fig. 12 (opposite). Autocorrelation function used to characterise fabrics in shear zones. a) Image of granite sample (described in Rey *et al.* 1994) showing marked strain gradient towards shear zone. Two test areas indicated by squares. b) Microstructure of high strain area. c) Microstructure of low strain area. d),e) Autocorrelation transform of microstructure grey scale image in d) low strain area, e) high strain area. f), g) Thresholded images of central part of autocorrelation transform image for f) high strain area, g) low strain area, showing mean preferred orientation of f) 5°, g) 21°.

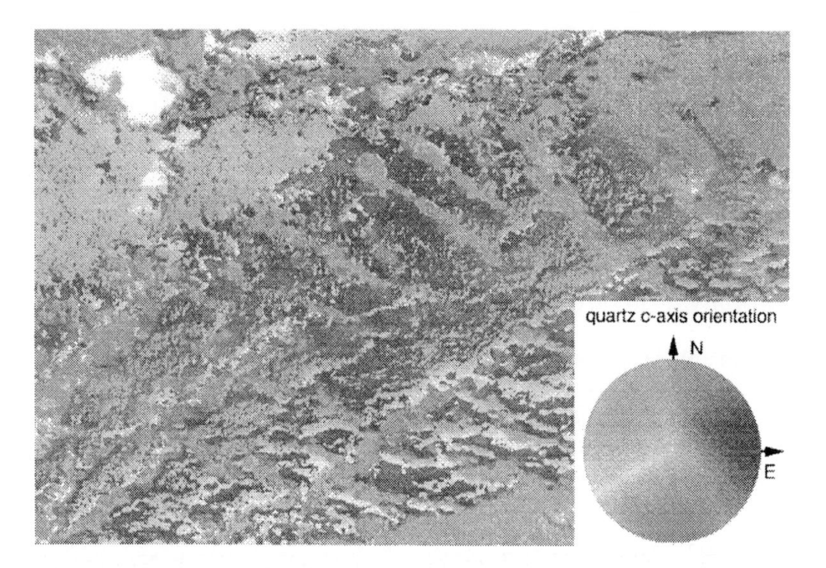

Fig. 13. Computer integrated polarization image of quartz *c*-axis orientations for naturally deformed sample. Inset in lower right hand corner shows colour coded lower hemisphere equal area stereonet. *c*-axes for each pixel independently determined, so grain boundaries can be accurately mapped. From van Daalen *et al.* (1995).

displacement paths of material points or particles. The sample material is often adorned with a grid or other markers (clay-box type experiments) or mixed with small distinct marker particles.

Image analysis can be used in several ways for strain analysis:

(i) to enhance sharpness or contrast to facilitate selection of markers, after which marker positions are digitised and further processed (Mancktelow 1991, Bons *et al.* 1993)

(ii) to locate positions of markers. An example is the method proposed by Allais *et al.* (1994), where image analysis is used to automatically locate nodes of a rectangular grid on a deforming sample at several stages during progressive deformation. For the case of marker particles, one can define particles in the image at one stage and then use pattern matching to locate these particles at subsequent stages (Jessell, unpubl. data). The output of this analysis is a list of positions of the grid-nodes or marker-particles for further processing.

(iii) to measure strain on images themselves. For example, Dong *et al.* (1992) describe a method where a multiple exposure image is made of a deforming sample mixed with distinct marker particles. The delay between the exposures is such that each particle is shown as a streak or array of dots on the image. The

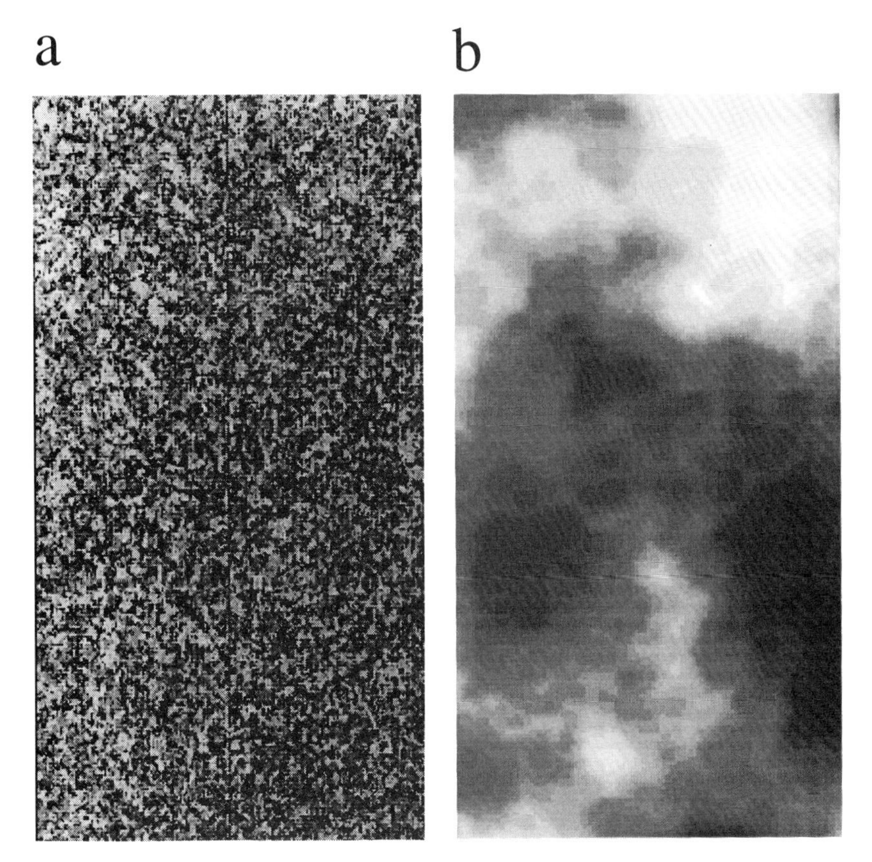

Fig. 14. Surface reconstruction using pattern matching. a) One of stereo pair of images of granite flake produced by quarry blasting. Second image (not shown) taken after camera moved to right. b) Grayscale representation of surface topography (bright areas are relative highs), determined by calculating apparent offset in features in image pair which is caused by parallax resulting from camera shift. Range in heights is 13.7 mm. From Jessell *et al.* (1995).

image is then subdivided into small regions. For each region the auto-correlation function is then calculated, which gives the average length and orientation of the particles in the region during the exposure sequence. This is essentially the displacement vector of the particles in the region. Also, Bons & Jessell (1995) describe a method derived from Butterfield (1971) and Butterfield *et al.* (1970). Here images of a deforming material mixed with marker particles are captured at regular intervals. Two images at a time are then compared to determine the displacement field. The local region around a point in the image of one (relatively undeformed) stage is then used as a template

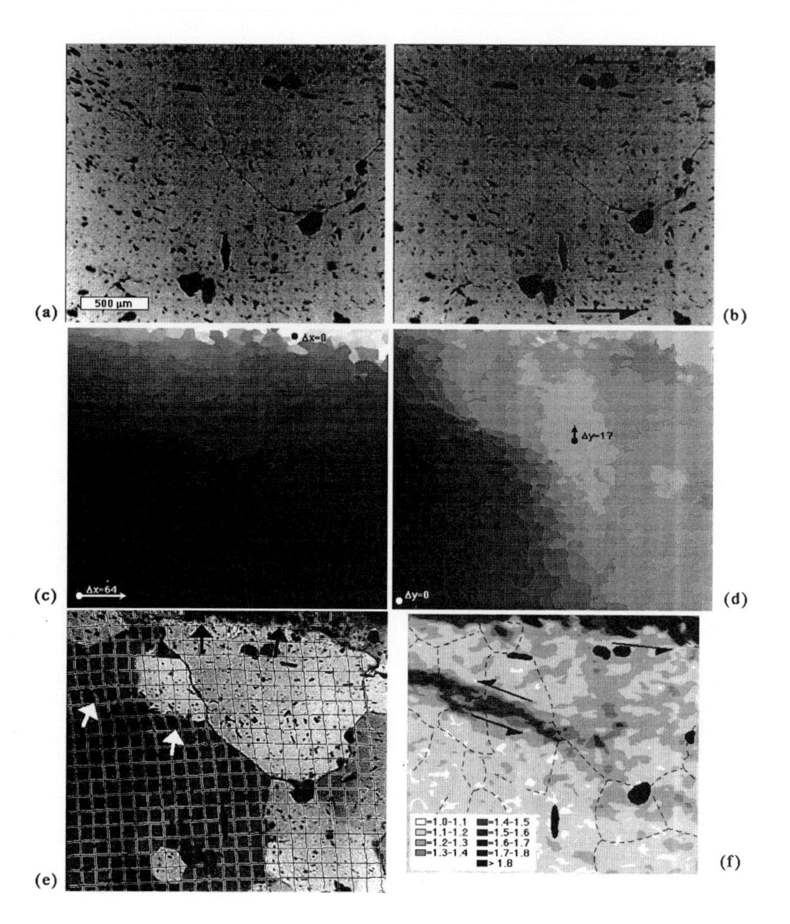

Fig. 15. Application of strain analysis with pattern matching to deformation experiment with OCP. a),b) Digital image of thin sheet of OCP grains before and after sinistral shear strain of 0.15 lasting 4 minutes. Shear zone edge at top of image. Black dots: particles of aluminium powder mixed with OCP. Plane polarised light. c),d) Grayscale contour maps of *X*- and *Y*-component of displacement field calculated with PPL images (a & b). Arrows = displacement magnitude. Weighted cross-correlation routine (Russ 1994, p. 342) with 30×20 oval shaped mask was used. e) Deformation visualised by imaginary originally rectangular grid passively deformed according to displacement field shown in c), d) superimposed on image of deformed grains. Note localisation of shear at edge of shear zone (black arrows) and in narrow zone on left (white arrows). f) Grayscale map of axial ratios of finite strain ellipse. Two localised shears stand out; one coincides with grain boundaries and indicates grain boundary sliding/shear was a deformation mechanism. (From Bons & Jessell 1995).

for pattern matching. The template is moved over the other image to find the best matching region in this image. The position of this region is the position of the point in the second, deformed stage. The routine is repeated for each point (pixel) in the first image to determine the whole displacement field, which can be visualised in the form of two grayscale images of the x- and y-component of the displacement (Fig. 15).

Conclusions

In this paper, we have outlined the basics of image analysis and listed some of the more specific applications to microstructural analysis. These techniques can either provide information that is unavailable through normal microscopy or is time-consuming or subject to bias when collected manually. An important consideration when selecting a data processing technique is the purpose of the analysis. During experiments, or when surveying variation in microstructures across a region, it may be more important to be able to contrast the variations based on an unbiased estimate of a parameter than to spend the considerable extra time to measure the actual values of the parameter. For example, automated grain size measurements are without doubt flawed but if the bias is systematic and understood, and the variation between microstructures is the important question, then image processing will provide a useful data reduction technique. In contrast, if precise grain size distribution and details of grain boundary shapes are required, then manual digitising may be more appropriate.

The user should always be aware of the limitations and possible errors in automatic image analysis and should never use image analysis as a 'black box' to quickly produce pseudo-quantitative results. The references given in this chapter, especially those for the textbooks, should help the interested reader to get more acquainted with the methods outlined.

Acknowledgments

We would like to thank Mirjam van Daalen, Renée Panozzo-Heilbronner, Holges Stiinitz, and Phillipe Gaillot whose work we have borrowed for some of the figures in this chapter, and we hope that our descriptions of their work do them justice. We would also like to thank Patrice Rey for the loan of his shear zone.

References

Alliers, L., Champenois, M., Macaudiere, J., & Bertrand, J.M. 1995. Use of image analysis in the measurement of finite strain by the normalized Fry method: geological implications for the 'Zone Houillière' (Briançonnais zone, French Alps). *Min. Mag.* **59**: 179-187.

Allais, L., Bornert, M., Brethau, T., & Caldemaison, D. 1994. Experimental characterization of the local strain field in a heterogeneous elastoplastic material. *Acta Metall. Mater.* **42:** (11) 111-222.

Baker, D. W., Wenk, H. R., & Christie, J. M. 1969. *X*-ray analysis of preferred orientation in fine grained quartz aggregates. *J. Geol.* **77**: 144-172.

Beyna, G., Leymarie, P., Buffet, G., & Nault, L. 1990. Principe de l'analyse des fabriques d'axes C du quartz par le traitement d'images numeriques. *Compt. Rend., Serie-2.* **310**: 1233-1239.

Bons, P. D., Jessell, M. W., & Passchier, C. W. 1993. The analysis of progressive deformation in rock analogues. *J. Struct. Geol.* **15:** 403-411.

Bons, P. D. & Jessell, M. W. 1995. Strain analysis in deformation experiments with pattern matching or a stereoscope. *J. Struct. Geol.* **17**: 917-921.

Burkhard, M. 1990. Ductile deformation mechanisms in micritic limestones at low temperatures (150-350°C). In: Deformation mechanisms, rheology and tectonics. Knipe, R.J. & E.H. Rutter (eds.). *Spec. Publ. Geol. soc. Lond.* **54**: 241-257.

Butterfield, R. & Andrawes, K. Z. 1971. The visualization of planar displacement fields. In: *Stress-strain behaviour of soils* . Parry, R. H. G. (ed.). *Proc. Roscoe Mem. Sym. Camhridge Univ.*, G. T. Fonlis & Co, Oxfordshire, pp. 467-475.

Butterfield, R., Harkness, R. M., & Andrawes, K. Z. 1970. A stereo-photogrammetric method for measuring displacement fields. *Géotechnique.* **20**: 308-314.

Carlson, W. D. & Denison, C. 1992. Mechanisms of porphyroblast crystallization: results from high-resolution computed *X*-ray tomography. *Science* **257**: 1236-1239.

Colletta, B., Letouzey, J., Pinedo, R., Ballard, J.-F. & Bale, P. 1991. Computerized *X*-ray tomography analysis of sandbox models: examples of thin skinned thrust systems. *Geology* **69**: 1063-1067.

De Paor, D. G. 1988. Strain determination from three known stretches; an exact solution. *J. Struct. Geol.* **10**: 639-642.

De Paor, D. G. 1990. Determination of the strain ellipsoid from sectional data. *J. Struct. Geol.* **12**: 131-137.

Dong, R., Chu, S., & Katz, J. 1992. Quantitative visualization of the flow within the volute of a centrifugal pump. Part A: technique. *J. Fluids Eng.* **114**: 390-395.

Dougherty, E. R. (ed.) 1994. *Digital Image Processing Methods.* Marcel Dekker Inc., New York, 472 pp.

Dutruge, C., Chery, J. & Hurtrez, J. E. 1993. Une approche numerique des effets de taille de grain sur la localisation de la deformation ductile. *Compt. Rend. Serie-2.* **317**: 195-201.

Fabbri, A. G. 1984. *Image Processing of Geological Data.* Van Nostrand Reinhold Co. Inc, New York, 244 pp.

Gaillot, P. 1995. Etude de la fabrique des roches par le trallement d'image: apport des ondelettes. Unpubl. DEA Thesis, Universite de Toulouse III, June 1995.

Gonzalez, R. C. & Wintz, P. 1987. *Digital Image Processing.* Addison-Wesley, Reading MA, 403 pp.

Jessell, M. W., Cox, S. J. D., Schwarze, P. & Power, W. 1995. The anisotropy of surface roughness measured using a digital photogrammetric technique. In: Ameen, S.M. (ed.). Fractography: fracture topography as a tool in fracture mechanics and stress analysis. *Geol. Soc. Spec. Publ.* **92**: 27-37.

Launeau, P., Bouchez, J.-L., & Benn, K. 1990. Shape preferred orientation of object populations: automatic analysis of digitized images. *Tectonophysics* **180**: 201-211.

Lloyd, G. E. & Ferguson, C. C. 1986. A spherical electron-channeling pattern map for quartz petrofabric analysis. *J. Struct. Geol.* **8**: 517-526.

Mancktelow, N. S. 1991. The analysis of progressive deformation from an inscribed grid. *J. Struct. Geol.* **13**: 859-864.

Means, W. D. 1989. Synkinematic microscopy of transparent polycrystals. *J. Struct. Geol.* **11**: 163-174.

Mazerolle, F., Geraud, Y., & Latiere, H. J. 1989. Analysis of the internal structure of rocks and characterization of mechanisms of deformation by a non-destructive method. *Tectonophysics* **159**: 149-159

Ohashi, Y. 1992. Three-dimensional reconstructions of pore geometry from serial sections- image algebraic approach. In: Pflug, R. & J. W. Harbaugh (eds.). *Computer Graphics in Geology*. Freiburg, Germany, Univ. Freiburg. pp. 29-39.

Oltra, P.-H. 1988. Une nouvelle méthode de quantification de la déformation subie par un échantillon. Appors du traitement numérique d'images et du filtrage de convolution. *Compt. Rend., Série 2* **306**: 1493-1499.

Panozzo, R. 1983. Two-dimensional analysis of shape-fabric using projections of digitized lines in a plane. *Tectonophysics*. **95**: 279-294.

Panozzo, R. 1984. Two-dimensional strain from the orientation of lines in a plane. *J. Struct. Geol.* **6**: 215-221.

Panozzo-Hellbronner, R. 1992. The autocorrelation function: an image processing tool for fabric analysis. *Tectonophysics* **212**: 351-370.

Panozzo-Heilbronner, R. & Pauli, C. 1993. Integrated spatial and orientation analysis of quartz *c*-axes by computer-aided microscopy. *J. Struct. Geol.* **15**: 369-382.

Price, G. P. 1973. The photometric method in microstructural analysis. *Am. J. Sci.* **273**: 523-537.

Prince, C. M. 1991. DECODE and DFOUR: 2-D Fourier processing of petrographic images. *Comp. & Geosci.* **17**: 505-525.

Rey, P. F., Fountain, D. M., & Ciement, W. P. 1994. *P*-wave velocity across a noncoaxial ductile shear zone and its associated strain gradient: Consequences for upper crustal reflectivity. *J. Geophys. Res.* **99**: 4533-4548.

Russ, J. C. 1990. *Computer-assisted Microscopy - The Measurement and Analysis of Images.* Plenum Press, New York, 453 pp.

Russ, J. C. 1994. *The Image Processing Handbook.* (2nd. ed.) CRC Press, Florida, 674 pp.

Schmid, S.M. & Casey, M. 1986. Complete fabric analysis of some commonly observed quartz *c*-axis patterns. *A. G. U. Monogr.* **36**: 263-286.

Shtuka, A. & Royer, J. J. 1994. *IMGEO*: Interactive image processing software for geological and geophysical data. In: Fabbri, A. G. & Royer, J. J. (eds) *3rd CODATA Conference on Geomathematics and Geostatistics.* Sci. de la Terre, Sér. Inf. Nancy, **32**: 539-552.

Treagus, S. H., & Sokoutis, D. 1992. Laboratory modelling of strain variation across rheological boundaries. *J. Struct. Geol.* **14**: 405-424.

van Daalen, M., Panozzo-Heilbronner, R., & Stiinitz, H. 1995. Strain localisation and initiation of dynamic recrystallisation in naturally deformed quartz rocks. (abs.) *J. Czech Geol. Soc.* **40**: C 125-126.

Wilson, C. J. L. 1986. Deformation induced recrystallization of ice: the application of in situ experiments. *Geophys. Monogr.* **36:** 233-232.

Zhu, Z. 1992. A study of quantitative crystal optics. (abs.) *Geol. Soc. Rust.* **32**: 220.

Calculation of Rock Properties from Pole Figures Using *LabView*

Johann Lapierre, David Mainprice, and Walid Ben Ismail
Laboratoire de Tectonophysique, Université Montpellier II, Place E. Bataillon, 34095 Montpellier cédex 05, France. johann@dstu.univ-montp2.fr david@dstu.univ-montp2.fr walid@dstu.univ-montp2.fr

Abstract– Calculating physical properties of rocks requires an Orientation Distribution Function (ODF) of crystalline aggregates derived from individual orientation data (*i.e.*, in general, universal stage measurements). Though frequently projected and contoured on stereograms to yield pole figures for tectonic interpretations, orientation datasets are rarely published and are often lost. To calculate physical properties, we developed an image processing method to recover orientation density data from published pole figures. An inversion program used in materials science was modified to derive partial ODFs from recovered orientation density dataset. We describe the full method implemented in the *LabView*™ program *SCANSTEREO* for orientation density recoverage and in the *FORTRAN77*™ program *INVPOLE* for ODF calculation. We give a test of consistency between seismic properties derived from inverted and measured data for a peridotite sample from Lanzo massif (Alps, Italy).

Introduction

Many rock-forming minerals are intrinsically anisotropic as single crystals, and aggregates often contain an anisotropic distribution of crystals. This leads to the concept of rock texture (commonly called petrofabric) which may be mathematically described by the orientation distribution function (ODF). An ODF is the essential parameter to determine physical properties of anisotropic rocks (*e.g.*, Mainprice *et al.* 1993, Mainprice & Humbert 1994) and is critical for the interpretation of geophysical data such as seismic anisotropy or electric and thermal properties.

Petrofabrics have been used classically as indicators of shear sense on rocks deformed at P/T conditions permitting plastic behaviour (Nicolas *et al.* 1973, Schmidt & Casey 1986); they are now being used increasingly to calculate rock physical properties. Recent studies of teleseismic shear-wave splitting (Mainprice & Silver 1993, Barruol & Mainprice 1993) have shown that relationships between petrofabric and anisotropic seismic properties allow the determination of the directions of plastic flow in the upper mantle. Furthermore, a classification of mantle textures and related seismic anisotropy has been initiated (Ben Ismaïl 1995) on the basis of 110 measured petrofabrics from various geodynamical contexts. Elastic properties of crustal rocks derived from petrofabrics are also useful to characterize seismic reflectors detected during deep continental seismic reflection profiling (*e.g.,* COCORP, ECORS, DECORP, *etc.*).

The above studies are restricted to a small number of petrofabrics and the desirability of establishing a more comprehensive database is clear. Such a revival of interest in texture analysis has prompted the acquisition of new data sets as well as the use of existing ones. However, the problem of retrieving lost data sets from petrofabric projections such as pole figures has not been addressed. For example, numerous existing petrofabric data were stored on IBM punch cards, but almost complete obsolescence of such data storage methods means the majority of this data has been lost. Moreover, original petrofabric data sets related to published pole figures are often unavailable in the literature (*e.g.*, in theses). We solve this problem by introducing a method combining image processing and numerical techniques derived from texture goniometry to recover the ODF from published pole figures.

We review the main features of pole figures to show which types of data are recoverable. Then we present an image processing procedure (*SCANSTEREO*) to extract point orientation density values from digitized images of pole figures. Next, we explain briefly the inversion procedure (*INVPOLE*) which allows the calculation of the ODF coefficients on the basis of data provided by *SCANSTEREO*. Finally, we develop an application on a known peridotite sample from Lanzo massif (Alps, Italy) to compare seismic properties derived from original and inverted datasets in order to estimate the bias introduced by the inversion method.

The Pole Figure Geometry and Its Implications

A pole figure is a representation of the Lattice Preferred Orientations (LPO) of a rock-forming mineral population (see Nicolas & Poirier 1976 for a review). Typically, it is plotted as a scatter diagram on lower hemisphere equal area projection. Individual orientation data as derived from universal stage measurements are a population of individual poles. The data are contoured to obtain an orientation density diagram. Data derived from volume diffraction measurements (*e.g.*, *X*-ray or neutron texture goniometry) are routinely presented as contoured diagrams. A pole figure can be defined analytically by the density of crystallographic orientation $P_{(h)}(y)$ in a y direction of the structural reference frame, and describes the fundamental relationship of texture:

$$P_{h_i}(y) = \frac{1}{2\Pi} \int_{h_i // y} f(g) d\chi \qquad (1)$$

with:
h_i: A crystallographic direction corresponding to the normal to a low-index lattice plane with Miller indices $h1$, $h2$, $h3$.

$f(g)$: The ODF which may be defined in terms of generalized spherical harmonics (Bunge 1982).

$d\chi$: An element of solid angle in the unit hemisphere of the projection.

In practice, orientation density counting is often achieved using the Schmidt grid on the projected data set. Counting and contouring methods are many (*e.g.*, Vollmer 1995 for a recent review) but, essentially, they all smooth and interpolate the individual poles. Thus, it is impossible to recover the original data set of individual measurements from the pole figure and the only alternative is to produce a lower accuracy gridded set for given increments in azimuth and inclination such as those used in contouring texture goniometry measurements. This is detailed in the two next sections where we describe how digitized images of pole figures are processed. A first pre-processing stage is necessary to preserve only orientation density information and to convert density areas into an appropriate format. Then pole figure center coordinates and radius are calculated to produce a circular grid in order to extract grayscale density values.

Setting Up the Image of the Pole Figure

The images of pole figures are digitized with a flat-bed scanner and saved as TIFF 256 gray level images. Objects not related to density information (labels, scanning noise, *etc.*) are removed from the images. Because the pole figures considered here are contour plots, we need to produce an image with homogeneous gray level areas corresponding to each different density interval. We have done this with Adobe *Photoshop*™ 2.5.1, but it can be achieved with any drawing software. It is simply necessary to join the contour lines delimiting the density areas, and then fill them with a gray level representing the density value. The gray level scale can be chosen arbitrarily within the range of 1 to 254 gray value. The values 0 and 255 are reserved for diagram background and zero density areas respectively. After these operations the image can be processed with the *SCANSTEREO* program.

Working Procedure of the Program *SCANSTEREO*

Reading the Image

SCANSTEREO reads uncompressed 256 gray level TIFF or PICT images. The user can choose between processing only one image or a folder (directory) of images which are all identically grayscaled. In the first case the image is displayed full size if the resolution is less than 300×300 pixels, otherwise it is reduced to 300×300 pixels and displayed. In the folder option only the first image is displayed to save time and memory.

Processing the Image

SCANSTEREO first calculates the center coordinates and the radius of the pole figure as follows. The image is first thresholded in order to obtain a binary image of the whole pole figure: levels between 1 and 255 are replaced by the value 1 and background remains as 0. Then the center coordinates of the diagram and the size of the rectangle which surrounds the pole figure are calculated. The diameter of the pole figure is obtained considering only the minimum of either the height or the width of the surrounding rectangle so that the scanned data points are as much as possible contained within the pole figure even if its scanned image is not perfectly circular.

Next a sub-routine called *GREYDENS* allows the user to select the density value corresponding to each gray level automatically selected by the program. Then the *CALCGRID* subroutine scans the image assuming the pole figure to be an equal area projection and using a grid calculated with the following formulas:

$$X_{Gi} = \mathrm{INT}\left(X_c + 2\sin\left(INC_i/2\right)\sin\left(AZ_i\right) R/\sqrt{2}\right)$$
$$Y_{Gi} = \mathrm{INT}\left(Y_c + 2\sin\left(INC_i/2\right)\cos\left(AZ_i\right) R/\sqrt{2}\right)$$

where X_{G_i} and Y_{G_i} are the coordinates of the grid points (in pixels), X_C and Y_C are the center coordinates of the pole figure (in pixels), R is the radius of the pole figure (in pixels), INC_i is the inclination (from 0° to 90° every 5°) and AZ_i is the azimuth (from 0° to 355° every 5°). Note: as in most images and screen formats, the origin is in the top left hand corner.

For each point, X_{G_i}, Y_{G_i}, the gray level is extracted and the corresponding density value is stored in a 2D array. In order to avoid measuring points outside the diagram, when the inclination reaches 90° the *CALCGRID* subroutine checks if each point has a gray level of zero (*i.e.*, the point is outside of the pole figure). If this is true, it decreases the radius and recalculates the coordinates of the measured point until its gray value becomes different from zero. Finally the image is displayed full size with the grid superimposed.

Output of Results

The program *SCANSTEREO* produces a matrix of 19 rows by 72 columns containing the crystallographic orientation density values of the pole figure in an ASCII spreadsheet format file. Such a dataset is analogous to those obtained from *X*-ray or neutron goniometry. Consequently, a partial ODF may be calculated according to the inversion method implemented in the *INVPOLE* program, presented in the following section.

The Inversion Method

Inversion of pole figures has been widely developed in materials science (Bunge 1982) for *X*-ray or neutron goniometry. Recent work (Bunge & Wenk 1977, Casey 1981, Wagner & Humbert 1987) introduced inversion methods for the low symmetry materials which prevail in Earth sciences. These procedures make use of harmonic methods which lead to a spherical harmonic form of the ODF. We outline only the principles of the inversions here; further details of the theory and formalization can be found in the above references.

The discrete development of the relation in eqn. (1) in terms of symmetrical generalized spherical harmonics may be expressed as follows:

$$P_{h_i}(y) = \sum_{l=0}^{\infty} \sum_{\nu=1}^{N(l)} F_l^{\nu}(h_i)\ \dot{k}_l^{\nu}(y)$$

where $N(l)$ is the number of specimen symmetry harmonics, $\dot{k}_l^{\nu}(y)$ are the spherical harmonics satisfying sample symmetry, and $F_l^{\nu}(h_i)$ denotes the theoretical coefficients of the pole figure which are related to the coefficients $C_l^{\mu\nu}$ of the serial development of the ODF:

$$^{Th}F_l^{\nu}(h_i) = \frac{4\pi}{2l+1} \sum_{\mu=1}^{M(l)} C_l^{\mu\nu}\ \dot{\dot{k}}_l^{*\mu}(h_i) \tag{2}$$

Here, $M(l)$ is the number of linearly independent harmonics (*i.e.*, the number of independent measured pole figures), and $\dot{k}_l^{*\mu}(h_i)$ are complex conjugates of spherical harmonics satisfying crystal symmetry. The coefficients $F_l^{\nu}(h_i)$ can also be related to measured pole density values $P_{h_i}(y)$ by:

$$^{Meas}F_l^{m}(h_i) = \int_{\alpha=o}^{\pi/2} \int_{\beta=0}^{2\pi} P_{h_i}(y)\, \dot{k}_l^{*\nu}(y) \sin\alpha\, d\alpha\, d\beta$$

where α and β are the spherical coordinates in the pole figure reference frame. Using eqns. (2) and (3) it is possible, for a given set of pole figures, to calculate the ODF coefficients $C_l^{\mu\nu}$ using a least square method:

$$\sum \left[{}^{Meas}F_l^{m}(h_i) \ - \ {}^{Th}F_l^{\nu}(h_i) \right]^2 \ = \ \min.$$

In the examples presented here, petrofabrics are derived from an individual measurement technique, the universal stage, hence the consistency of data between different pole figures of the same sample is better than those obtained by *X*-ray analysis. Therefore, we can expect a more accurate inversion.

Universal stage measurements that are used here are for olivine and are non-vectorial, that is directions are measured without a positive and negative sense, therefore pole figures are centrosymmetric. Hence the inversion method only permits the determination of the coefficients of the ODF for even values of l (expansion range of the spherical harmonics).

Relationships between *M(l)* and *l* (equation (2)), depending on crystal symmetry given by Bunge (1982) and Humbert & Diz (1991), show that it is

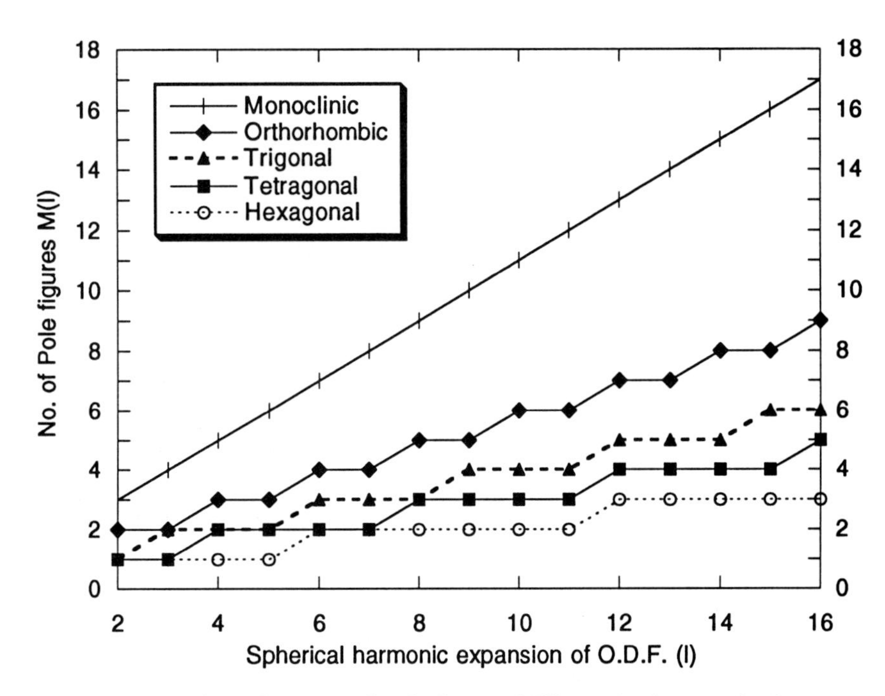

Fig. 1. Number of measured pole figures *M(l)* required to reach given spherical harmonic expansion *l* of orientation distribution function for the crystal symmetries treated by program *INVPOLE*. For seismic properties $l = 4$ expansion is minimum requirement and therefore 1, 2, 2, 3, and 5 pole figures would be needed for hexagonal, tetragonal, trigonal, orthorhombic, and monoclinic crystal and triclinic specimen symmetries. Spherical harmonic expansions of 10 to 16 are typical for quantitative studies of lattice preferred orientation (texture) where many pole figures are required for low symmetry minerals.

possible to calculate physical properties of symmetrical tensors of orders 2 and 4 from ODFs characterized by spherical harmonics with $l = 2$ and $l = 4$, respectively. Because tensors related to physical properties of geophysical interest are centrosymmetric, the bias introduced by the centrosymmetry of the pole figure does not affect the present analysis.

The method of pole figure inversion can be applied to any crystal symmetry (Bunge 1982). In the present application two factors restrict its application, firstly the number of pole figures available, and secondly the numerical inversion scheme used. Published LPO are typically given for only three principal crystallographic directions in the structural reference frame (*i.e.*, three pole figures). This constitutes a limitation for the inversion method because Bunge (1982) showed that with lowering symmetry class (i.e. cubic towards triclinic) the number of independent pole figures required increases. For example, for hexagonal, tetragonal, trigonal, orthorhombic and monoclinic symmetry classes one requires 1,2,2,3, and 5 independent pole figures, respectively, to calculate an ODF of order 4 for seismic properties. If only three independent pole figures ($M(l) = 3$) are published, tensors

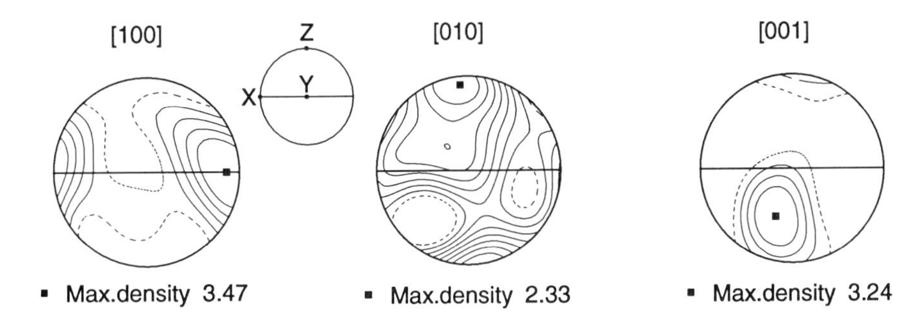

Fig. 2. Olivine pole figures of lherzolite for [100], [010], and [001]. Equal-area projection, lineation (*X*) east-west, foliation plane (*XY*) perpendicular to projection plane. Contour intervals 0.5 times uniform distribution for [100] and [001], 0.25 times for [010]; lowest contour is dashed line, higher values - solid lines. Highest value - black square.

of orders 2 and 4 (electric, thermal, and seismic properties) cannot be calculated for monoclinic or lower symmetry minerals (see Fig.1). In the current application we have used the *INVPOLE* program developed in *FORTRAN* 77 by Wagner & Humbert (1986). To make the inversion scheme as memory efficient as possible, this program is restricted to non-cubic crystal classes containing a two fold axis, hence excluding cubic and triclinic classes as shown in Fig. 1.

Calculation of Seismic Properties

To calculate the seismic properties of a polycrystal, one needs to evaluate the elastic properties of the aggregate. In the case of an aggregate with a crystallographic fabric or texture, the anisotropy of the elastic properties of the single crystal must be taken into account. For each orientation $\boldsymbol{g}$ the single crystal properties have to be rotated into the specimen coordinate frame using the orientation matrix g_{ij},

$$C_{ijkl}(\boldsymbol{g}) = g_{im} \cdot g_{jn} \cdot g_{ko} \cdot g_{lp}\ C_{mnop}(\mathbf{g^o})$$

where $C_{ijkl}(\boldsymbol{g})$ is the elastic property in sample coordinates, $\boldsymbol{g} = g_{ij} = g(\varphi_1\ \phi\ \varphi_2)$ the measured orientation in sample coordinates and $C_{mnop}(\boldsymbol{g}^o)$ is the elastic property in crystal coordinates, with $\boldsymbol{g}^o = g^o{}_{ij} = g^o(\varphi_1{=}0\ \phi{=}0\ \varphi_2{=}0)$.

The elastic properties of the polycrystal may be calculated by integration over all possible orientations of the ODF. The ODF may be derived from individual measurements and pole figure inversion as in the current application. Bunge (1982) has shown that integration is given as:

$$\langle C_{ijkl} \rangle = \int C_{ijkl}(\boldsymbol{g}) . f(\boldsymbol{g})\ d\boldsymbol{g}$$

where $\langle C_{ijkl} \rangle$ is the elastic tensor of the aggregate and $d\boldsymbol{g} = 1/8p^2 \sin\phi\ d\varphi_1\ d\phi\ d\varphi_2$. The final step is the calculation of the three seismic phase velocities from the solution of the Christoffel equation:

$$\mathrm{Det} \mid \langle C_{ijkl} \rangle X_i X_j - \delta_{ik} \rho V^2 \mid = 0$$

where $X_i X_j$ are the direction cosines of the wave propagation direction, ρ is density, δ_{ik} is the Kronecker delta, and V is one of the three seismic phase velocities. The calculation is repeated for every propagation direction on a hemisphere in 6 degree intervals and contoured (*e.g.,* Fig. 2). The Voigt average has been used in all calculations presented here (see Mainprice & Humbert 1994 for further details).

Example

In order to validate results of the inversion method we compared seismic properties calculated after inversion of the pole figure and those derived from universal stage measurements. We could have presented the intermediate step by comparing the ODF calculated from the universal stage measurements and from the inverted pole figures at an expansion of order 4, but we chose to illustrate the method with calculated physical properties as this was the objective of our study. We used data from Boudier (1969), collected on a lherzolite sample from Lanzo massif (see Fig. 2 for related pole figures). We calculated the seismic properties of olivine because it is orthorhombic and therefore amenable to our method, which cannot be used for monoclinic and triclinic minerals. Contoured projections of *P*-wave velocities (*V*p) presented in Figs. 3a and 3b show the same velocity distribution with a maximum close to the lineation (*X*), and a minimum perpendicular to the foliation plane (*XY*). Such coherence between the two methods of calculation is also qualitatively verified for projection of *S*-wave birefringence (d*V*s) and fastest *S*-wave polarization plane displayed in Figs. 3d, 3e 3g, and 3h, respectively. Quantitatively, the errors on *P*-wave velocity and *S*-wave birefringence for the principal directions of the structural reference frame are summarized in Table 1, which shows only values less than 1%. These errors have also been projected and contoured on pole figures displayed in Fig. 3c for *Vp* and Fig. 3f for *dVs*. Both stereograms show a minimum close to the foliation plane and perpendicular to the lineation related to the *SCANSTEREO* grid resolution which is more important in the center of the pole figure. Generally, the distributions of these errors are related to a combination of the lack of accuracy of the *SCANSTEREO* scanning circular grid and the smoothing and interpolation necessary to construct the original pole figure.

Conclusion

Although this inversion method does not provide optimal determination of the ODF, it allows satisfying calculations of seismic properties. Qualitatively the results are very similar to those derived from original measurements. Quantitatively, *Vp* and *dVs* determination is very accurate in all directions of the structural reference frame (generally, error is less than 5%). As a result, orientation distribution of seismic anisotropy in the structural reference frame may be derived from the published pole figures without individual lattice orientation measurements. This method will allow the establishment of an interpretive

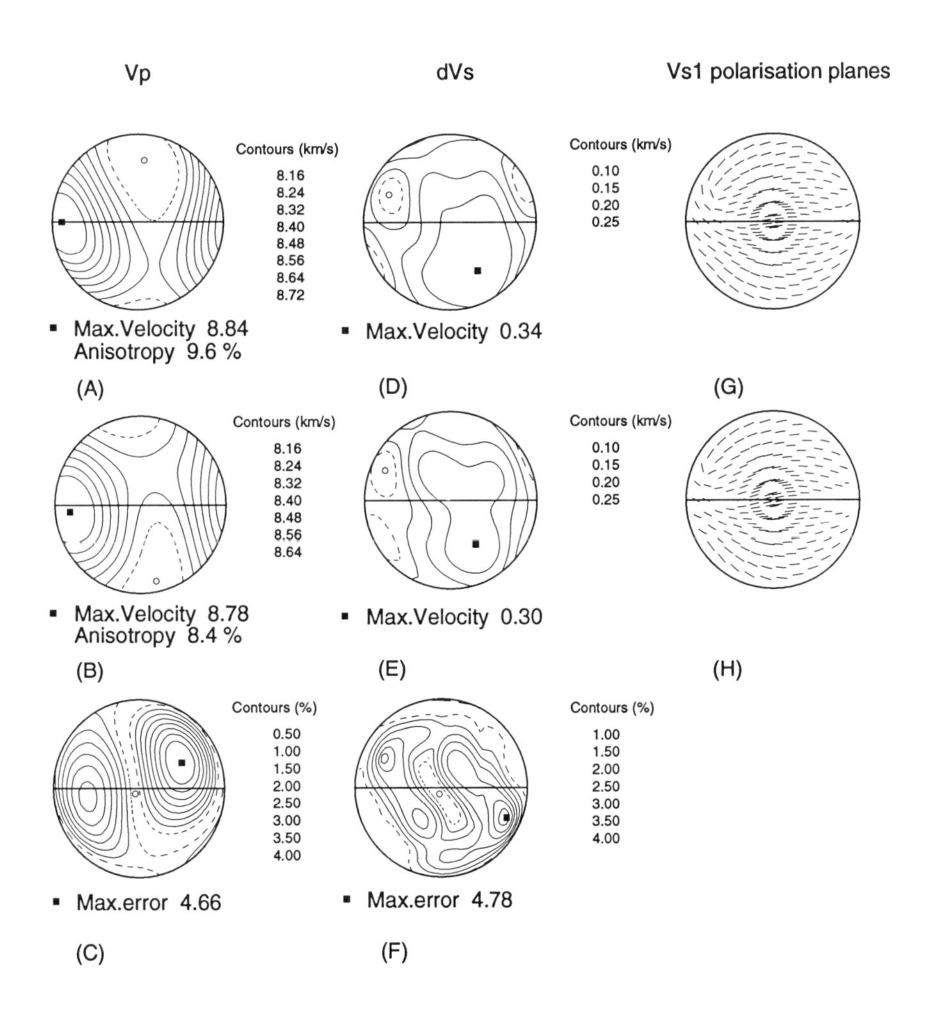

Fig. 3. Seismic properties of lherzolite sample with olivine calculated directly and after inversion, and related errors. Maximum values – black square, minimum values – open circle. a) Contoured *P*-wave distribution calculated from individual measurements (km/s). b) Contoured *P*-wave distribution calculated from recovered data (km/s). c) Contoured error on *Vp* estimate (%). d) Contoured *S*-wave birefringence distribution calculated from individual measurements (km/s). e) Contoured *S*-wave birefringence distribution calculated from recovered data (km/s). f) Contoured error on *dV*s estimate (%). g) Fastest *S*-wave polarization plane derived from individual measurements. h) Fastest *S*-wave polarization plane derived from recovered data.

Table 1. Seismic wave velocities and birefringence and related errors in the principal directions of the structural reference frame.

Azimuth / Inclination	0°/90° *Y*	0°/0° *Z*	90°/0° *X*
*V*p direct (km/s)	8.104	8.203	8.796
*V*p inv. (km/s)	8.100	8.244	8.722
***V*p error**	**0.5%**	**0.049%**	**0.84%**
*V*s1 direct (km/s)	4.820	4.879	4.913
*V*s1 inv. (km/s)	4.798	4.912	4.886
***V*s1 error**	**0.22%**	**0.23%**	**0.27%**
*V*s2 direct (km/s)	4.595	4.618	4.744
*V*s2 inv. (km/s)	4.630	4.640	4.756
***V*s2 error (km/s)**	**0.35%**	**0.22%**	**0.12%**
*dV*s direct (km/s)	0.225	0.261	0.169
*dV*s inv. (km/s)	0.168	0.262	0.130
***dV*s error**	**0.57%**	**0.45%**	**0.39%**

framework for teleseismic studies of the upper mantle, which consists mostly of orthorhombic minerals.

Acknowledgements

We thank G. Barruol, B. Ildefonse, R. M. Russo, and A. Wendt for useful discussion and review of this work . Special thanks go to F. Boudier who provided the original universal stage measurements as part of her Thesis in 1969. We would also like to thank F. Wagner (University of Metz) who provided a copy of *INVPOLE*. Finally we thank the reviewers for their helpful comments.

References

Barruol, G. & Mainprice, D. 1993. A quantitative evaluation of the contribution of crustal rocks to shear-wave splitting of teleseismic SKS waves: *Phys. Earth Planet. Inter.* **78**: 281-300.

Ben Ismaïl, W. 1995. *Variation de l'anisotropie sismique du manteau supérieur par l'étude de pétrofabriques d'olivine*. Rapport DEA: Université de Montpellier II, 47 pp.

Boudier, F. 1969. *Le massif lherzolitique de Lanzo (Alpes Piemontaises) Etude structurale et petrologique*. Thèse de doctorat: Université de Nantes, 163 pp.

Bunge, H. J. & Wenk, H. R. 1977. Three-dimensional texture analysis of three quartzites (trigonal crystal and triclinic specimen symmetry): *Tectonophysics* **40**: 257-285.

Bunge, H. J. 1982. *Texture analysis in materials science*. Butterworths, London, 593 pp.

Casey, M. 1981. Numerical analysis of *X*-ray texture data: an implementation in *FORTRAN* allowing triclinic or axial specimen symmetry and most crystal symmetries. *Tectonophysics* **78**: 51-64.

Humbert, M. & Diz, J. 1991. Some practical features for calculating the polycrystalline elastic properties from texture. *J. Appl. Cryst.* **24**: 978-981.

Mainprice, D. & Humbert, M. 1994. Methods of calculating petrophysical properties from lattice prefered orientation data. *Survey Geophys.* **15**: 575-592.

Mainprice, D., Lloyd, G. E. & Casey, M. 1993. Individual orientation measurements in quartz polycrystals: advantages and limitations for texture and petrophysical property determinations. *J. Struct. Geol.* **15**: 1169-1187.

Mainprice, D. & Silver, P. 1993. Interpretation of SKS-waves using samples from the subcontinental lithosphere. *Phys. Earth Planet. Inter.* **78**: 257-280.

Nicolas, A., Boudier, F., & Boullier, A. M. 1971. Mechanisms of flow in naturally and experimentally deformed peridotites. *Am. J. Sci.* **273**: 853-876.

Schmidt, S. M. & Casey, M., 1986. Complete fabric analysis of some commonly observed quartz *c*-axis patterns. In : H. C. Heard, H. C. & Hobbs, B. E. (eds.). Mineral and rock deformation: Laboratory studies-The Paterson volume. *AGU Monogr.* **36.** AGU, Washington, D.C., pp. 263-286.

Vollmer, F. W. 1995. *C* program for automatic contouring of spherical orientation data using a modified Kamb method. *Comp & Geosci.* **21**: 31-49.

Wagner, F. & Humbert, M. 1987. Texture analysis from incomplete pole figures in low symmetry cases: *Textures & Microstruct.* **7**: 115-129.

Appendix

SCANSTEREO Algorithm

Because *SCANSTEREO* has been implemented under *LabView* 3.0 with the *Concept*™ VI image analysis package which is copyright protected, we just give a *C*-like pseudo code easily adaptable in other programming environments (*e.g.*, *NIH Image* macro). For those who own the licenses, the *LabView* VI is available by contacting D. Mainprice (david@dstu.univ-montp2.fr)

```
Read_Image (Original_Image);
Display(Image_Window_0,Original_Image);
Threshold_Image(Low_Limit=1,High_Limit=255,Replace_Value=1,Image_In=Original_Image,Image_Out=Thresholded_Image);
Get_Parameters(Image_In=Thresholded_Image,Center_X,Center_Y,Rectangle_Width,Rectangle_Height);
Diameter = Min(Rectangle_Width,Rectangle_Height);
Radius = Diameter/2;
Array_Histogram=Get_Image_Histogram(Original_Image);
For (i=1;i<256;i++)
|       {
|       Density[i]=0;
|       }
For (j=1;i<256;i++)
|       {
|       If(Array_Histogram[i] <> 0)
|       |       {
|       |       /* BEGINNING OF GREYDENS SUB-ROUTINE*/
|       |       Open_Image_Window(Image_Window_1);
|       |       j=0;
|       |       While(Get_Button_Value('OK',Default_Value=FALSE)==FALSE)
|       |       |       {
|       |       |       If j==0;
|       |       |       |       Threshold_Image(Low_Limit=i,High_Limit=i,
|       |       |       |       Replace_Value=255-i, Image_In= Original_Image,
|       |       |       |       Image_Out= Thresholded_Image_1);
|       |       |       /*HIGHLIGHTS THE AREAS OF INTEREST */
|       |       |       If int(j/2)==j/2
|       |       |       |       Display(Image_Window_1,Original_Image);
|       |       |       else
|       |       |       |       Display(Image_Window_1,Thresholded_Image_1)v
|       |       |       Density[i]=Get_Slider_Value('DENSITY',Default_Value=
|       |       |       Density[i-1]);
|       |       |       j+=1;
|       |       |       }
|       |       Close_Image_Window(Image_Window_1);
|       |       /* END OF GREYDENS SUB-ROUTINE*/
|       |       }
|       }
/* BEGINNING OF CALCGRID SUB-ROUTINE*/
For(i=0;i<19;i++)
|       {
|       INC=i*5*3.14/180;
|       For(j=0;j<72;j++)
|       |       {
|       |       AZ=j*5*3.14/180;
|       |       While(temp==0)
|       |       |       {
|       |       |       Grid_X=Center_X+sin(INC/2)*sin(AZ)*Radius/SQRT(2);
|       |       |       Grid_Y=Center_Y-sin(INC/2)*cos(AZ)*Radius/SQRT(2);
|       |       |       temp=Get_Pixel_Value(Original_Image, Grid_X,Grid_Y);
|       |       |       /*AVOIDS MEASURING OUTSIDE OF THE POLE FIGURE*/
|       |       |       if i==18
|       |       |       |       {
|       |       |       |       if temp==0
|       |       |       |       |       Radius-=1;
|       |       |       |       else
|       |       |       |       |       {
|       |       |       |       |       Set_Pixel_Value(Original_Image, Grid_X,
|       |       |       |       |       Grid_Y,255-temp);
|       |       |       |       |       Array_Out [i,j] = Density[temp];
|       |       |       |       |       }
|       |       |       |       }
|       |       |       else
|       |       |       |       {
|       |       |       Set_Pixel_Value(Original_Image, Grid_X,Grid_Y,255-temp);
|       |       |       Array_Out [i,j] = Density[temp];
|       |       |       |       }
|       |       |       }
|       |       }
|       }
/* END OF CALCGRID SUB-ROUTINE*/
Display(Image_Window_0,Original_Image);
Write_To_File(Array_Out);
```

III: ANALYSIS OF ORIENTATION DATA

SpheriCAD: An *AutoCAD* Program for Analysis of Structural Orientation Data

Carl. E. Jacobson
Department of Geological and Atmospheric Sciences,
Iowa State University, Ames IA 50011-3210, U.S.A.
cejac@iastate.edu

Abstract– *SpheriCAD* is an *AutoCAD*™ add-in program developed to facilitate regional structural analysis. Its culminating feature is the ability to plot equal-area projections ("stereonets") of structural orientation data by outlining an area on a map. *SpheriCAD* also handles the preliminary steps of digitizing field stop locations from paper maps and plotting attitude symbols from lists of structural measurements. The program is particularly useful for areas of complex structure, where many iterations are commonly required to define domains of locally uniform structural orientation.

Introduction

Geologists have long used equal-area projections ("stereonets") of foliations, folds, lineations, and other features to help interpret regional structural geology (Turner & Weiss 1963, Ramsay 1967, Hobbs *et al.* 1976). In areas of complex structure, it is often desirable to define subdomains in which structural orientations are locally uniform (Weiss & McIntyre 1957). For large data sets, this process is tedious and time consuming. Various computer programs are available to plot the stereonets, but data must first be collated into files organized by location. This is usually an iterative process, as plotting the stereonets will commonly indicate the need to refine subarea boundaries.

This paper describes an *AutoCAD* application, named *SpheriCAD*, that facilitates defining structural subregions for any type of planar or linear structure. Its principal feature is the ability to quickly plot equal-area nets of orientation data by outlining regions of interest on a map (*i.e.*, an *AutoCAD* drawing). The orientation measurements are contained in database files external to the *AutoCAD* drawing. *SpheriCAD* also handles the preliminary steps of digitizing field stop locations from paper maps and plotting attitude symbols from lists of structural measurements. The program is currently available for both the

DOS and *Windows* versions of *AutoCAD* Release 12. It is substantially revised from a previous version that worked only with *AutoCAD* for *DOS* and which required the user to own *Quattro Pro*™ for *DOS* (Jacobson 1994).

The *AutoCAD* Drafting and Programming Environment

Before describing *SpheriCAD*, it is useful to provide an overview of *AutoCAD* and its utility for geological applications. *AutoCAD* is the most widely used drafting and design program for the *DOS* and *Windows* platforms. It is also available for the *Macintosh OS* and *UNIX* workstations. It can be used to produce both conventional, two-dimensional geologic maps and rendered, three-dimensional surface models; for example, to illustrate topography and the trace of geologic contacts across topography.

In contrast to some drawing programs that utilize coordinate systems based on paper sizes, *AutoCAD* drawings are generally created using real-world coordinates. This results in automatic registry when maps of adjacent areas are combined into a single drawing. *AutoCAD* does require the use of a Cartesian coordinate system, so latitude and longitude cannot be used. A suitable reference frame is the Universal Transverse Mercator (UTM) grid system (Merrill 1986). An advantage of the UTM system is that it is widely used for digital geographic data. For example, most information on United States Geological Survey quadrangle maps (topographic elevations, streams, political boundaries, cultural features) can be obtained in digital format keyed to the UTM grid system. *AutoCAD* add-in programs are available to contour the elevation data or, for that matter, any type of *X, Y, Z* data.

Drawing entities in *AutoCAD* are created on "layers," which can be turned on or off independently. Attitude symbols for different types and generations of structures can be placed on separate layers, as can geologic contacts, formation hatch patterns, sample locations, *etc*.

The success of *AutoCAD* is due in large measure to the fact that it is easily customized. Add-in applications can be developed using the *AutoLISP* programming language built into *AutoCAD* (Gesner & Smith 1992, Head 1992) or the *AutoCAD Development System* (*ADS*), which provides an interface to a variety of commercial *C* compilers (Hampe & Boyce 1993). For the *Windows* version of *AutoCAD*, programs can also be developed using Microsoft *Visual BASIC*. Commands introduced by add-in applications are accessed exactly the same way as standard *AutoCAD* commands: via the command-line interface, pull-down menus, dialog boxes, or *AutoCAD*'s unique screen menus. Thus, they appear to be an integral part of *AutoCAD*.

SpheriCAD is written in *AutoLISP*, which is an interpreted (non-compiled) language. *AutoLISP* is slower than *C*, but easier to use. *AutoLISP* programs are stored in ASCII text files that must be loaded every time *AutoCAD* is started. In *SpheriCAD*, this is handled transparently by a pull-down menu (Fig. 1; this and subsequent figures are screen captures that show the actual *AutoCAD* working environment.)

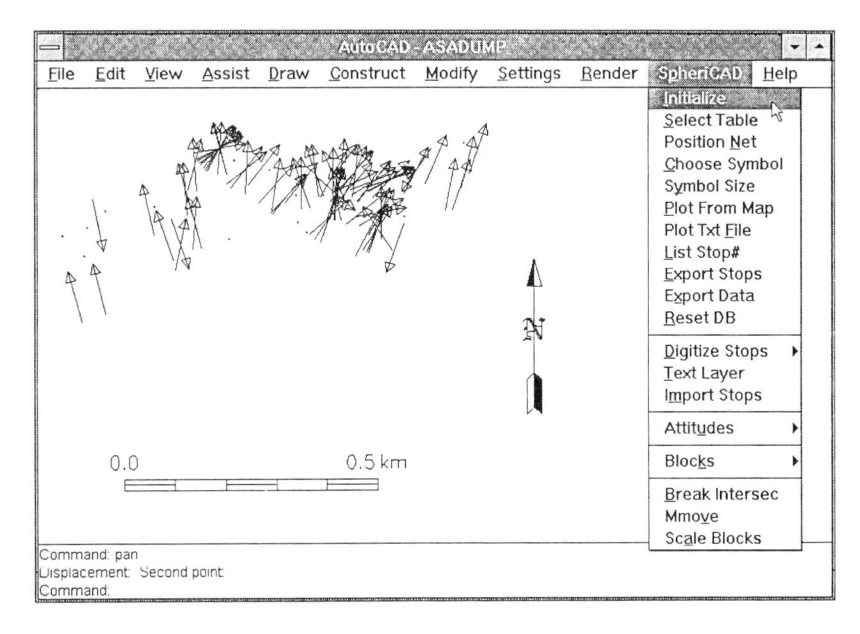

Fig. 1. *AutoCAD* map of F_2 fold orientations in portion of Orocopia Mountains, S. California (Jacobson & Dawson 1995). *SpheriCAD* pull-down menu is shown. Hard-copy output gives much higher resolution. Toolbar and toolbox have been turned off for simplicity.

AutoLISP is based on the general *LISP* programming language (Winston & Horn 1989). It contains a subset of the *LISP* commands, as well as commands developed specifically for the CAD environment. Although *LISP* is not a common language, it is well suited for a drafting program because lists of items, such as pairs of *X-Y* coordinates, can be assigned to a single variable name. This avoids the need to define arrays.

SpheriCAD utilizes a set of functions and drivers, called the *AutoCAD SQL Interface* (*ASI*), that allows *AutoCAD* programs to query external database files. *SQL* is an abbreviation of "Structured Query Language," which was introduced by IBM in 1974 to provide a platform-independent, English-like language to access databases (Hampe & Boyce, 1993). *ASI* was originally developed for *AutoCAD* applications written in *C* (*ADS* programs), but was later extended to *AutoLISP*. The use of *ASI* in *AutoLISP* is not documented in the *AutoCAD* reference manuals, but is described in detail by Hampe & Boyce (1993). *ASI* was added to *AutoCAD* in Release 12. Thus, *SpheriCAD* will not operate with earlier versions of *AutoCAD*.

The *ASI* drivers provide access to a variety of database file formats. The drivers for *Paradox*™ and *dBASE III Plus*™ are particularly convenient, because they allow data files to be manipulated, even if the parent database program is not installed. *SpheriCAD* uses the *dBASE* driver. Files of orientation measurements formatted for *dBASE III Plus* can easily be created with spreadsheet programs such as *Excel*™ and *Quattro Pro*™.

The *SpheriCAD* Program

Details of operating *SpheriCAD* are explained in the user's manual supplied with the program. This paper simply outlines the general capabilities of the program.

Digitizing Stop Locations

In order to create stereonets, a special symbol must be present at each field stop in the *AutoCAD* drawing. The symbol is an *AutoCAD* block, where a block is simply a group of entities treated as a single unit. The stop-symbol block consists of two entities, a point (*i.e.*, one pixel) and an attribute. An attribute is a tag that contains information about a block. In this case, the information is the stop number. When an area on the map is selected, *SpheriCAD* finds all the enclosed stop symbols and reads their stop numbers (attributes). The stop numbers are then used to search a database file of orientation measurements. In the present version of *SpheriCAD*, stop numbers must have an integer numeric value.

A point is used for the stop symbols because it is inconspicuous. For example, Fig. 1 shows the orientations of open-to-tight (F_2) folds in Orocopia Schist from a small portion of the Orocopia Mountains in southern California, USA (Jacobson & Dawson, 1995). The points indicate stop locations where no folds were measured. Stop symbols are also present, but not visible, at the center of each attitude symbol. If the stop symbol were larger, it would interfere with viewing the attitudes. However, a larger symbol can be added if desired.

Stop-location symbols are inserted into the *AutoCAD* drawing by digitizing paper maps or reading a list of *X-Y* coordinates from an ASCII text file. In either case, a text label with the stop number is created next to each stop symbol. The layer that contains the text labels is usually turned off, again in order to avoid clutter.

The stop-location symbols are used to create a database file that lists all the stop numbers and their *X-Y* locations (Table 1; referred to as the "stop *X-Y* file"). This process involves an *AutoCAD* command that extracts attribute values and block information to ASCII text files. The ASCII file is imported into a database or spreadsheet program and saved to *dBASE III* format. Stop *X-Y* files can also be created by hand-entering tables of *X-Y* coordinates. The stop *X-Y* file is used to plot attitude symbols and blocks that show sample locations, mineral distributions, *etc*.

Plotting Attitude Symbols

The first version of *SpheriCAD* (Jacobson 1994) plotted attitude symbols by reading lists of measurements from ASCII text files. The current version provides an additional method that utilizes the relational database capabilities of *ASI*. The newer method is simpler and quicker for file preparation, but considerably slower at actually plotting the symbols.

Table 1.

Database file structure for stop locations.

X	Y	Stop
602.666	3719.306	25
602.689	3719.437	26
602.808	3719.471	27
602.899	3719.556	28
603.001	3719.550	29
603.137	3719.544	30

The database method involves two files, the stop *X-Y* file (Table 1) and a file of attitude measurements (Table 2). Any given research area will have only one stop *X-Y* file. There will be a separate attitude file for each different type of structure. The attitude file (Table 2) is read one line at a time. The stop number is used as a "key" to query the stop *X-Y* file to determine where to plot the attitude symbol.

The ASCII method uses a single file with all necessary information (Table 3). It is relatively inefficient in terms of file storage, because *X* and *Y* must be included on every line, even if successive measurements are from the same location. Furthermore, most users store attitude data in files containing trend, plunge, and stop number, but not *X* and *Y*. Adding *X* and *Y* can be tedious, because there generally is a different number of measurements at each stop. Therefore, the database method is recommended, despite its slow plotting speed.

For either method, an attitude symbol can be chosen from a graphical menu (Fig. 2) or by typing the symbol name at the *AutoCAD* command line. The attitude symbols are blocks. From the command line, any valid block name can be entered (i.e., it is possible to create custom symbols). Symbol size is entered at the command line. Selection of data files, whether of *dBASE* or ASCII format, is via a dialog box (Fig. 3). The value of dip or plunge can optionally be plotted at an arbitrary position relative to the attitude symbol.

Plotting Stereonets

Plotting a stereonet involves five steps: (1) selecting the database table for the structure of interest (*e.g.*, S_1, L_1, L_2, *etc.*), (2) choosing a plotting symbol, (3) setting the symbol size, (4) placing a stereonet circle in the drawing, and (5) delineating an area from which to plot data. The first three steps are analogous to those for plotting attitude symbols. A set of plotting symbols is supplied with *SpheriCAD*, and custom ones can be added. A stereonet circle can be placed anywhere in the *AutoCAD* drawing (Fig. 4). The center and radius are indicated either by pointing with the cursor or by typing values at the command line. Specifying the area to plot is equivalent to selecting the stop symbols within the area. It is most easily accomplished by drawing an irregular polygon around the region (stops) of interest (Fig. 4). The stop numbers within the polygon are used to query the

Table 2.
Database file structure for attitude measurements.

Stop	Trend	Plunge
25	344	29
27	347	3
28	15	30
28	355	24
29	4	6
29	39	13
29	53	14

Table 3.
ASCII file structure for plotting attitude symbols.

X	*Y*	Trend	Plunge	Stop
602.666	3719.306	344	29	25
602.808	3719.471	347	3	27
602.899	3719.556	15	30	28
602.899	3719.556	355	24	28
603.001	3719.550	4	6	29
603.001	3719.550	39	13	29
603.001	3719.550	53	14	29

selected attitude table. The query result is a list of all the attitude measurements at the selected stops. The resultant measurements are then plotted on the stereonet (Fig. 5) using the standard equations for an equal-area projection (Hobbs *et al.* 1976, p. 501).

Additional nets can be created elsewhere in the drawing, or more data can be added to the first net. For example, in Fig. 6, the layer containing symbols for F_2 folds has been turned off and the one with foliation symbols turned on. The open squares indicate poles to foliation for the same area used to plot the F_2 folds (outlined region in Fig. 4). To plot the second data set, it was necessary to choose a new symbol and to switch from the database file of fold axes to the file of foliation measurements. However, it was not necessary to redraw the selection polygon. The reason is that *AutoCAD* can recall the most recently selected group of objects. This saves work and guarantees that the fold axis and foliation data come from exactly the same stops.

In the above example, turning the trend/plunge symbols off and the strike/dip symbols on was superfluous. This is because *SpheriCAD* ascertains the data to plot based on the selected stops and current database table, not on the attitude symbols visible.

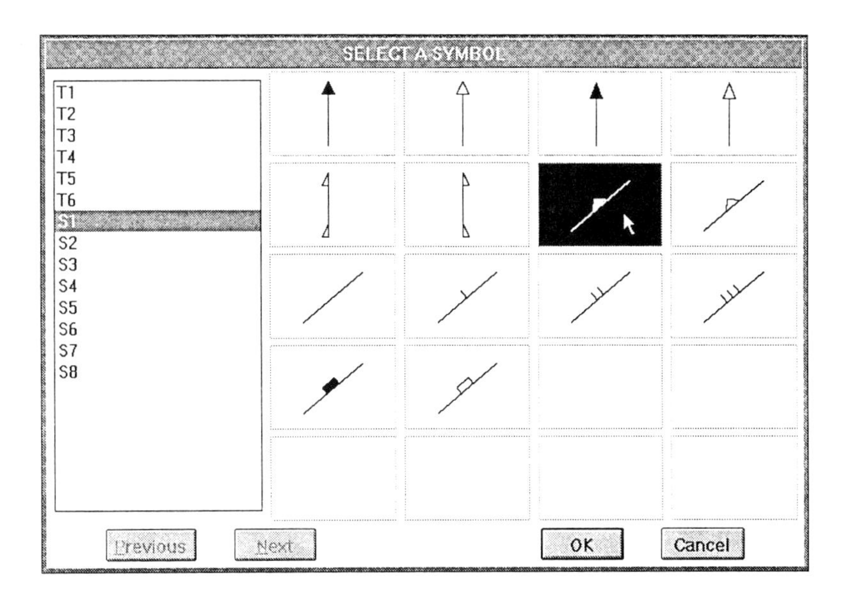

Fig. 2. Graphical menu for selecting an attitude symbol. A similar menu is used to choose plotting symbols for stereonets.

The utility of *SpheriCAD* is emphasized by a closer look at the Orocopia Mountains. The F_2 folds in that mountain range exhibit a wide variation in axial trend (Jacobson *et al.* 1988, Jacobson & Dawson 1995). When measurements from the entire area are plotted on a single stereonet, the spread in axial trend appears relatively continuous. This gives the impression that the folds were produced during a single progressive deformation that involved reorientation of axes (*cf.* Escher & Watterson 1974). However, a detailed analysis of subareas indicates that the F_2 folds probably belong to two separate generations, one with approximately NE-SW trend, the other with approximately NW-SE trend. In some subregions, only one of the two sets is developed (Fig. 7, subarea A). Elsewhere, both sets are present and can be differentiated (Fig. 7, subarea B). In yet other regions, refolding and interference results in a continuous range of trends (Fig. 7, subarea C). Elucidating relationships such as these can be very tedious without an automated tool such as *SpheriCAD*.

Any net can be saved in the drawing file alongside the map. Nets can also be written to separate files without the map.

SpheriCAD does not contour stereonets or perform any other type of statistical analysis. Nor can it represent planes by great-circle arcs rather than by poles. Such functionality may be incorporated in future versions. In the meantime, *SpheriCAD* can export data sets as ASCII text files for manipulation

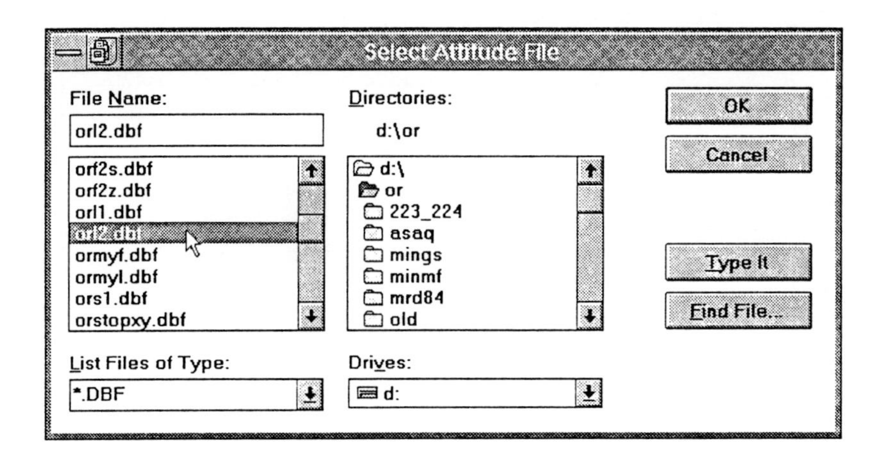

Fig. 3. Dialog box for file selection.

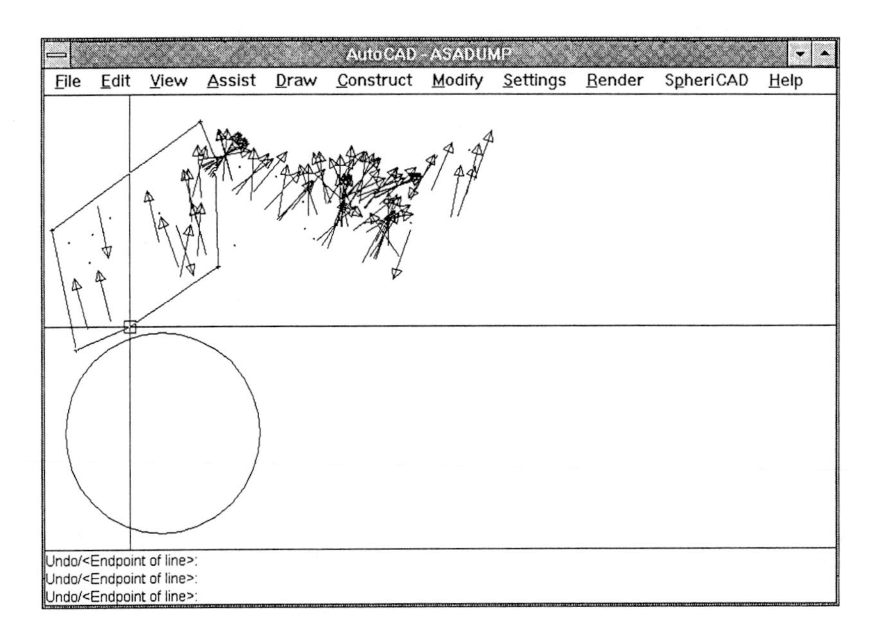

Fig. 4. An empty stereonet circle has been added to the drawing. The region of interest is being outlined, with the final point (at the crosshairs) about to be selected.

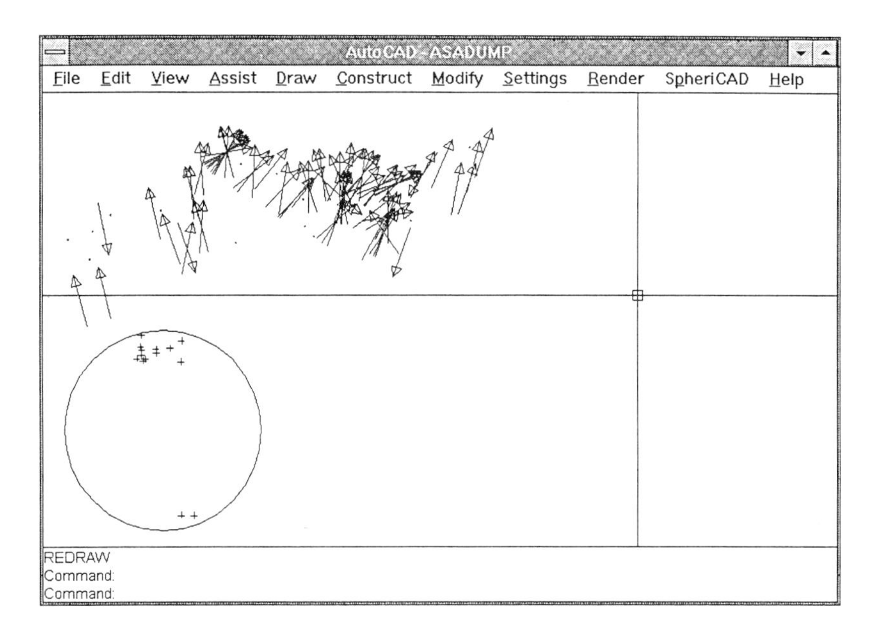

Fig. 5. Completed stereonet for the area defined in Fig. 4.

in other stereonet programs. The data are selected in exactly the same manner as for plotting a stereonet; *i.e.*, by outlining an area. Conversely, *SpheriCAD* can plot stereonets in *AutoCAD* directly from ASCII files without having to select an area on the map.

SpheriCAD can be used to create separation-arc diagrams (Hansen 1967) to determine transport direction in shear zones from fold asymmetry. This requires that folds with sinistral ("S") and dextral ("Z") asymmetry be stored in separate database files. The two files are then plotted individually, each with a different symbol.

Plotting Blocks

SpheriCAD was created primarily to work with structural data, but can be used to insert any type of block into an *AutoCAD* drawing from a list of either stop numbers in a *dBASE* file or *X-Y* pairs in an ASCII file. The *dBASE* method requires that a stop *X-Y* file be present in order to define the coordinates of the stop numbers. There are three primary reasons to insert blocks:

- The standard stop symbol is a point. It is sometimes desirable to add a more visible symbol at all stop locations.

- Symbols can be used to distinguish a subgroup of stops, such as those where a certain type of sample was collected or where a key index mineral is present.

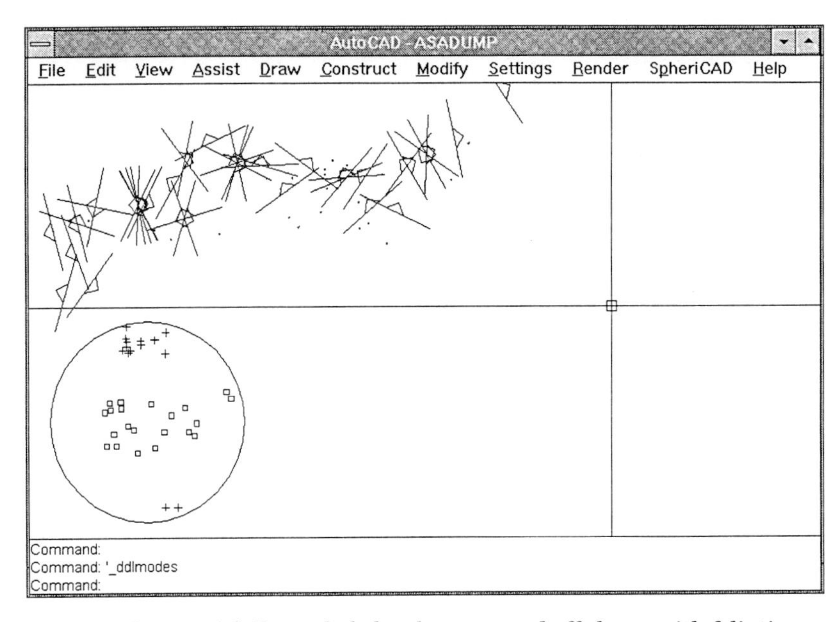

Fig. 6. Layer with F_2 symbols has been turned off; layer with foliation symbols turned on. Poles to foliation (open squares) plotted for area previously selected for F_2 folds (area outlined in Fig. 4).

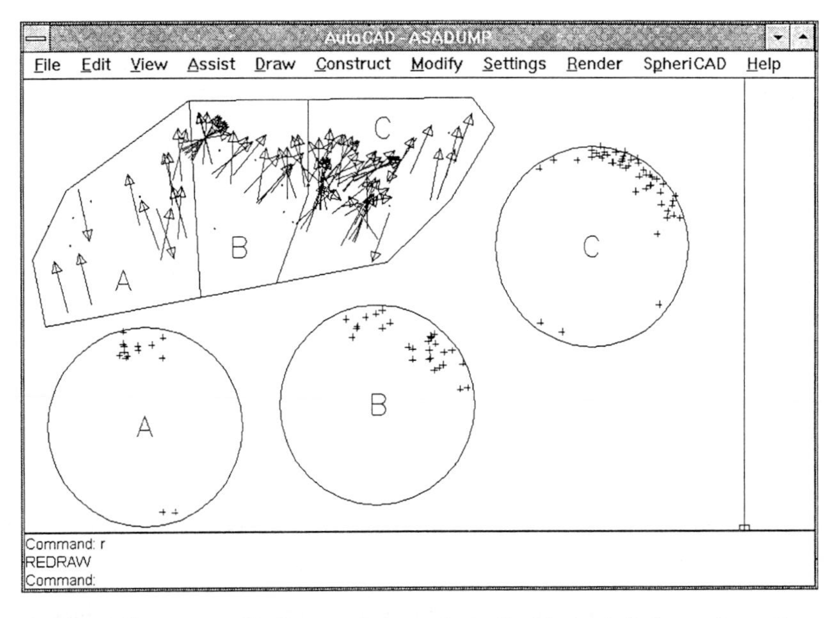

Fig. 7. Stereonets for three subareas described in text. Subarea boundaries were added with *AutoCAD* pline (polyline) command.

- Pairs of *X-Y* values can be plotted on a graph. In general, however, producing graphs is best accomplished with a dedicated graphing or spreadsheet program. In *AutoCAD*, graph axes must be created manually.

SpheriCAD also includes several block utilities that are useful even for non-geological applications; for example, to scale the size of all or selected blocks independently of other entities in the drawing.

Availability of *SpheriCAD*

SpheriCAD can be downloaded free of charge via either ftp or the World Wide Web. The package includes a readme file (ASCII text) and two self-extracting zip files (binary files), one containing the program files and the other the user's manual formatted for Microsoft *Word* for *Windows* 6.0. For ftp use the following parameters:

ftp site:	isum1.iastate.edu
userid:	anonymous.cejac
password:	your e-mail address (including domain name; *e.g.*, cejac@iastate.edu)

On the World Wide Web access URL:

http://www.public.iastate.edu/~cejac/anonymous

The program can alternatively be ordered by mail from:

Carl Jacobson,
Geological Sciences,
Iowa State University,
Ames IA 50011-3210, U.S.A.

The cost is $7 US for both the disk and hard-copy user's manual. The disk alone is $5, the manual alone is $4. Add $3/disk and $2/manual for overseas shipments. The disk includes the file version of the user's manual. The disk will be 3.5" unless specified otherwise. Make check or money order payable to "Geological Sciences."

Acknowledgements

The project was funded by NSF grants EAR-9004354 and EAR-9316916. Felix Oyarzabal and John Baldwin helped test the program and John Baldwin assisted in formatting the user's manual. The manuscript benefited from the helpful comments of Brendan Murphy and an anonymous reviewer. Neil Alen of Autodesk and Robert McNeel & Associates of Seattle provided invaluable assistance for demonstrating the program at the 1994 Annual meeting of the Geological Society of America.

References

Escher, E. & Watterson, J. 1974. Stretching fabrics, folds and crustal shortening. *Tectonophysics* **22**: 223-231.

Gesner, B. R. & Smith, J. J. 1992. *Maximizing AutoLISP.* New Riders Publishing, Carmel, Indiana.

Hampe, K. & Boyce, J. 1993. *The AutoCAD Professional's API Toolkit.* New Riders Publishing, Carmel, Indiana.

Hansen, E. 1967. Methods of deducing slip-line orientations from the geometry of folds. *Year Book Carnegie Inst. Washington* **65**, 387-405.

Head, G. O. 1992. *AutoLISP in Plain English.* Ventana Press, Chapel Hill, North Carolina.

Hobbs, B. E., Means, W. D. & Williams, P. F. 1976. *An Outline of Structural Geology.* John Wiley & Sons, Inc., New York, New York.

Jacobson, C. E. 1994. Structural analysis using *AutoCAD* and *Quattro Pro. Geol. Soc. Am. Abs. w. Prog.* **26**: A183.

Jacobson, C. E. & Dawson, M. R. 1995. Structural and metamorphic evolution of the Orocopia Schist and related rocks, southern California: Evidence for late movement on the Orocopia fault. *Tectonics* **14:** 933-944.

Jacobson, C. E., Dawson, M. R. & Postlethwaite, C. E. 1988. Structure, metamorphism, and tectonic significance of the Pelona, Orocopia, and Rand Schists, southern California. In: *Metamorphism and Crustal Evolution of the Western United States, Rubey Volume VII* (edited by Ernst, W. G.). Prentice-Hall, Englewood Cliffs, New Jersey, 976-997.

Merrill, G. K. 1986. Map location literacy - How well does Johnny Geologist read? *Bull. geol. Soc. Am.* **97**: 404-409.

Ramsay, J. G. 1967. *Folding and Fracturing of Rocks.* McGraw-Hill, New York, New York.

Turner, F. J. & Weiss, L. E. 1963. *Structural Analysis of Metamorphic Tectonites.* McGraw-Hill, New York, New York.

Weiss, L. E. & McIntyre, D. B. 1957. Structural geometry of Dalradian rocks at Loch Leven, Scottish Highlands. *J. Geol.* **65**: 575-602.

Winston, P. H. & Horn, B. K. P. 1989. *LISP.* Addison-Wesley Publishing Co., Reading, Massachusetts.

Appendix: Overview of *SQL* Syntax

This section describes a small portion of the *SpheriCAD* code to illustrate *SQL* syntax. When an area is outlined for plotting a stereonet, *SpheriCAD* determines all the stop numbers in the area and writes them to a database file named `subset.dbf` (`dbf` is the standard file extension used by *dBASE III*). This file has a single field (column) entitled `stop`. The values in the `stop` field are used to query the attitude file, which has three fields: `stop`, `trend`, and `plunge` (this is true even for files of planar features, which must contain the trend and plunge of the pole to the plane). The *AutoLISP* code that composes the query statement is:

```
(strcat
    "SELECT * FROM "
    tblnam
    ", subset WHERE "
    tblnam
    ".stop = subset.stop"
)
```

The `strcat` command concatenates the five following lines into a single text string. The `tblnam` term is a text variable that contains the name of the most recently selected database file of attitude measurements. All the other terms are string constants. If the current attitude file is `orf2.dbf`, then the resultant query will be:

```
SELECT * FROM orf2, subset WHERE orf2.stop = subset.stop
```

By convention, terms that are part of the *SQL* language are written in upper case, whereas file and field names are shown in lower case, although this is not required. The `dbf` file extension is not included in the *SQL* statement, but is handled by the *ASI* driver. Terms such as `orf2.stop` refer to the `stop` field in the `orf2.dbf` database file. The * indicates that the query result should include all fields (`stop`, `trend`, and `plunge`) in the queried files (`orf2.dbf` and `subset.dbf`). The query criterion is that a given stop number be present in both `orf2.dbf` and `subset.dbf`. The query result is stored as an array in memory called the "cursor." *ASI* commands are available to extract values, in this case trend and plunge, from the cursor.

A Computer Program to Print Inclined Spherical Projections

John Starkey
Department of Earth Sciences, University of Western Ontario,
London, Ontario N6A 3B7 Canada. jstarkey@julian.uwo.ca

Abstract– A computer program, *Net*, is described which constructs nets based on the Stereographic, Lambert Equal Area, and Orthographic Projections. The nets are drawn to the computer screen, printer, or plotter. The size of the net and the plane of projection are specified, within the size limits imposed by the output device. The program also provides control over the spacings of the great and small circles on the net. The computational procedures have been described elsewhere (Starkey 1970). The program is written in Turbo Pascal for use on an IBM PC compatible microcomputer under *DOS*. The program is capable of producing publication quality nets and is available as shareware from the author.

Introduction

In the earth sciences three dimensional orientation data are often represented on the surface of a reference sphere. The data are located on the sphere by means of a grid of great and small circles. The great circles represent planes which intersect along one diameter of the sphere, which thus defines an axis. The small circles represent planes perpendicular to this axis. These great and small circles correspond to the familiar lines of longitude and latitude on a geographical globe. The grid of great and small circles is projected onto a plane as a net of intersecting, usually curved lines. The net is used to locate and manipulate orientation data.

Three spherical projections are in common use. Crystallographers traditionally use the Stereographic Projection for the representation of pole figures. Structural geologists and geophysicists more commonly use the Lambert Equal Area Projection, which offers particular advantages for the preparation of contoured pole figures. Block diagrams are constructed with the help of the Orthographic Projection (Mclntyre & Weiss, 1956), although it has also been used to prepare pole figures (Buerger 1963).

Nets based on these three spherical projections are not readily available, especially when specific sizes are required. Further, the nets which are available

```
          NET version 1.0 - A program to prepare spherical projections
  by John Starkey, Department of Earth Sciences, University of Western Ontario,
********************** London, Ontario, Canada. N6A 3B7 ************************

     <*> Output to ....................................... LPT1    LaserJet 3

     <1> Projection .......................................... Stereographic

     <2> Diameter (in millimetres) ................................... 100.0

     <3> Tilt of Projection .......................................... 35.26

     <4> Angular spacing for fine small circles ......................     2

     <5> Angular spacing for fine great circles ......................     2

     <6> Annulus free of fine small circles ..........................    20

     <7> Angular spacing for bold small circles ......................    10

     <8> Angular spacing for bold great circles ......................    10

     <9> Annulus free of bold small circles ..........................     2

     <E>xecute the program or <Q>uit --->          ENTER SELECTION >
```

Fig. 1. The main menu for program *Net.*

are usually only for equatorial projections, in which the plane of projection is tangential to the reference sphere at the equator, and therefore parallel to the axis of the sphere. A polar projection is more convenient for the plotting of linear data defined by an azimuth and plunge because it avoids the need to rotate the net. A polar net represents the projection of the reference sphere onto a plane tangential to the sphere at its pole, which is defined as the point of emergence of the axis. For the analysis of data which are distributed about a direction inclined to the horizontal, use of a net obtained by projection onto a plane tangential to the reference sphere such that the axis of the sphere coincides with the direction is most convenient. This allows for the direct manipulation of the data and improves on the usual techniques which use an equatorial projection and successive rotations about three horizontal axes to emulate the inclined direction. The solutions of many problems benefit from the use of nets based on equatorial, polar, and inclined projections in combination. Linear data are plotted using the polar projection, planar data are plotted using the equatorial projection, and an inclined projection is used to solve problems of rotation of the data about some general axis.

The Program *Net*

The program *Net* is completely menu driven. The main menu is illustrated in Fig. 1. The menu lists each of the available options and the currently selected parameters. Selection of an option from the main menu displays sub-menus for the entry of relevant parameters. The sub-menus are hidden after the parameters are chosen so that the main screen remains uncluttered.

To select the output device an asterisk is entered. This displays a sub-menu with the following five options:

1. The computer screen. The program recognizes the following graphics adapters CGA, EGA, VGA, Hercules™, IBM 8514™, ATT 400™ and PC 3270™.

```
<P>erimeter ....... 0.35 ┐
<B>old Lines ...... 0.20 │  Line Thickness (in mm.).
<F>ine Lines ...... 0.10 ┘

<M>ain Title ...... LAMBERT EQUAL AREA PROJECTION

<S>ub Title ....... J. Starkey
                    Dept. of Earth Sciences
                    University of Western Ontario
                    London, Ontario, Canada

<R>eturn

ENTER > _
```

Fig. 2. The sub-menu in *Net* used to define parameters for the LaserJet III printer.

2. A dot matrix printer. Choice of this option displays sub-menus for the selection of the vertical pitch, either 60 or 72 dots per inch, and the operative parallel or serial port.

3. A Hewlett Packard Laserjet III™ compatible printer. Selection of this option displays a sub-menu for the identification of the parallel or serial port to which the printer is attached.

4. A Hewlett Packard HP7474™ compatible plotter. Choice of this option displays a sub-menu for selection of the parallel or serial port to which the plotter is attached.

5. A computer file in *Hewlett Packard Graphics Language* (*HP-GL*) format. A filename must be entered to which the extension `.HPG` is added automatically.

Selection of output options 3, 4, or 5, displays additional sub-menus which permit the setting of drafting parameters. Choice of option 3, `Laserjet III`, produces the `LaserJet Parameters` menu shown in Fig. 2, which displays the current parameters. By pressing `P`, `B` or `F`, the thickness of the lines used to draw the perimeter of the net and the thick and thin lines on the net are entered, in millimetres. Pressing `M` displays the `Main Title` menu shown in Fig. 3. From this menu, `M` is pressed to enter the main title and *X* and *Y* are pressed to position it relative to the centre of the net. The *X* and *Y* coordinates are in millimetres, measured from the centre of the net to the centre of the bottom of the main title. The remaining parameters define the characteristics of the font used in the main title. They are the parameters required by the font definitions in the *Hewlett Packard Graphics Language* (*HP-GL/2*). They define the symbol set, font spacing, pitch, height, posture, stroke weight, and typeface. These parameters are explained in the *HP-GL/2* manual but a brief explanation can be obtained by pressing `?` which displays the information shown in Fig. 4. Pressing `R` returns to the `LaserJet Parameters` menu.

Pressing `S` on the `LaserJet Parameters` menu displays the `Sub-Title` menu, which is shown in Fig. 5. The sub-title can contain up to 4 lines of text. These are edited by pressing `S`. Pressing *X* and *Y* permits entry of the coordinates in

```
<M>ain Title ...... LAMBERT EQUAL AREA PROJECTION

<X>-coordinate ....   0.00 ┐ Position left (-) or right (+) of centre
<Y>-coordinate ....  60.00 ┘ Position above (+) or below (-) centre
                             (Measured to bottom middle of title)

<C>haracter Set ... 277    ┐
<F>ont Spacing .... 1      │
<P>itch ........... 5.900  │ Refer to the HP-GL manual for the "KIND"
<H>eight .......... 16.00  │ under the AD or SD commands.
<A>ttitude ........ 0      │
<W>eight .......... 3      │
<T>ypeface ........ 4148   ┘

<R>eturn

ENTER > _
```

Fig. 3. The sub-menu in *Net* to set up the main title.

Parameters used in the Hewlett Packard Graphics Language to Define the Font and its Attributes.

Character set	:	designates the symbol set, e.g. 277 = Roman-8
Font Spacing	:	0 = fixed spacing, 1 = proportional spacing
Pitch	:	pitch for fixed-space fonts, in characters per inch
Height	:	point size for proportional fonts
Attitude	:	posture of the characters, 0 = uprights, 1 = italic
Weight	:	line thickness, -7 (Ultra Thin) to 7 (Ultra Black)
Typeface	:	font designation, *e.g.* 4101 = CG Times

Fig. 4. Explanation of `KIND` parameters used in *HPGL* to define font attributes.

```
<S>ub Title ....... J. Starkey
                    Dept. of Earth Sciences
                    University of Western Ontario
                    London, Ontario, Canada.

<X>-coordinate .... -55.00 ┐ Position left (-) or right (+) of centre
<Y>-coordinate .... -50.00 ┘ Position above (+) or below (-) centre
                             (Measured to top left corner of title)
<L>ine Spacing .... 0.00

<C>haracter Set ... 277    ┐
<F>ont Spacing .... 1      │
<P>itch ........... 9.000  │ Refer to the HPGL manual for the "KIND"
<H>eight .......... 7.00   │ under the AD or SD commands.
<A>ttitude ........ 1      │
<W>eight .......... 0      │
<T>ypeface ........ 4101   ┘

<R>eturn

ENTER > _
```

Fig. 5. The sub-menu in *Net* to set up the sub-title.

```
<P>erimeter ....... 1 ┐
<B>old Lines ...... 2 │  Plotter pen number.
<F>ine Lines ...... 3 ┘

<M>ain Title  : LAMBERT EQUAL AREA PROJECTION

<S>ub Title   : J. Starkey
                Dept. of Earth Sciences
                University of Western Ontario
                London, Ontario, Canada

 <R>eturn

ENTER > _
```

Fig. 6. The sub-menu in *Net* to define plotter parameters.

```
<M>ain Title ...... LAMBERT EQUAL AREA PROJECTION

<X>-coordinate ....   0.00 ┐ Position left (-) or right (+) of centre
<Y>-coordinate ....  95.00 ┘ Position above (+) or below (-) centre
                             (Measured to bottom middle of title)

<H>eight .......... 4.00  ┐  Height and width of characters (in mm.)
<W>idth ........... 2.70  │  used to write the title.
<P>en Number ...... 1     ┘  Plotter pen number.

<R>eturn

ENTER > _
```

Fig. 7. The sub-menu in *Net* to set up the main title for plotter output.

millimetres, of the top left corner of the sub-title to be positioned relative to the centre of the net. The spacing of the lines of the sub-title is entered after pressing *L*. The remaining parameters characterize the font used for the sub-title. Their meaning is the same as on the `Main Title` menu and the explanation shown in Fig. 4 can be displayed by pressing `?`.

The selection of output options 4 or 5, plotter or file output, displays the `Plotter Parameters` menu shown in Fig. 6, or the essentially identical *HP-GL* `File Parameters` menu. The parameters are similar to those on the `LaserJet Parameters` menu (Fig. 2), except that the widths of the lines used to draw the perimeter of the net and the thick and thin lines on the net are specified by pen number rather than in millimetres. Pressing `M` or `S`, to select the `Main Title` or the `Sub-Title` menus, displays sub-menus as shown in Figs. 7 and 8. These are similar to those for the LaserJet, (Figs. 3 and 5), except that the font characteristics are restricted to height and width, in millimetres, and the number of the pen used, which controls the line thickness. The remaining options on the main menu (Fig. 1), are largely self-explanatory, as follows:

1. The spherical projection used. Nets are drawn based on the Lambert Equal Area, Stereographic, or Orthographic projection.

2. The diameter of the projection, in millimetres. This does not apply to screen output.

```
<S>ub Title ....... J. Starkey
                    Dept. of Earth Sciences
                    University of Western Ontario
                    London, Ontario, Canada.

<X>-coordinate .... -90.00 ┐ Position left (-) or right (+) of centre
<Y>-coordinate .... -90.00 ┘ Position above (+) or below (-) centre
                             (Measured to top left corner of title)
<L>ine Spacing .... 0,00

<H>eight .......... 2.00  ┐ Height and width of characters (in mm.)
<W>idth ........... 1.30  ┘ used to write the title.

<P>en Number ...... 2     } Plotter pen number.

<R>eturn

ENTER > _
```

Fig. 8. Sub-menu in *Net* to set up the sub-title for plotter output.

3. The angle between the plane of projection and the axis of the reference sphere. (*e.g.*, 0° for equatorial projections and 90° for polar projections).

4. The angular spacing of small circles on the net drawn as thin lines. See the discussion of the line thickness parameters on the output menus above (Figs. 2 and 6). For screen output all lines have the same thickness.

5. The angular spacing of great circles on the net drawn as thin lines.

6. The angular annulus about the axis of the projection within which the small and great circles represented by thin lines are omitted. This avoids congestion by lines too closely spaced near to the projected pole.

7. The angular spacing of small circles on the net drawn as thick lines.

8. The angular spacing of great circles on the net drawn as thick lines.

9. The angular annulus about the axis of the projection within which the small and great circles represented by thick lines are omitted. This avoids congestion by lines too closely spaced near to the projected pole.

`E` executes the program and `Q` terminates the program.

Upon termination, the program writes the parameters which are in effect to a file, `Net.inp`. This file provides the default settings the next time the program executes. Figure 9 illustrates an example of Laserjet III printer output from *Net*. It is a net based on stereographic projection onto a plane inclined at 35.26° to the axis of the reference sphere. Such a projection is useful for the manipulation of crystal directions in cubic crystals involving rotation about [111]. Figure 1 lists the input parameters, the *LaserJet* parameters are indicated in Figs. 2, 3, and 5.

The output options offered by *Net* give control over the output quality. The LaserJet III printer produces high quality output and provides formatting

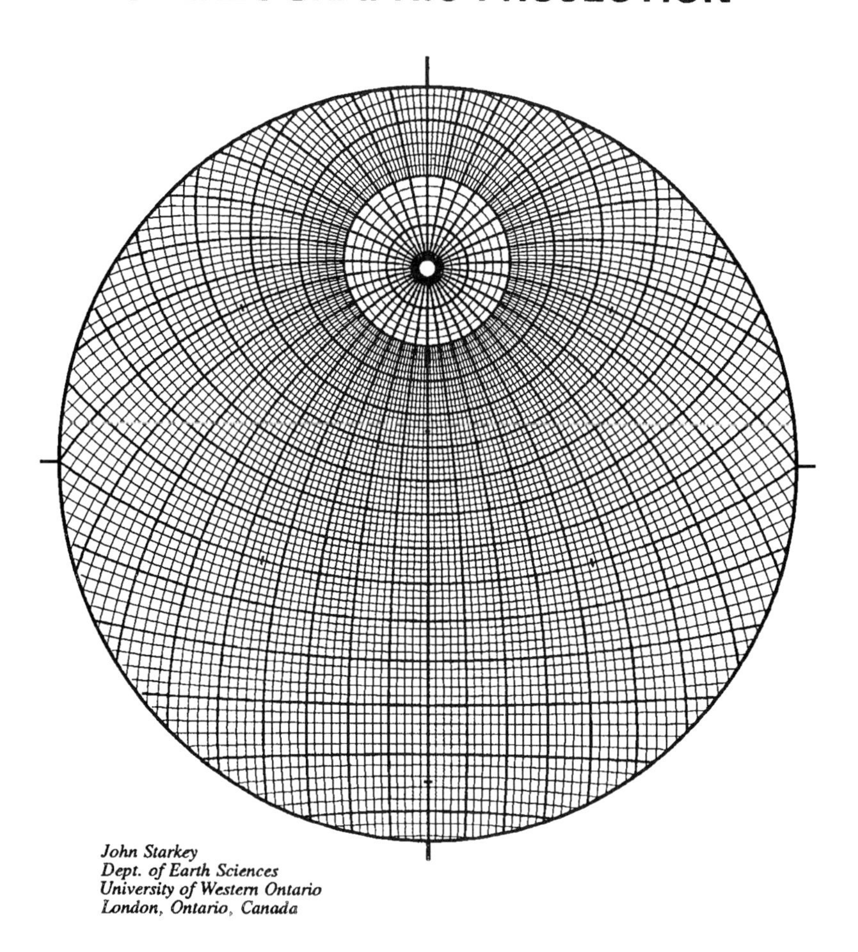

Fig. 9. A net produced by stereographic projection onto the plane tangential to the reference sphere at 35.26° to the axis.

flexibility for the titles (see, for example, Fig. 9). A pen plotter also produces high quality output but formatting options for the titles are limited. The precision of dot matrix printer output is limited by the spacing of the dots; however, usable working copies may be produced. Use of the option to write a data file in *HP-GL* format allows for final output by other programs, including presentation graphics programs. This provides the maximum output flexibility and permits incorporation of nets into other documents. The screen display is useful for proofing the net before committing it to another output device.

References

Buerger, M. J. 1963. *Elementary Crystallography*. John Wiley & Sons. 528 pp.

Mclntyre, D. B. & Weiss, L. E. 1956. Construction of block diagrams to scale in orthographic projection. *Proc. Geol. Assoc.* **67**: 142-155.

Starkey, J. 1970. A computer program to construct spherical projections. In: Paulitsch., P. (ed.). *Experimental and Natural Rock Deformation*. Springer, pp. 38-50.

Presentation of Orientation Data in Spherical Projection

John Starkey
Department of Earth Sciences, University of Western Ontario,
London, Ontario N6A 3B7, Canada. jstarkey@julian.uwo.ca

Abstract– The presentation of three dimensional orientation data in the Earth Sciences is commonly accomplished by representing the data on the surface of a reference sphere from which orientation diagrams are prepared using spherical projections. Where the orientation data are to be contoured, an equal area spherical projection is normally used. The programs described here prepare orientation diagrams using the Lambert Equal Area Projection.

Introduction

A program, *Fabfile*, is used to enter data into a microcomputer and prepare computer files for input to a plotting program, *Fabric*. The orientation data entered into *Fabfile* can include any combination of linear data, defined by azimuth and plunge, and planar data, defined either by strike, dip and the approximate direction of dip, or by azimuth and dip. The data can represent orientations on either the upper or lower reference hemisphere, or on both. Scalar and vector data can be entered. The orientations of reference directions and planes can also be entered; these appear in the final orientation diagrams but do not form part of the data set when the data are contoured, or where their inclusion could be confused with the other orientation data.

The program *Fabric* accepts data files created by *Fabfile* and manipulates the data to produce orientation diagrams. The computations have been described elsewhere (Starkey 1970). Orientation diagrams which can be prepared include pole figures, great circle diagrams, and contoured pole figures. The diagrams can represent projections of either the upper or lower-reference hemisphere. Vectorial data may be represented as pole figures plotted over the entire sphere, in which case their direction is preserved by the use of distinctive symbols to indicate the hemisphere on which they occur. The orientation diagrams can be displayed on the computer screen or printed on dot matrix printers, laser printers, or pen plotters. They can also be written as computer files for input to other programs.

Contoured pole figures are prepared by counting the number of data within an area corresponding to a circle on the surface of the reference sphere,

```
         FABFILE version 3.0 - A program to prepare input files for FABRIC
  by John Starkey, Department of Earth Sciences, University of Western Ontario,
*********************** London, Ontario, Canada. N6A 3B7 ***********************

                     Enter a name for the data file SAMPLE.FAB

                   This file already exists in this directory:
                     to Continue with this file type .... C
                     to Replace this file type .......... R
                     to Enter another file name type .... E
                     to List existing files type ........ L
                     to Quit the program type ........... Q _
```

Fig. 1. Menu used in *Fabfile* to initialize data file.

located at the nodes of a square grid placed over the projected hemisphere. The area of the counting circle is 100/*n* % of the reference hemisphere, where *n* is the sample size. This results in contours which are independent of sample size and are expressed in multiples of a uniform distribution (Starkey 1977, 1993).

The programs *Fabfile* and *Fabric* have been written in *Turbo Pascal*. The current version, 3.0, has been compiled under version 5.5 for use with an IBM PC compatible microcomputer under *DOS*.

Fabfile

This program prepares data files to be read by *Fabric*. Orientation data can be entered for lineations or planes. The number of data which can be entered is limited only by the amount of available disk storage. The input data can include orientation data for up to 300 special lineations, to be represented on the orientation diagrams by designated symbols, and special planes, to be represented by great circles. If the data are subsequently contoured the special directions are not included in the contoured data set. The special planes are excluded from great circle diagrams, where they could be confused with the measured orientation data. (See the discussion of the data file `Planes.fab` below).

The opening screen display of *Fabfile* requests entry of a filename to receive data, to which the extension `.fab` is added automatically. This file is a text file, it can therefore subsequently be viewed or edited with a text editor. However, it is recommended that existing files be edited using the editing facilities which are incorporated into *Fabfile*, since this program ensures that the entered data are acceptable.

If the filename which is entered already exists in the current directory, a menu is displayed with a series of options. These are selected by entering the appropriate, highlighted character on the screen display (Fig. 1). This assists in the selection of a nonconflicting filename.

Entry of a new filename produces a request to enter the hemisphere on which the measured data are plotted, upper or lower, and a header. Indication of the plotting hemisphere ensures consistency with the projected hemisphere subsequently used by *Fabric*. If a new filename is entered, a header containing up to 75 characters of text can be added to the new file. The header is not used by either *Fabfile* or *Fabric*, it simply permits entry of a comment which may be read

```
Input can be either linear or planar data.
Linear data are entered as azimuth and plunge.
Planar data are entered either as strike, dip and approximate direction of dip
                        OR as azimuth and dip.
Special data can be entered to represent particular directions or surfaces
(e.g. lineation, bedding etc.). These can be entered as either linear or planar
data and can be plotted either as poles, using special symbols, or as great
circles.
                  Linear data ......................... 1
                  Planar data (strike and dip).......... 2
                  Planar data (azimuth and dip) ........ 3
                  Special directions ................... 4
                  Special surfaces (strike and dip) .... 5
                  Special surfaces (azimuth and dip) ... 6
                  Edit the file ........................ E
                  Terminate data entry ................. T

                  Enter your selection ................. _
```

Fig. 2. Main *Fabfile* menu from which data entry mode is selected.

using a text editor. The header can be left blank. If a new file, or addition of data to an existing file, is selected, the main menu is displayed, from which the action to be taken by the program is chosen (Fig. 2). As each option is selected, an appropriate display is presented to guide the entry of data.

During data entry, the set of orientation data for each measured element is numbered consecutively on the screen. At any time during data entry, data sets can be eliminated by entering `R`, followed by the number of the data set. Data entry is terminated by entering `Q`, which returns to the main menu. The terms `dip` and `plunge`, displayed during data entry, refer to inclinations relative to the horizontal, either up or down. Internally the programs use the following convention. If upper hemisphere input is selected, elevations measured on the upper hemisphere are indicated by positive dips and plunges, data on the lower hemisphere are indicated by negative values. Conversely, if lower hemisphere input is selected, data on the lower hemisphere have positive dips and plunges and data on the upper hemisphere have negative dips and plunges. The data can be entered on one hemisphere in *Fabfile* and presented as orientation diagrams based on projection of the opposite hemisphere, or both hemispheres, in *Fabric*.

The following options are available from the main menu: Entered data values are automatically restricted to their acceptable range, azimuths, and strikes from 0° to 360°, and dips and plunges from –90° to 90°

1: Linear data specified by azimuth and plunge (Fig. 3). Invisible to the operator, the program adds two more data to the data set for each lineation. These are an 'L', which indicates that the data pertain to a lineation, and an 'N', which indicates that the data are not to be represented by a special symbol.

2: Planar data specified by strike and dip (Fig. 4). The approximate direction of dip is indicated by 'N', 'E', 'S', or 'W', representing the cardinal direction towards which the plane is dipping. Invisibly, the program adds an 'N' to the data set, which indicates that the data do not pertain to a special plane.

3: Planar data specified by azimuth and dip (Fig. 5). The azimuth is the direction towards which the plane is dipping. Invisibly, the program adds 'P' to the data set, to indicate that the data represent the azimuth and dip of a plane, and an 'N', to indicate that the data are not for a special plane.

```
                    *** LINEAR DATA ***
To remove data after they are entered type R followed by their NUMBER
To return to the main menu type Q
                  NUMBER   AZIMUTH     PLUNGE
                    1        45.0       30.0
                    2       _____
```

Fig. 3. Data entry screen displayed by *Fabfile* which guides input of orientation data for directions defined by azimuth and plunge.

```
                    *** PLANAR DATA ***
To remove data after they are entered type R followed by their NUMBER
To return to the main menu type Q
              NUMBER    STRIKE      DIP    DIRECTION
                2        45.0       30.0       N
                3       _____
```

Fig. 4. Data entry screen displayed by *Fabfile* which guides input of orientation data for planes defined by strike, dip and direction of dip.

```
                    *** PLANAR DATA ***
To remove data after they are entered type R followed by their NUMBER
To return to the main menu type Q
                  NUMBER   AZIMUTH      DIP
                    3        45.0       30.0
                    4       _____
```

Fig. 5. Data entry screen displayed by *Fabfile* which guides input of orientation data for planes defined by azimuth and dip.

4: Special linear data, such as geographical coordinates or specimen directions, specified by azimuth and plunge (Fig. 6). A symbol is selected to represent the lineation in the orientation diagram. The choice is 'X', 'Y', 'Z', '+', '*', 'S' (square), 'D' (diamond), or 'T' (triangle) (*n.b.*: these symbols can be illustrated by drawing a scatter diagram with *Fabric* using the data file `Symbols.fab` as input. See the description of data files below.) Invisible to the operator, the program adds an 'L' to the data set; this denotes that the data pertain to a special lineation.

5: Special planar data, such as a plane of reference, specified by strike and dip (Fig. 7). The approximate direction of dip is indicated by 'N', 'E', 'S', or 'W', representing the cardinal direction towards which the plane is dipping. Invisibly the program adds a 'G', which indicates that the data are to be represented by a great circle.

6 : Special planar data, such as a plane of reference, specified by azimuth and dip (Fig. 8). The azimuth is the direction towards which the plane is dipping. Invisibly, a 'P' is added, to indicate that the data are the azimuth and dip of a plane, and a 'G', to indicate that the data are to be represented by a great circle.

E : This displays the data file for editing (Fig. 9). The sets of orientation data in the data file are numbered consecutively and displayed in groups of 20. The data listed in Fig. 9 are the input data illustrated in Figs. 3 to 8. To edit data, a number from 0 to 20 is entered. A '0' indicates that the displayed data are not to be changed and the next group of 20 is displayed. Entry of the number of a data set provides two options.

> **M**: allows selected data set to be modified. Data in the `Strike` or `Azimuth` and `Dip` or `Plunge` columns are self-explanatory.

```
                    *** SPECIAL LINEAR DATA ***
To remove data after they are entered type R followed by their NUMBER
To return to the main menu type Q
                 NUMBER    AZIMUTH      PLUNGE      SYMBOL
                    4        45.0        30.0         X
                    5       _____
```

Fig. 6. Data entry screen displayed by Fabfile which guides input of orientation data for special directions defined by azimuth and plunge. Note symbol used to represent direction in orientation diagrams.

```
                    *** SPECIAL PLANAR DATA ***
To remove data after they are entered type R followed by their NUMBER
To return to the main menu type Q
                 NUMBER     STRIKE        DIP     DIRECTION
                    5         45.0        30.0        N
                    6       _____
```

Fig. 7. Data entry screen displayed by *Fabfile* which guides input of orientation data of special planes defined by strike, dip and direction of dip. These are represented by great circles on orientation diagrams.

```
                    *** SPECIAL PLANAR DATA ***
To remove data after they are entered type R followed by their NUMBER
To return to the main menu type Q
                    NUMBER    AZIMUTH         DIP
                       6        45.0          30.0
                       7       _____
```

Fig. 8. Data entry screen displayed by *Fabfile* which guides input of orientation data of special planes defined by azimuth and dip. These are represented by great circles on orientation diagrams.

In the `Direction of Dip` column an 'L' indicates that the data pertain to a lineation (see options 1 and 4 above), 'P' indicates a plane defined by its azimuth and dip (see 3 and 6 above) and 'N', 'S', 'E', and 'W' are the cardinal directions towards which a plane, defined by its strike and dip, is dipping (see 2 and 5 above). In the `Special Symbol` column, 'N' indicates that the data do not pertain to a special direction or plane (see 1, 2 and 3 above), 'G' indicates a special plane to be represented by a great circle (see 5 and 6 above), other characters are codes for symbols indicating special directions (see 4 above).

R: removes the selected data set. Upon completion of editing the program returns to the Main Menu.

The option is presented to prepare another data file or to return to *DOS*.

Fabric

The program *Fabric* uses the data in files prepared by *Fabfile* to prepare orientation diagrams. Orientation diagrams can be drawn to the computer screen in graphics mode. The program can autodetect and write to CGA, EGA, VGA, Hercules™, IBM 8514™, ATT 400™, or PC 3270™ graphic adapters. *Fabric* can also draw to a Hewlett Packard 7475™ compatible plotter, an Epson™ compatible dot matrix printer with either 60 or 72 dots per inch vertically and 60 dots per

```
                         *** EDIT THE FILE ***
                Strike or     Dip or   Direction     Special
     Number      Azimuth      Plunge    of Dip        Symbol
       1          45.0         30.0       L             N
       2          45.0         30.0       N             N
       3          ____         30.0       P             N
       4          45.0         30.0       L             X
       5          45.0         30.0       N             G
       6          45.0         30.0       P             G

       The NUMBER selected = 3      to Modify type M to remove type R
```

Fig. 9. *Fabfile* edit screen permits modification or removal of records in data file. *c.f.* Figs. 3 – 9. Data set 3 has been selected for modification.

```
             FABRIC version 3.0 - A program to prepare orientation diagrams.

  by John Starkey, Department of Earth Sciences, University of Western Ontario,
*********************** London, Ontario, Canada. N6A 3B7 ***********************

                    Enter the name of the data file SAMPLE.FAB

This file does not exist in this directory. To enter another file name type <E>
                                                 To list available files type <L>
                                                     To quit the program type <Q>
```

Fig. 10. *Fabric* menu from which input data file is initialized.

inch horizontally, or to a Hewlett Packard Laserjet III™ compatible printer. A file can also be written in *HP Graphics Language* (*HP-GL*) format which can be imported to other programs, such as *WordPerfect*™ or *CorelDraw*™, for more elaborate formatting suitable for publication. If a mathematic co-processor chip is available, this is detected and used, else a mathematic co-processor is emulated in software.

The number of orientation data which can be entered is limited only by the disc space available, since the data are stored in a temporary file and not in memory. The number of special directions is limited to 300.

The name of the input data file is entered. Only the root of the filename is required, the program automatically adds the extension `.fab`. If the file does not exist in the current directory a menu is displayed which presents three options (Fig. 10).

E: another filename may be entered
L: list the filenames in the current directory with the extension `.fab`. This aids in the selection of a file name.
Q: terminate the program.

Once a valid filename has been entered, a title for the orientation diagram may be specified. After the file name and title have been entered, the main menu is displayed from which the desired options are selected (Fig. 11).

P: select the hemisphere to be projected to prepare the orientation diagram. A sub-menu is displayed with the following options:

L: lower hemisphere. **U**: upper hemisphere. **B**: both hemispheres.

If both hemispheres are to be projected only a scatter diagram (*i.e.* a pole figure) can be drawn. In this case poles on the upper and lower hemisphere are represented by different symbols, by default a + and diamond respectively (see note 2 below). Special symbols are plotted on the appropriate hemisphere, and special planes are plotted as great circles on both hemispheres.

O: select the output device. After typing `O` the current assignments of output devices and ports are displayed. These are obtained either internally or from a file called `Fabric.inp` (see below under option `K` and note 2). A sub-menu is displayed offering the following options

> **S**: screen display.
>
> **P**: plotter output to a device which can read *HP-GL* files. The plotter can be either an HP7475 compatible pen plotter or an HP Laserjet III compatible printer. Selecting plotter output displays another sub menu with two options.
>
> > **I**: titles can be included or excluded by selecting 'I', which toggles between yes and no. Titles are written in *HP-GL* in an unattractive stick font (Fig. 12). Suppressing the titles allows them to be added manually.
> >
> > **D**: The diameter of the orientation diagram must be specified, in millimetres. The program accepts values between 20 and 200. However, there may be other mechanical constraints imposed by the output device.
>
> **R**: printer output, this refers to a dot matrix printer.
>
> **F**: write the output to an *HP-GL* file to which the extension `.hpg` will be added automatically. An attempt to save to an existing filename results in prompts being displayed as follows.
>
> > **E**: enter another file name.
> > **R**: replace the existing file.
> > **L**: list the existing files with the `.hpg` extension to assist in assigning a nonconflicting filename. Once an acceptable filename is entered, a sub-menu is displayed with two options.
> >
> > > **I**: titles can be included or excluded by selecting I, which acts as a toggle. This allows suppression of the titles, which can then be added by importing the file into a presentation graphics drawing program.
> > >
> > > **D**: the diameter of the orientation diagram is specified between 20 to 200 millimetres, inclusive.

F: select the type of figure to be drawn. A sub-menu is displayed with the following options.

> **S**: scatter diagram (pole figure). Input linear data are represented by poles. Input planar data are represented by the normals to the planes.
>
> **G**: great circle diagram. Input planar data are presented by great circles. Input linear data are represented by great circles

```
TITLE =                 Sample data - rotated

<P>rojected Hemisphere ..................................... Upper

<O>utput to ................................................ Plotter
          <I>nclude Titles ................................. Yes
          <D>iameter in millimetres (20.0 - 200.0) ........ 100

<F>igure ................................................... Contoured
          <G>rid size (even number, 20 - 100) ............. 40
          <C>ontours (up to 10)  1  2  3  5
          <A>verage the matrix using 8 elements ........... Yes
          <W>rite the contoured matrix to a file .......... No

<R>otate the data .......................................... Yes
          Azimuth = 20       Plunge = 30      Rotation = 40
          <S>ave the rotated data to a file ............... No

<K>eep these parameters as future defaults ................. No

<E>xecute the program of <Q>uit.
                                            ENTER SELECTION :
```

Fig. 11. The main menu displayed by *Fabric*. When certain options are selected additional sub-menus are displayed from which parameters are set. At any given time the main menu displays the options which are currently in effect.

corresponding to the planes normal to the lineations.

C: a contoured diagram. This represents a contoured scatter diagram. If this option is selected the following four additional options are offered on a sub-menu.

G : the grid size, from 20 to 100. This is the number of cells along one side of a square array of grid points on which the counting circle will be centred. The area of the counting circle is $100/n\%$ of the projected hemisphere. If the spacing between the grid points is significantly greater than the diameter of the counting circle adjacent counting circles may not overlap to cover the surface of the hemisphere. Consequently, there may be data that will not be counted. Geometrical considerations suggest that the following minimum values for the grid spacing are appropriate for the sample sizes indicated; sample size 1000 +, grid spacing 50; sample size 500 +, grid spacing 25; sample size 100 +, grid spacing 20.

A further consideration is that on some output devices the interaction between certain grid spacings and the resolution of the output-device can produce Moiré interference fringes in the output. Experimentation with the grid size, the diameter of the projection and the scale set on the output device may be required.

C: allows entry of values for up to ten contours. The area of the diagram below the lowest specified contour will be stippled, the area above the highest contour will be crosshatched. The ornamentation can be suppressed by specifying a minimum contour value of 0 (zero) or a

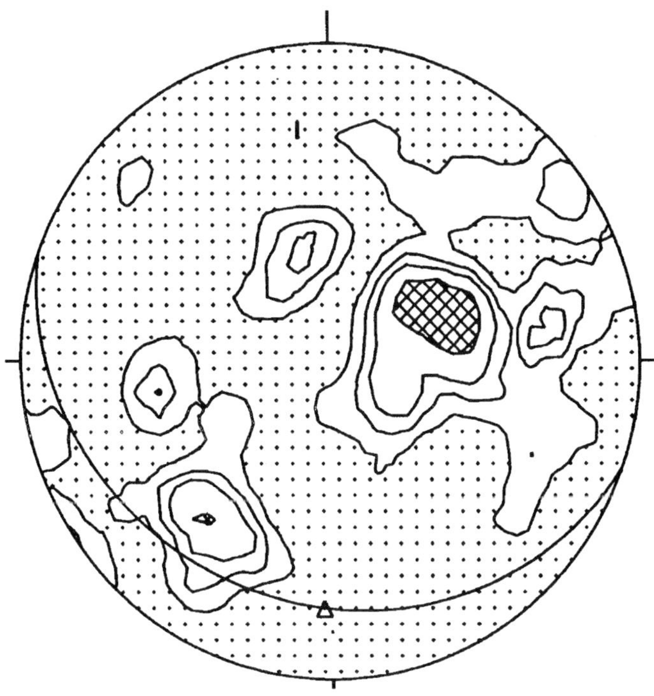

Fig. 12. Orientation diagram prepared by *Fabric* using the options and parameters displayed in Fig. 11.

maximum value higher than the maximum concentration present.

A: toggles the answer `Yes/No` to the question "Average the matrix using eight elements?". Averaging the matrix is accomplished by replacing each element in the matrix by the weighted mean of its eight nearest neighbours. The four nearest neighbours above, below, and to each side, are given unit weight. The next four nearest neighbours, above to right and left, and below to right and left, are weighted by $1/\sqrt{2}$. This averaging is for cosmetic purposes; it yields a more visually pleasing diagram.

W: toggles the answer `Yes/No` to the question "Write the contoured matrix to a file?". `Yes` causes the contoured matrix to be written to a text file which can be used by other programs. A filename is entered and the extension `.mtx` is added to it automatically.

R: toggles the answer `Yes/No` to the question "Rotate the data?", answer `Yes` requires entry of the azimuth and plunge of a rotation axis and the amount of rotation about the axis (a positive rotation is clockwise as viewed in the direction of the axis from the centre of the sphere). The option is provided of saving the rotated data as follows:

S: toggles the answer `Yes/No` to the question "Save the rotated data?". `Yes` requests the name for a file in which to save the rotated data, the extension `.fab` is added automatically. The saved file is a text file, written in the same format as the files produced by *Fabfile*.

By saving the rotated data, a series of rotations can be performed incrementally. For instance, a general rotation can be performed by rotating the data about sequential horizontal axes. This procedure is analogous to that commonly followed when manually rotating data on a spherical projection.

K: toggles the answer `Yes/No` to the question "Keep these parameters as future defaults?". If `Yes`, the parameters on the screen will be written to a file, `Fabric.inp`, and they will appear as the default parameters the next time the program is run (see note 2 below). If "No", the default parameters in place at the start of the current session will remain in effect.

E: executes the program and causes the diagram to be drawn. Alternatively, the program can be terminated without drawing the diagram or saving the selected options by typing `Q`.

When *Fabric* is executed the orientation diagram is output to the device selected (Fig. 12). If output is to the screen, the program halts when the diagram is displayed and a key must be pressed before the program continues. If a contoured diagram has been selected the magnitude of the maximum concentration and the areas of the projected hemisphere occupied by the different point concentrations are displayed on the screen (Fig. 13).

Data Files

The distributed file, `FabricXX.exe`, includes three sample data files, `Sample.fab`, `Planes.fab` and `Symbols.fab`. These illustrate some of the features of *Fabric*. Their contents can be viewed by using the `TYPE` command from *DOS*. `Sample.fab` consists of a header and 147 data sets. The first data set pertains to a reference plane which will be represented by a great circle on the orientation diagram. The second data set pertains to a reference direction which will be represented by a triangle. The remaining 145 data sets define lineations. Figure 12 illustrates a contoured orientation diagram prepared from this file.

`Planes.fab` is a small file intended primarily to illustrate a great circle diagram. It contains six data sets (Fig. 14). The first two pertain to linear data, the second two to planes, the fifth to a special lineation and the sixth to a special plane. The great circle diagram drawn from these data shows only 4 great circles, for data sets 1 to 4, and a square representing the special lineation (Fig. 15a). Data set 6 is not plotted, since the resulting great circle, representing a special plane, would be confused with the other four. A pole figure drawn from these data shows four poles, representing the normals to the planes defined by data sets 1 to 4, a square, representing data set 5, and a great circle representing data set 6 (Fig. 15b).

`Symbols.fab` is a file containing data to display the 8 available special symbols. They can be displayed by drawing a scatter diagram to the output device.

Notes

1) During execution, *Fabric* writes a temporary file for intermediate data, called `Yekrats.onj`, into the current directory. This is a binary file. This file is erased by the program if it terminates normally. If the program does not terminate normally this file might be left in the directory. However, subsequent execution of *Fabric* will remove it.

2) Keeping the parameters displayed on the opening menu for future default values causes them to be written to a file called `Fabric.inp`. This is a text file and therefore it can be edited using a text editor. In general this is inadvisable since it can be more easily and safely modified the next time *Fabric* is executed. However, this file allows for some customization of *Fabric* by editing lines 1, 2, and 5. These lines are listed below with their initial default values.

Line 1: `LPT1 60 72`. These are parameters for a dot matrix printer. They include the active printer port, LPT1, LPT2, *etc*., and the horizontal and vertical pitches of the attached printer, in dots per inch. The default settings are 60 and 72, respectively, which are appropriate for Epson-compatible 9-pin printers. With some printers, for instance Epson 24-pin printers, this may result in the projected circle becoming an ellipse, with a long axis oriented vertically on the page. This can be corrected by changing the pitches to 60 and 60.

Line 2: `COM2 HP7475`. These are parameters for the plotter or laser printer. The first parameter is the port to which the device is attached, either a serial port COM1, COM2, *etc*., or a parallel port, LPT1, LPT2, *etc*. The second parameter indicates the plotting device, either a HP-compatible pen plotter HP7475 or a HP LaserJet III- compatible printer LASER3.

Line 5: `40.0 60.0 + D`. These parameters pertain to the symbols used in the diagrams. The first value specifies the size of the Special Symbols, as a fraction of the radius of the diagram. The default value is 1/40. The second value similarly specifies the size of the symbols which will indicate data on the upper and lower hemispheres of the projection when a pole figure is to be drawn on both hemispheres. The default is 1/60. The symbols to be used in the latter case are selected by the last two parameters, which have the same significance as

```
Sample size = 145
The Maximum Concentration = 8 times uniform.

 0 times uniform occupies 64.60% of the area
 1 times uniform occupies 22.12% of the area
 2 times uniform occupies  6.36% of the area
 3 times uniform occupies  3.74% of the area
 4 times uniform occupies  1.51% of the area
 5 times uniform occupies  0.72% of the area
 6 times uniform occupies  0.56% of the area
 7 times uniform occupies  0.24% of the area
 8 times uniform occupies  0.16% of the area

    Upper Hemisphere Input Data
    Upper Hemisphere Plot

                             Press any key to continue
```

Fig. 13. Orientation statistics for pattern shown in Fig. 12.

```
Upper * Data to illustrate the drawing of great circles. (see documentation).

 135.0  20.0 L N
 280.0  60.0 L N
  90.0  45.0 S N
 300.0  80.0 S N
  80.0  45.0 L S
 170.0  70.0 W G
```

Fig. 14. Contents of data file `Planes.fab`.

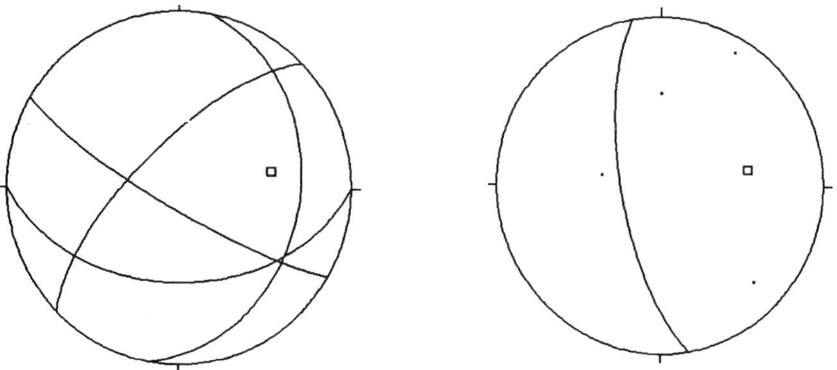

Fig. 15. a) Great circle diagram and b) scatter diagram from input file `Planes.fab` (Fig. 14).

explained in the instructions for *Fabfile*, option 4, above. The defaults are + and *D*, indicating a diamond, for the upper and lower hemisphere respectively.

If *Fabric* is executed and `Fabric.inp` does not exist in the active directory, then *Fabric* generates default parameters which are displayed on the main menu, where they can be changed. The parameters can then be written to the file `Fabric.inp` by electing to keep the parameters from the menu. It is then possible to use a text editor to modify lines of `Fabric.inp`, as discussed above, using the file as a model. If the file is corrupted it can be erased and *Fabric* will recreate it next time it is executed.

Conclusion

The interaction between the data input menus and editing in *Fabfile* is illustrated in Figs. 3 to 9. In Figs. 3 to 8 one data set for each of the types of input data is entered. The program keeps count of the number of data sets entered. On the editing screen, Fig. 9, the data sets are listed in the order in which they were entered and they are displayed in their entirety, including the parameters which are added automatically by the program. The data files produced by *Fabfile* are text files which are therefore readily accessible to other programs. In particular, since *Fabric* reads only the first 4 values in each record in the file, additional data, which will be ignored by *Fabric*, can be added to the records. Such additional data might include information about the individual data sets which could be used by other programs to sort or select subsets of data.

During the entry of input parameters to *Fabric* only data and options pertinent to the current selections are displayed. Thus, the main menu screen (Fig. 11), changes according to the kind of orientation diagram and output device selected.

An example of a contoured orientation diagram prepared by *Fabric* is illustrated in Fig. 12. The input file was `Sample.fab` and the input parameters are those shown in Fig. 11. Fig. 12 was prepared by sending the output of *Fabric* to a Hewlett Packard Laserjet III printer, thus the titles are written in the HP-GL stick font. As noted earlier, the titles can be suppressed in *Fabric* and added independently. *Fabric* generates statistical information about the contoured orientation diagram, the data corresponding to Fig. 12 are shown in Fig. 13. Such data are useful in the analysis of orientation patterns (Starkey 1993).

Fabfile and *Fabric* are general purpose programs, written to accept the kinds of orientation data typically measured by earth scientists, and to prepare the commonly used types of orientation patterns. Thus, the use of the programs should prove to be intuitive. Interaction with the programs has been kept as direct and simple as possible, and thus only limited control over the quality of the output is provided within *Fabric*. However, flexibility can be increased by writing the data to HP-GL data files which can be read by presentation graphics programs. [Control over the quality of the output has been further enhanced since the manuscript was submitted].

Fabfile and *Fabric* are available by anonymous ftp from ftp.csn.net, in the COGS directory. The file is fabricXX.exe, where XX is the version number. It is a self extracting ZIP file which includes documentation and data files.

References

Starkey, J. 1970. A Computer Programme to Prepare Orientation Diagrams. In: Paulitsch., P. (ed.). *Experimental and Natural Rock Deformation*. Springer, pp 51-74.

Starkey, J. 1977. The Contouring of Orientation Data Represented in Spherical Projection. *Can. J. Earth Sci.* **14**: 268-277.

Starkey, J. 1993. The analysis of three-dimensional orientation data. *Can. J. Earth Sci.* **30**: 1355-1362.

Microcomputers and the Optical Universal Stage

John Starkey
Department of Earth Sciences, University of Western Ontario,
London, Ontario N6A 3B7, Canada. jstarkey@julian.uwo.ca

Abstract– A petrographic microscope equipped with a universal stage permits determination of the orientations of optical directions and morphological features of minerals in thin section. An interactive computer program is described which converts universal stage measurements into the data required to represent orientations in spherical projection. It also allows the internal consistency of the measured data to be tested. Further, the program computes the orientations of directions which cannot be measured directly. The program is written in Turbo Pascal.

Introduction

The following discussion pertains to the use of a Leitz 4-axis universal stage to measure morphological features and optical directions in uniaxial and biaxial minerals. Only three of the rotation axes are required. Following the nomenclature of Emmons (1943; see p. 13, fig. 8) these are:

1. The inner vertical axis (I.V). This is vertical with respect to the plane of the thin section and is the axis about which the thin section is rotated. The rotation is measured on a scale that surrounds the specimen stage.

2. The north-south axis (N-S). This axis lies in the plane of the thin section and, when the microscope stage is in the zero position, it also lies in the plane of polarization of the polarizer, which is oriented north-south by convention. Rotations about this axis are measured using hinged scales (known as Wright arcs) which are attached to the support for the specimen stage.

3. The east-west axis (E-W). This is identified as the outer east-west axis (O.E-W) by Emmons, whose discussion relates to a 5-axis universal stage (for a discussion of the 4-axis stage see

Phillips 1971). With the microscope stage at zero, this axis is oriented east-west, in the plane of polarization of the analyzer. Rotations on this axis are measured on a graduated drum on the right side of the universal stage.

In addition, the microscope stage is used to rotate the universal stage about the microscope axis (M).

A 5-axis universal stage allows a mineral to be uniquely oriented optically, in a single combination of universal stage settings. However, in practice, the 4-axis stage is most commonly used and the complete optical orientation is achieved by plotting intermediate data on a spherical projection.

Ultimately, the universal stage measurements are converted into coordinates relative to the thin section for representation in spherical projection. The plane of the thin section provides a horizontal reference plane onto which the orientation data, represented on the surface of a reference sphere, are projected. Within the reference plane, the usual geographic coordinates are used – with north at the top. The orientation of the universal stage on the microscope, when the microscope stage is set at zero, is such that the 0° mark on the I.V scale is to the south, and the scale for the E-W axis is to the east (see Turner & Weiss 1963, p.198). Note, some of these conventions are peculiar to the Leitz universal stage; other brands may differ. Also, the algorithms described below pertain to the Leitz universal stage. The orientation of a measured direction is specified by an azimuth, measured clockwise from north in the plane of the thin section, and an angle of elevation above the plane of the thin section. Following crystallographic convention, which is necessary if directions are to be identified crystallographically, the data are represented on a spherical projection of the upper, reference hemisphere.

Morphological Features

Morphological features which can be measured using the universal stage include crystal faces, cleavage, and twin planes. Crystal faces and planes of cleavage are usually measured in plane polarized light, and twin planes in cross polarized light, but the procedure remains the same. The thin section is rotated about the I.V axis to bring the trace of the plane parallel to the east-west cross wire of the microscope. The measured I.V rotation corresponds to the azimuth of the plane. The thin section is rotated about the E-W axis to bring the plane vertical, in which orientation the plane is viewed on edge and appears sharp. The inclination of the plane is measured from the scale on the E-W axis; with the Leitz universal stage, a value greater than, but close to 0°, indicates a tilt to the north. A value less than, but close to, 360° indicates a tilt to the south. From the universal stage measurements, the computer program obtains the data required to plot the normal to the plane on a spherical projection, using the following algorithm:

```
{To convert U-stage measurements of a plane to the azimuth and elevation of the
                                normal to the plane}

begin
 if (E_W < 180) then                    {E_W is read from the scale on the E-W axis}
 begin
```

```
  Elevation := E_W;
  Azimuth := 360 - I_V              {I_V is read from the scale on the I.V axis}
 end
 else
 begin
  Elevation := 360 - E_W;
  Azimuth := 180 - I_V;
  if (Azimuth < 0) then Azimuth := Azimuth + 360;        {Keep az in range 0-360°}
end;
```

Optical Axis of Uniaxial Minerals

With optically uniaxial minerals, the only optic direction that can be located with the universal stage is the optic axis. This is identified between cross polars by observing extinction. As the thin section is rotated about the I.V axis, the mineral goes to extinction when the optic axis lies in either the north-south or east-west planes of polarization of the polarizer and analyzer respectively. If the optic axis lies in the north-south plane, rotation about the E-W axis will not cause it to depart from this plane and extinction will be maintained. The extinction position must be chosen for which the optic axis lies in the east-west plane. This is identified when the mineral departs from extinction on rotation about the E-W axis. An arbitrary tilt, approximately 30°, is applied to the E-W axis and, while the tilt is maintained, the mineral grain is returned to extinction by rotation on the N-S axis. The E-W axis is returned to zero. In this orientation, principal sections of the uniaxial indicatrix are oriented vertically north-south and east-west, parallel to the planes of polarization, and the optic axis is either parallel to the E-W axis or parallel to the microscope axis, M. The two orientations are differentiated by rotating the universal stage about the M axis. The grain remains at extinction if the optic axis is parallel to M and departs from extinction if it is parallel to the E-W axis. The inclination of the optic axis corresponds to the tilt on the N-S axis, which is measured using the right or left Wright arcs on the universal stage.

The measurements of the rotation on the I.V axis, and the tilt on N-S, define the orientation of the optic axis with respect to the universal stage. These data are converted to an azimuth and elevation relative to the thin section. The appropriate algorithm depends on whether the optic axis is vertical (parallel to the microscope axis M) or horizontal (parallel to the E-W axis) and whether the tilt on the N-S axis is measured on the right or left Wright arc:

```
        {To convert U-stage data for measured dir. to azimuth & elevation}
begin
 if Horizontal then
 begin                                         {Direction is parallel E- W axis}
  Elevation := N_S;
  if Right_Arc then
  Azimuth := 270 - I_V                         {N-S rot. measured on right arc}
 else
  Azimuth := 90 - I_V;                          {N-S rot. measured on left arc}
 end
 else
 begin
  Elevation := 90 - N_S;                          {Direction is parallel M axis}
  if Right_Arc then
   Azimuth := 90 - I_V                         {N-S rot. measured on right arc}
  else
   Azimuth := 270 - I_V;                        {N-S rot. measured on left arc}
  end;
end;
```

Optic Directions of Biaxial Minerals

With biaxial minerals, it is possible to measure two of the three principal optic directions, *X*, *Y*, and *Z*, in a single grain. The grain is rotated to extinction on the I.V axis. Two light rays pass through the mineral grain vibrating parallel to the planes of polarization of the polarizer and analyzer. An arbitrary tilt on the E-W axis causes the grain to depart from extinction. Extinction is restored by tilting on the N-S axis. The tilt on the E-W axis is reversed through zero and, if necessary, extinction is restored by rotation on the I.V axis, without disturbing the N-S and E-W axes. Repeated back and forth tilts on the E-W axis, with adjustments on the N-S and I.V axes to restore extinction, may be required before extinction is maintained for all positions of the E-W axis. These adjustments bring one of the principal sections of the biaxial indicatrix parallel to the plane of vibration of the polarizer. Therefore one of the optical directions is parallel to the E-W axis. To identify this optic direction the u-stage is rotated 45° about the microscope axis, M, in response to which the mineral grain departs from extinction. The section is rotated back and forth approximately 45° about the E-W axis. If the grain goes through extinction it is because one of the optic axes is brought into coincidence with the microscope axis and the mineral is rotating about the optic normal, *Y*. If no extinction is observed the grain is rotating about either *X* or *Z* and the direction is identified using an accessory optical plate. Alternatively, in the case of pleochroic minerals, the pleochroic scheme may be used to identify the optic direction which is parallel to the E-W axis.

The angle of rotation about the I.V axis, and the tilt on the N-S axis, define the orientation of the measured optical direction. The azimuth and elevation are calculated in a manner analogous to that described for uniaxial minerals where the optic axis is brought into coincidence with the E-W axis. The second optic direction is measured by rotating the grain 90° about I.V to the second extinction position and repeating the procedure.

Confirmation of Measured Data and Location of Non-measured Directions

The computer program confirms the internal consistency of data as they are measured and calculates the orientations of non-measured directions. This is illustrated with reference to measurements of hornblende. The data required to completely describe the crystallographic and optical orientations of hornblende are illustrated in Fig. 1, in which *a*, *b*, and *c* are the crystallographic axes and *X*, *Y*, and *Z* are the axes of the optical indicatrix. Hornblende is monoclinic, with two planes of cleavage parallel to the crystal form {110} that intersect parallel to the crystallographic *c*-axis. The angle between the normals to the two cleavages is 124°, and is bisected by the *b*-axis. *Y* is parallel to *b*. In hornblende, *Z* lies within 30° of *c*, between +*a* and +*c*. Universal stage measurements reveal this angle to be 16° in this particular hornblende. The crystallographic angle, *β*, between the +*a* and +*c* axes is 105° and therefore is approximately 1° from *a*. These crystallographic data are incorporated into the computer program to manipulate the universal stage measurements obtained for hornblende. An example of the screen display of the computer program is illustrated in Fig. 2. The orientation data for two measured optic directions are entered as measured on the universal stage.

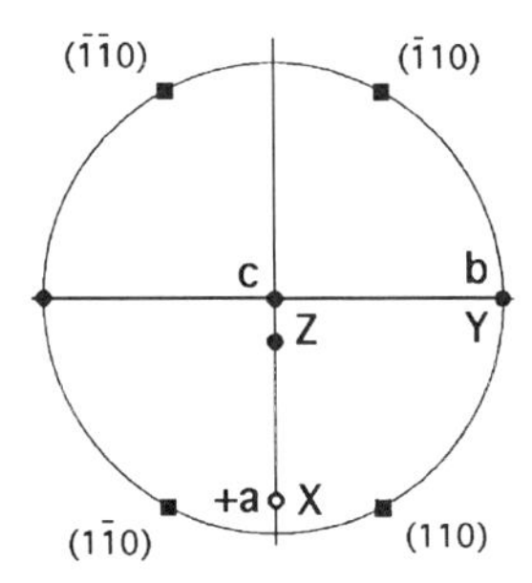

Fig. 1. The crystallographic and optical orientations of hornblende, represented in Lambert Equal Area spherical projection. Solid circles represent the orientations of crystallographic axes and the axes of the optical indicatrix on the upper hemisphere. Open circle represents *+a* and *X* axes on lower hemisphere. Squares represent normals to {110} cleavage.

The program translates these data into an azimuth and elevation relative to the thin section. The two measured directions define a plane, and the normal to the plane is the third optic direction for which the azimuth and elevation are calculated by spherical trigonometry, as illustrated in Fig. 3. If the absolute difference in the azimuths of the two measured directions is less than 180° then the geometry is that shown in Fig. 3a and the direction with the smallest azimuth is arbitrarily identified as *J*. If the difference in the azimuths is greater than 180° the direction with the smallest azimuth is chosen as *K* (Fig. 3b). The directions *J* and *K*, and the centre of the projected hemisphere, *M*, define a spherical triangle. The sides *MJ* and *MK* are the distances of the two directions from the vertical, obtained by subtracting the elevations of *J* and *K* from 90°. The normal to the great circle containing *J* and *K*, corresponding to the third, unmeasured, optic direction is represented by *L*. The azimuth of *L* is the azimuth of *K* plus the angle *KML*. The elevation of *L* is 90°-*LM*:

```
            {Determine orientation of plane normal given 2 directions}
begin
 Sign := +1;                          {Initialize Sign for L on upper hemisphere}
 Angle_JMK := abs(Azimuth_J-Azimuth_K);                    {Abs. diff. in azs.}
 if (Angle_JMK = 0) or (Angle_JMK = Pi) then                       {Pi = 180°}
 begin
  Side_LM := 0                                  {Special case, JK is vertical}
  Azimuth_L := Azimuth_K + Ninety;                 {Ninety is 90° in radians}
 end
 else
 begin
  if (Angle_JMK < Pi) then
  begin                                                         {See Fig. 3a}
   if Azimuth_K < Azimuth_J then
   begin                        {Select direction with smallest azimuth as J}
    Interchange(Side_JM,Side_KM);                          {interchange data}
    Interchange(Azimuth_J,Azimuth_K);
   end;
  end
  else
  begin
  Angle_JKM := Two_Pi - Angle_JMK;                  {Two_Pi radians = 360°}
  if Azimuth_K > Azimuth_J then
  begin                         {Select direction with smallest azimuth as K}
   Interchange(Side_JM, Side_KM);
   Interchange (Azimuth_J,Azimuth_K);
  end;
 end;
```

a

```
#1     Measured           Calculated                 X to Y =   90.0(0.0)        [90]
   Angle Tilt Arc    Azimuth Elevn.             => X to Z =   80.1(9.9) ***  [90]

X = 268     21  R        2.0    21.0
                                                   Y to Z =   90.0(0.0)        [90]
Y =                    171.7    68.7

Z = 170      7  L      280.0     7.0

1 =



2 =


                                                   b to X =   90.0(0.0)        [90]
                                                   b to Y =    0.0(0.0)        [0]
                                                   b to Z =   90.0(0.0)        [90]

                  b = 171.7    68.7



SELECTION >    (Q to Quit  :  S to Save Data)
```

b

```
#1     Measured           Calculated                X to Y =  90.0(0.0)        [90]
   Angle Tilt Arc    Azimuth Elevn.            => X to Z =  80.1(9.9)  ***  [90]
                                               => X to 1 =  31.3             [30 to 40]
X = 268     21  R        2.0    21.0              X to 2 =  30.1             [30 to 40]
                                                  Y to Z =  90.0(0.0)        [90]
Y =                    171.7    68.7              Y to 1 =  60.7(1.3)        [62]
                                                  Y to 2 =  62.0(0.0)        [62]
Z = 170      7  L      280.0     7.0           => Z to 1 =  71.4             [64 to 80]
                                                  Z to 2 =  71.1             [64 to 80]
1 = 190      8                                    1 to 2 =  57.3(1.3)        [56]
    190      7                                    a to X =   3.5             [0 to 15]
    190    311                                    a to Y =  90.0(0.0)        [90]
                                                  a to Z =  83.5             [75 to 90]
2 =                    349.1    49.3              a to 1 =  32.6(0.6)        [32]
                                                  a to 2 =  31.5(0.5)        [32]
                                                  b to X =  90.0(0.0)        [90]
                                                  b to Y =   0.0(0.0)        [0]
               +a = 185.7   -20.7                 b to Z =  90.0(0.0)        [90]
                                                  b to 1 =  60.7(1.3)        [62]
                b = 171.7    68.7                 b to 2 =  62.0(0.0)        [62]
                                                  c to X =  78.5        ***  [60 to 78]
               +c =  79.9     0.7                 c to Y =  90.0(0.0)        [90]
                                                  c to Z =  21.5             [12 to 30]
                                                  c to 1 =  90.0(0.0)        [90]
SELECTION >    (Q to Quit  :  S to Save Data)     c to 2 =  90.0(0.0)        [90]
```

Fig. 2. Computer screen display a) during and b) after data entry.

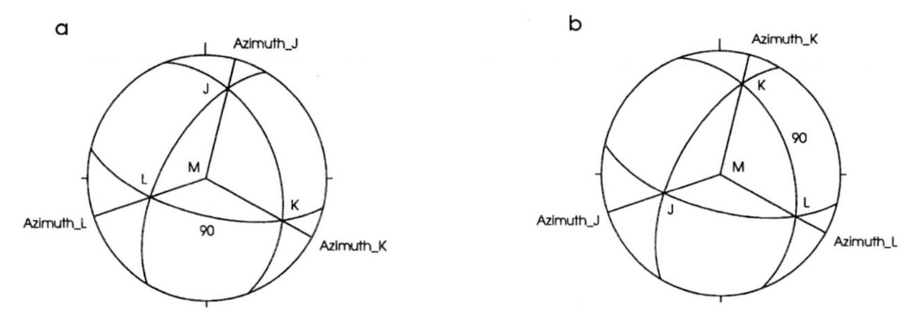

Fig. 3. The spherical triangles to be solved to compute the orientation of the direction, L, normal to the plane defined by two measured directions, J and K. In a) the azimuths of the two measured directions differ by less than 180°, in b) the difference is greater than 180°.

```
Angle_JKM := AngleB(Side_KM,Side_JM,Angle_JMK);{Trig.Func.}
Angle LKM := Ninety - Angle_JKM;
if (Angle_LKM = 0 then
begin                                    {Special Case, plane KL is vertical}
 Side_LM := Ninety - Side_KM;
 Azimuth_L := Azimuth_K + Pi;
end
else
begin
 if (Angle_LKM < 0) then
 begin                                   {L is on lower hemisphere, change Sign}
  Sign : = -Sign;
  Angle_LKM:= - Angle_LKM;
 end;
 Side_LM := SideC(Side_KM,Ninety,Angle_LKM);                      {Trig. Func.}
                  Angle_KML := AngleB(Side_KM,Ninety,Angle_LKM!; {Trig. Func.}
 Azimuth_L := Azimuth_K + (Sign * Angle_KML);      {modify azimuth for hemisphere
                                                              where L occurs}
end;
if (Azimuth_L > Two_Pi) then Azimuth_L := Azimuth_L - Two_Pi;
if (Azimuth L < 0) then Azimuth_L := Azimuth_L + Two_Pi;
if (Side_LM > Ninety) then
begin                                        {Restore L to upper hemisphere}
 Side_LM := Pi - Side_LM;
 Azimuth_L := Azimuth_L + Pi;
  if (Azimuth_L > Two_Pi) then Azimuth_L := Azimuth_L - Two_Pi
 end;
 end;
end;
```

The trigonometric functions *AngleB* and *SideC* are implementations of standard trigonometric solutions of spherical triangles. They have been incorporated into a *Turbo Pascal* unit which is presented as an Appendix.

Additional data appear on the computer screen at this time, see Fig.2a. They include the azimuth and elevation of the *b*-crystallographic axis, which is identical to *Y*, and the angles between all combinations of *X, Y, Z,* and *b*. The angles between measured directions are indicated by =>. The ideal angles between the indicated pairs of directions are listed in the column on the right of the screen. The differences between the calculated and ideal values are indicated in parentheses after the calculated values. Differences greater than 5° are marked by ***.

The angular separation between two directions is obtained from a spherical triangle defined by the two directions, *J* and *K*, and the centre of the projection, *M*, see Fig. 4. The sides *JM* and *KM* are obtained from the elevations of the two directions, and the angle *JMK* is the absolute difference in their azimuths. The angular separation is obtained by solving the spherical triangle for side *JK*.

```
        {To calculate the angular separation between two directions}
begin
 if (Angle_JMK = 0) then
  Separation := abs(Side_JM - Side_KM)           {J and K have same azimuth}
 else
 begin
  if (Angle_JMK = Pi) then
   Separation := Side_JM + Side_KM                  {J and K are 180° apart}
  else
  begin
   if (Angle_JMK < Pi) then
     Separation := SideC (Side_JM, Side_KM,Angle_JMK)  {Angle JMK acute; Fig. 4}
   else
      Separation := SideC (Side_JM,Side_KM,(Two_Pi - Angle_JMK));{Angle JMK is
                                                                   obtuse}
  end;
 end;
end;
```

The measured orientation data for cleavages are entered by selecting 1, for the first cleavage, or 2, for the second cleavage (Fig. 2). The rotations on the I.V and E-W axes required to bring the cleavage east-west and vertical are entered. The azimuth and elevation of the normal to the cleavage are obtained as indicated above under the discussion of morphological features. Orienting the cleavage trace parallel to the east-west crosswire of the microscope is easy but recognizing when the cleavage is vertical is often difficult, especially in strongly coloured minerals like hornblende. Thus the measured rotation on I.V is likely to be more than the measured tilt on the E-W axis. An indication of the precision of the measurements on the E-W axis is provided by the program. The program calculates two planes which have the measured I.V rotation and for which the normals are 62° from *Y*. One of these two calculated orientations should coincide with the measured cleavage orientation, thus confirming the internal consistency of the data.

The I.V measurement of the observed cleavage defines a vertical plane in which the normal to the cleavage, *S*, must lie (Fig. 5). The optic *Y* direction may also lie in this plane (Fig. 5a) but generally the *SY* plane is inclined (Fig. 5b). Where the *SY* plane is vertical (Fig. 5a) the azimuths and elevations of the normals to the two possible cleavages, *S* and *T*, 62° away from *Y*, are readily computed from the known location of *Y*. Where the *SY* plane is inclined (Fig. 5b), computation of the locations of the two possible cleavage normals is less direct. The plane through *Y*, which is perpendicular to the vertical plane defined by *S*, intersects the latter plane at *P*. This defines a right spherical triangle *MPY*. Solution of this triangle, and the interfacial angle 62°, allows calculation of the azimuths and elevations of the two consistent cleavage normals, *S* and *T*. From the azimuths and elevations of *S* and *T*, the program calculates the rotations about the E-W axis required to bring the normals to each cleavage to the horizontal. Subtraction of the calculated rotation for *T* from 360° achieves conformity with the graduations on the Leitz universal stage. The calculated universal stage settings for the two possible cleavages are listed on the computer screen beneath the measured data (Fig. 2b). The similarity between the measured tilt and one of the calculated tilts indicates the likely precision of the measured data.

```
        {Check orientation of two cleavages consistent with measured I.V}
begin
 EW_Tilt_S := -1;                              {Set EW_Tilt_S negative as a flag}
 if (Side_MY = 0) then
 begin
  EW_Tilt_S := Ninety - Sixty_Two;                  {Sixty_Two = 62° in radians}
  EW_Tilt_T := Two_Pi - EW_Tilt_S;
 end
 else
 begin
  Angle_SMY := Azimuth_Y + I_V;                 {Measured cleav.,S, on N-radius}
  if (Angle_SMY > Two_Pi) then Angle_SMY := Angle_SMY - Two_Pi;
  if (Angle_SMY = 0) then
  begin                          {Special case, Y on north radius - Fig. 5a}
   Side_MS := Side_MY + Sixty_Two;                      {Calc. 1st. cleavage, S}
   if (Side_MS < Ninety) then
    EW_Tilt_S := Ninety - Side_MS
   else
    EW_Tilt_S := Two_Pi + (Side_MS - Ninety);
   Side_MT : = Sixty_Two - Side_MY;                     {Calc. 2nd. cleavage, T}
   if (Side_MT < 0) then
    EW_TilT := Ninety + Side_MT
   else
    EW_Tilt_T := Two_Pi - (Ninety - Side_MT);
```

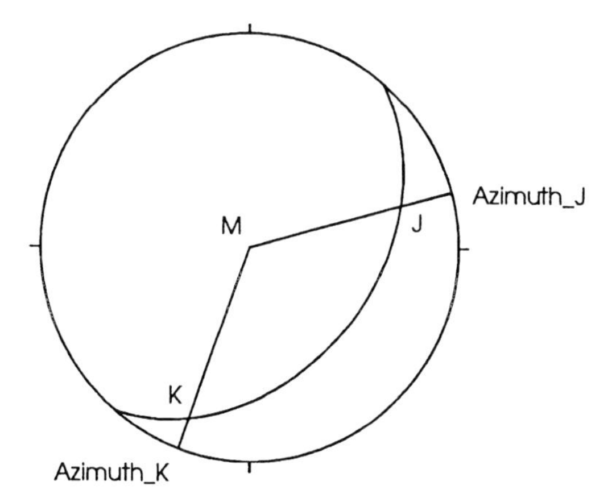

Fig. 4. The spherical triangle to be solved to determine the angular separation of two measured directions, J and K.

```
end
else
begin
 if (Angle_SMY = Pi) then
 begin                          {Special case, Y on south radius - Fig. 5a}
  Side_MS := Sixty_Two - Side_MY;                  {Calc. 1st. cleavage, S}
  if (Side_MS < 0) then
   EW_Tilt_S := Two_Pi - (Ninety + Side_MS)
  else
   EW_Tilt_S := Ninety - Side_MS
  Side_MT := Side_MY + Sixty_Two;                  {Calc. 2nd. cleavage, T}
  if (Side_MT > Ninety) then
   EW_Tilt_T := Side_MT - Ninety
  else
   EW_Tilt_T := Two_Pi - (Ninety - Side_MT);
 end
 else
 begin                                    {General solution - Fig. 5b}
  if (Angle_SMY > Pi) then Angle_SMY := Two_Pi - Angle_SMY;
  if(Angle_SMY = Ninety) and (Side_MY <= Sixty_Two) then
  begin                               {Special case, Y on east radius}
   Side_MS := SideB(Sixty_Two,Side_MY,Angle_SMY);
   if (Side_MS > Ninety) then
   begin
    EW_Tilt_S := Two_Pi - (Side_MS- Ninety);
    EW_Tilt_T := Side_MS - Ninety;
   end
   else
   begin
    EW_Tilt_S := Ninety - Side_MS;
    EW_Tilt_T := Two_Pi -(Ninety - Side_MS);
   end;
  end
  else
  begin
   if(Angle_SMY < Ninety) then
   begin              {General case, Y in north-east quadrant - Fig. 5b}
    Side_PY := arcSin(sin(Side_MY) *sin(Angle_SMY));   {Solve right spherical
                                                              triangle, MPY}
    if (Side_PY < = Sixty_Two) then
    begin
     Side_MS := SideB(Sixty_Two,Side_MY,Angle_SMY);
     if (Side_MS > Ninety) then
       EW_Tilt_S := Two_Pi - (Side_MS-Ninety)
     else
       EW_Tilt_S := Ninety- Side_MS;
     Side_MP := arcTan(tan(Side_MY) * cos(Angle_SMY));  {Solve right spherical
```

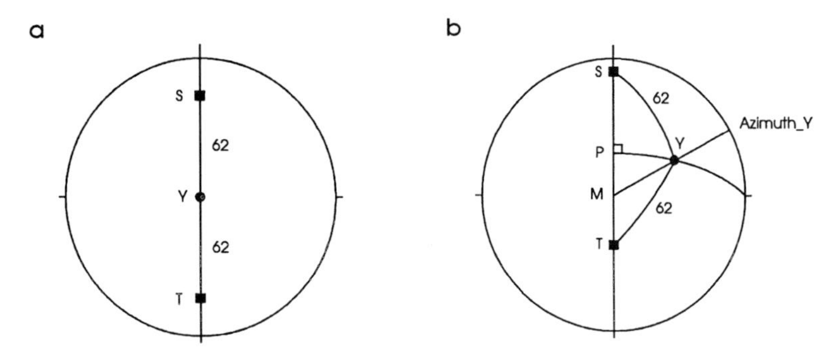

Fig. 5. Confirmation of universal stage measurements of cleavage in hornblende. a) Special case where optic direction, *Y*, lies in same vertical plane as the normal to the measured cleavage, *S*. b) General case where plane containing the normal to the measured cleavage, *S*, and the *Y* direction is inclined. Tilts measured on the universal stage are rotations required to bring cleavage normals to horizontal.

```
                                                              triangle, MPY}
      Side_MT := Side_MP - (Side_MS - Side_MP);
      if (Side_MT < 0) then
        EW_Tilt_T := Two_Pi - (Ninety + Side_MT)
      else
        EW_Tilt_T := Ninety - Side_MT;
    end;
  end
  else
  begin                    {General case, Y in south-east quadrant - Fig. 5b}
    Side_PY := arcSin(sin(Side_MY) * sin(Pi - Angle_SMY));        {Solve right
                                                   spherical triangle, MPY}
    if (Side_PY <= Sixty_Two) then
    begin
      Side_MT := SideB(Sixty_Two,Side_MY,(Pi - Angle_SMY));
      if (Side_MT > Ninety) then
        EW_Tilt_T := Side_MT- Ninety
      else
        EW_Tilt_T := Two_Pi - (Ninety - Side_MT);
      Side_MP := arcTan(tan(Side_My) * cos(Pi - Angle_SMY));      {Solve right
                                                    spherical triangle MPY}
      Side_MS := Side_MP - (Side_MT - Side_MP);
      if (Side_MS < 0) then
        EW_Tilt_S := Ninety + Side_MS
      else
        EW_Tilt_S := Two_Pi - (Ninety-Side_MS);
    end;
  end;
  if (EW_Tilt_S >= 0) then                            {Test EW_Tilt_S flag}
    Write solutions to screen            {EW_Tilt_S positive, solution found}
  else
    Write warning message to screen        {EW-Tilt_S negative, no solution}
  end;
 end;
end;
end;
end;
```

The program calculates the orientation of the normal to the second cleavage from the orientation of the first measured cleavage and *Y*. The geometry differs depending on whether the angle between *Y* and the normal to the measured cleavage is approximately 62° or the complimentary angle 118° (see Figs. 6a and 6b, respectively). In the spherical triangle defined by the *Y* axis and the first cleavage, *S*, two sides and the included angle are known. The sides *MY* and *MS* are obtained by subtracting the elevations of *Y* and the normal to the cleavage, *S*, from 90°. The angle *SMY* is the difference in the azimuths of *Y* and *S*. Thus angle *MYS* can be calculated. In the spherical triangle defined by *Y* and the normal to the second cleavage, *T*, the sides *MY* and *TY* are known, and the included angle, *MYT*, is the compliment of angle *MYS*. Thus, side *MT* and angle *TMY* are obtained. The elevation of the second cleavage is 90° - *MT* and its azimuth is the azimuth of *Y* plus or minus angle *TMY*, depending on whether *SY* is close to 62° or close to 118° (see Figs. 6a and 6b).

```
                              {Locate the second cleavage}
begin
 CalculateSeparation(Side_MS,Side_MY,abs(Azimuth_S - Azimuth_Y),Side_SY); {Calcu-
                              late separation between two directions - Fig. 4}
 Angle_SMY := Azimuth_S - Azimuth_Y;
 if (Angle_SMY < 0) then Angle_SMY := Angle_SMY + Two_Pi;
 if (Side_SY < Ninety) then Sign := -1 else Sign := +1;   {Sign initialized to -1
                                                   for Fig. 6a; +1 for Fig. 6b}
 if (Angle_SMY > Pi) then
 begin
  Sign : = -Sign;
  Angle_SMY := Two_Pi - Angle_SMY;
 end;
 Angle_MYS := AngleB(Side_MY, Side_MS,Angle_SMY);
 if (Side_SY < Ninety) then
 Angle_MYT := Pi - Angle_MYS;                                          {Fig. 6a}
 Side_MT : = SideC (Sixty_Two, Side_MY, Angle_MYT);
 Angle_TMY := AngleB (Side_MY,Sixty_Two,Angle_MYT);
 Azimuth_T := Azimuth_Y + (Sign * Angle_TMY);      {Sign adjusts for Fig. 6a or b}
 Elevation_T := (Ninety - Side_MT)* Radians;
 if (Elevation_T < 0) then
 begin                         {Restore cleavage normal to upper hemisphere}
  Elevation_T := abs(Elevation_T);
  Azimuth_T := Azimuth_T + Pi;
  if (Azimuth_T >= Two_Pi) then Azimuth_T := Azimuth_T - Two_Pi;
 end;
end;
```

If the orientation of the second cleavage is subsequently measured, the program performs the same calculations as for the first cleavage (see Figs. 5 and 6), and the azimuth and elevation obtained are substituted for those calculated from the first cleavage. The attitude of the crystallographic *c*-axis is calculated from the normal to the first, measured cleavage and *Y*. These directions define a plane and the normal to this plane is the *c*-crystallographic axis. The computation is the same as that described in connection with Fig. 3.

Computation of the location of the +*a*-crystallographic axis relies on the facts that *a* lies in the plane containing the *c*-crystallographic axis and *Z*, that the crystallographic angle β is 105°, and, in addition, that *Z* lies between +*a* and +*c* and less than 30° from +*c*. The special cases, where the *c*-axis occurs at the centre of the projection, or where the *c*-axis and *Z* lie in the same vertical plane, are readily computed since the azimuth of +*a* is the same as that of *Z* in the direction in which the angle between *Z* and *c* is acute. In this case the elevation of +*a* is obtained from β. The more general case is illustrated in Figs. 7a and 7b. In Fig. 7b, the +*a* axis is on the lower hemisphere, which is indicated by a negative elevation.

```
                              {Compute the orientation of +a}

begin                                                   {C lies at the centre}
 if (Side_CM = 0) then
 begin
  Azimuth_A := Azimuth_Z;
  Side_AM := Beta;                {Beta is the crystallographic angle 105°}
```

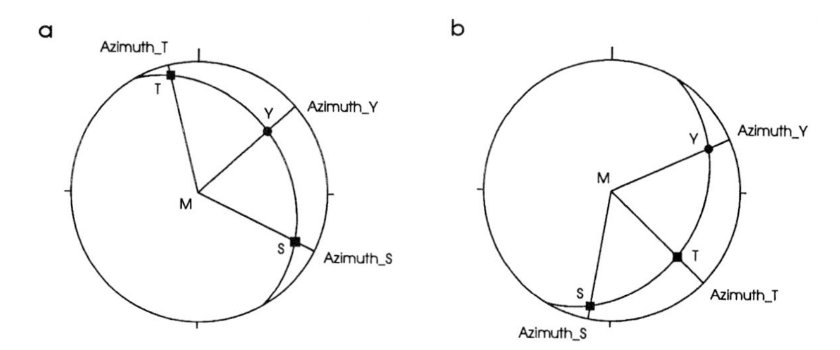

Fig. 6. Spherical triangles to compute the orientation of the second cleavage, *T*, from the measured cleavage, *S*, and the optic direction *Y*. Angle between measured cleavage and *Y* is ≈ 62° in a) and ≈ 118° in b).

```
end
else
begin
 Angle_CMZ := Azimuth_Z - Azimuth_C;
 if (Angle_CMZ < 0) then Angle_CMZ := Angle_CMZ + Two_Pi;
 if (Angle_CMZ = 0) then
 begin                                      {C and Z lie on north radius}
  if (Side_MZ < Side_CM) then
  begin                            {Z lies closest to centre of projection}
   Azimuth_A := Azimuth_C + Pi;
   Side_AM := Beta - Side_CM;
  end
  else
  begin                            {C lies closest to centre of projection}
   Azimuth_A := Azimuth_C;
   Side_AM := Side_CM + Beta;
  end;
 end
 else
 begin
  if (Angle_CMZ = Pi) then
  begin                          {C and Z lie on opposite sides of centre}
   C_to_Z := Side_CM + Side_MZ;
   if (C_to_Z < Ninety) then
   begin
    Azimuth_A := Azimuth_C + Pi;
    Side_AM:= Beta - Side_CM;
   end
   else
   begin
    Azimuth_A := Azimuth_C;
    Side_AM:= Side_CM + Beta;
   end;
  end
  else
  begin
   if (Angle_CMZ < Pi) then
   begin                        {Z in NE quadrant relative to C - Fig. 7a}
    C_to_Z := SideC(Side_CM,Side_MZ,Angle_CMZ)
                                                        {determine C_to_Z}
    Angle_MCZ := AngleA (Side_MZ,Side_CM,C_to_Z);
    if (C_to_Z < Ninety) then
    begin
     Azimuth_A := Azimuth_C + AngleB(Side_CM,Beta,Angle_MCZ);
     Side_AM := SideC(Side_CM,Beta,Angle_MCZ);
    end
    else
    begin
     Azimuth_A := Azimuth_C - AngleB(Side _CM,Beta,(Pi - Angle_MCZ));
     Side_AM := SideC(Side_CM,Beta,(Pi - Angle_MCZ));
    end;
   end
   else
```

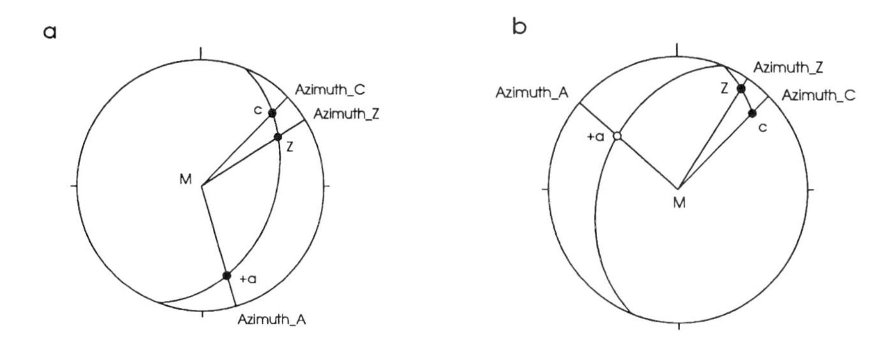

Fig. 7. Computation of the orientation of the $+a$ crystallographic axis from the orientations of the c crystallographic axis and the optic direction Z. The directions $+a$, c, and Z lie on a great circle with Z less than 30° from c and $+a$ 105° from c measured in the same direction. In a) the a axis is on the upper hemisphere, in b) it is on the lower hemisphere.

```
        begin                         {Z in NW quadrant relative to C - Fig. 7b}
          C_to_Z := SideC(Side_CM,Side_MZ,(Two_Pi - Angle_CMZ));
          Angle_MCZ := AngleA(Side_MZ,Side_CM,C_to_Z);
          if (C_to_Z < Ninety) then
          begin
            Azimuth_A := Azimuth_C - AngleB(Side_CM,Beta,Angle_MCZ);
            Side_AM := SideC(Side_CM,Beta,Angle_MCZ);
          end
          else
          begin
            Azimuth_A := Azimuth_C + AngleB(Side _CM,Beta,(Pi - Angle_MCZ));
            Side_AM := SideC(Side_CM,Beta,(Pi - Angle_MCZ));
          end;
        end;
      end;
    end;
  end;
  Elevation_A := Ninety - Side_AM;
end;
```

The program calculates the remaining angles between pairs of directions listed in the columns to the right of Fig. 2b, as described in connection with Fig. 4. If unacceptably large errors appear, data can be remeasured and reentered at any time. Finally, the orientations of X, Y, Z, $+a$, b, c, and the normals to the two cleavages are saved to a data file from which they can be read by other programs, which plot the data or perform statistical analyses.

Conclusion

The interactive computations outlined above can be applied to measurements of any mineral using the universal stage. The algorithms are general and provide models which are applicable to any universal stage measurements. The computer program provides continuous feedback as data are measured, and thus allows the internal consistency of the measured data to be checked. By computing the orientations of other, non-measured crystallographic directions, and checking these against the known crystallography, the program can warn of gross errors. It is therefore possible to continually refine the measurements until the desired precision is obtained at which time the data are saved to a data file.

References

Emmons, R. C. 1943. The Universal Stage. *Geol. Soc. Amer. Mem.* **8**, 205pp.

Phillips, W. R. 1971. *Mineral Optics.* W. H. Freeman and Co., 249pp.

Turner, F. J. & Weiss, L. E. 1963. *Structural Analysis of Metamorphic Tectonites.* McGraw-Hill Inc., 545pp.

Appendix

```
Unit Trig;
{The nomenclature conforms to that normally used in the solution of spherical
triangles. The three angles are identified as AngleA, AngleB, and AngleC; the
sides of the triangle opposite these angles are SideA, SideB, and SideC,
respectively.}

interface

const
 Ninety = Pi/2;

 Function Tan(x : real): real;
 Function Cot(x : real): real;
 Function  ArcCos(x: real): real;
 Function ArcSin(x: real): real;
 Function AngleA (SideA,SideB,SideC: real): real;
 Function SideC (SideA,SideB,AngleC: real): real;
 Function AngleB (SideA,SideB,AngleC: real): real;
 Function SideB (SideA,SideC,AngleA: real): real;

implementation

                         {~~~~~~~~~~~~~~~~~~~~~~~~~~~~~~~~}

Function Tan;                                     {Returns the tangent of angle x}
var
 CosX: real;

begin
 CosX := cos(x);
 if CosX = 0 then CosX := 0.00001;                          {avoids division by 0}
 Tan := sin(x)/CosX;
end; {Function Tan}

                         {~~~~~~~~~~~~~~~~~~~~~~~~~~~~~~~~}

Function Cot;                                          {Returns the cotangent of x}
var
 TanX: real;

begin
 TanX := tan(x);
 if TanX = 0 then TanX := 0.00001;                          {Avoids division by 0}
 Cot := 1/TanX;
end; {Function Cot}

                         {~~~~~~~~~~~~~~~~~~~~~~~~~~~~~~~~}

Function ArcCos; {Returns the arcCosine of angle x. Values range from +1 through
```

```
0 to -1. Angles 0 through 90 to 180 returned in radians.}

begin
 if x = 0 then
 begin
  ArcCos := Pi/2;
 end
 else
 begin
  if x < 0 then
  begin
   if x > -1 then ArcCos := Pi + arctan(sqrt(1 - sqr(x))/x)
  else
   ArcCos := Pi;
  end
  else
  begin
   if x < 1 then ArcCos := arctan(sqrt(1 - sqr(x))/x) else ArcCos := 0;
  end;
 end;
end; {Function ArcCos}

                         {~~~~~~~~~~~~~~~~~~~~~~~~~~~~~~~}

Function ArcSin; {Returns the arcSine of angle x. Values range from -1 though O
to + 1. Angles -90 through O to + 90 returned in radians}

begin
 if x <= -1 then
  ArcSin := -(Pi/2)
 else
  if x >= 1 then
   ArcSin := Pi/2
  else
   ArcSin := arctan(x/sqrt(1 - sqr(x)));
end; {Function ArcSin}

                         {~~~~~~~~~~~~~~~~~~~~~~~~~~~~~~~}

Function AngleA; {Solves for the angle opposite the first side of an oblique
spherical triangle given three sides. As the opposite side increases from O
towards the sum of the other two sides the angle increases from O towards 180}

begin
 AngleA := ArcCos((cos(SideA) - (cos(SideB) * cos(SideC)))/ (sin(SideB) *
                                                                sin(SideC)));
end; {Function AngleA}

                         {~~~~~~~~~~~~~~~~~~~~~~~~~~~~~~~}

Function SideC; {Solves for the unknown side of an oblique spherical triangle
when two sides and the included angle are known. As the included angle varies
from O towards 180 the unknown side varies from O towards the sum of the other
two sides}
begin
 SideC := ArcCos ((cos(SideA) * cos(SideB)) + (sin(SideA) * sin(SideB *
                                                               cos(AngleC)));
end; {Function SideC}

                         {~~~~~~~~~~~~~~~~~~~~~~~~~~~~~~~}

Function AngleB; {Solves for the angle opposite the second side of an oblique
spherical triangle when two sides and the included angle are known - solves for
angles greater than 90. The sum of the two given sides of the triangle must be <
= 180.}

var
TempSideC : real;

begin
 TempSideC := SideC(SideA,SideB,AngleC);
 AngleB := ArcCos ((cos(SideB) -
(cos(SideA) * cos(TempSideC))) / (sin(SideA) * sin(TempSideC)));
end;  {Function AngleB}

                         {~~~~~~~~~~~~~~~~~~~~~~~~~~~~~~~}
```

```
Function SideB; {Solves for the unknown side of an oblique spherical triangle
when two sides and the angle adjacent to one of the sides are known}

var
AngleB : real;

begin
 AngleB := ArcSin((sin(AngleA) * sin(SideC)) / sin(SideA));
 SideB := ArcCos(((cos(SideA) * cos(SideC)) - (sin(SideA) * sin(SideC)*
          cos(AngleA) * cos(AngleB)))/(1 - sqr(sin(SideC)) * sqr(sin(AngleA))));
end; {Function SideB}

                        {~~~~~~~~~~~~~~~~~~~~~~~~~~~~~~~}

                                                                          end.
```

Stereonet Applications for *Windows* and Macintosh

Declan G. De Paor
Department of Earth & Planetary Sciences, 20 Oxford Street,
Cambridge MA 02138, U.S.A. depaor@eps.harvard.edu

Abstract– Stereonet applications with advanced graphical user interfaces and many custom features are now available on a commercial basis from a number of sources. The best *Windows*™ application in this reviewer's opinion is *StereoNet* by Per Ivar Steinsund whilst the leading Macintosh application is Neil Mancktelow's *StereoPlot*.

Introduction

One of the first applications that structural geologists found for their new personal computers in the early 1980s was the construction of stereograms (*alias* stereoplots, or stereonets). Since that time, net drawing applications have come a long way. The purpose of this review is to alert readers to the existence of relatively new and advanced stereonet applications.

StereoNet

StereoNet for Windows is the creation of Per Ivar Steinsund (perivar@ibg.uit.no) and is sold by Rockware Inc. for $299. It requires an 80286 or better processor, 2MB of RAM, and *Windows* 3.1 or '95. There is no limit to the number of data points that can be plotted given enough memory; for example, 1MB of RAM can hold 125,000 points, considerably more than the average structural geologist needs! *StereoNet* supports 24 bit color and can send output to any Windows-compatible printer. Data can be exchanged with other Windows applications such as Microsoft's *Excel*™ by cutting and pasting to and from the clipboard. Planes can be plotted as great circles or as poles using a variety of symbols. Data can be contoured and analyzed statistically. Two dimensional orientation data can be presented as a rose diagram (Fig.1).

StereoNet comes with a very good user manual. The one area of confusion concerns plotting and file formats. The manual states that "the dip direction format uses 360 degrees...and the strike should be in the dip direction." What is meant is that dip direction data is to be used instead of strike data. The manual refers to the "plunge" of a lineation in a plane when "pitch" would be a more

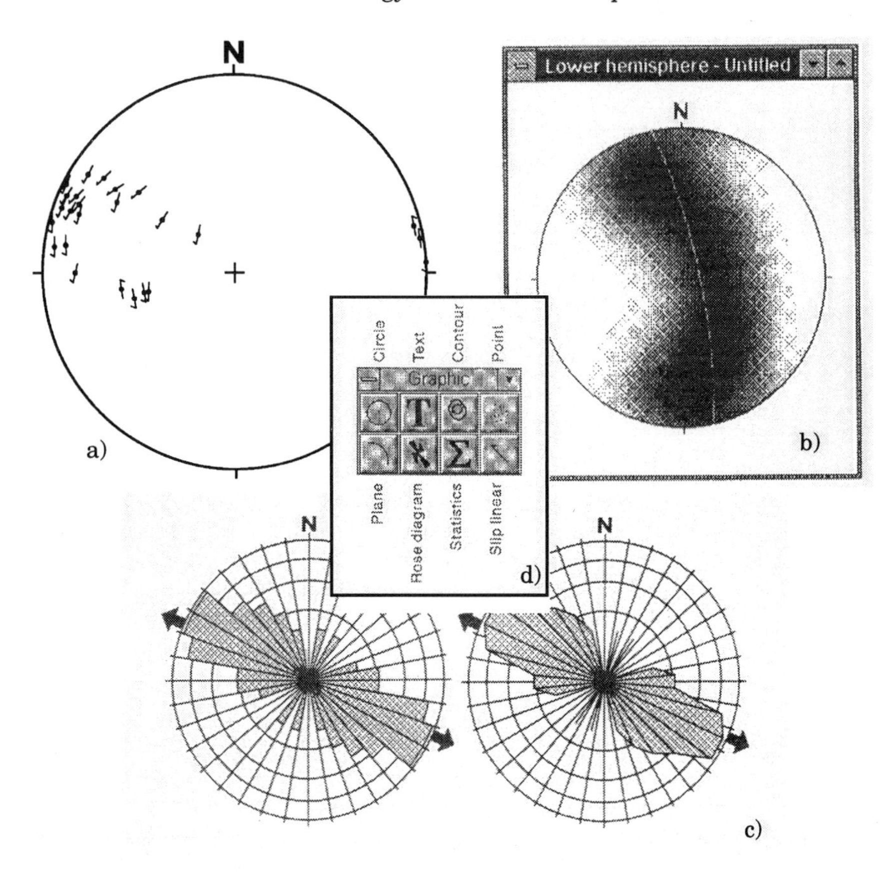

Fig. 1. Sample output from *StereoNet* for *Windows*. a) Slip linear plot. b) Density plot. c) Two types of rose diagram. d) Sample tool palette with explanation. Scanned from manual.

appropriate term and the user is instructed a) to use a negative striation plunge "if the striation is in the opposite direction to the strike" and b) to indicate "the movement direction with a U (up) or D (down)". One example given is "-16° D" for a lineation on a plane with strike 120° and dip 25°. Many students will have a difficult time figuring out the slip system intended by this convention! However, a little practice with test data should clear up any confusion.

In addition to its friendly graphical user interface (Fig. 2), *StereoNet*'s great strength lies in the supported computations, which include calculations of plane intersections, three dimensional rotations of data sets, and eigenvector analysis.

StereoPlot

StereoPlot is distributed as shareware by its author, Neil Mancktelow (neil@erdw.ethz.ch). One may examine the application free of charge but if you

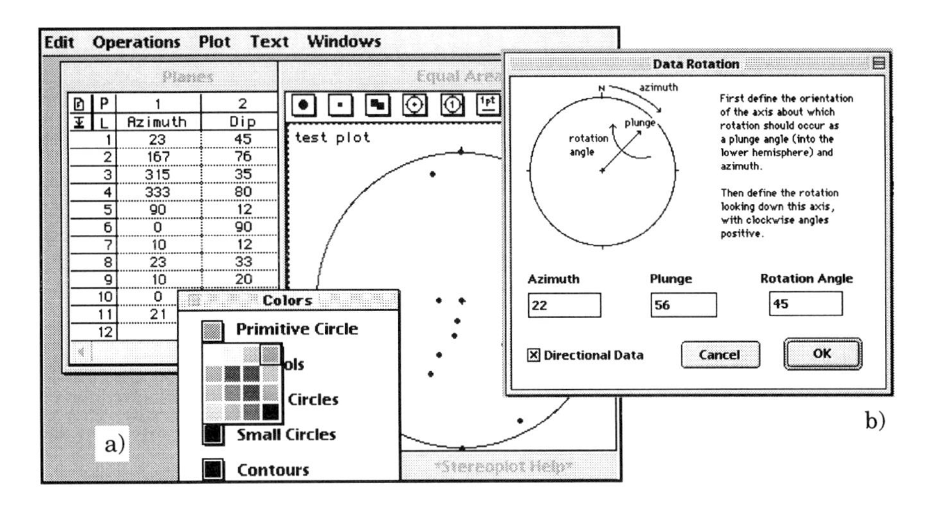

Fig. 2. Sample output from *StereoPlot* for the *MacOS*. a) View of three program windows; data spreadsheet, graphic window, and tear-off palette with pop-up menus. b) Dialog box for rotation of data set about an arbitrary axis.

decide to keep it for research or teaching use, then you are requested to pay the license fee of 100 Swiss Francs.

StereoPlot is available in versions optimized for the PowerMac, for any Macintosh with a floating point processor, and for older Macintosh computers. The maximum number of data supported on a single plot is 1,000 though a larger 5,000 point, memory-hogging version can be requested from the author. The 1,000 point version should suffice for most users and requires only 1MB of RAM. The program uses color *QuickDraw*™ routines for rapid refreshing on screen and *Postscript*™ printing technology for optimum hard copy resolution.

The program makes full use of the standard Macintosh user interface. Balloon Help may be turned on in order to see explanations of interface items. The interface (Fig. 2) consists of a spreadsheet window and a graphics window. Data is entered either by opening a tab-delimited TEXT file, or by typing entries at the keyboard. Data can be cut and pasted and mistakes can be undone. Menu items are selected in order to create a plot and plotting details such as color and symbols are controlled by standard tear-off palettes and pop-up menus.

A significant feature of *StereoPlot* is that plots can be overlain, so that points from separate subareas, for example, can be plotted on a single diagram. However, these cannot be contoured or statistically analysed as a single entity. The ability to create multiple overlays is a great advantage for those occasions when such a feature is needed and a nuisance at other times. Because a click and drag of the mouse creates a new overlay, one must be careful not to click the mouse at the wrong time. The text entry process is also a little irritating. Overall, however, this is a very useful program with lots of extra features such as the ability to rotate data about an inclined axis (Fig. 2b).

Manipulation of Orientation Data Using Spreadsheet Software

Gustavo Tolson
Instituto de Geología, UNAM, Ciudad Universitaria,
México D.F. 04510. tolson@servidor.unam.mx

Francisco Correa-Mora
Instituto de Geofísica, UNAM, Ciudad Universitaria,
México D.F. 04510. pancho@tonatiuh.igeofcu.unam.mx

Abstract– The mathematical treatment of orientation data generally requires specific, rather specialized, software. In this paper we show how simple spreadsheet functions can be combined in macros to manipulate orientation data. These macros can in turn be combined in a modular manner to create sophisticated programs which have the advantage that they can be easily modified to suit a particular user's needs. The example we present is from an application that calculates the orientation of the *c* crystallographic axis of orthoamphiboles based on cleavage plane orientation data derived from universal stage measurements.

Introduction

As geologists we are often faced with the need to work with orientation data of different kinds. This work generally consists of capturing, reducing, manipulating, and visualizing the data. To this end, and in this day of widespread microcomputer availability, there are several commercial, shareware, or freeware programs available to most of us in the earth science community. However, despite this diversity of software to which we have access, most users find that they want to do something with their data that no single program can do. For example, we are sometimes forced to use one program to perform rotations or other calculations on our data and a second to produce visually satisfactory output.

Orientation data in particular are more difficult to manage using different computer programs because of the variability in written notation. There is also considerable flexibility concerning the mathematical rendering of orientation data. The choice of notation conventions is often the result of belonging to some "school", be it the "European School" or the "American School", while the mathematical representation is usually chosen on the basis of its suitability for a certain calculation. Thus N51E 43NW, 43° 321°, and 231° 43° could all be written equivalents for the orientation of a plane as measured in the field, while the direction cosines of the unit normal, the stereographic projection of the pole, or the (x, y, z) coordinates of the dip direction unit vector of the same plane could be mathematical expressions suited for different calculation or visualization purposes.

The objective of this paper is to show that off-the-shelf commercial spreadsheet software is a very useful and mathematically quite powerful tool for the capture, reduction, manipulation, and visualization of orientation data. With such software, users can write functions and macros which together can be assembled into modular programs. This allows tool-kits or utilities to be put together or enhanced by different people without any worry about compatibility. Most of the large commercial spreadsheet packages available on the market are capable of importing or exporting data and associated functions and macros in different file formats, which are practically platform independent. Furthermore, most spreadsheet packages allow the definition of databases, which allows the user to capture a large amount of data and then filter it or selectively display it on the basis of user-defined criteria.

The spreadsheet presented here illustrates utilities written to translate different orientation notations, to calculate the intersection of two planes, to calculate the plane defined by two lines, and to apply an arbitrary rotation of a given sense and a given magnitude about a given axis to an arbitrary line or plane. The spreadsheet can also be used as a visualization tool since lines and planes can be represented on an equal area or stereographic projection net.

The Spreadsheet

The spreadsheet presented here was written using the Microsoft *Excel*™ (v. 4.0) spreadsheet package running under Microsoft *Windows*™ (a compatible version of *Excel* is available for the *MacOS*). This spreadsheet environment provides users with powerful built-in mathematical and text functions as well as the opportunity to write user-defined functions. Such an environment has the added advantage of providing a row-column tabular format which offers an intuitive structure for data capture and processing. *Excel* also provides the possibility of grouping related spreadsheets in workbooks which permits a logical organization of interacting files under a single filename. The workbook written for the orientation application consists of four spreadsheets. The first is the Macro spreadsheet, where all the user-defined functions are kept. The second is the Main spreadsheet where the data are input and where the user-defined functions are called to process the data. The third is the Projection spreadsheet, where the data can be visualized on a projection net. The last sheet is an Output spreadsheet, which allows selective output of processed data along with any other textual information needed by the next application.

Main Spreadsheet

The Main spreadsheet is the one used for data input, reductions and manipulation (Fig. 1). It has no initial restrictions regarding structure, and these can be

Universal Stage data reduction | Sample No.: | Date:

#	Universal Stage orientation: Azimuth 1	Co-latitude1	Azimuth 2	Co-latitude2	Projection net orientation: Azimuth 1	Inclination 1	Azimuth 2	Inclination 2	c-Axis
1	357	31	221	23	003	59	139	67	067 36
2	061	23	275	14	=FIG1.XLS!U_StageAz_NetAz(B6)			76	007 41
3	309	15	082	43	051	75	278	47	331 33
4	001	45	143	17	359	45	217	73	298 26
5	031	12	174	29	329	=90-C9	186	61	249 39
6	089	14	231	38	271	76	129	52	190 32
7	278	41	127	27	082	49	=FIG1.XLS!U_StageAz_NetAz(D11)		
8	332	25	184	16	028	65	176	74	098 36
9	051	65	288	40	309	25	072	=90-E13	357 17
10	271	22	174	08	089	68	186	82	114 66
11	344	45	335	15	016	45	025	75	=Plane_Int
12	102	15	141	23	258	75	219	67	(G15;H15;
13	077	17	302	12	283	73	058	78	I15;J15)
14	184	38	056	31	176	52	304	59	236 32
15	329	20	024	28	031	70	336	62	344 62
16	067	31	126	30	293	59	234	60	266 56

Fig. 1. A portion of the main spreadsheet showing the capture and output fields. Forumale in selected cells shown in shaded boxes.

modified by the user to suit his/her purposes. We have found, however, that the most logical and intuitive structure is to have the data elements in one or more columns on the left, and the results of operations in succeeding columns to the right. In this way several operations on the same datum can be performed incrementally with the advantage that intermediate results can be used in other calculations, output to other applications, or rendered graphically.

The example shown here is used for the reduction of universal stage data collected from orthoamphiboles. The orientation of the *c*-axis of orthoamphiboles is difficult to measure directly but is parallel to the intersection of the two prominent cleavage planes, whose measurement is straightforward (Shelley 1992). Thus the orientation of the two cleavage planes is recorded in the first four columns, and the intersection (in the u-stage frame of reference) is calculated in the fifth column. Frame of reference corrections, including rotation of the thin-section back to its orientation in the field, are accomplished in the next few columns.

Projection Spreadsheet

The Projection spreadsheet (Fig. 2) is perhaps the most varied and complex of the spreadsheets. It takes as input columns of orientation data and provides visual output in the form of an equal area net. As implemented here it consists of two regions; one visible and one hidden from view. The hidden region is where the mathematical calculations for the projection are carried out and the visible region

Microsoft Excel - FIG1.XLS

File Edit View Insert Format Tools Data Window Help

G12

	A	B	C	D	E
5	Rad / °	0.017			
6	x Values	Top of Circle y	Bottom y	y-Data	Trend / Plunge
7	-1.000	0.000	0.000		067 36
8	-0.990	0.141	-0.141		007 41
9	-0.980	0.199	-0.199	=IF(Main!$K7<>"";Main!$K7;"")	
10	-0.970	0.243	-0.243		298 26
11	-0.960	0.280	-0.280		249 39
12	-0.950	0.312	-0.312		190 32
13	-0.940	0.341	-0.341		154 20
204	0.970	0.243	-0.243		
205	0.980	=SQRT(1-(A205)^2)			
206	0.990	0.141	-0.141		
207	1.000	0.000	=-1*B207		
208	0.591			0.251	
209	=IF(E8<>"",SQRT(2)*SIN((90-RIGHT(E8,2))/2*$B$5)*SIN(LEFT(E8,3)*$B$5),NA())				
210	-0.327			0.590	
211	-0.662			=IF(E10<>"",SQRT(2)*SIN((90-RIGHT(E10,2))/2*$B$5)*COS(LEFT(E10,3)*$B$5),NA())	
212	-0.568				
213	-0.119			-0.675	
214	0.356			-0.729	
215	0.636			-0.089	

Equal Area Projecti

Macros / Main / Projection / Output

Fig. 2. A portion of the Projection spreadsheet. Shaded region is normally hidden from view, since there is not generally any need to change it. Leftmost column stores x values for chart window where data are graphically displayed. The next two columns are y values of top and bottom halves of net's primitive. Fourth column has y values of orientation data. The x and y values to be graphed are functions of the text fields where the data are tabulated.

is where the data to be projected are tabulated and graphically rendered. The (x,y) coordinates to be graphed are in columns in the hidden region, which in turn are functions of the trend and plunge data in the visible data table (Fig. 2).

Macro spreadsheet

The Macro spreadsheet contains all the user-defined functions and their documentation. Thus, individual users can add more functions at will or use the existing functions to build more complex ones. The Macro command language is relatively easy to learn and one has access to on-line help at the press of a key. The user-defined functions used in the Macro spreadsheet are a mixture of text, mathematical and logical spreadsheet-functions. The mathematical and logical functions handle the calculations, while the text functions are used to manage input and output. Table 1 summarizes the different user-defined functions and their relative dependencies.

The function `Plane_Int` illustrates how the logical, mathematical, and character manipulation of data elements can be integrated. The function takes as

Table 1. Summary of user defined functions implemented in the Macro spreadsheet, showing required input, a brief description, output, and called functions. Functions to convert among notation conventions are also implemented in the spreadsheet but are not described.

Function	Arguments	Description	Output	Calls
FindSeparator	Text String	Given a text string with orientation data, finds the position of the comma, slash or blank space.	Integer value	
GetAzimuth	Text String	Gets the azimuth from a string.	Text	FindSeparator
GetInclination	Text String	Gets the inclination from a string.	Text	FindSeparator
AzimuthTxt	One numerical or text value	Uses text functions to format output to three-digit azimuth values.	Three-digit text	
PlungeTxt	One numerical or text value	Uses text functions to format output to two-digit dip or plunge value.	Two-digit text	
PlaneToPole	Orientation text string the orientation of a plane (using the right hand rule convention) into a text string with the orientation of the corresponding pole (trend and plunge).	Converts the text string which describes AzimuthTxt PlungeTxt	Orientation text string	GetAzimuth GetInclination
PoleToPlane	Orientation text string the orientation of a pole to a plane (trend and plunge) into a text string with the orientation of the corresponding plane (using the right hand rule convention).	Converts the text string which describes AzimuthTxt PlungeTxt	Orientation text string	GetAzimuth GetInclination
CrossProduct	Two orientation text strings	Calculates the cross product of two lines.	Orientation text string	GetAzimuth GetInclination AzimuthTxt PlungeTxt
Plane_Int	Two orientation text strings planes.	Calculates the line of intersection of two	Orientation text string	PlaneToPole CrossProduct
CalculatePlane	Two orientation text strings	Calculates the plane defined by two lines.	Orientation text string	CrossProduct PoleToPlane
LineRotation	Two orientation text strings and an integer.	Performs the rotation of a line about an axis.	Orientation text string	GetAzimuth GetInclination AzimuthTxt PlungeTxt
PlaneRotation	Two orientation text strings and an integer.	Performs the rotation of a plane about an axis.	Orientation text string	GetAzimuth GetInclination PlaneToPole AzimuthTxt PlungeTxt LineRotation PoleToPlane

Microsoft Excel - FIG1.XLS

File Edit View Insert Format Tools Data Window Help

	A	B	C
44	**CrossProduct**	=ARGUMENT("LineOne")	Function which calculates the cross product of two
45		=ARGUMENT("LineTwo")	unit vectors expressed by their orientations
46		Aone=GetAzimuth(LineOne)*PI()/180	orientación.
47		Fone=GetInclination(LineOne)*PI()/180	
48		Atwo=GetAzimuth(LineTwo)*PI()/180	This function is called by the Plane_Int and
49		Ftwo=GetInclination(LineTwo)*PI()/180	CalculatePlane functions.
50		Sone=SIN(Aone)	
51		Stwo=SIN(Atwo)	
52		Sthree=SIN(Fone)	
53		Sfour=SIN(Ftwo)	
54		Cone=COS(Aone)	
55		Ctwo=COS(Atwo)	
56		Cthree=COS(Fone)	
57		Cfour=COS(Ftwo)	
58		PX=Sone*Cthree*Sfour-Sthree*Stwo*Cfour	
59		PY=Sthree*Ctwo*Cfour-Cone*Cthree*Sfour	
60		PZ=Cone*Cthree*Stwo*Cfour-Sone*Cthree*C	
61		PX=IF(PZ<0,-PX,PX)	
62		PY=IF(PZ<0,-PY,PY)	
63		PZ=IF(PZ<0,-PZ,PZ)	
64		P=SQRT(PX^2+PY^2+PZ^2)	
65		PP=PZ/P	
66		FF=ATAN(PP/SQRT(1-PP^2))	
67		AA=IF(PX=0,IF(PY>0,PI()/2,AA=3*PI()/2),ATA	
68		F=ROUND(FF*180/PI(),0)	
69		A=ROUND(AA*180/PI(),0)	
70		A=IF(PX<0,A+180,IF(AND(PX>0,PY<0),A+360	
71		Answer_=AzimuthTxt(A)&" "&BuzTxt(F)	
72		=RETURN(Answer_)	

Macros / Prueba

Fig. 3. A portion of the Macro spreadsheet illustrating the `CrossProduct` function. Note use of built-in mathematical and logical functions.

arguments two text values, namely the strike and dip of two planes in Right Hand Rule convention. Following the algorithm described by Wallbrecher (1986), the orientations of the planes are converted to 3-dimensional, right-handed Cartesian coordinates and expressed as unit vector plane normals. The cross product of these normals, parallel to the intersection direction, is calculated by the `CrossProduct` function (Fig. 3), after which it is converted to geographic coordinates with a series of IF statements. Two user-defined text functions, `AzimuthTxt` and `PlungeTxt`, provide the last formatting step, returning a text value in the form *xxx yy*, where *xxx* is a three digit azimuth and *yy* is a two digit plunge.

Output Spreadsheet

This spreadsheet is initially empty and to it are copied data from the Main spreadsheet in order to export them in some suitable format (*e.g.,* ASCII) to other applications. In the case of the example shown, the format is suitable for exporting the orthoamphibole *c*-axis orientations to a projection net software package capable of reading *DOS*-text (ASCII) files.

Discussion and Conclusions

When asked informally about computer programming, most geologists would admit to a certain hesitancy about learning a programming language. Commercially available spreadsheet software, however, is a viable alternative to traditional programming languages and indeed offers certain advantages. Spreadsheet packages offer a vast repertoire of text, mathematical, and logical functions and operators, which make them suitable for complicated mathematical manipulation of data, including matrix-oriented calculations. The greatest advantage offered by spreadsheet software is, however, its ability to easily provide graphical output of data, something which is not trivial in traditional programming environments.

Furthermore, a spreadsheet environment encourages the use of good programming practices, in that there is a very close relationship between the data and the procedures operating on them. A modular, procedural approach to program development is also practically intrinsic to the structure of a spreadsheet, given its row and column structure. By putting one's data in columns on the left and applying various operations to them in columns to the right, the result is a procedural manipulation of the data. The database management capabilities of today's spreadsheets are also an extremely useful feature, allowing one to capture all data in a single sheet and later filter it on the basis of different criteria in order to apply different manipulations or visualization techniques to different subsets of the data.

The spreadsheet described above is available via anonymous FTP from ftp.unam.mx in subdirectory pub/ciencia/geologia.

Acknowledegments

We would like to thank Gabriela Solís Pichardo and an anonymous reviewer for helpful comments which improved earlier versions of the manuscript. G.T. is also grateful to Dr. Ludwig Masch of the Mineralogisch- und Petrographisches Institut for computer and universal-stage facilities provided during a stay at the University of Munich.

References

Shelley, D. 1992. *Igneous and Metamorphic Rocks under the Microscope*. Chapman & Hall, London, 445 pp.

Wallbrecher, E. 1986. *Tektonische und Gefügeanalytische Arbeitsweisen*. Ferdinand Encke Verlag, Stuttgart, 244 pp.

Discussion and Conclusions

IV: STRAIN AND KINEMATIC ANALYSIS

Modeling Growth and Rotation of Porphyroblasts and Inclusion Trails

Eric C. Beam
Department of Geological Sciences, University of Texas at Austin, Austin TX 78712, U.S.A.
eric@maestro.geo.utexas.edu

(Current address: Exxon, P.O. Box 4778, Houston TX 77210-4778, U.S.A.)

Abstract– Trails of mineral inclusions within metamorphic porphyroblasts can provide useful kinematic information, but interpretation of these structures is plagued by ambiguity.

Spiral trails or differences in orientation between internal and external foliations can be interpreted as reflecting either differential rotation of the porphyroblast and cleavage during growth, or as static overgrowth upon earlier foliations. These contrasting interpretations have conflicting implications for the behavior of deforming rocks.

This paper presents a computer program which uses previously published descriptions of the behavior of structural elements to simulate the growth and rotation of porphyroblasts and cleavages. This program models synkinematic porphyroblast growth as a series of steps of growth and rotation. The porphyroblast is considered to be a rigid inclusion in a viscous matrix. As it grows it includes one of three model foliations. Foliation may be the plane of maximum finite strain, maximum infinitesimal strain, or a preexisting passive marker. Both the finite and passive foliations may rotate as deformation proceeds. The program incorporates different growth laws in order to simulate growth in different metamorphic conditions. The porphyroblast may grow by adding fixed increments of radius, surface area, or volume. Porphyroblasts may be equant or elongate.

Resultant inclusion trails are complex because of the variable relative rates of rotation of inclusions and cleavage. Different growth laws cause the curvature of trails to be distributed differently within the porphyroblast. In some cases foliations are generated which could easily be interpreted as an included crenulation cleavage; other cases give an apparent sense of rotation opposite to the actual rotation.

Introduction

An important goal for structural geologists is the determination of strain paths. This information can often provide insight additional to that from finite strain data, which can be the end result of an infinite variety of strain histories. Rigid porphyroblasts can preserve the foliation at intermediate stages in the strain history of a rock and have been used to determine the kinematics of deformation. Consequently the attempt to obtain information about strain paths has become entwined with a debate over the behavior of rigid objects in a deforming medium.

When foliation within a porphyroblast is not parallel to external foliation a rotation is inferred. Often the internal foliation shows a sigmoidal or spiral pattern, which is assumed to be the result of simultaneous growth and rotation. Apparent rotations of porphyroblasts have been interpreted as representing two different deformation histories. One view (Spry 1963, Rosenfeld 1970, Williams & Schoneveld 1981) is that apparent rotation of porphyroblasts results from actual rotation of porphyroblasts during growth in simple shear. The other view (Bell 1985, Bell *et al.* 1992) is that apparent rotation of porphyroblasts results from the reorientation of external foliation by pure shear.

Because the principles which govern the motion of particles in a deforming viscous medium have been quantified, this problem is suitable to computational experiments and several have been published (Masuda & Mochizuki 1989, Bjørnerud & Zhang 1994, Gray & Busa 1994). All of these previous studies consider spherical porphyroblasts.

Masuda & Mochizuki (1989) model the incorporation into a growing rigid spherical porphyroblast of passive marker planes in the surrounding matrix, which is a Newtonian viscous fluid deforming in simple shear. This model incorporates deflection of the marker planes adjacent to the porphyroblast. The sphere rotates at one-half the rate of simple shearing ($\dot{\gamma}/2$), and grows by addition of increments of constant volume (dv/dt = constant). The matrix deforms according to the solution of Masuda & Ando (1988) for a rigid spherical inclusion in a Newtonian viscous matrix undergoing simple shear. The flow is deflected around the inclusion leading to distortion of material lines in the adjacent matrix. In simple cases with the marker planes initially parallel to the shear plane, the modeled porphyroblasts show an increase in curvature of the trails towards the outside of the sphere, due to the volumetric growth law. Also shown are cases in which the initial orientation of the marker planes is oblique to the shear plane. Because the rotation rate of the passive marker planes is variable, and the rotation rate of the porphyroblast constant, this can form complex inclusion trails, including reversals in the curvature of the resultant inclusion trail. This result is not surprising given the existing knowledge of the rotation of passive markers; a passive marker may rotate faster or slower than a spherical porphyroblast depending on the passive marker orientation relative to the shear plane (Ghosh & Ramberg 1976, Ghosh 1993).

Bjørnerud & Zhang (1994) performed an analysis similar to Masuda & Mochizuki (1989) but extended it to investigate the effect of the degree of mechanical coupling between the rigid porphyroblast and viscous matrix in simple shear.

The matrix deforms according to the solution of Bjørnerud (1989) for a rigid spherical inclusion in a Newtonian viscous matrix undergoing simple shear. This solution seems to be essentially the same at that of Masuda & Ando (1988; see Gray & Busa 1994, p. 576, for a discussion of the development of descriptions of deformation kinematics). As in Masuda & Mochizuki (1989) the flow is deflected around the inclusion, leading to distortion of material lines in the adjacent matrix. The porphyroblast in Bjørnerud & Zhang (1994) overgrows passive marker planes which are initially parallel to the shear plane. They show that curved inclusion trails may form even with no coupling, as the porphyroblast overgrows foliation that is deflected around it. Episodic growth produces abrupt changes in inclusion trail curvature in their model.

Gray & Busa (1994) extend the analysis of simulated inclusion trails formed during simple shearing to three dimensions. The matrix deforms according to the solution of Einstein (1956) for a rigid spherical inclusion in a Newtonian viscous matrix. In the case of simple shearing this solution is the same as those used in Masuda & Mochizuki (1989) and Bjørnerud & Zhang (1994) and again the flow is deflected around the inclusion, leading to distortion of material lines in the adjacent matrix. Gray & Busa (1994) consider a spherical porphyroblast fully coupled to its matrix overgrowing passive marker planes initially parallel to the shear plane. Growth is by increments of constant radius, area, or volume. This model incorporates deflection of the marker planes adjacent to the porphyroblast. Slices through the modeled porphyroblast show complex patterns, including some that look like the 'millipede' structures of Bell & Rubenach (1980), which the authors interpret as precluding a simple shear deformation history.

All of the models above have some features in common. They all consider spherical porphyroblasts in simple shear with passive marker foliations. Consideration of other shapes and foliations, however, can provide more fruitful and less ambiguous results. The model presented here considers some more general cases.

Model Design

This model is purely kinematic, relying on the descriptions of particle and foliation movement in both simple shear and combinations of pure and simple shear published in Ghosh & Ramberg (1976) and Ghosh (1982, 1993). This model is thus able to consider more general cases than some previous models, but at the expense of some realism. In particular, distortion of foliation by deflection of matrix flow around the porphyroblast is not modeled. This model assumes that the medium around the porphyroblast has a linear viscous rheology, that the porphyroblast is rigid, and that there is no slip along the interface between the two. Only the central inclusion trail is modeled.

This program simulates the formation of inclusion trails by calculating the orientation of porphyroblasts and foliations in progressive general shear. Foliation may be modeled as the plane of maximum finite strain, maximum infinitesimal strain, or as a pre-existing passive marker. The strain history is divided into increments and particles of the foliation are included at each increment. Once a particle is included it rotates with the porphyroblast (Fig. la).

In order to perform this simulation it is necessary to have quantitative descriptions of all the structural elements involved under all the conditions modeled. These are drawn from a variety of sources. The orientation ϕ of the major axis of the infinitesimal strain ellipse for combined pure and simple shear is given by

$$\theta = \frac{1}{2}\tan^{-1}\left(\frac{1}{2s_r}\right) \tag{1}$$

(Ghosh, 1982, eqn. 10) where s_r, is the ratio of the rate of pure shear to the rate of simple shear, that is

$$s_r = \frac{\dot{\varepsilon}}{\dot{\gamma}}. \tag{2}$$

In the case of simple shear only,

$$\lim_{s_r \to \infty} (\theta) = 45°. \tag{3}$$

The orientation (θ') of the major axis of the finite strain ellipse for combined pure and simple shear is given by

$$\tan 2\theta' = \frac{\frac{1}{s_r}\left(1 - e^{-2\gamma s_r}\right)}{\left(\frac{1}{s_r}\sinh(\gamma s_r)\right)^2 + \left(e^{2\gamma s_r} - e^{-2\gamma s_r}\right)} \tag{4}$$

(Ghosh 1982, eqn. 8) where γ is simple shear strain. In the case of simple shear only, this is

$$\tan 2\theta' = \frac{2}{\gamma} \tag{5}$$

(Ghosh 1993, eqn. 8.69). The rotation of a passive marker line in combined pure and simple shear is given by

$$\tan\phi = e^{2\gamma s_r}\tan\phi_0 + \frac{1}{2s_r}\left(e^{2\gamma s_r} - 1\right) \tag{6}$$

(Ghosh & Ramberg 1976, eqn. 16), where ϕ_0 is the initial and ϕ the final orientation of the line. In the case of simple shear only, this is

$$\cot\alpha' = \cot\alpha + \gamma \tag{7}$$

(Ghosh 1993, eqn. 8.73), where α is the initial and α' the final orientation of the line.

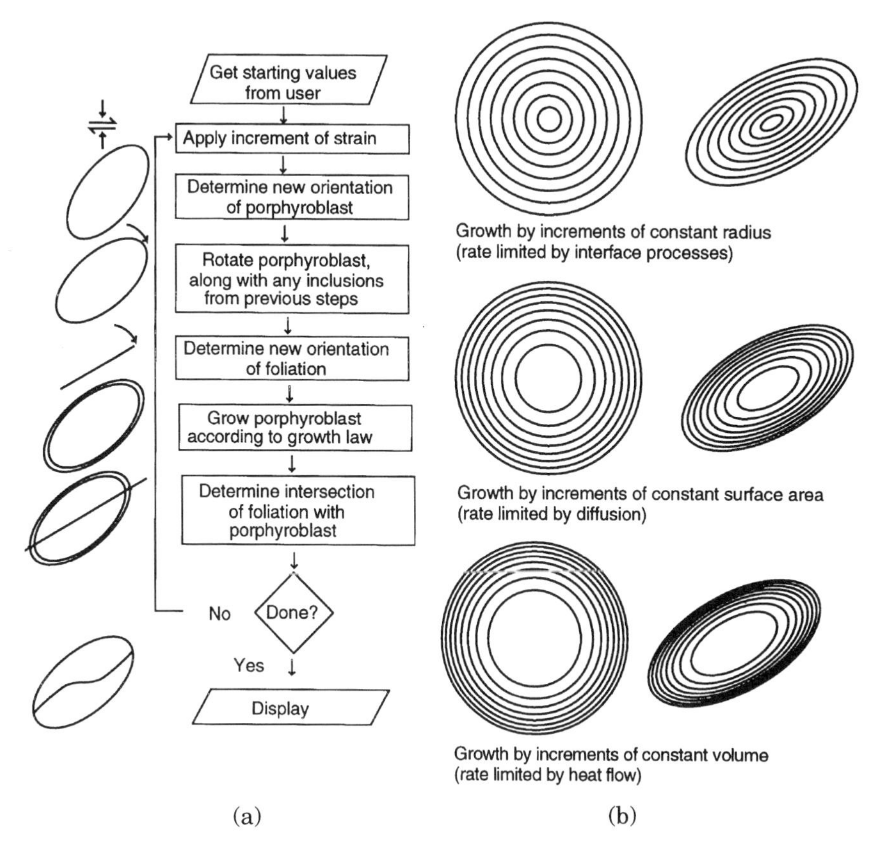

Fig. 1. a) Flow chart showing general operation of program, which determines orientation of foliation and porphyroblast, and the intersection between the two at successive increments. b) Effect of growth law illustrated by growth rings, for porphyroblasts with circular or elliptical cross section.

The motion of the rigid particles is described by equations from Ghosh & Ramberg (1976). There are two relevant equations which apply in different fields determined by the axial ratio R of the rigid inclusion, and the ratio of pure to simple shear strain rates s_r. If

$$R < \frac{1 + \sqrt{1 + 4s_r^2}}{2s_r} \tag{8}$$

then

$$\phi = \tan^{-1}\left[\frac{h}{C}\tan\left\{\gamma h + \tan^{-1}\frac{C\tan\phi_0 + B}{h}\right\} - \frac{B}{C}\right] \tag{9a}$$

where

$$h = \sqrt{AC - B^2} \tag{9b}$$

(Ghosh & Ramberg 1976, eqn. 12), where ϕ_0 is the initial and ϕ the final orientation of the line, and *A, B,* and *C* are defined in Ghosh & Ramberg (1976) as

$$A = \frac{R^2}{R^2 + 1} \tag{10}$$

$$B = s_r \frac{R^2 - 1}{R^2 + 1} \tag{11}$$

$$C = \frac{1}{R^2 + 1}. \tag{12}$$

When

$$R > \frac{1 + \sqrt{1 + 4s_r^2}}{2s_r^2} \tag{13}$$

then

$$\phi = \tan^{-1}\frac{P(B + k) - B + k}{C(1\text{-}P)} \tag{14}$$

(Ghosh & Ramberg 1976, eqn. 11), where *P* is

$$P = \frac{C \tan\phi_0 + B - k}{C \tan\phi_0 + B + k} e^{2\gamma k} \tag{15a}$$

and

$$k = \sqrt{B^2 - AC} \tag{15b}$$

with *A, B* and *C* defined in eqns. 10, 11, and 12, respectively. These equations are valid for simple and combined simple and pure shear and for equant and elongate inclusions.

The relative rate of porphyroblast growth and rotation determines the curvature of an inclusion trail. Variation of growth with time determines the geometry of that curvature. This program divides the strain history into equal increments. During each increment a shell of material must be added to the porphyroblast. This program allows an increment of constant radius, area, or volume. These options correspond to different geological scenarios. According to Carlson (1989) the constant radius growth corresponds to growth rate limited by interface processes, the constant area to growth rate limited by diffusion, and the

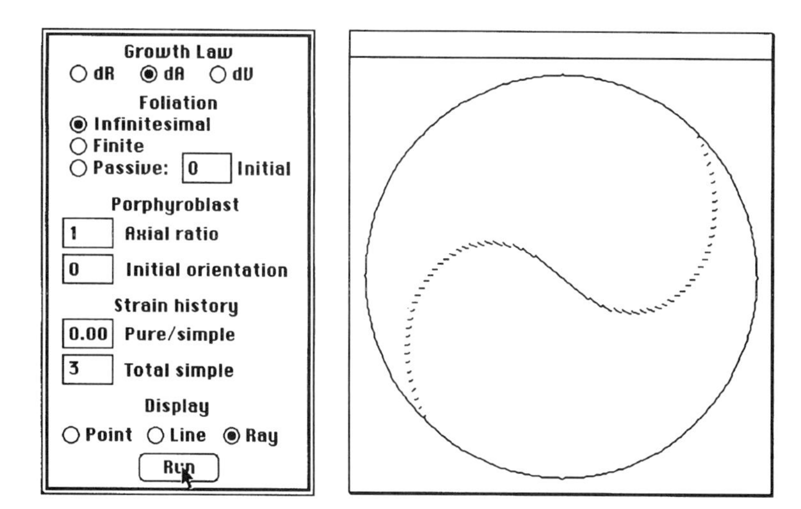

Fig. 2. User interface of program which allows users to rapidly investigate many combinations of input values and immediately see the results. In this case simple shear of spherical porphyroblast has produced smoothly curving sigmoidal inclusion trail.

constant volume to growth rate limited by heat flow. These different growth laws make substantial differences in the growth pattern of the porphyroblast (Fig. lb). All of these options may be input by the user in a dialog, it is not necessary to reenter all the choices in order to rerun the analysis. This makes it easy to perform exploratory analysis (Fig. 2). Results may be displayed as a series of points representing the intersection of the cleavage and porphyroblast at each growth increment, as a line connecting these points, or as a series of line segments corresponding to the cleavage segments captured during each growth increment.

Results

These calculations can produce simple patterns similar to classic porphyroblast textures (Fig. 2). In some cases the inclusion trails constructed using these simple governing equations are complex (Fig. 3). This is a result of the varying relative rates of rotation of the porphyroblast and foliation. Because the inclusion trail records the relative rotation of the porphyroblast and foliation, inclusion trails may show an apparent rotation larger or smaller than the true porphyroblast rotation (relative to a fixed external reference frame). Some cases preserve a sense of rotation opposite to the true rotation. In addition, the changing relative rate of rotation can cause reversals in the inclusion trail (Masuda & Mochizuki 1989); in this model this sometime produces patterns that could be interpreted as included crenulation cleavages. Varying the growth law produces substantial changes in inclusion trail geometry. A porphyroblast will undergo the same rotation history regardless of growth law, but curvature of the trail will be distributed differently within the porphyroblast (Fig. 4). When growth is by

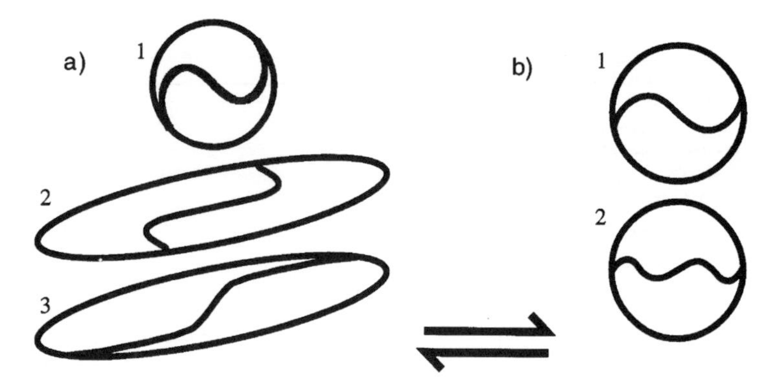

Fig. 3. a) Three porphyroblasts have undergone dextral simple shear γ=3, growth $\propto$ volume. (1) is equant and includes an infinitesimal strain fabric. (2) also includes infinitesimal strain fabric but is elongate (5:1) and initially 45° from shear plane. This could be misinterpreted as included crenulation cleavage. (3) is same as (2) but includes passive marker originally $\perp$ to shear plane which rotated faster than porphyroblast, giving apparent sense opposite to actual rotation. b) Two porphyroblasts underwent equal rates of pure shear and simple shear γ = 2, growth $\propto$ area. (1) overgrows infinitesimal strain foliation; (2) overgrows passive marker foliation initially 90° from shear plane.

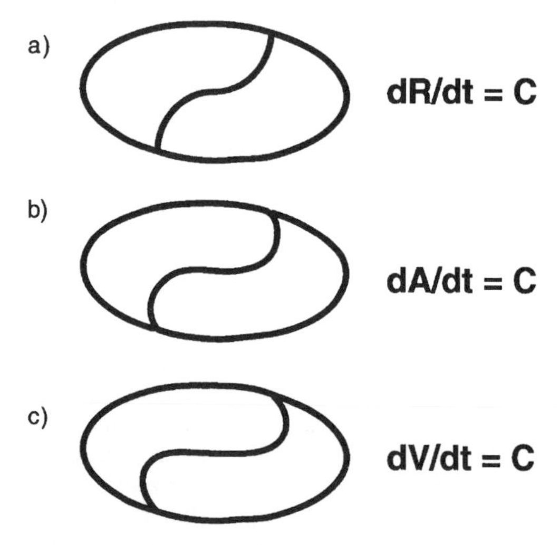

Fig. 4. a) Three porphyroblasts underwent dextral simple shear γ= 3, overgrowing infinitesimal strain foliation. All have axial ratio = 2 and initial orientation 45° from shear plane. (a) grew by increments of constant radius, (b) by increments of constant surface area, and (c) by increments of constant volume. Change in curvature from center to margin more abrupt in (c), although all have identical rotation histories.

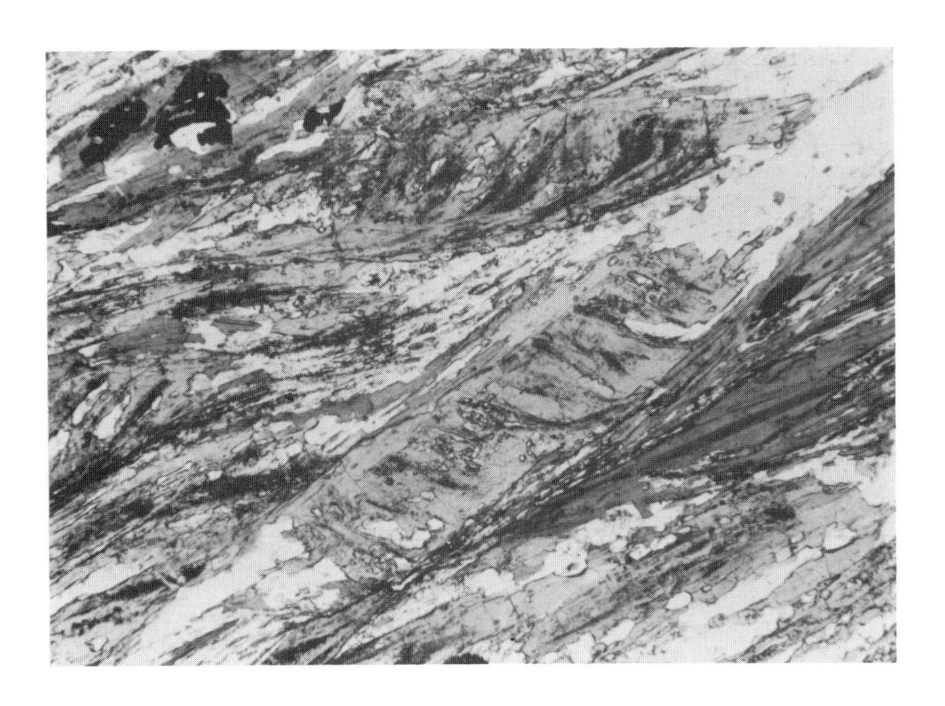

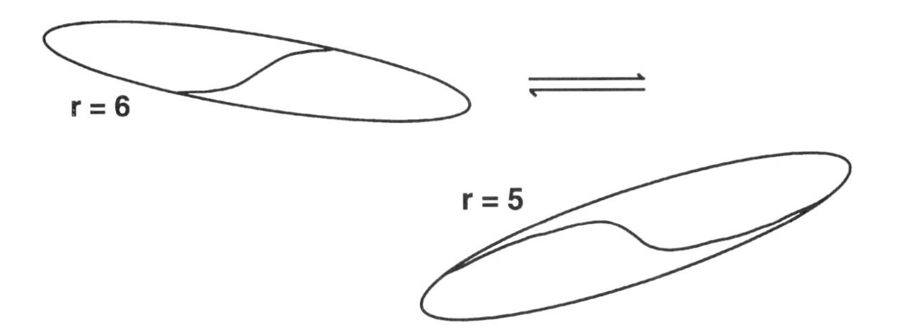

Fig. 5. a) Two amphiboles from Maclaren Glacier Metamorphic Belt, south-central Alaska, with trails of quartz and graphite. F.O.V : 2 mm. b) Possible explanation of observed texture; both porphyroblasts have undergone the same strain history. $s_r = 0.2$, total $\gamma = 2$, foliation is long axis of finite strain ellipse. Growth $\propto$ volume. Upper porphyroblast was initially oriented 10° clockwise of shear plane and thus rotated slower than foliation. Lower one was 90° counterclockwise of shear plane and rotated faster than foliation.

increments of constant volume, an abrupt change in curvature may be interpreted as a truncation. This has been used as a criteria for identifying non-rotated porphyroblasts (Bell *et al.*, 1992).

Because the user of this model can easily vary entered values, a rapid search of parameter space may be made, enabling the user to find the combination of values which best fits both the inclusion trail geometry and any other kinematic data in the area. For example (Fig. 5a), two amphibole porphyroblasts are shown from the Maclaren Glacier Metamorphic Belt, a shear zone characterized by top-to-the-south simple shear (Smith 1981, Davidson *et al.* 1994). The porphyroblasts pictured, however, show opposite senses of curvature of inclusion trails, a pattern that suggests a large component of pure shear. A series of models shows that these textures could have formed with a very small component of pure shear, if the porphyroblasts had different initial orientations and overgrew a finite strain foliation (Fig. 5b). This is not a unique solution, but it is very similar to values determined nearby (~1.5 km down section) by analysis of pressure shadows around rutile porphyroblasts (Beam & Fisher 1990).

Conclusions

Simple governing principles can result in complex inclusion trail geometries. Opposite senses of rotation can be explained by one deformation history. Abrupt changes in inclusion trail geometry are a natural result of porphyroblast growth laws, and of the shapes of porphyroblasts. Geometries which look like included crenulations are easily produced by continuous growth and rotation.

Often structural geologists find that they are able to produce a coherent kinematic history which is consistent with all their observations except for inclusion trails. In light of this conflict it is tempting to regard complex inclusion trails as a potential "... interpretative tool unrivaled by any other geometric or microstructural phenomenon..." (Bell 1985, p. 115). However, it is more likely that the complexity of inclusion trails represents a complex response to a kinematic history consistent with other structural data. Using this program geologists working in deformed, metamorphosed rocks may be able to reconcile complex inclusion patterns with somewhat more straightforward structural histories determined by other methods, and in doing so make inclusion trails a tool rather than an obstacle in kinematic analysis.

Acknowledgements

The author received fellowship support from the University of Texas Geology Foundation. Comments from Sarah Penniston-Dorland, Mark Cloos, and an anonymous reviewer improved the manuscript. A copy of a Macintosh version of the program described is available by anonymous ftp to muon.geo.utexas.edu, in the directory /pub/beam.

References

Beam, E. C. & Fisher, D. M. 1990. Rotation of elongate porphyroblasts in a shear zone, Kahiltna terrane, south central Alaska: *Geol. Soc. Amer. Abstracts with Prog.*, **22**: A183.

Bell, T. H. 1985. Deformation partitioning and porphyroblast rotation in metamorphic rocks: a radical reinterpretation: *J. Met. Geol.* **3**: 109-118.

Bell, T. H., Forde, A., & Hayward, N. 1992. Do smoothly curving, spiral-shaped inclusion trails signify porphyroblast rotation? *Geology* **20**: 59-62.

Bell, T. H., Johnson, S. E., Davis, B., Forde, A., Hayward, N., & Wilkins, C. 1992. Porphyroblast inclusion-trail orientation data: eppur non son girate! *J. Met. Geol.* **10**: 2-95-307.

Bell, T. H. & Rubenach, M. J. 1980. Crenulation cleavage development-evidence for progressive bulk inhomogeneous shortening from "millipede" microstructures in the Robertson River Metamorphics: *Tectonophysics* **68**: T9-T15.

Bjørnerud, M. G. 1989. Mathematical model for folding of layering near rigid objects in shear deformation: *J. Struct. Geol.* **11**: 245-254.

Bjørnerud, M. G. & Zhang, H. 1994, Rotation of porphyroblasts in non-coaxial deformation: insights from computer simulations: *J. Met. Geol.* **12**: 135-139.

Carlson, W. D., 1989. The significance of intergranular diffusion to the mechanisms and kinetics of porphyroblast crystallization: *J. Met. Geol.* **103**: 1-24.

Davidson, C., Schmid, S. M. & Hollister, L. S. 1994. Role of melt during deformation in the deep crust. *Terra Nova* **6**: 133-142.

Einstein, A. 1956, *Investigations on the theory of the Brownian movement.* Dover, New York, 119 pp.

Ghosh, S. K. 1982. The problem of shearing along axial plane foliations. *J. Struct. Geol.* **4**: 63-67.

Ghosh, S. K. 1993. *Structural Geology Fundamentals and Modern Developments.* Pergamon Press, Oxford, 598 pp.

Ghosh, S. K. & Ramberg, H. 1976. Reorientation of inclusions by combination of pure shear and simple shear. *Tectonophysics* **34**: 1-70.

Gray, N. H. & Busa, M. D. 1994. The three-dimensional geometry of simulated porphyroblast inclusion trails: inert marker, viscous-flow models: *J. Met. Geol.* **12**: 575-587.

Masuda, T. & Ando, S. 1988. Viscous flow around a rigid spherical body: a hydrodynamical approach. *Tectonophysics* **148**: 337-346.

Masuda, T. & Mochizuki, S. 1989, Development of snowball structure: numerical simulation of inclusion trails during synkinematic porphyroblast growth in metamorphic rocks. *Tectonophysics* **170**: 141-150.

Rosenfeld, J. L. 1970. *Rotated garnets in metamorphic rocks*. Geol. Soc. Amer., 105 pp.

Smith, T. E. 1981. *Geology of the Clearwater Mountains, South-central Alaska*. Anchorage, Alaska Div. Geol. & Geophys. Surv., 72 pp.

Spry, A., 1963. The origin and significance of snowball structure in garnet. *Journ. Petrol.* **4**: 211-222.

Williams, P. F. & Schoneveld, C. 1981. Garnet rotation and the development of axial plane crenulation cleavage. *Tectonophysics* **78**: 307-334.

Simulated Pressure Fringes, Vorticity, and Progressive Deformation

Kyuichi Kanagawa
Department of Earth Sciences, Chiba University,
Chiba 263, Japan. kyu@earth.s.chiba-u.ac.jp

Abstract– An alternative approach to numerically simulating displacement controlled rigid fiber growth in pressure fringes takes account of rigid-body rotation of an object and its surrounding pressure fringe according to their shape and flow in the matrix. Provided that the process of fiber growth is known, both syntaxial and antitaxial rigid fiber growths in pressure fringes can be numerically simulated by this method for any plane, isochoric, Newtonian viscous flow with constant flow parameters. Three syntaxial and two antitaxial growth models are suggested, but it is uncertain yet which model is appropriate for natural fiber growth in pressure fringes. The method is illustrated for the rigid fiber models of syntaxial and antitaxial fiber growths around a circular rigid object. The lengths and curvatures of fibers in a simulated pressure fringe are constant except for a few fibers in these models and they are shown to vary systematically according to finite strain and kinematic vorticity number. Although the growth process of pressure fringes is required to be known, simulated fibers provide a powerful tool for quantitative estimates of vorticity and progressive deformation.

Introduction

Fibrous pressure fringes are commonly found around rigid objects in rocks deformed predominantly by pressure solution. If the growth of fibers is displacement controlled, their trajectories should contain important information on the deformation path (*e.g.*, Elliott 1972, Durney & Ramsay 1973, Wickham 1973, Ellis 1986, Ramsay & Huber 1983). Many studies have thus attempted to use fiber trajectories in pressure fringes for analysis of progressive deformation. The most commonly used method in these studies is based on the displacement controlled rigid fiber model (Durney & Ramsay 1973, Gray & Durney 1979, Ramsay & Huber 1983), which assumes rigid fibers whose growth trajectories are parallel to the maximum incremental extension direction. Fibers are divided into small, straight segments, each orientation and length of which are regarded as the direction and magnitude of the maximum incremental extension. But the orientations of fiber segments do not necessarily preserve the incremental strain directions because the rigid object and its surrounding attached rigid fibers would have rotated with respect to their matrix during general non-coaxial flow or even during coaxial flow for an elongate rigid object (*c.f.* Ghosh & Ramberg 1976, Ishii 1995).

Another attempt has been made to estimate shear sense as well as the amounts of finite strain and rotation from the comparison of natural and computer-simulated pressure fringes (Etchecopar & Malavieille 1987, Malavieille & Ritz 1989, Ishii 1995). This simulation method is based on the geometric best-fit between a rigid object and its matrix. For each strain increment the matrix is homogeneously deformed. The rigid object and its previously formed fringe are rotated and translated such that gaps and overlaps between them and their matrix are minimized, the still remaining gaps being filled with newly grown fibers (Etchecopar & Malavieille 1987). The simulation rotates the object and pressure fringe with respect to their matrix but it is still uncertain whether their rotation follows known rules of rigid-body rotation (*e.g.*, Ghosh & Ramberg 1976).

This paper presents an alternative simulation method for displacement-controlled rigid fiber growth in pressure fringes, which is based on the rotation of a rigid object in a flowing, viscous matrix. Provided that the growth process of fibers is known, the method can be applied to both syntaxial and antitaxial rigid fiber growths around a rigid object in any plane, isochoric, Newtonian viscous flow with constant flow parameters. The method is illustrated for the displacement controlled rigid fiber models of syntaxial and antitaxial fiber growths. It is shown that the length and curvature of simulated fibers vary systematically according to finite strain and kinematic vorticity number, and therefore that they can be useful parameters for quantitative estimates of vorticity and progressive deformation as suggested by Simpson & De Paor (1993).

Description of Flow and Progressive Deformation

Any homogeneous flow can be described by a velocity gradient tensor **L** which relates the velocity of a particle and its position (Malvern 1969). Let the coordinates for pure shear flow be x_ε and y_ε where longitudinal strain rate along the x_ε axis is $\dot{\varepsilon}\ (\geq 0)$ and that along the y_ε axis is $-\dot{\varepsilon}$ (Fig. 1). The velocity gradient tensor for pure shear flow $\mathbf{L}_\varepsilon$ with respect to $x_\varepsilon y_\varepsilon$ is then

$$\mathbf{L}_\varepsilon = \begin{pmatrix} \dot{\varepsilon} & 0 \\ 0 & -\dot{\varepsilon} \end{pmatrix} \tag{1}$$

Let the coordinates for simple shear flow be x_γ and y_γ, where the simple shear direction is parallel to x_γ and the shear strain rate $\dot{\gamma}$ is positive for dextral motion (Fig. 1). The velocity gradient tensor for simple shear flow $\mathbf{L}_\gamma$ with respect to $x_\varepsilon y_\varepsilon$ is

$$\mathbf{L}_\gamma = \begin{pmatrix} 0 & \dot{\gamma} \\ 0 & 0 \end{pmatrix} \tag{2}$$

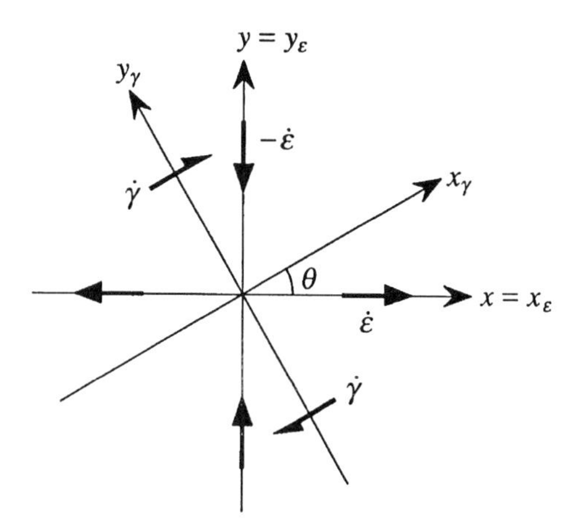

Fig. 1. Pure shear coordinates $x_\varepsilon y_\varepsilon$ ($= xy$) and simple shear coordinates $x_\gamma y_\gamma$ where x_γ axis makes an angle θ with x_ε axis. Longitudinal strain rate along x_ε axis is $\dot{\varepsilon}$ (≥ 0); along y_ε axis is $-\dot{\varepsilon}$. Simple shear direction is parallel to x_γ axis, and the shear strain rate is $\dot{\gamma}$ where dextral shear is taken as positive.

Any plane isochoric flow can be described by a simultaneous superposition of pure shear flow and simple shear flow (Ramberg 1975). Now let the x_ε and y_ε axes coincide with x and y axes respectively which are fixed in space, and let the x_γ axis subtend an angle θ with the x axis as in Fig. 1. Then $\mathbf{L}_\gamma$ with respect to xy becomes (Ramberg 1975)

$$\mathbf{L}_\gamma = \begin{pmatrix} -\dot{\gamma}\sin\theta\cos\theta & \dot{\gamma}\cos^2\theta \\ -\dot{\gamma}\sin^2\theta & \dot{\gamma}\sin\theta\cos\theta \end{pmatrix} \tag{3}$$

Adding equations (1) and (3) we obtain the velocity gradient tensor $\mathbf{L}$ for general plane isochoric flow

$$\mathbf{L} = \mathbf{L}_\varepsilon + \mathbf{L}_\gamma = \begin{pmatrix} \dot{\varepsilon}-\dot{\gamma}\sin\theta\cos\theta & \dot{\gamma}\cos^2\theta \\ -\dot{\gamma}\sin^2\theta & -\dot{\varepsilon}+\dot{\gamma}\sin\theta\cos\theta \end{pmatrix} \tag{4}$$

Internal vorticity, the non-coaxiality of flow, can be described by the kinematic vorticity number W_k (Truesdell 1954, Means *et al.* 1980). Any flow may comprise one of the following five types; (1) pure shear flow ($W_k = 0$), (2) simple shear flow ($W_k = 1$), (3) sub-simple shear flow ($0 < W_k < 1$), (4) super-simple shear flow ($1 < W_k < \infty$) (Ramberg 1975, Means *et al.* 1980, De Paor 1983). (5) Rigid rotation with $W_k = \infty$ is an end member of super-simple shear flow. W_k can be expressed in terms of the ratio $\dot{\varepsilon}/\dot{\gamma} = s_r$ and θ as follows (Ghosh 1987):

$$W_k = 1/\sqrt{\cos^2 2\theta + (2s_r - \sin 2\theta)^2} \tag{5}$$

Combinations of s_r and θ which satisfy eqn.(5) give identical particle paths.

Any homogeneous finite deformation can be described by a position gradient tensor **P** which relates the position of a particle in undeformed and deformed states. For plane isochoric (area-constant) deformation the determinant $|\mathbf{P}| = 1$. The strain ellipse shape tensor is then given as $\mathbf{PP}^T$ whose eigenvectors and eigenvalues give the orientation and squared axial lengths of the strain ellipse (Wheeler 1984, Kanagawa 1993). When the components of a position gradient tensor are the functions of time t, then the particle paths for progressive deformation can be obtained from the position gradient tensor. Such position gradient tensors can be derived from velocity gradient tensors for the above five types of flow (Ramberg 1975, Passchier 1988). Once either strain rate $\dot{\varepsilon}$ or $\dot{\gamma}$ is assumed, particle paths and the evolution of the strain ellipse with t for any W_k can be computed from the corresponding position gradient tensor.

Pure Shear Flow and Progressive Deformation ($W_k = 0$, $s_r = \infty$)

The position gradient tensor is given by Ramberg (1975) as

$$\mathbf{P} = \begin{pmatrix} e^{\dot{\varepsilon}t} & 0 \\ 0 & e^{-\dot{\varepsilon}t} \end{pmatrix} \tag{6}$$

The particle paths are rectangular hyperbolas with x and y axes being their asymptotes, and the strain is irrotational (Fig. 2a).

Simple Shear Flow and Progressive Deformation ($W_k = 1$, $s_r = 0$)

When $\theta = 0°$, the position gradient tensor is

$$\mathbf{P} = \begin{pmatrix} 1 & \dot{\gamma}t \\ 0 & 1 \end{pmatrix} \tag{7}$$

The particle paths are straight lines parallel to x axis, and the strain is moderately rotational (Fig. 2b).

Sub-simple Shear Flow and Progressive Deformation ($0 < W_k < 1$)

We choose $\theta = 0°$ for simplicity. Substituting this value into eqn. (4) we obtain

$$\mathbf{L} = \begin{pmatrix} \dot{\varepsilon} & \dot{\gamma} \\ 0 & -\dot{\varepsilon} \end{pmatrix} \tag{8}$$

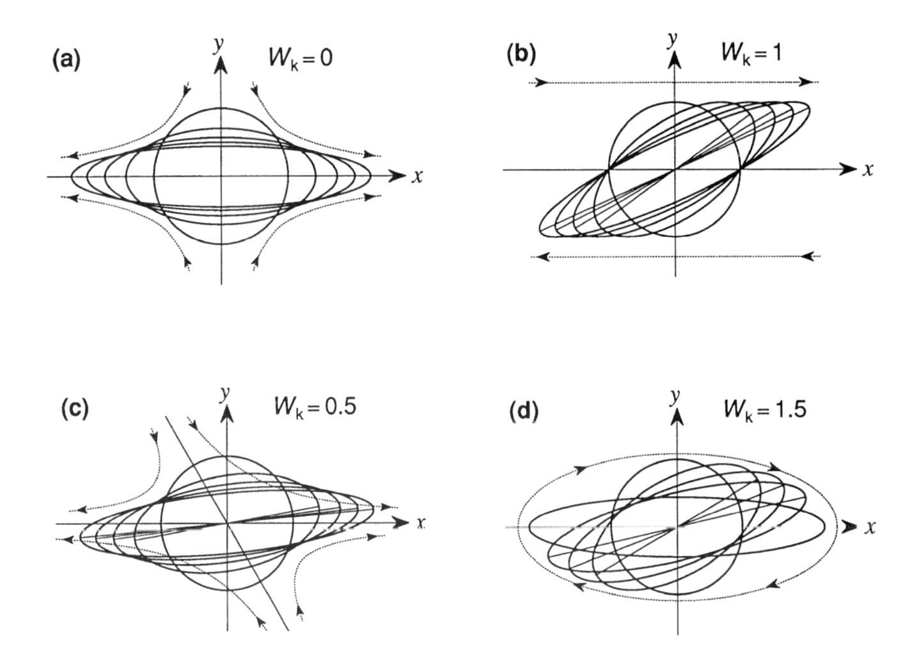

Fig. 2. Strain ellipse evolution for four types of flow. a) Pure shear flow ($W_k = 0$), b) Simple shear flow ($W_k = 1$). c) Sub-simple shear flow ($W_k = 0.5$). d) Super-simple shear flow ($W_k = 1.5$). Initial circle and strain ellipses with axial ratios $R_s = 2, 3, 4$, and 5 shown for each flow type. Dotted curves indicate representative particle paths.

Substituting $\theta = 0°$ into eqn. (5) and solving for s_r

$$s_r = \frac{1}{2}\sqrt{\frac{1}{W_k^2} - 1} \tag{9}$$

Since $0 < W_k < 1$, $s_r > 0$. s_r can thus be determined from eqn. (9) for any particular value of W_k. The position gradient tensor is given by Ramberg (1975) as

$$\mathbf{P} = \begin{pmatrix} e^{\dot{\varepsilon}t} & \dfrac{\sinh(\dot{\varepsilon}t)}{s_r} \\ 0 & e^{-\dot{\varepsilon}t} \end{pmatrix} \tag{10}$$

The particle paths are hyperbolas with their asymptotes being the x axis and $y = -2s_r x$, and the strain is slightly to moderately rotational (Fig. 2c).

Super-simple Shear Flow and Progressive Deformation ($W_k > 1$)

In this type of flow the velocity gradient tensor has complex eigenvalues. Since eqn. (8) always gives real eigenvalues $\dot{\varepsilon}$ and $-\dot{\varepsilon}$ we must choose θ other than 0°. Following Ramberg (1975) we choose $\theta = 45°$ and put this value in eqn. (4) to get

$$\mathbf{L} = \begin{pmatrix} \dot{\varepsilon} - \frac{\dot{\gamma}}{2} & \frac{\dot{\gamma}}{2} \\ -\frac{\dot{\gamma}}{2} & -\left(\dot{\varepsilon} - \frac{\dot{\gamma}}{2}\right) \end{pmatrix} \tag{11}$$

where $0 < \dot{\varepsilon} < \dot{\gamma}$ is required for complex eigenvalues. Substituting $\theta = 45°$ into eqn. (5) and solving for s_r

$$s_r = \frac{1}{2}\left(1 - \frac{1}{W_k}\right) \tag{12}$$

Since $W_k > 1$, $0 < s_r \leq 1/2$ (the equality holds when $W_k = \infty$). The position gradient tensor is given by Ramberg (1975) as

$$\mathbf{P} = \begin{pmatrix} \cos(a\dot{\gamma}t) + \frac{2s_r - 1}{2a}\sin(a\dot{\gamma}t) & \frac{1}{2a}\sin(a\dot{\gamma}t) \\ -\frac{1}{2a}\sin(a\dot{\gamma}t) & \cos(a\dot{\gamma}t) - \frac{2s_r - 1}{2a}\sin(a\dot{\gamma}t) \end{pmatrix} \tag{13}$$

where $a = \sqrt{s_r - s_r^2}$. The particle paths derived from equation (13) are ellipses whose long axes are oriented 45° from the x axis. We rotate coordinate axes ±45° before and after deformation respectively here in order to bring the simple shear direction parallel to the x axis as for other non-coaxial types of flow. The elliptical particle paths thus become oriented parallel to the x axis (Fig. 2d). The strain is moderately to strongly rotational with strain ellipses pulsating between a circle and an ellipse oriented parallel to the x axis (Fig. 2d).

Rotation of a Rigid Object in a Flowing Viscous Matrix

The rotation rate of a rigid elliptical object embedded in a Newtonian viscous matrix for any particular plane isochoric flow depends on the object's axial ratio R and long-axis orientation ϕ (Ghosh & Ramberg 1976; Fig. 3). In pure shear flow, the rotation rate $\dot{\phi}_\varepsilon$ of a rigid ellipse is given as follows (Ghosh & Sengupta 1973, Ghosh & Ramberg 1976)

$$\dot{\phi}_\varepsilon = -\dot{\varepsilon}\frac{R^2-1}{R^2+1}\sin 2\phi \tag{14}$$

In simple shear flow, the rotation rate $\dot{\phi}_\gamma$ with respect to $x_\gamma y_\gamma$ coordinates is given by the following eqn. (Jeffery 1922, Ghosh & Ramberg 1976, Bilby & Kolbuszewski 1977)

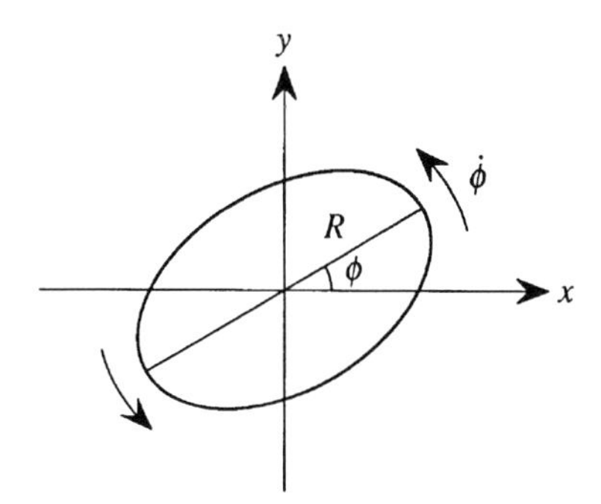

Fig. 3. Rotation rate $\dot{\phi}$ of a rigid ellipse with axial ratio R and long-axis orientation ϕ. Anticlockwise rotation is taken as positive.

$$\dot{\phi}_\gamma = -\dot{\gamma}\frac{R^2\sin^2\phi + \cos^2\phi}{R^2+1} \tag{15}$$

With respect to xy coordinates where the x_γ axis makes an angle θ from the x axis as in Fig. 1, eqn. (15) becomes

$$\dot{\phi}_\gamma = -\dot{\gamma}\frac{R^2\sin^2(\phi-\theta) + \cos^2(\phi-\theta)}{R^2+1} \tag{16}$$

The rotation rate of a rigid ellipse in general plane isochoric flow, $\dot{\phi}$, is therefore obtained by the sum of two rotation rates given by eqns. (14) and (16) (Ghosh & Ramberg 1976)

$$\dot{\phi} = \dot{\phi}_\varepsilon + \dot{\phi}_\gamma \tag{17}$$

$$= -\dot{\varepsilon}\frac{(R^2-1)\sin 2\phi}{R^2+1} - \dot{\gamma}\frac{R^2\sin^2(\phi-\theta) + \cos^2(\phi-\theta)}{R^2+1} \tag{17a}$$

Using $\dot{\varepsilon}/\dot{\gamma} = s_r$ and rearranging, eqn. (17) is rewritten as follows

$$\dot{\phi} = -\dot{\gamma}\left(A\cos^2\phi + B\sin 2\phi + C\sin^2\phi\right) \tag{18}$$

where

$$A = \frac{\cos^2\theta + R^2\sin^2\theta}{R^2+1} \tag{18a}$$

$$B = \frac{(s_r - \sin\theta\cos\theta)(R^2-1)}{R^2+1} \tag{18b}$$

$$C = \frac{R^2 \cos^2\theta + \sin^2\theta}{R^2+1} \tag{18c}$$

Equations (14), (15), and (18) can also be written in the following form:

$$-\dot{\varepsilon} \text{ or } -\dot{\gamma} = \frac{\dot{\phi}}{f(\phi)} \tag{19}$$

The relation between orientations of a rigid ellipse before and after a deformation increment is derived by integrating both sides of eqn. (19) with the boundary conditions that ε or $\gamma = 0$ and $\phi = \phi_0$ when $t = 0$, and that $\varepsilon = \dot{\varepsilon}t$ or $\gamma = \dot{\gamma}t$ and $\phi = \phi$ when $t = t$. Once either strain rate $\dot{\varepsilon}$ or $\dot{\gamma}$ and a particular value of s_r are assumed, the progressive rotation of a rigid ellipse with axial ratio R and initial orientation ϕ_0 is determined from ϕ as a function of t (*e.g.* Fig. 4).

Progressive Rotation of a Rigid Ellipse in Pure Shear Flow ($W_k = 0$)

The rotation rate is given by eqn. (14) (Fig. 4a). Rearranging this equation as eqn. (19) and integrating both sides from $t = 0$ to $t = t$ yields

$$-\dot{\varepsilon}t = \frac{R^2+1}{2(R^2-1)} \ln\left(\frac{\tan\phi}{\tan\phi_0}\right) \tag{20a}$$

therefore

$$\phi = \tan^{-1}\left\{\tan\phi_0 \, e^{-2\dot{\varepsilon}t(R^2-1)/(R^2+1)}\right\} \tag{20b}$$

Progressive Rotation of a Rigid Ellipse in Simple Shear Flow ($W_k = 1$)

The rotation rate is given by eqn. (15) (Fig. 4b) where $\theta = 0°$. Rearranging this equation as eqn. (19) and integrating both sides from $t = 0$ to $t = t$ gives

$$-\dot{\gamma}t = \frac{R^2+1}{R}\left\{\tan^{-1}(R\tan\phi) - \tan^{-1}(R\tan\phi_0)\right\} \tag{21a}$$

therefore

$$\phi = \tan^{-1}\left[\frac{1}{R}\tan\left\{\tan^{-1}(R\tan\phi_0) - \dot{\gamma}t\frac{R}{R^2+1}\right\}\right] \tag{21b}$$

Progressive Rotation of a Rigid Ellipse in Sub-simple Shear Flow ($0 < W_k < 1$)

We put $\theta = 0°$ in eqn. (18) (Fig. 4c) as for flow and progressive deformation, so its coefficients A, B, and C given by eqns. (18a)-(18c) become

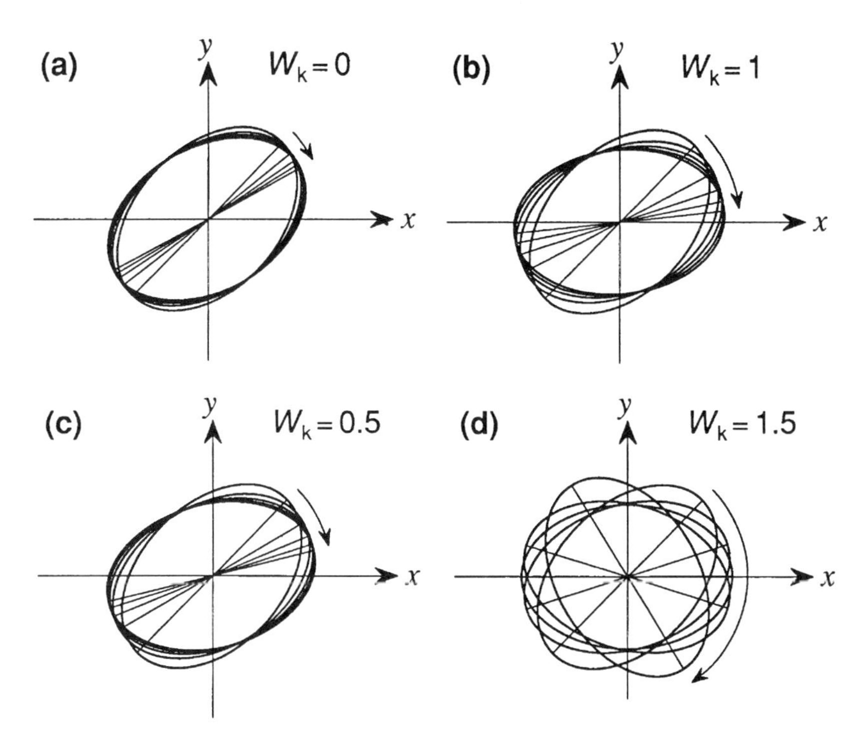

Fig. 4. Progressive rotation of rigid ellipse. R = 1.5, ϕ_0 = 45°. a) Pure shear flow, $W_k = 0$. b) Simple shear flow, $W_k = 1$. c) Sub-simple shear flow, $W_k = 0.5$. d) Super-simple shear flow, $W_k = 1.5$. Initial orientation and orientations after matrix strain ratios $R_s = 2, 3, 4, 5$ shown for each case.

$$A = \frac{1}{R^2+1} \tag{22a}$$

$$B = s_r \frac{R^2-1}{R^2+1} \tag{22b}$$

$$C = \frac{R^2}{R^2+1} \tag{22c}$$

Rearranging eqn. (18) as eqn. (19) and integrating both sides from $t = 0$ to $t = t$ we get

$$-\dot{\gamma}t = \int_{\phi_0}^{\varphi} \frac{d\phi}{A\cos^2\phi + B\sin 2\phi + C\sin^2\phi} \tag{23}$$

The integral on the right-hand side of eqn. (23) has three different solutions as follows depending on the values of B^2 and AC (Ghosh & Ramberg 1976).

(1) Case $B^2 < AC$

$$-\dot{\gamma}t = \left[\frac{1}{h}\tan^{-1}\left(\frac{C\tan\phi + B}{h}\right)\right]_{\phi_0}^{\phi} \tag{24}$$

where

$$h = \sqrt{AC-B^2}. \tag{24a}$$

Solving this equation for ϕ we obtain

$$\phi = \tan^{-1}\left[\frac{h}{C}\tan\left\{\tan^{-1}\left(\frac{C\tan\phi_0+B}{h}\right) - \dot{\gamma}t\,h\right\} - \frac{B}{C}\right] \tag{24b}$$

(2) Case $B^2 = AC$

$$-\dot{\gamma}t = \left[-\frac{1}{C\tan\phi + B}\right]_{\phi_0}^{\phi} \tag{25}$$

therefore

$$\phi = \tan^{-1}\left[\frac{1}{C}\left\{\frac{C\tan\phi_0 + B}{1+\dot{\gamma}\,t\,(C\tan\phi_0 + B)} - B\right\}\right] \tag{25a}$$

(3) Case $B^2 > AC$

$$-\dot{\gamma}t = \left[\frac{1}{2k}\ln\left(\frac{C\tan\phi + B - k}{C\tan\phi + B + k}\right)\right]_{\phi_0}^{\phi} \tag{26}$$

where

$$k = \sqrt{B^2-AC}, \tag{26a}$$

therefore

$$\phi = \tan^{-1}\left(\frac{D(B + k) - B + k}{C(1-D)}\right) \tag{26b}$$

where

$$D = \frac{C\tan\phi_0 + B - k}{C\tan\phi_0 + B + k} e^{-2\dot{\gamma} t k} \tag{27}$$

Progressive Rotation of a Rigid Ellipse in Super-simple Shear Flow (W_k>1)

We put $\theta = 45°$ in eqn. (18) (Fig. 4d) as for flow and progressive deformation, so its coefficients *A, B,* and *C* given by eqns. (18a)-(18c) become

$$A = \frac{1}{2} \tag{28a}$$

$$B = \left(s_r - \frac{1}{2}\right)\frac{R^2-1}{R^2+1} \tag{28b}$$

$$C = \frac{1}{2} \tag{28c}$$

Since $0 < s_r \leq 1/2$ and $\frac{R^2-1}{R^2+1} < 1$, $-\frac{1}{2}<B\leq 0$, it therefore always holds that

$$B^2 < \frac{1}{4} = AC \tag{28d}$$

Hence ϕ is given by eqn. (24). ϕ_0 is rotated 45° before deformation and ϕ is rotated -45° after deformation here (Fig. 4d), corresponding to rotating coordinate axes ±45° before and after deformation, respectively.

An Alternative Numerical Simulation Method for Displacement-controlled Rigid Fiber Growth Around a Rigid Object

A rigid object and its pressure fringe composed of rigid fibers, embedded in a flowing matrix, will have a rate of rotation as given by eqn.(17) depending on their total shape and orientation. A deformation increment will therefore result in an incremental rotation of the rigid object and its pressure fringe, as well as in an incremental growth of the pressure fringe fiber. We can numerically simulate these processes for any plane, isochoric, Newtonian flow with constant flow parameters as follows, provided that the process of fiber growth is assumed.

We know the position gradient tensor **P** and the rotation of a rigid ellipse for any particular value of W_k. These are the functions of strain rate multiplied by time, $\dot{\varepsilon}t$ or $\dot{\gamma}t$. For incremental deformation, $\dot{\varepsilon}t$ or $\dot{\gamma}t$ is replaced by $\dot{\varepsilon}\Delta t$ or $\dot{\gamma}\Delta t$, which is in turn expressed as $\Delta\varepsilon$ or $\Delta\gamma$, because the strain rates are assumed constant. The strain rate ratio $s_r = \dot{\varepsilon}/\dot{\gamma}$ can be rewritten as $\Delta\varepsilon/\Delta\gamma$ because

$$s_r = \frac{\dot{\varepsilon}}{\dot{\gamma}} = \frac{\dot{\varepsilon}\Delta t}{\dot{\gamma}\Delta t} = \frac{\Delta\varepsilon}{\Delta\gamma} \tag{28e}$$

For each incremental strain $\Delta\varepsilon$ or $\Delta\gamma$, the incremental fiber growth can be numerically simulated by the following three steps (Fig. 5). Initial position vectors of the outline of a rigid object are set as $\boldsymbol{x}_{00k}$ ($k = 1, 2, .., n_p$), where n_p is the number of points. Position vectors of fibers after i–1 strain increments are $\boldsymbol{x}_{i-1jk}$ ($j = 0, 1, ..., i-1$; $k = 1, 2, ..., n_p$), where $\boldsymbol{x}_{i-10k}$ ($k = 1, 2, ..., n_p$) defines the rigid object at this stage. A kth fiber is thus represented by $\boldsymbol{x}_{i-1jk}$ ($j = 0, 1, ..., i-1$).

For the ith increment of strain, the best-fit ellipse of the outline of the previously formed pressure fringe is calculated by the least-squares method, and its axes l_{1i} and l_{2i}, and long-axis orientation ϕ_i are obtained (Fig. 5a). If i=1 (first strain increment), then this ellipse-fitting is applied to the rigid object. Next, incremental rotation, $\Delta\omega_i = \phi_i' - \phi_i$, is calculated by substituting R (l_{1i}/l_{2i}) and ϕ_i and replacing $\dot{\varepsilon}t$ or $\dot{\gamma}t$ by $\Delta\varepsilon$ or $\Delta\gamma$ in eqns. (20), (21) or (24)-(26). The rigid object and pressure fringe are then rotated such that

$$[\mathbf{l}_{1i}'\,\mathbf{l}_{2i}'] = \mathbf{R}[\mathbf{l}_{1i}\,\mathbf{l}_{2i}] \tag{29a}$$

(Fig. 5b) where

$$\mathbf{R} = \begin{pmatrix} \cos\Delta\omega_i & -\sin\Delta\omega_i \\ \sin\Delta\omega_i & \cos\Delta\omega_i \end{pmatrix} \tag{29b}$$

The third step consists of incremental syntaxial or antitaxial fiber growth, Δl_{ik} ($k = 1, 2, ..., n_p$), depending on the fiber growth process, which results in fibers represented by $\boldsymbol{x}_{ijk}$ ($j = 0,1,...,i$) (Fig. 5c).

Syntaxial Fiber Growth Models

Syntaxial fiber growth from a rigid object toward its matrix occurs in crinoid-type pressure fringes (Ramsay & Huber 1983). Since fibers are assumed to be rigid, the object and previously formed fringe behave as a combined rigid body during incremental deformation. For incremental syntaxial fiber growth there are three possible models. Fibers may grow parallel to incremental particle paths in the matrix (Fig. 6a) which are directly derived from the incremental position gradient tensor $\Delta\mathbf{P}$, which is in turn obtained by replacing $\dot{\varepsilon}t$ or $\dot{\gamma}t$ by $\Delta\varepsilon$ or $\Delta\gamma$ in eqns. (6), (7), (10) or (13). Such growth is equivalent to the passive fiber growth proposed by Ellis (1986). Fibers may also grow parallel to the maximum incremental extension $\Delta\boldsymbol{e}_1$ which is derived from the incremental strain ellipse whose shape tensor is given by $\Delta\mathbf{P}(\Delta\mathbf{P})^T$. Two other models are possible. One is incremental fiber growth so as to fill a gap formed by incrementally strained matrix (Fig. 6b), while the other is growth equivalent to the magnitude of the maximum incremental extension, $\Delta\boldsymbol{e}_1$

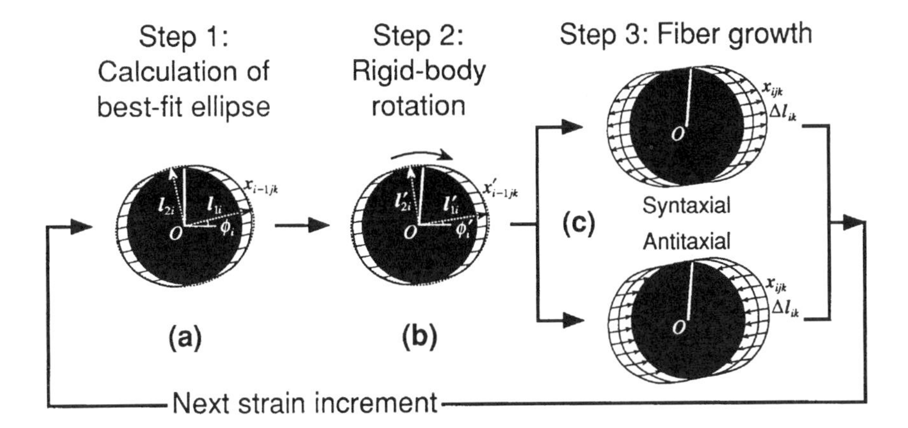

Fig. 5. Three simulation steps of rigid pressure fringe growth around rigid object. After $i-1$ strain increments, fiber is represented by $\boldsymbol{x}_{i-1jk}$ (see text). a) Best-fitting ellipse of previous formed fringe calculated by least-squares in ith strain increment; axial lengths l_{1i}, l_{2i}, long-axis orientation ϕ_i (step 1). b) Rigid object and pressure fringe rotated by $\Delta\omega_i$ (step 2). c) Incremental syntaxial/antitaxial fiber growth depending on growth process results in fiber represented by $\boldsymbol{x}_{ijk}$ (step 3).

(Fig. 6c). The latter corresponds to the rigid fiber model of Durney & Ramsay (1973). It is uncertain yet which of the above three models better simulates natural syntaxial fiber growth.

Antitaxial Fiber Growth Models

Antitaxial fiber growth from the matrix toward a rigid object occurs in pyrite-type pressure fringes (Ramsay & Huber 1983). Incremental fiber growth occurs between a rigid object and its previously formed rigid pressure fringe, filling the space formed by their separation due to incremental strain. It is reasonable to consider incremental antitaxial fiber growth parallel to, and equivalent to the maximum incremental extension Δe_1, as in the rigid fiber model of Durney & Ramsay (1973) and Ramsay & Huber (1983). There are two possible models for incremental rotation. One is the coupled rotation model in which incremental rotation of the previously formed pressure fringe is coupled with that of a rigid object (Fig. 7a). The incremental rotation of the object and its previously formed pressure fringe is obtained in this model from the axial ratio and long-axis orientation of the best-fit ellipse of the outline of the previously formed pressure fringe. The other model is decoupled rotation, where incremental rotations of the object and pressure fringe are different (Fig. 7b). We therefore calculate in this model the best-fit ellipse for each outline of object and two sides of the pressure fringe, then obtain their incremental rotations separately. We do not know yet whether either of them is appropriate for natural antitaxial fiber growth.

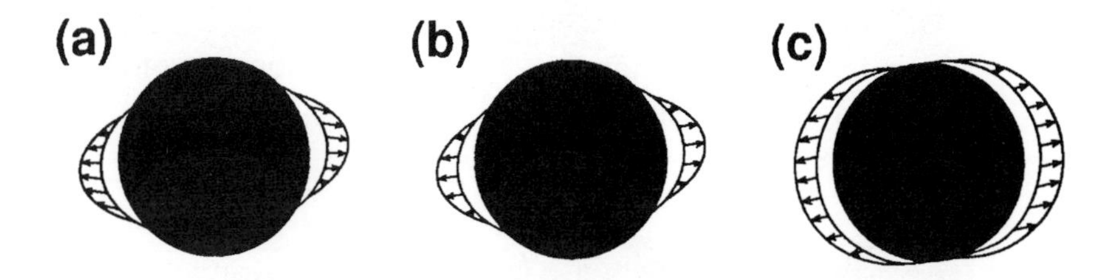

Fig. 6. Possible models for displacement-controlled syntaxial rigid fiber growth. Growth parallel to: a) incremental particle paths in matrix, b) direction of maximum incremental extension to fill gap between previously formed pressure fringe and incrementally strained matrix, (c) maximum incremental extension direction.

Examples of Syntaxial and Antitaxial Fiber Growths Around a Circular Rigid Object

The numerical simulation method of displacement-controlled rigid fiber growth in pressure fringes proposed above is illustrated below for the rigid fiber model of syntaxial fiber growth and the coupled rotation model of antitaxial fiber growth around a circular rigid object. In both models, an object and its previously formed pressure fringe rotate as a combined rigid body during incremental deformation, and fibers grow parallel to the maximum incremental extension $\Delta\boldsymbol{e}_1$ by the amount equivalent to $\Delta\boldsymbol{e}_1$.

We set the origin O at the center of a rigid circle, and n_p points define the circle whose position vectors are x_{00k} ($k = 1, 2, ..., n_p$). W_k is initially given. A strain increment is then given by $\Delta\varepsilon$ or $\Delta\gamma$, from which the incremental position gradient tensor $\Delta\mathbf{P}$ is obtained. For the ith increment of strain, the ellipse-fitting of the previously formed pressure fringe (step 1 in Fig. 5) and the subsequent incremental rotation (step 2 in Fig. 5) are performed as described above. The base length for incremental fiber growth, l_i, can be assumed to be the radial length of the rotated best-fit ellipse in the direction of $\Delta\boldsymbol{e}_1$ (Fig. 8a). The length of incremental fiber growth, Δl_{ik} ($k = 1, 2, ..., n_p$), is then

$$\Delta l_{ik} = l_i \Delta\boldsymbol{e}_1. \tag{30}$$

Incremental syntaxial fiber growth occurs outward from the end points of fibers, $\boldsymbol{x}'_{i-1i-1k}$ ($k = 1, 3 ..., n_p$). Incremental syntaxial fiber growth vector, Δl_{ik}, depends on the location of $\boldsymbol{x}'_{i-1i-1k}$ with respect to $\Delta\boldsymbol{e}_1$, and is given as follows (Fig. 8b):

$$\Delta l_{ik} = \left\{ \begin{array}{ll} l_i \Delta e_1 & x'_{i-1i-1k}.\Delta e_1 \geq 0 \\ -l_i \Delta e_1 & x'_{i-1i-1k}.\Delta e_1 < 0 \end{array} \right\} \tag{31}$$

Position vectors of fibers after the ith incremental syntaxial fiber growth, $\boldsymbol{x}_{ijk}$ ($j = 0, 1, ..., i$; $k = 1, 2, ..., n_p$), are then (Fig. 8b)

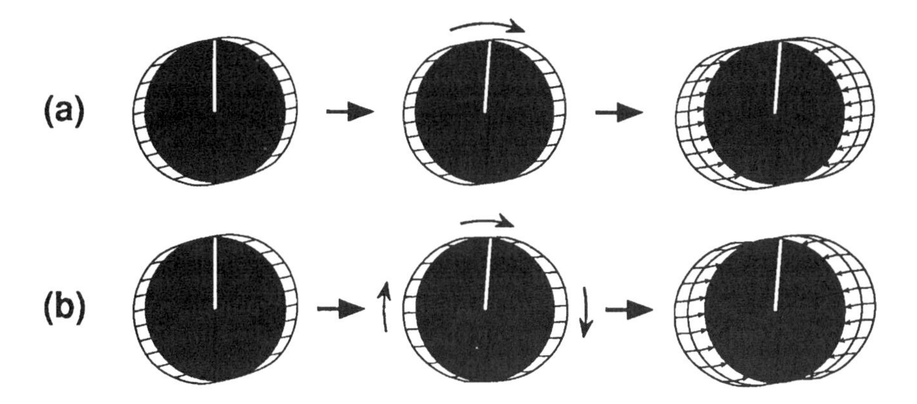

Fig. 7. Possible models for displacement-controlled antitaxial rigid fiber growth. Rigid body rotation of previously formed pressure fringe is coupled with that of rigid object in a) – coupled rotation model – whereas it is decoupled in b) – decoupled rotation model. In both models, fibers grow parallel to, and with extension equal to, maximum incremental extension.

$$x_{ijk} = \begin{Bmatrix} x'_{i\text{-}1jk} \ \ (j = 0,1,...,i\text{-}1) \\ x'_{i\text{-}1j\text{-}1k} + \Delta l_{ik} \ \ (j = i) \end{Bmatrix} \tag{32}$$

Incremental antitaxial fiber growth occurs inward from the points of fibers on the rigid object outline, $\boldsymbol{x}'_{i-10k}$ ($k = 1, 2,..., n_p$). The incremental antitaxial fiber growth vector, Δl_{ik}, depends on the location of $\boldsymbol{x}'_{i-10k}$ with respect to $\Delta \boldsymbol{e}_1$, and is given as follows (Fig. 8c):

$$\Delta l_{ik} = \begin{Bmatrix} -l_i \Delta e_1 \ \ (x'_{i\text{-}10k} . \Delta e_1 \geq 0) \\ l_i \Delta e_1 \ \ (x'_{i\text{-}10k} . \Delta e_1 = 0) \end{Bmatrix} \tag{33}$$

Position vectors of fibers after the ith incremental antitaxial fiber growth, x_{ijk} ($j = 0, 1,..., i$; $k = 1,2,..., n_p$), are then (Fig. 8c)

$$x_{ijk} = \begin{Bmatrix} x'_{i-1jk} \ \ (j = 0) \\ x'_{i-1j-1k} - \Delta l_{ik} \ \ (j = 1,2,...,i) \end{Bmatrix} \tag{34}$$

The incremental rotation of the rigid circle and pressure fringe followed by the incremental fiber growths given above necessarily produces a few exceptional fibers. This occurs where Δl_{ik} changes its sign by k as shown in eqns (31) and (33), and results in an intersection of two adjacent fibers for syntaxial fiber growth (Fig. 8d) or a sharp bend in a fiber for antitaxial fiber growth (Fig. 8e), which are unrealistic or unusual in naturally formed pressure fringes. For a syntaxial fiber whose incremental growth vector intersects the adjacent fiber, we therefore stop the incremental growth of the fiber where it touches the adjacent fiber (Fig. 8d).

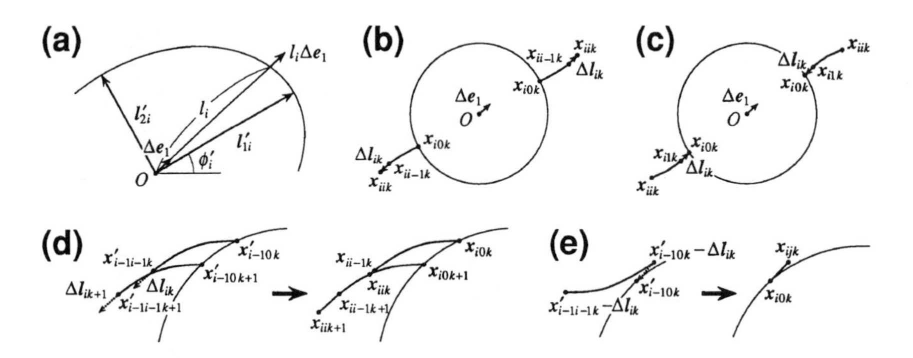

Fig. 8. Detail of ith incremental fiber growth (step 3) for syntaxial and antitaxial fiber growths. a) Radial length of rotated best-fitting ellipse in direction of $\Delta\mathbf{e}_1$. l_i is base length for fiber growth. b) Syntaxial growth occurs outward from end points of fibers (see text). c) Incremental antitaxial fiber growth occurs inward from points of fibers on rigid object outline (see text). d) If incremental growth vector from end point of syntaxial fiber intersects adjacent fiber, incremental growth of fiber is stopped where it touches adjacent fiber. e) If antitaxial fiber changes growth direction due to rotation resulting in sharp bend, previously formed fiber is removed.

For antitaxial fiber growth a sharp bend in a fiber occurs when Δl_{ik} has the opposite sign to Δl_{i-1k}. In such a case we remove the previously formed fiber (Fig. 8e).

The position gradient tensor for cumulative strain after n increments, $\mathbf{P}_f$, can be obtained by n times sequential multiplication of $\Delta\mathbf{P}$ (Elliott 1972)

$$\mathbf{P}_f = (\Delta\mathbf{P})^n \tag{35}$$

The orientation and axial ratio of the finite strain ellipse in the matrix is then derived from the eigenvectors and eigenvalues of $\mathbf{P}_f\mathbf{P}_f^T$.

Some results of syntaxial and antitaxial pressure fringe growths thus simulated for four types of progressive deformation (W_k = 0, 0.5, 1 and 1.5) are shown in Figs. 9 and 10. Progressive deformations with different W_k values result in quite different pressure fringes. For coaxial progressive deformation (W_k =0), both syntaxial and antitaxial fiber growths result in the same pressure fringes with symmetric straight fibers. For non-coaxial progressive deformations (W_k > 0), syntaxial and antitaxial fibers grow as asymmetric curves having different convex sides with respect to the matrix. For a particular value of W_k, fiber lengths increase with increasing finite strain. For the same amount of finite strain, the amount of rotation and curvature of fibers systematically increase with increasing W_k. The number of fibers that are stopped from growing or partly removed during progressive deformation increases with increasing W_k as well as with increasing finite strain, due to the increasing amount of rotation. Except for such fibers, the lengths and curvatures of fibers are constant in each pressure fringe.

Curvature Analysis of Simulated Fibers

Since the lengths and curvatures of fibers in the simulated syntaxial and antitaxial pressure fringes around a rigid circle are constant and apparently dependent on W_k and finite strain (Figs. 9 and 10), as suggested by Simpson & De Paor (1993), they should be sensitive to varying W_k and finite strain.

We use fiber length normalized by the radius of a rigid circle, r. The normalized total length of a fiber x_{njk} ($j = 0, 1, ..., n$), l'_{nk}, is then given as

$$l'_{nk} = \sum_{i=1}^{n} \frac{\Delta l_{ik}}{r} \tag{36}$$

where Δl_{ik} is the length of the ith incremental fiber growth. The curvature of the ith to i+1th segments of the fiber (Fig. 11), κ_{ik}, can be represented by the angle $\Delta\theta_{ik}$ between Δl_{ik} and Δl_{i+1k} measured in radians, divided by the sum of their lengths,

$$\kappa_{ik} = \frac{\Delta\theta_{ik}}{\Delta l_{ik}+\Delta l_{i+1k}} \tag{37}$$

The sign of κ_{ik} depends on $\Delta\theta_{ik}$ as is evident from eqn. (37), and therefore κ_{ik} is positive for an anticlockwise deflection of a fiber during growth. κ_{ik} changes with fiber growth, and therefore is plotted against the normalized cumulative length of fibers, l'_{ik}, where

$$l'_{ik} = \sum_{1}^{i} \frac{\Delta l_{ik}}{r} \tag{38}$$

Thus Fig. 11 represents a progressive curvature change with growth.

Progressive curvature changes on the simulated syntaxial and antitaxial fibers for variable W_k ($0 \leq W_k \leq 2$) are shown in Fig. 12 and Fig. 13, respectively. For coaxial progressive deformation ($W_k = 0$), curvature is always 0 as is evident from straight fibers in Figs. 9 and 10. For non-coaxial progressive deformations ($W_k > 0$), curvatures of both syntaxial and antitaxial fibers are always positive (anticlockwise deflection from the growth direction) for dextral shear as in this study, whereas it is always negative (clockwise deflection) in sinistral shear. As inferred from Figs. 9 and 10, absolute values of curvature systematically increase with increasing W_k. Although they monotonously decrease with finite strain, syntaxial fibers for $W_k > 1.75$ have a minimum curvature at a late stage of their growth (Fig. 12). Due to the pulsating nature of super-simple shear progressive deformation ($W_k > 1$), the finite strain ratio never exceeds a limiting value of 5 for $W_k = 1.5$ and 3 for $W_k = 2$.

The angle $\Delta\theta_{ik}$ between Δl_{ik} and Δl_{i+1k} (Fig. 11) is equal to $-\Delta\omega_{i+1}$, where $\Delta\omega_{i+1}$ is the i+1th incremental rotation of a pressure fringe (see Fig. 8a-c). The progressive curvature change of fibers obtained from eqns. (37) and (38) therefore represents the change of incremental rotation $\Delta\omega$ with respect to incremental fiber growth Δl. For a particular value of W_k, $\Delta\omega$ depends on the axial ratio R and

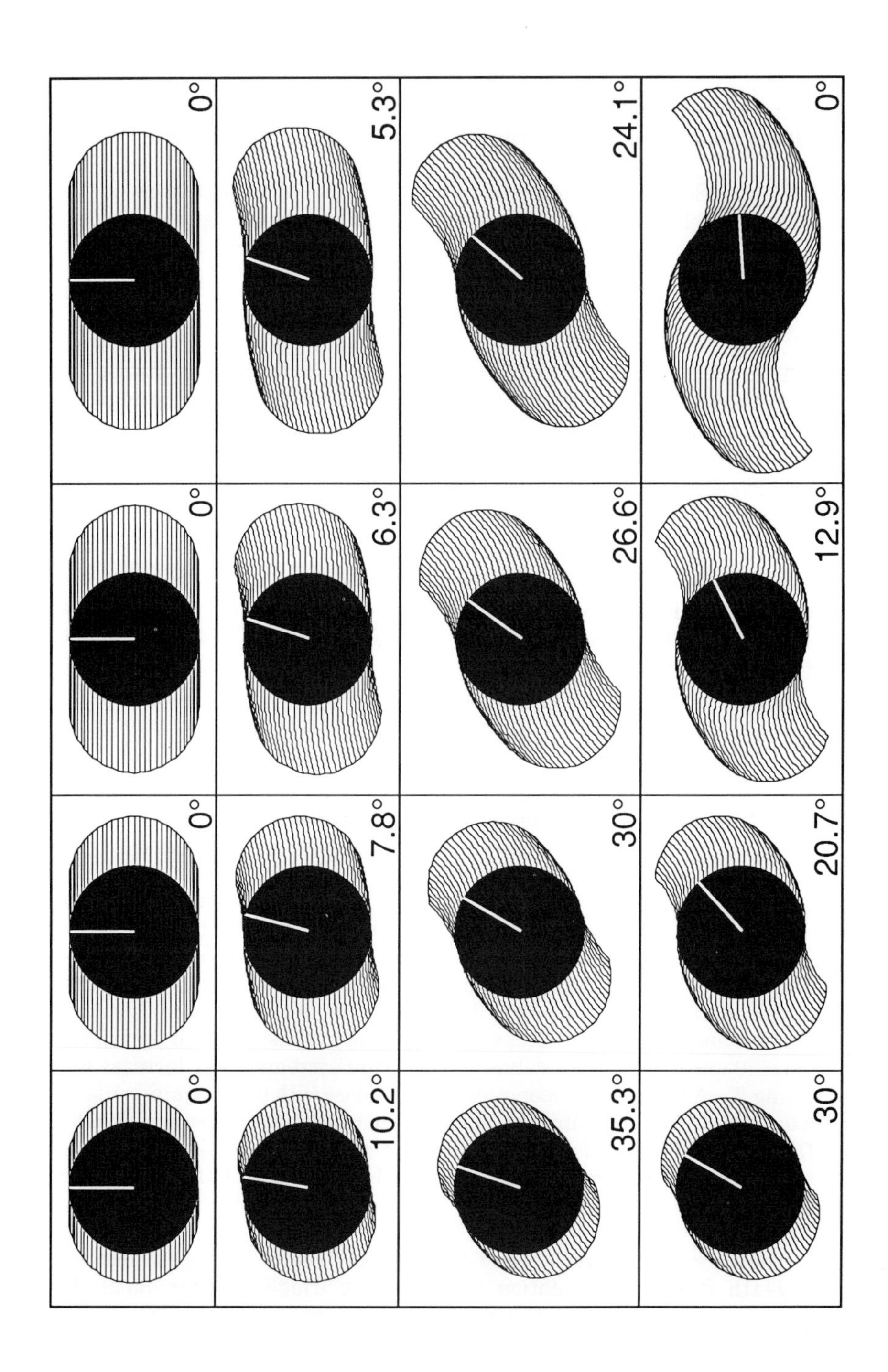
0°
0°
0°
0°
5.3°
6.3°
7.8°
10.2°
24.1°
26.6°
30°
35.3°
0°
12.9°
20.7°
30°

long-axis orientation ϕ of the best-fit ellipse of a pressure fringe (eqns. (20), (21) or (24-26)). $\Delta\omega$ can also be evaluated from the rotation rate $\dot{\phi}/\dot{\gamma}$ (eqn. 18). Δl depends on the radial length l of the rotated best-fit ellipse in the direction of the maximum incremental extension Δe_1 (eqns. (31) and (33)), which in turn depends on the axial lengths (l_1 and l_2) and orientation ϕ of the best-fit ellipse.

In order to understand the progressive curvature change of simulated fibers, the progressive change of R, ϕ, l and $-\dot{\phi}/\dot{\gamma}$ (clockwise rotation rate) are plotted against the normalized cumulative fiber length l' for three-types of non-coaxial progressive deformation ($W_k = 0.5$, 1, and 2) (Fig. 14). Although $-\dot{\phi}/\dot{\gamma}$ is initially 0.5 irrespective of W_k, the strain rate ratio $s_r = \dot{\varepsilon}/\dot{\gamma}$ varies with W_k and so does the rotation rate $\dot{\phi}$. R and l monotonously increase with l' except for the syntaxial fiber growth with $W_k = 2$ (Fig. 14c), and therefore Δl also generally increases. In contrast, ϕ monotonously decreases by clockwise rotation of a pressure fringe. $-\dot{\phi}/\dot{\gamma}$ decreases with decreasing ϕ (also see Ghosh & Ramberg 1976, Figs. 2 and 3) and so would $\Delta\omega$. Therefore the curvature generally decreases monotonously with growth of fibers (Figs. 12 and 13).

For syntaxial fiber growth with $W_k = 2$ (Fig. 14c), l reaches a maximum at $l' \approx 0.7$, then slightly decreases with l'. A rapid decrease of ϕ due to a large rotation results in a large angle between ϕ and Δe_1 and a decrease in l even though l_1 increases. In addition, $\dot{\phi}$ becomes negative at $l' \approx 1.1$, reaching a minimum in $-\dot{\phi}/\dot{\gamma}$. The decrease in l with l' as well as reaching a minimum in $-\dot{\phi}/\dot{\gamma}$ must be responsible for an appearance of minimum curvature (Fig. 12) for the syntaxial fiber growth with $W_k \geq 1.75$.

The difference in ϕ between the simulated syntaxial and antitaxial pressure fringes for $W_k \geq 1$ (*e.g.* Fig. 14c and f) is probably due to their different shapes (Figs. 9 and 10). A smaller ϕ for syntaxial pressure fringes than that for antitaxial ones results in smaller $-\dot{\phi}/\dot{\gamma}$ (*e.g.*, Figs, 14c and f), hence a smaller angle of rotation. This difference in rotation angle between syntaxial and antitaxial pressure fringes is demonstrated when we compare the simulated pressure fringes with $W_k = 1.5$ and $R_s = 5$ in Figs. 9 and 10.

Fig. 9 (opposite). Simulated syntaxial pressure fringes around rigid circular object for 4 types of progressive deformation ($W_k = 0, 0.5, 1, 1.5$) by displacement-controlled rigid fiber model of syntaxial fiber growth. Four stages of pressure fringe corresponding to $R_s = 2, 3, 4, 5$ shown for each type. Long-axis orientation of matrix strain ellipse is indicated on lower right of each pressure fringe. Number of strain increments = 10, 20, 30, 40, respectively. White line in rigid object was initially at 90°.

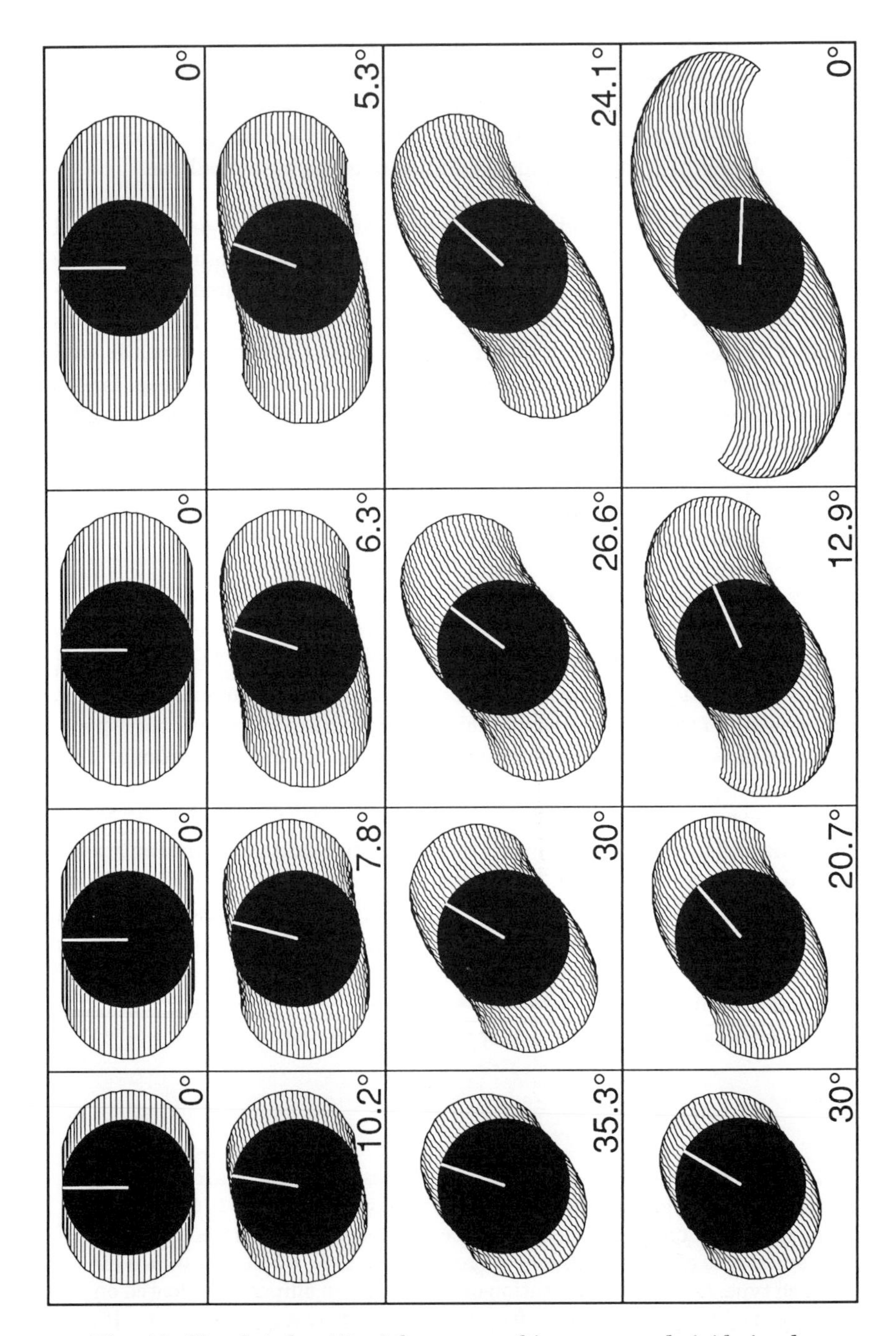

Fig. 10. Simulated antitaxial pressure fringes around rigid circular object for 4 types of progressive deformation (W_k = 0, 0.5, 1, 1.5) by coupled rotation model of displacement-controlled antitaxial rigid fiber growth. See text for further explanation.

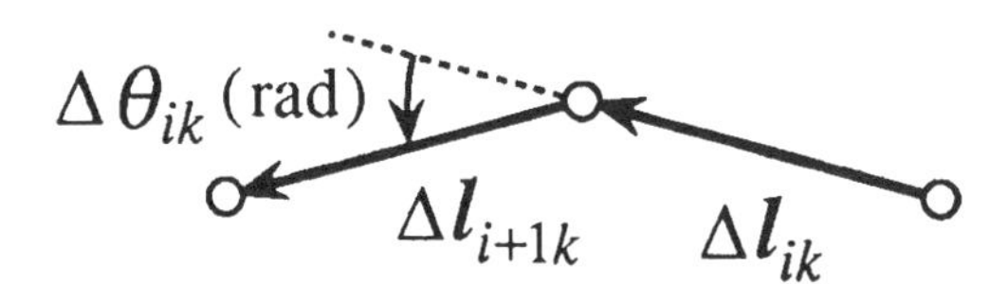

Fig. 11. Curvature analysis of pressure fringe fibers. See text for explanation.

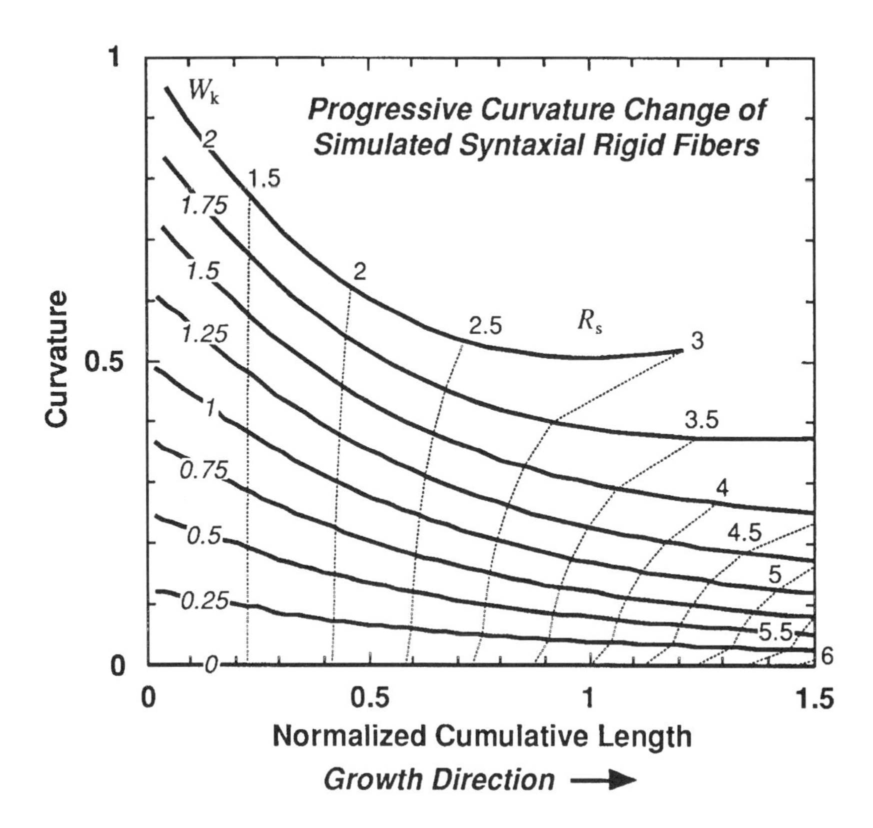

Fig. 12. Progressive curvature change (solid lines) of simulated syntaxial fibers by displacement-controlled rigid fiber model for variable kinematic vorticity numbers ($0 \leq W_k \leq 2$). Curvature is plotted against cumulative fiber length normalized by radius of object. Dotted lines link points of equal R_s in matrix. Curvature positive for dextral shear, negative for sinistral shear.

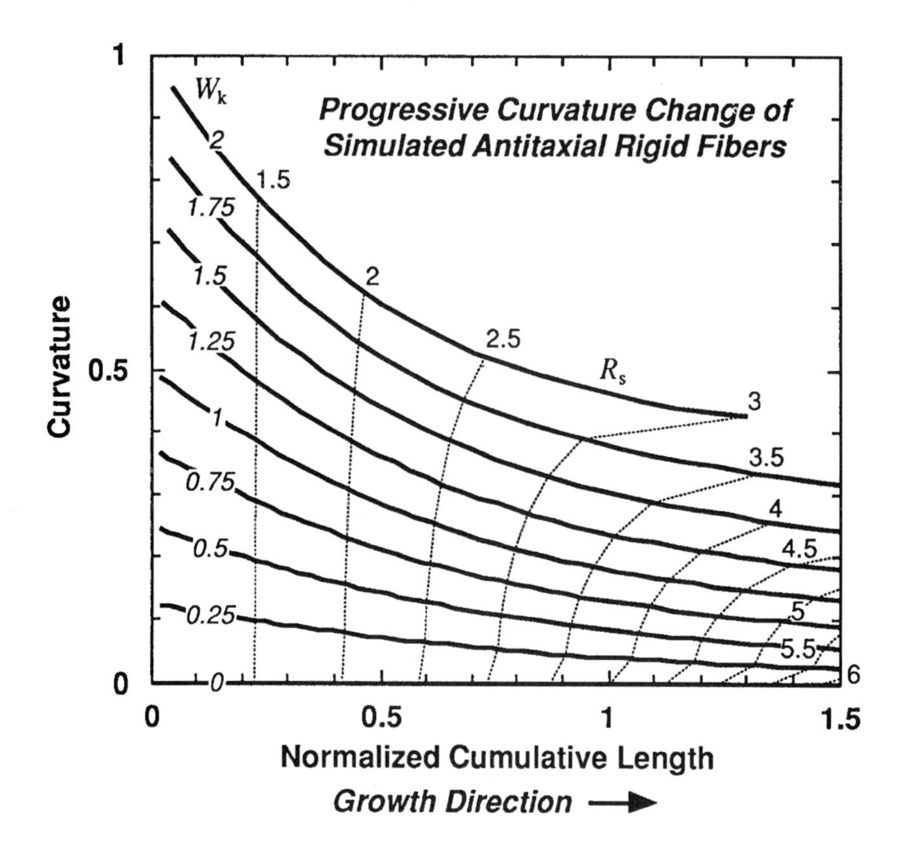

Fig. 13. Progressive curvature change (solid lines) of simulated antitaxial fibers by coupled rotation model of displacement-controlled antitaxial rigid fiber growth. See text for further explanation.

Conclusions

In contrast to the previous numerical simulation method of Etchecopar & Malavieille (1987), which is based on a geometric best-fit between a rigid object, its pressure fringe and the matrix, the alternative numerical simulation method proposed here is based on the rotation of a rigid object and its pressure fringe according to their shape and flow in a viscous matrix. The method can be applied to both syntaxial and antitaxial rigid-fiber growths for any plane, isochoric, Newtonian viscous flow with constant flow parameters, provided that the process of fiber growth is known. Three models are possible for syntaxial rigid fiber growth, whereas two models are possible for antitaxial rigid fiber growth.

The simulation method is illustrated here for the rigid fiber models of syntaxial and antitaxial fiber growths around a circular rigid object, where incremental rotation of the previously formed pressure fringe is coupled with that of a rigid object. Lengths and curvatures of most fibers in a simulated pressure fringe are constant (Figs. 9 and 10), and progressive curvature change of

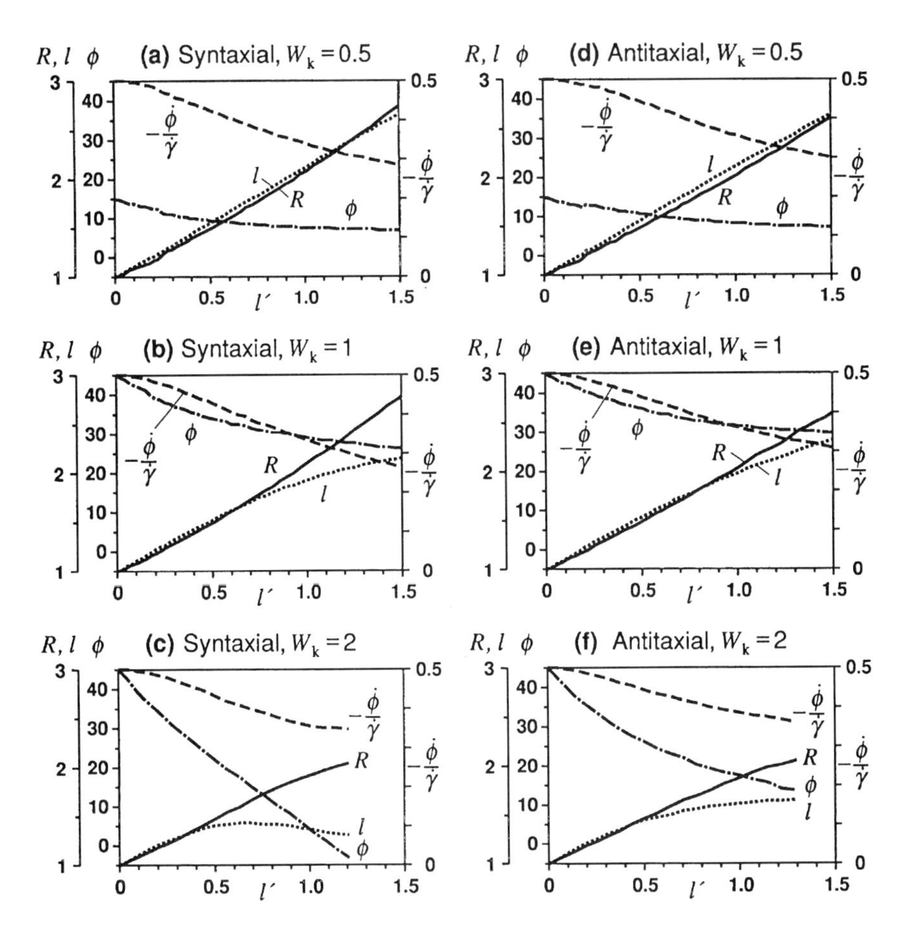

Fig. 14. Progressive changes of R (solid line) and long-axis orientation (ϕ dashed-dotted line) of best-fitting ellipse of simulated pressure fringe, its radial length l (dotted line) in maximum incremental extension direction after rigid rotation, and its rate of clockwise rotation $-\dot{\phi}/\dot{\gamma}$ (dashed line). Values for 3 types of non-coaxial progressive deformation (W_k = 0.5, 1, 2) plotted against normalized cumulative fiber length l' in syntaxial and antitaxial fiber growth models.

simulated fibers systematically vary according to kinematic vorticity number and finite strain (Figs. 12 and 13). Both curvature and normalized cumulative fiber length defined by equations (37) and (38) are measurable parameters from natural pressure fringes. Comparison of progressive curvature changes obtained from natural fibers and those of simulated fibers would therefore enable us to estimate internal vorticity and finite strain quantitatively, if a natural rigid object is circular in section normal to the vorticity vector, if natural pressure fringes are rigid and grow as in the simulation models illustrated in this study, and if vorticity is constant during progressive deformation. Many framboidal pyrites are spherical in shape and suitable for this analysis. Progressive curvature change analysis is also applicable to the decoupled rotation model of antitaxial rigid fiber growth which yields constant length and curvature of most fibers in a simulated pressure fringe. In these cases simulated pressure fringes would provide a powerful tool for quantitative estimates of vorticity and progressive deformation. In order for such estimates to be practical, however, it is required that the process of fiber growth is known. This is not possible from numerical simulations only; we need to examine natural pressure fringes in more detail from the viewpoint of fiber growth.

Acknowledgements

I thank Declan De Paor for his encouragement and editorial efforts, an anonymous reviewer for helpful comments and improving the manuscript, and Kazuhiko Ishii for providing me his manuscript in press, discussion on pressure fringe simulation, and especially for pointing out an error in an earlier algorithm of fiber growth.

References

Bilby, B. A. & Kolbuszewski M. L. 1977. The finite deformation of an inhomogeneity in two-dimensional slow viscous incompressible flow. *Proc. R. Soc. Lond.* **A355**: 335-353.

De Paor, D. G. 1983. Orthographic analysis of geological structures—I. Deformation theory. *J. Struct. Geol.* **5:** 255-277.

Durney, D. W. & Ramsay, J. G. 1973. Incremental strains measured by syntectonic crystal growths. In: *Gravity and Tectonics* (ed. De Jong, K. A. & Scholten R.). Wiley, New York, 67-96.

Elliott, D. 1972. Deformation paths in structural geology. *Bull. geol. Soc. Am.* **83**: 2621-2638.

Ellis, M. A. 1986. The determination of progressive deformation histories from antitaxial syntectonic crystal fibers. *J. Struct. Geol.* **8**: 701-709.

Etchecopar, A. & Malavieille, J. 1987. Computer models of pressure shadows: a method for strain measurement and shear-sense determination. *J. Struct. Geol.* **9**: 667-677.

Ghosh, S. K. 1987. Measure of non-coaxiality. *J. Struct. Geol.* **9**: 111-113.

Ghosh, S. K, & Ramberg, H. 1976. Reorientation of inclusions by combination of pure shear and simple shear. *Tectonophysics* **34**: 1-70.

Ghosh, S. K. & Sengupta, S. 1973. Compression and simple shear of test models with rigid and deformable inclusions. *Tectonophysics* **17**: 133-175.

Gray, D. R. & Durney, D. W. 1979. Investigations on the mechanical significance of crenulation cleavage. *Tectonophysics* **58**: 35-79.

Ishii, K. 1995. Estimation of non-coaxiality from crinoid type pressure fringes: comparison between natural and simulated examples. *J. Struct. Geol.* **17:** 1267-1278.

Jeffery, G. B. 1922. The motion of ellipsoidal particles immersed in a viscous fluid. *Proc. R. Soc. Lond.* **A102**: 161-179.

Kanagawa, K. 1993. Competence contrasts in ductile deformation as illustrated from naturally deformed chert-mudstone layers. *J. Struct. Geol.* **15**: 865-885.

Malavieille, J. & Ritz, J. F. 1989. Mylonitic deformation of evaporites in décollements: examples from Southern Alps, France. *J. Struct. Geol.* **11:** 583-590.

Malvern L. E. 1969. *Introduction to the Mechanics of a Continuous Medium.* Prentice-Hall, Englewood Cliffs.

Means, W. D., Hobbs, B. E., Lister G. S. & Williams, P. F. 1980. Vorticity and non-coaxiality in progressive deformation. *J. Struct. Geol.* **2**: 371-378.

Passchier, C. W. 1988. The use of Mohr circles to describe non-coaxial progressive deformation. *Tectonophysics* **149**: 323-338.

Ramberg, H. 1975. Particle paths, displacement and progressive strain applicable to rocks. *Tectonophysics* **28**: 1-37.

Ramsay, J. G. & Huber, M. I. 1983. *The Techniques of Modern Structural Geology. Volume 1: Strain Analysis.* Academic Press, London.

Simpson, C. & De Paor, D. G. 1993. Strain and kinematic analysis in general shear zones. *J. Struct. Geol.* **15**: 1-20.

Truesdell, C. 1954. *The Kinematics of Vorticity.* Indiana University Press, Bloomington.

Wheeler, J. 1984. A new method to display the strain of elliptical markers. *J. Struct. Geol.* **6**: 417-423

Wickham, J. S. 1973. An estimate of strain increments in a naturally deformed carbonate rock. *Am J. Sci.* **273**: 23-47.

Flinn Diagram Construction on *Macintosh* Computers

Jay Zimmerman
Department of Geology, Southern Illinois University
at Carbondale, Carbondale IL 62901-4324, U.S.A.
zimmerman@geo.siu.edu

Abstract– Flinn diagrams, two-dimensional data plots that offer graphical means of quantitatively describing and comparing ellipsoids, have long been part of the standard strain analysis repertoire and are receiving increased attention by instructors and students in structural geology courses. This paper describes and discusses two Flinn diagram plotting programs written in *BASIC* for Macintosh computers. The two programs utilize *XYZ*-type and R_f- or R_s - type data, respectively.

Introduction

Flinn diagrams offer graphical means to quantitatively describe and compare the shapes of triaxial ellipsoids of any type: material, geometric, or optical. In structural geology, the diagrams have typically been used to portray sets of strain (geometric) ellipsoids of use in analyses of total or progressive deformation of rocks.

Undergraduate-level and advanced structural geology texts and treatises (*e.g.*, Ramsay 1967, Means 1976, Hobbs, Means, & Williams 1976, Ramsay & Huber 1983, Davis 1984, Twiss & Moores 1992, Ghosh 1993) address the construction and use of Flinn diagrams, and in specific instances (Rowland & Duebendorfer 1994, Hatcher 1995) offer practical exercises to be completed by students. The programs described in this paper may, therefore, be useful to instructors of, and students in, introductory or advanced structural geology courses, as well as to researchers interested in object/fabric shape characterization or in the analysis of deformation.

The Flinn Diagram and Its Modifications

In 1956, Flinn published his modification of an axial ratio diagram originally introduced by Zingg (1935) to describe and classify the shapes of sedimentary clasts. The Flinn method requires measurement of the three principal axes ($X \geq Y \geq Z$) of an ellipsoid and the plotting of axial ratios X/Y *vs.* Y/Z as ordinate and abscissa, respectively, each ellipsoid being represented by a single point in Cartesian "Flinn space". Spheres, for which

$$X = Y = Z \tag{1}$$

plot at the origin of the diagram, and ellipsoids become increasingly eccentric with distance from the origin. If

$$X/Y > Y/Z \tag{2}$$

the point will locate in a part of the diagram populated by prolate ellipsoids of various shape, whereas for

$$X/Y < Y/Z \tag{3}$$

it will locate in a field of oblate ellipsoids. In cases where all three principal axes of individual ellipsoids cannot be measured directly (not atypical in geology), X/Y and Y/Z axial ratios of appropriate objects can be determined from suitable outcrops or from specimens cut parallel to XY and YZ planes. The set of measured axial ratios can then be averaged to a single point, R_f, in Flinn space that represents collective (population) object shape data from the outcrop or specimen.

Numerous methods for determining principal axes of the strain ellipse and strain ellipsoid have been developed (see, for example, Ramsay 1967, Ramsay & Huber 1983). When the strain ellipsoid, R_s, is plotted on the Flinn diagram, coordinate axis values represent ratios of principal stretches

$$1{+}e_1/1{+}e_2 = S_1/S_2 = R_{xy} \tag{4}$$

for the ordinate. The field of prolate ellipsoids is then interpreted as a field of constrictional strain, and the field of oblate ellipsoids is considered one of actual or apparent (in the case of volume loss preceding or during deformation) flattening strain (Ramsay & Wood 1973). In such cases, the Flinn diagram can provide insight into the character of total strain, or into the history of progressive deformation.

In 1962 Flinn defined and added to his diagram a parameter

$$k = (a-1)/(b-1), \tag{5}$$

where, in effect,

$$a = (1+e_1)/(1+e_2) \tag{6}$$

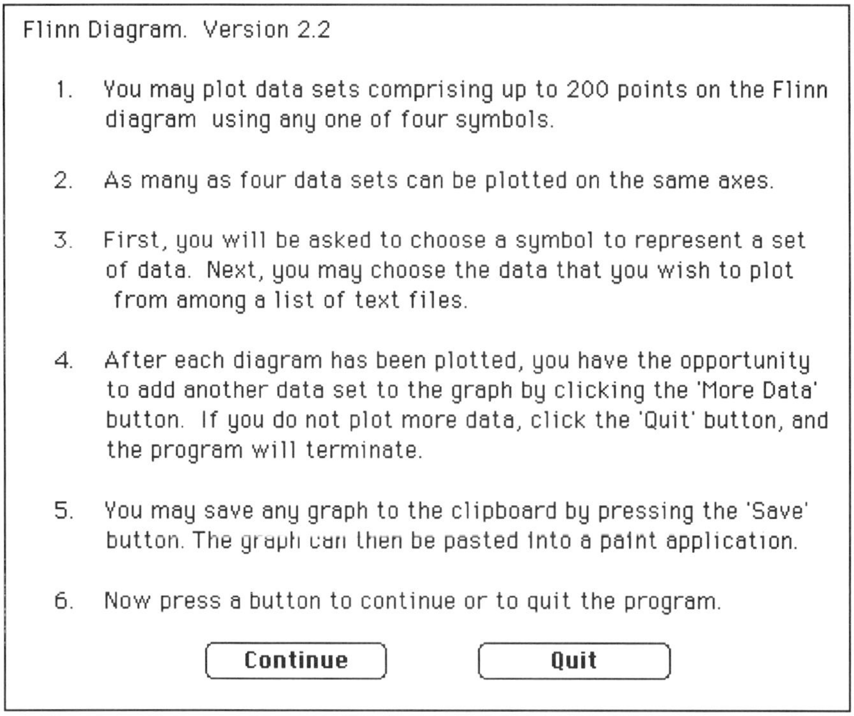

Flinn Diagram. Version 2.2

1. You may plot data sets comprising up to 200 points on the Flinn diagram using any one of four symbols.

2. As many as four data sets can be plotted on the same axes.

3. First, you will be asked to choose a symbol to represent a set of data. Next, you may choose the data that you wish to plot from among a list of text files.

4. After each diagram has been plotted, you have the opportunity to add another data set to the graph by clicking the 'More Data' button. If you do not plot more data, click the 'Quit' button, and the program will terminate.

5. You may save any graph to the clipboard by pressing the 'Save' button. The graph can then be pasted into a paint application.

6. Now press a button to continue or to quit the program.

Continue   Quit

Fig. 1. This explanation window appears when either application is opened. Users may choose to continue or exit the program.

and

$$b = (1+e_2)/(1+e_3) \tag{7}$$

to better describe simple deformation paths: k being the slope of a line from the coordinate origin to the point representing any ellipsoid in Flinn space. This parameter was later redefined for incremental deformation as

$$k' = \ln a / \ln b \tag{8}$$

(Flinn 1965). For most routine purposes, it is sufficient to keep in mind that

$$k = k' = \infty \tag{9}$$

defines uniaxial prolate ellipsoids lying along the ordinate of the Flinn diagram; that

$$k = k' = 0 \tag{10}$$

defines uniaxial oblate ellipsoids along the abscissa; and that

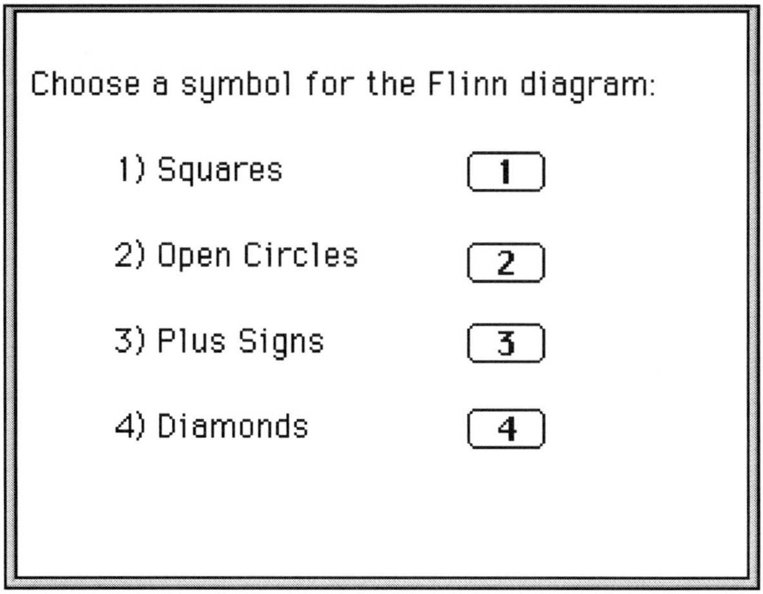

Choose a symbol for the Flinn diagram:

1) Squares [1]

2) Open Circles [2]

3) Plus Signs [3]

4) Diamonds [4]

Fig. 2. Users are presented with a choice of four symbols to represent data on the Flinn diagram.

$$k = k' = 1 \qquad (11)$$

describes a line inclined at 45° to the abscissa that separates the fields of prolate and oblate ellipsoids (this line also defines the equal-volume plane strain path). Elsewhere in Flinn space, $k \neq k'$. Flinn (1965, 1978), Ramsay (1967), Ramsay & Wood (1973), and Ramsay & Huber (1983) explain the uses of these parameters in characterizing deformation paths.

In adapting the Flinn diagram for the specific purposes of strain analysis (that is, recording strain ellipsoids rather than ellipsoidal objects) Ramsay (1967) suggested that

$$\log_e (1+e_1)/(1+e_2) = \ln R_{xy} \qquad (12)$$

be plotted against

$$\log_e (1+e_2)/(1+e_3) = \ln R_{yz}. \qquad (13)$$

In this way, the origin of the logarithmic Flinn diagram (0,0) represents a state of zero strain, and all points within the body of the diagram represent deformations rather than fabric or object shapes. The conceptual link between the original Flinn diagram – origin at (1,1) – and Ramsay's logarithmic modification – origin

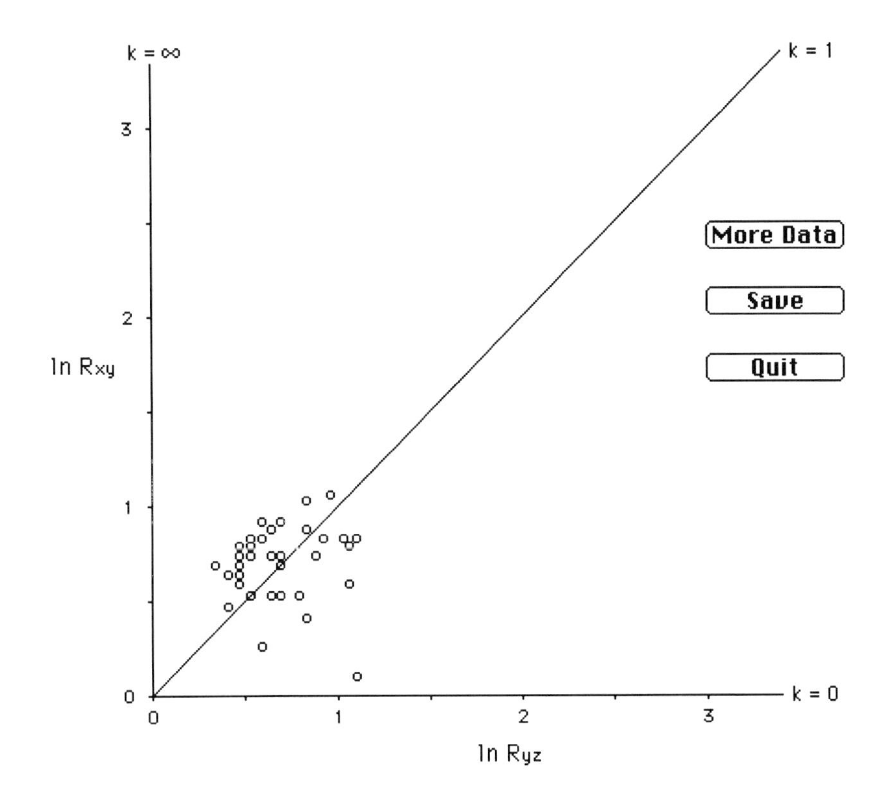

Fig. 3. Preliminary Flinn diagram. Users can elect to add data sets, save the current diagram to the clipboard, or exit the program without saving the diagram. Data: deformed cobbles from the Funzie conglomerate, Shetland Islands (Flinn 1956).

at (0,0) – lies in the reasonable assumption that populations of statistically spherical objects in rocks, such as oöids or reduction spots, indicate zero net deformation.

Owing to my professional interest in the analysis of deformation, the computer programs discussed here incorporate the Ramsay modification and are constructed to plot ln R_{xy} *vs.* ln R_{yz}. It should be emphasized, however, that they can be used to record and describe the other types of ellipsoids mentioned earlier.

Program Descriptions and Execution

In current versions of the programs the user must initially decide between two separate routines depending on the type of raw data to be used. One program utilizes *XYZ* data (individual measurements of the long, intermediate, and short

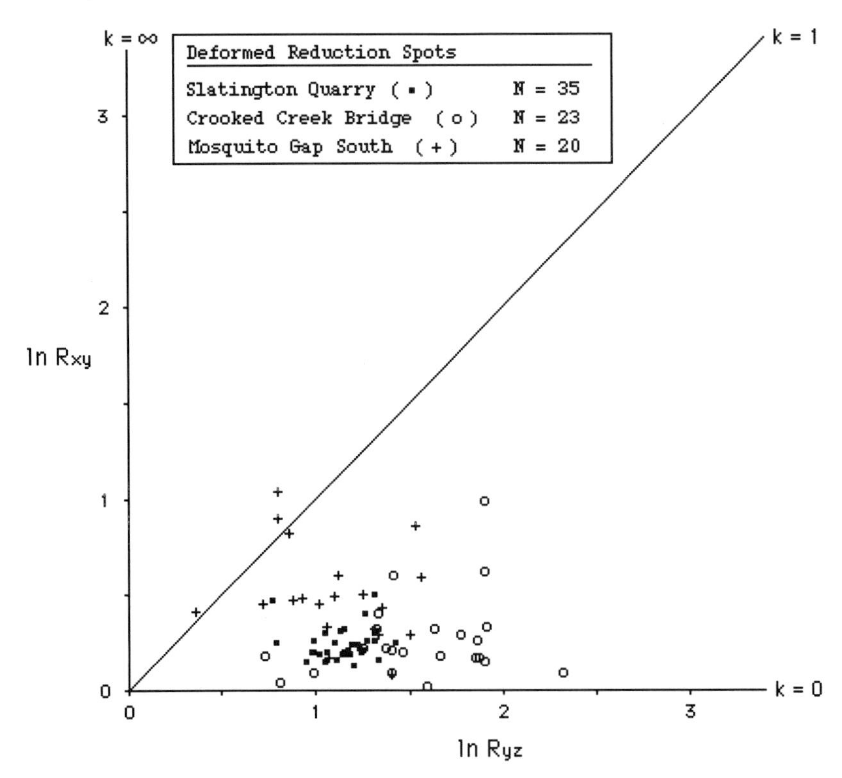

Fig. 4. Flinn diagram containing three data sets. This diagram was saved to the clipboard and pasted into a paint application for editing and labeling. Data: reduction spots from three locations in the Missouri Mountain Formation (Silurian), Ouachita Mountains, Arkansas.

axes of ellipsoids), and the other operates on R_f (R_s) data (ratios of long to intermediate and intermediate to short ellipsoid axes). Once opened and running, both programs require identical responses.

After opening the chosen application the user is presented with a text window (Fig. 1) that cites limits on the total permissible number of data entries and data sets, choice of symbols, and options for disposition of the plotted diagram. Buttons provide the choice of continuing or exiting the program at this stage.

If the user continues, a second window (Fig. 2) prompts a choice between four symbols (squares, open circles, plus signs, and filled diamonds) for the first data set. Following selection of a symbol, the program presents a list of data files. The user must select and open the desired file in the usual manner by using the system's `File` menu, the appropriate key strokes (cmd-O), or by double-clicking the selection. At this point it is most important that the format of the filed data entries (*XYZ* or R_f) match the program being used. A mismatch will result in an error message or data points plotted outside the coordinate axes.

Opening an appropriately formatted data file will automatically produce a fully plotted Flinn diagram (Fig. 3). At this stage, the user can choose among three button options: adding another data set to the plot, saving the current plot, or exiting the program. If one chooses to add a data set, the symbol window will reappear followed by a list of data files to select and open, and the steps described in the preceding paragraph are repeated. Fig. 4 is a diagram on which three independent sets of data have been plotted.

When all desired data sets have been plotted, the user can elect to exit the program by clicking the `Quit` button (Fig. 3), in which case the plot will not be saved. Alternatively, activating the `Save` button will copy the completed diagram to the clipboard (prior to automatically terminating the program run) from which it can be pasted into a graphics program such as Aldus *SuperPaint*™, *Canvas*™, or Claris *MacPaint*™, for clean-up, labeling, and final formatting. A finished, labeled diagram is shown in Fig. 4.

Constructing Data Files

Several standard word processing applications including Microsoft *Word*™, *ClarisWorks*™ (*MacWrite*™), and *SimpleText*™, can be used to create *XYZ* and $R_f(R_s)$ data files. In all cases tested to date the file must be saved in "Text" (ASCII) format in order to be successfully read by the plotting programs.

Data file construction is straightforward. In *XYZ* format the three measured lengths of the principal axes of an ellipsoid, separated by commas, occupy each line. The last number on each line should not be followed by a comma. Each $R_f(R_s)$ data line comprises two numbers, the X/Y and Y/Z ratios, separated by a comma. Note that in both formats the file must end with the final numerical entry in a column of data. If a carriage return is inadvertently placed in the line following the final entry, an error message will appear when the programs attempt to plot the data.

Acknowledgements

The author gratefully acknowledges help and advice from Michael Reese, Michael Kruge, and Dhananjay Ravat.

References

Davis, G. H. 1984. *Structural Geology of Rocks and Regions*. John Wiley & Sons, New York, 492 pp.

Flinn, D. 1956. On the deformation of the Funzie conglomerate, Fetlar, Shetland. *Journal of Geology* **64**: 480-505.

Flinn, D. 1962. On folding during three-dimensional progressive deformation. *Q. J. Geol. Soc.* **118**: 385-433.

Flinn, D. 1965. Deformation in metamorphism. In: Pitcher, W. S. and Flinn, G. W. (eds.), *Controls of Metamorphism*. John Wiley & Sons, New York, pp. 46-72.

Flinn, D. 1978. Construction and computation of three-dimensional progressive deformation: *J. Geol. Soc.Lond.* **135**: 291-305.

Ghosh, S. K. 1993. *Structural Geology: Fundamentals and Modern Developments.* Pergamon Press, Oxford, 598 pp.

Hatcher, R. D., Jr. 1995. *Structural Geology: Principles, Concepts, and Problems* (2nd Ed). Prentice Hall, New Jersey, 525 pp.

Hobbs, B. E., Means, W. D., & Williams, P. F. 1976. *An Outline of Structural Geology*. John Wiley & Sons, New York, 571 pp.

Means, W. D. 1976. *Stress and Strain*. Springer-Verlag, New York, 339 pp.

Ramsay, J. G. 1967. *Folding and Fracturing of Rocks*. McGraw-Hill Book Co., New York, 568 pp.

Ramsay, J.G. & Huber, M. I. 1983. *The Techniques of Modern Structural Geology, Vol. I: Strain Analysis*. Academic Press, London, 307 pp.

Ramsay, J.G. & Wood, D. S. 1973. The geometric effects of volume change during deformation processes. *Tectonophysics* **16**: 263-277.

Roland, S. M. & Duebendorfer, E. M. 1994. *Structural Analysis and Synthesis* (2nd ed.). Blackwell Scientific Publications, Boston, 279 pp.

Twiss, R. J. & Moores, E. M. 1992. *Structural Geology*. W. H. Freeman and Company, New York, 532 pp.

Zingg, T. 1935. Beitrag zur Schotteranalyze. *Schweizer. Mineral. und Petrograph. Mitteil.* **15**: 39-140.

A Modified Data Input Procedure for the *Fry* 5.8 Strain Analysis Application

Jay Zimmerman
Department of Geology, Southern Illinois University
at Carbondale, Carbondale IL 62901-4324, U.S.A.
zimmerman@geo.siu.edu

Abstract– *Fry* version 5.8 is an application for doing Fry strain analysis (Fry 1979) on the Macintosh computer. This short contribution describes an improved data entry procedure which resolves problems in identifying object centers when data is derived from digitized documents. The user's ability to accurately locate object centers is greatly improved if the template is converted to color. The digitizing cursor then shows up in a complementary color. The procedure described requires a color monitor and the Aldus *SuperPaint*™ application.

Introduction

One of the major advantages of the Macintosh application called *Fry* version 5.8 by De Paor and Simpson is that it enables the user to digitize positions of objects such as centers of ooids with the computer mouse. Like me, however, other users may have experienced difficulty seeing the black or white (b&w) spot made by the mouse against the background of a b&w template. The digitizing spot tends to get lost in the clutter of the template which is often a scanned photograph with a wide range of gray tones as well as b&w. Because the cursor is programmed to XOR with the screen (that is to show black against white and *vice versa*), I find that I often fail to select centers of some objects while marking the centers of others more than once. Fortunately, the fit of a strain ellipse to data in Fry analysis is totally unaffected by multiple markings of the same center but it may be adversely affected by omissions. One solution is to convert the template from b&w to color. An undocumented feature of the program is that the digitizing cursor then marks the spot in a complementary color that is much more visible than the original b&w. I have had the best results with a green cursor against a deep red or magenta template and with blue against orange. A green template also works well.

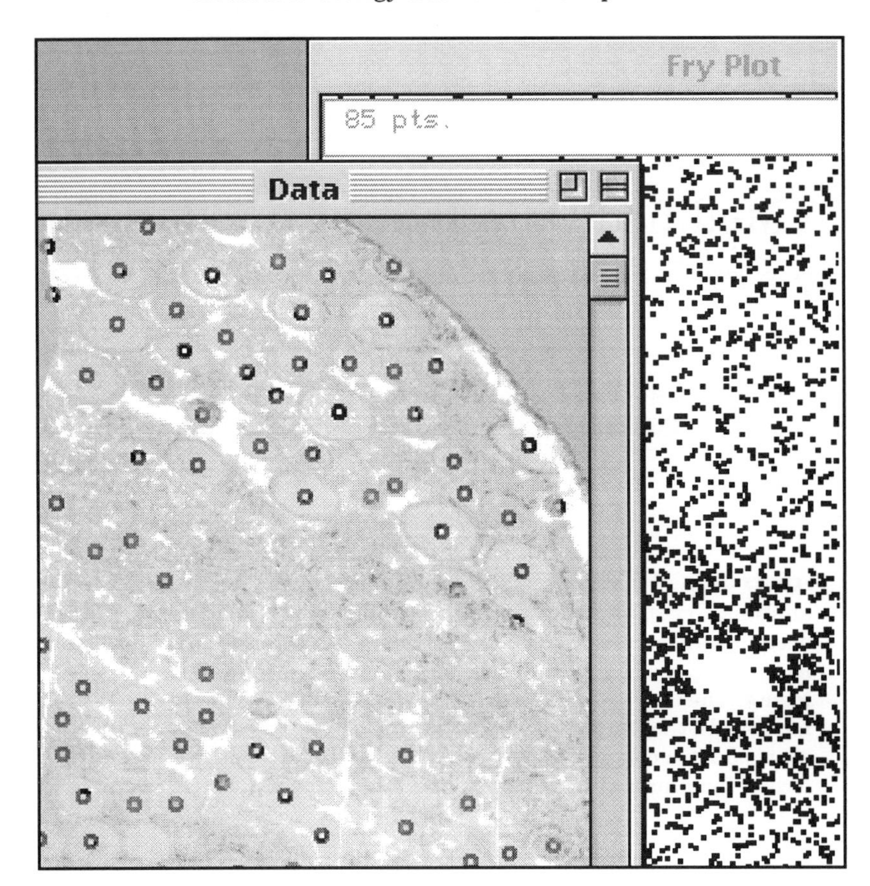

Fig. 1. Screen shot of two windows from the *Fry* 5.8 application. Front window shows scanned template of data. Colors appear as grays. Thick dark rings are imprinted on the template when the mouse is clicked on an object center. Note how shade varies with background. Window in rear is Fry plot for this data. Template prepared in *Photoshop*.

Provided that the user has a color computer and monitor, the following procedure for converting a b&w template to color may be of value. I have used Aldus *SuperPaint* throughout.

Procedure

1. Only 1-bit b&w documents can be transformed to color. Multiple-bit documents such as the Ashton PICT (8-bit) which accompanies *Fry* 5.8, must be converted to 1-bit depth. To do this, open the document in *SuperPaint*, then open `Document Info...` under the `File` menu. Click the `Black & White` (1-bit) button in the `Document is...` sub-window and then click the `OK` button. A warning message about loss of pixel depth will appear but just click `Continue`. This action will produce the 1-bit conversion but note that in many cases, marked (but not fatal) image degradation will result.

2. With the 1-bit b&w document opened in *SuperPaint*, click to activate the color choice palette. Six colors in addition to b&w will be offered. Pick one and click the appropriate box.

3. Now pull down the `View` menu and click `Color Preview`. A colored version of the document will appear on the monitor. `Save` it in PICT format.

4. This document can now be opened in color in the *Fry* 5.8 application using the standard Macintosh `Open` menu and dialog box. However, if opened in *SuperPaint*, it will remain b&w.

Equivalent steps may be taken with *Canvas*™ (and probably with other drawing applications) instead of *SuperPaint*. With sophisticated image processing applications such as *Photoshop*™, a color "rinse" may be added to an 8-bit grayscale image, thus avoiding loss of image quality (Fig. 1).

Reference

Fry, N. 1979. Random point distributions and strain measurement in rocks. *Tectonophysics* **60:** 69-105.

V: MATHEMATICAL AND PHYSICAL MODELING

Review of *Theorist*: a Symbolic Mathematics and Graphics Application

Andy R. Bobyarchick
Department of Geography and Earth Sciences
University of North Carolina at Charlotte, Charlotte NC 28223, U.S.A.
fgg00arb@email.uncc.edu http://anb-mac.uncc.edu

Abstract– This note presents a critical review of *Theorist*™, a symbolic mathemathics and graphing application of potential interest to the field of structural geology.

Introduction

Theorist is a symbolic mathematics and graphing application for microcomputers. The software is available for the *MacOS*™, *Windows*™, and *Windows* NT™ operating systems. Minimum requirements on the Macintosh are a Mac Plus or above (including Quadras and PowerPCs), 2 MB of RAM, 2 MB hard disk space, and *System* 6, 7, or *A / UX*. Suitable performance on PowerMacs and for graphics intensive applications benefit from additional RAM. On the *Windows* side, you need a minimum 386 processor, 4 MB RAM, 2 MB hard disk space, and *Windows* 3.1, NT, or 95. The version reviewed here is 2.01 running on a Macintosh 900 with a PowerPC upgrade card installed under *System* 7.5.1. Earlier versions of *Theorist* were marketed by Prescience Inc., but the program is now distributed by Waterloo Maple Software Inc., makers of the popular *Maple*™ mathematical system for microcomputers and workstations. Evidently, much of the core code in *Theorist* resembles *Maple* anyway.

Theorist supports all basic mathematical procedures you would expect in such software including: roots, factorials, and powers; integer and floating point numbers; mathematical constants and trigonometric functions; polynomial simplification and expansion; simultaneous solutions for systems of equations; vector and matrix algebra; and differentiation and integration. The maximum numerical precision is 15 digits.

In addition, *Theorist* comes with reasonable graphing abilities in two and three dimensions. Graphics can be animated in real time, and the program supports several projection systems (Cartesian, polar, cylindrical, and spherical). Options for coloring and shading plots allow creation of presentation-quality graphics that can also be exported in MetaFile format (*Windows*), or PICT format (Macintosh). Files copied onto the Macintosh clipboard within *Theorist* can be pasted into other Macintosh applications that accept the PICT format. EPS (Encapsulated Postscript) files can also be generated. These resolution-independent files can then be placed in page processors for publication. It is also possible to create a text version of the EPS file on the Macintosh; this file, with some editing, may then be opened in an modifiable form in a program such as Adobe *Illustrator*™.

The basic user environment in *Theorist* is the notebook. You can create notebooks optimized for your work that will open automatically every time the program is started. A notebook' s format is independent of the operating system, so it can be readily transferred between platforms with e-mail or over a network. Notebooks contain hot graphics windows that update automatically when an expression is changed within the notebook. A "statement" is the fundamental expression in a notebook. Statements are either "assumptions" or "conclusions" Statements entered by the user are assumptions; results from the program are conclusions. Statements may be equations or expressions composed of numbers, names, or wildcard variables. You can also define functions to be used within the notebook.

Theorist uses, in part, an iconic interface. You can create expressions and equations by clicking on a palette that contains either variable names or functions and mathematical operators. The program automatically positions an entry point in an expression where you may enter information from the keyboard. Expressions may also be created from the keyboard directly. With the appropriate typed operators (especially the placement of parentheses), *Theorist* will interpret keyboard entry in a syntax similar to the point-and-click entry technique. A `Fortranish` keyboard entry option lets you enter data in a way that you might use in programming or in a spreadsheet application.

For example, if you type `x^2+2` with the `Fortranish` option on, the result in the notebook will be x^2+2. With the `Fortranish` option off, the result is x^{2+2}. These different formats give the interface some flexibility, but they may also be confusing to the first-time user.

A simple notebook might show the somewhat busy icon palette, an equation in the notebook, and a graph of the function defined in the notebook. Details of the graph can be displayed at the bottom of the notebook by clicking on the "graph details" button at the bottom of a small palette within the graph window. Pop-up menus in the graph details area allow you to adjust some of the display attributes such as line width, style, and color. Other options accessible through icons in the graph window let you change the number of data points used in creating the curve, zoom in or out of the current view, or isolate certain parts of the plot. The graph is dynamically linked to the notebook statement that created it; the graph will update as you type changes in the statement definition. You can also click and drag within the graph window to see different parts of a graph in real time.

Theorist also handles the creation and import of tabular data. Import may be as simple as generating a table in a spreadsheet or word processing program, copying the text, and using a paste command in the table details window to automatically make a new table. You can also load tables saved in text format into *Theorist*. *Theorist* defines a table as a "... model of a smooth, continuous function defined by function values at regular intervals." Both real and complex data tables can be created or imported. Arrays of data can also be imported or exported within *Theorist*, but you should note that large tables or matrices of data use RAM space rapidly.

Conclusion

Theorist is a fine program for learning to apply mathematics to problems in structural geology because of the program's ability to dynamically link notebook expressions and graphs. The real-time manipulation of graphs (including such things as rotation of three dimensional spaces) is not currently implemented in high-end symbolic math programs like *Mathematica*. The notebook metaphor, with embedded graphics, is also available in some versions of *Mathematica*, however.

Theorist does have its limitations, however. It does not have the extensive resources of freely available notebooks and libraries that *Maple* and *Mathematica*™ have. The application is also limited in precision if you work with high numerical precision or very large or with very small numbers.

If you are at an institution that has a site license for *Maple*, you should check with your computing services department; it may be that a site license for *Theorist* has already been purchased.

Structural Geophysics: Integrated Structural and Geophysical Modelling

Mark W. Jessell and Rick K.Valenta
Australian Geodynamics Cooperative Research Centre, Monash University, Clayton, Victoria, 3168, Australia. mark@artemis.earth.monash.edu.au

Abstract– We present a technique for the integrated forward modelling of the three-dimensional structure and geophysical response of multiply deformed terrains. This technique allows information collected by field geologists and geophysicists to be reconciled by developing a simplified kinematic structural history of the area. The deformation history of the area is modelled using a succession of structural episodes, such as folding, shear zone activity, and intrusive events. The interaction of these episodes with a starting stratigraphy allows the prediction of the geometry of the final structures. By specifying geophysical rock properties for units in the initial stratigraphy we can also predict the potential field anomalies for gravity and magnetics. The accuracy of the model can be gauged by comparing the predictions with the observed structural and geophysical data. This approach points to a new methodology for the reconstruction of the geometry of structures in the Earth's crust, and has potential as a tool for both training and regional interpretation.

Introduction

The development of three-dimensional structural models in poorly exposed or complex terrains has always presented a problem. As a result, potential-field (magnetic and gravity) surveys have become increasingly important tools for regional structural synthesis (Whiting 1986, Isles *et al.* 1988, Leaman 1991). In regional geophysical interpretation, as in structural mapping, the primary goal is to determine the three-dimensional geometry of structures in the Earth. Given the ever-improving spatial resolution of gravity and magnetic surveys, the latter of which can now resolve quite thin lithological units, it is perhaps surprising how few tools are available to interpret the observed structural and potential-field data in an integrated fashion. One can consider three approaches to the computer based interpretation of structural or geophysical information:

1) A Geographic Information System (GIS) approach, which can integrate any number of spatially located data by making spatial comparisons within one

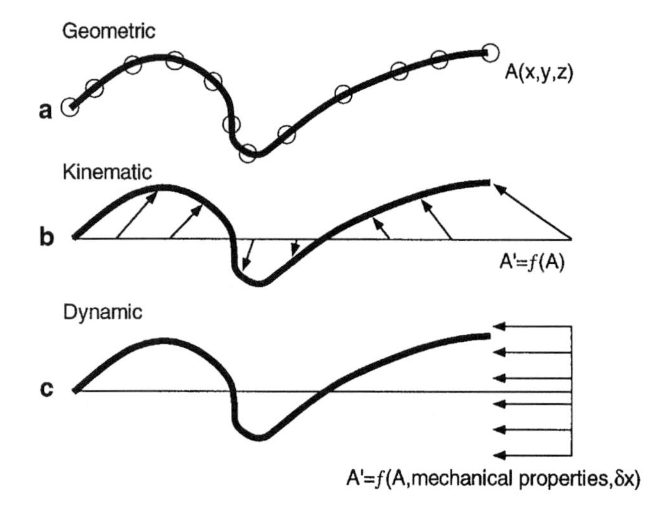

Fig. 1. Three classes of mechanical modelling. a) Geometric: each object (*e.g.* fault surface, bedding plane) completely described by its current *X,Y,Z* position. b) Kinematic: current geometry derived from initial geometry by superimposed displacement equations. c) Dynamic: current geometry calculated from initial geometry, mechanical properties, and imposed boundary conditions.

"layer" of data and logical comparisons among "layers" (see Bonham-Carter 1994 for a review).

2) A modelling approach, which generates a two or three-dimensional model of the distribution of rock types, and from this makes linked predictions of the geophysical response. Most potential-field modelling is of this type, including the work described in this paper.

3) An inversion approach, which can be taken to directly generate a two or three-dimensional model of the geology by inverting the observed geophysical field (Oldenburg 1993). Several systems have been developed to automatically remove the effects of deformation in basin settings, where detailed information is often available (Gratier *et al.* 1991, Rouby *et al.* 1993). Some interesting work has also been carried out to invert two dimensional geological sections in terms of their geological evolution (Flewelling *et al.* 1992, Simmons 1992).

In this paper we outline the principles of an integrated computer based system for the reconciliation of gravity, magnetic, and structural data sets based on a combined structural and potential-field forward modelling scheme.

Modelling 3-Dimensional Structures in the Earth's Crust

In order to model the potential-field response of geological structures we must first consider how to model the structures themselves and this depends in part on the nature of the data available. Mechanical modelling of geological structures can be carried out on three increasingly sophisticated levels: geometric, kinematic, and dynamic (Fig. 1) and the choice of technique depends on the preexisting knowledge of the structure and the reason for modelling.

Geometric modelling seeks to describe the current shape of a structure by defining its spatial extent so that a bedding plane, for example, may be described as a connected series of points. It is most like traditional map making in that it provides a synthesis of all the available knowledge about a region based on the interpretations made by geologists. A wide range of two and three-dimensional drafting and computer-aided design and visualisation packages is available, some of which are specifically tailored for geological use (Barchi *et al.* 1992, Barret & Bailey 1992, Luthi 1992, Mallet 1992, Mayoraz *et al.* 1992, Pflug 1992, Pflug, Klein *et al.* 1992, Prissang 1992, Verschuren 1992). These packages are well suited to interpolating the geometry of regions where there are dense three-dimensional data available, such as a three-dimensional seismic reflection survey, or the densely spaced boreholes at a mine site. In these cases, the flexibility afforded by being able to draw up the structures provides the easiest route to defining the geometry, although editing the model once it has been formulated is not necessarily as easy. A common approach is to develop a three-dimensional model by interpolating between two dimensional cross sections (Tipper 1977). Once constructed these models can then be used to model geophysical, geochemical, or hydrogeological properties.

Where the available three-dimensional geological data are sparse, as is generally the case in regional mapping programs, the geometric approach is less easily applied. In this case often only surface information is available and the problem is, instead, to extrapolate into the third dimension. As a consequence it is often difficult to distinguish between different hypotheses that match the data. In this case a kinematic approach may be preferable, where the structures are described by the effect of a series of deformations on a starting geometry. The kinematic modelling of structures has been developed for a number of different research and modelling goals (Thiessen & Means 1980, Jessell 1981, Charlesworth & McLellan 1986, Groshong & Usdansky 1987, Perrin *et al.* 1988, Contreras 1991, Guglielmo 1993) with many kinematic modelling systems concentrating on two dimensional balanced section analysis in basins and, therefore, restricted to the modelling of fault displacements and resulting folds. In this project we have developed a three-dimensional kinematic forward modelling approach for multiply deformed terrains.

Dynamic modelling is the most sophisticated type of modelling scheme, where not only the geometry but also the mechanical properties of rocks are considered. This type of modelling is well suited to improving our understanding of processes involved in geology at all scales (Dieterich 1969, Casey & Huggenberger 1993). However it is generally impractical to try to model from scratch the current three-dimensional geometry of specific multiply deformed terrains in terms of a mechanical model as there are too many uncertainties involved. It is, therefore, generally used to examine one particular stage in the structural evolution of an area, based on a geometry defining a preexisting geology (Zhang *et al.* 1994) so that in some respects the constraints of dynamic modelling are inherited from geometric modelling systems.

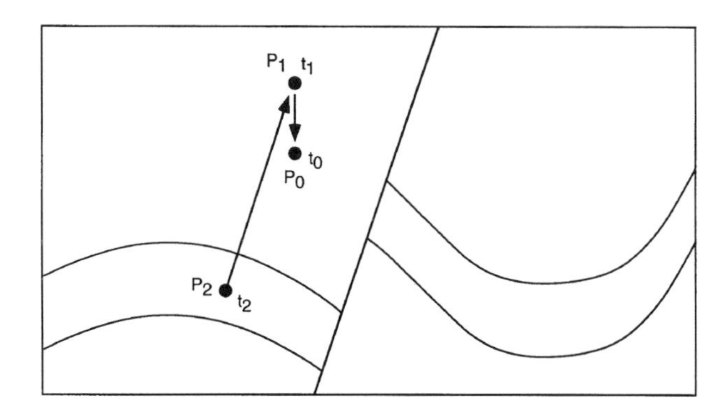

Fig. 2. Principles of Eulerian kinematic modelling. Final cross section of area after folding and normal faulting. To calculate rock type of a point at time t_2 and position P_2, we remove effects of faulting to restore to position P_1 at time t_1, then remove effects of folding to restore position P_0 at time t_0. Given layer cake stratigraphy at time t_0, rock type is a function of height.

Integrated Modelling of Structure and Geophysics

As a result of the difficulties in applying geometric or dynamic schemes to model complex regional structures, we have concentrated on the kinematic modelling approach. The language of structural geologists recognises the inherent sedimentary, structural, or metamorphic layering found through much of the Earth's crust, and often describes the observed structures in terms of the successive deformations which affect this layering, such as fold phases F_1, F_2, and F_3 (Tipper 1992). We have used a structural geologist's descriptive language to implement a kinematic geological forward modelling scheme combined with geophysical forward modelling and have developed a computer package called *Noddy* (Jessell 1981; see Appendix I) that allows the user to define the currently observed structures in terms of a simplified geological history. This history is based on the *a priori* information available to the geologist and could include: structural information such as outcrop patterns, age relationships, and bedding orientations; rock property information such as magnetic susceptibility; and potential-field information such as the observed gravitational and magnetic fields. Once a geological history has been specified, the modelling scheme allows the user to test the accuracy of the history by comparing the resulting structural and geophysical predictions with the observed data.

Kinematic Structural Modelling

Most three-dimensional geometric modelling schemes are either vector-based (so that a surface is described by a set of triangles, for example) or voxel-based (so that a geological volume is described by a set of volume elements). *Noddy* uses a model based on displacement equations superimposed on a given starting stratigraphy and, in principal, these equations could act on either raster or vector objects, although for most calculations a voxel-based system is used.

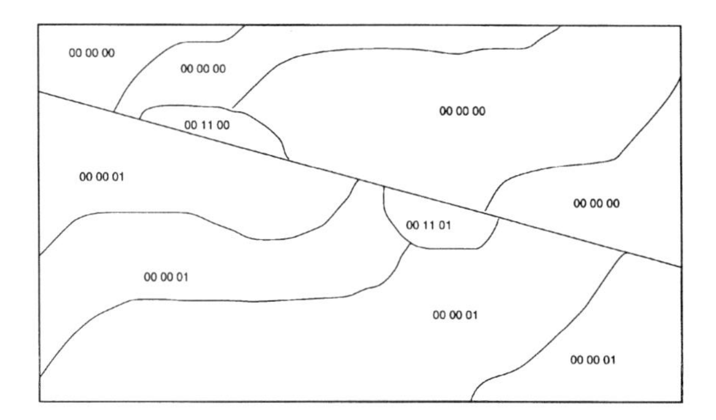

Fig. 3. Using discontinuity codes to characterise discrete contiguous geological volumes. See text.

Displacement equations are stored as a "history" which provides a parameterised definition of kinematics and rock properties. A voxel model (in some cases only one layer thick) is calculated by considering each voxel independently and is performed using the Eulerian (inverse) form of the defining Lagrangian displacement equations and by implementing them in reverse chronological order starting with the most recent deformation event (Fig. 2). The input to this calculation is the current *X,Y,Z* position of each voxel, and the output is the *X,Y,Z* position at the time the associated volume of rock was created. New rock can be treated either at time zero, when the base stratigraphy is defined, or when an unconformity, dyke, or plug emplacement event takes place. At present the events are assumed to be instantaneous, and thus only a relative time scale is needed. As well as the initial position of the point, a binary "discontinuity code" is stored, that records each time a voxel was affected by an event described by a discontinuous displacement equation (faults, unconformities, dykes, and plugs) but ignores events described by continuous displacement equations (folds, shear zones, strain, rotation, foliations and lineations) (Fig. 3). This code allows the accurate transformation of the voxel data set to a vector data set, since only voxels which have exactly the same sequence of discontinuity codes are part of a contiguous volume of rock. If two adjacent voxels have different codes, the difference in the discontinuity code that occurred most recently defines the specific discontinuity which separates them.

The starting point for the kinematic modelling is a layer cake stratigraphy, for which the heights of the stratigraphic contacts are defined, together with a geophysical rock property stratigraphy defining the densities and magnetic susceptibilities of each unit. The structural history is then developed from a set of events which produce folds, faults, unconformities, shear zones, dykes, igneous plugs, tilts, homogeneous strains, penetrative cleavages, and penetrative lineations (Fig. 4). Each event can be defined in terms of four classes of properties: form, position, orientation, and scale. For example, a fault is defined by its dip and dip direction, the pitch and magnitude of the slip vector, and the position of one point on its surface. The use of geological descriptions serves two purposes: it provides a natural framework for geologists to build a model, and once a model has been developed, it provides directly the type of information that the geologist would like to know (such as fault throw). Although the structural events in themselves

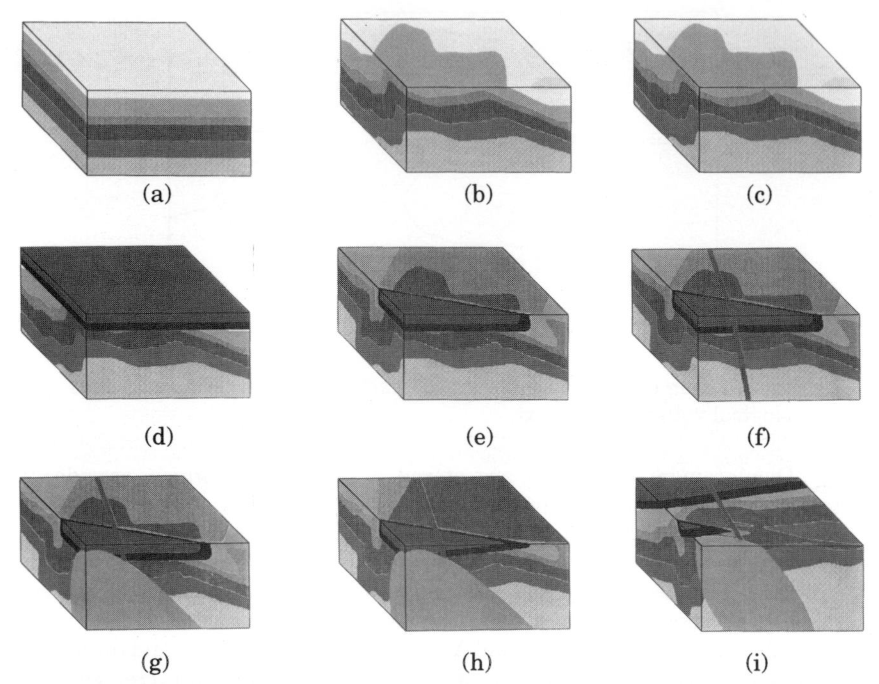

Fig. 4. Examples of principal classes of deformation events available within *Noddy*: a) simple layer cake stratigraphy, b) folds, c) faults, d) unconformities, e) shear zones, f) dykes, g) plugs, h) homogeneous strains, i) tilts. Two other events, penetrative foliation and lineation, do not affect outcrop patterns.

are relatively simple, complex geometries quickly develop as two or three events are superimposed on one another.

Calculations can be made for outcrop patterns and bedding, cleavage, and lineation orientation data, either for a flat land surface, or for a specified topography. In addition, block diagrams, sections, and the geometry of bedding surfaces can be easily calculated (Fig. 5). In order to calculate the orientations of specific features (bedding, foliations, fault planes, *etc.*) first the inverse and then the forward displacement equations need to be calculated (Fig. 6). Starting with the current *XYZ* location of a point, the position of this point at the time of formation of the structural feature (which may or may not be the time of formation of the rock) is calculated. Three points are defined close to this position which define a plane with the orientation of the feature prior to deformation. The current positions of these points are calculated as is the final orientation of the structural feature. The orientation of a linear feature is calculated from the intersection of two planes. Thus both Eulerian (inverse) and Lagrangian (forward) descriptions of the displacements must be available for a new deformation event to be included in the modelling scheme. We have tried to keep the displacement equations governing a particular type of deformation as simple as possible and rely on the ability to superimpose deformation events to produce structural complexity. A full description of the equations used is presented in Appendix II.

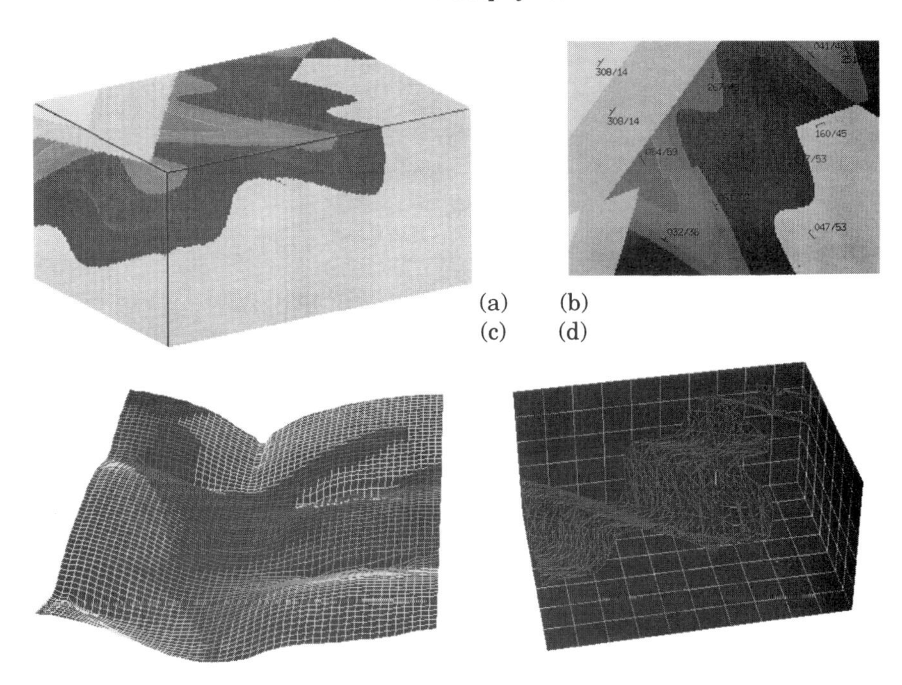

Fig. 5. Visualisation styles available for a simple fold-unconformity-tilt-fault history. a) Block diagram. b) Map. c) Geology draped over topography. d) Triangulated bedding surface.

Potential-field Modelling

An algorithm was developed to compute the magnetic and gravity anomalies due to any arbitrary 3-D structure that has been converted into a voxel form. The starting point of the calculations is the evaluation of the geometric coefficients which are the same for all elements in a given horizontal plane. This is performed using the analytic solutions to the magnetic and gravity anomalies of dipping prisms developed by Hjelt (1972, 1973), which were simplified to the case of cubes. The values of the gravity and induced magnetic field strength at any one X,Y location are obtained by convolving the coefficients with the respective susceptibility and density values of the element under consideration (Fig. 7). Although, in principle, each voxel in the model should be considered when calculating the field strength for a particular location, this resulted in extremely long calculation times. Instead we only consider voxels within a fixed range for each location, which can be justified because the coefficients diminish rapidly in magnitude as a function of distance. This lowers the accuracy of the calculation but the range parameter can always be increased if higher accuracy is required. The process of calculation can be summarised as follows:

1. For each horizontal layer, the geometric coefficients of a single cubic element are computed over a given calculation

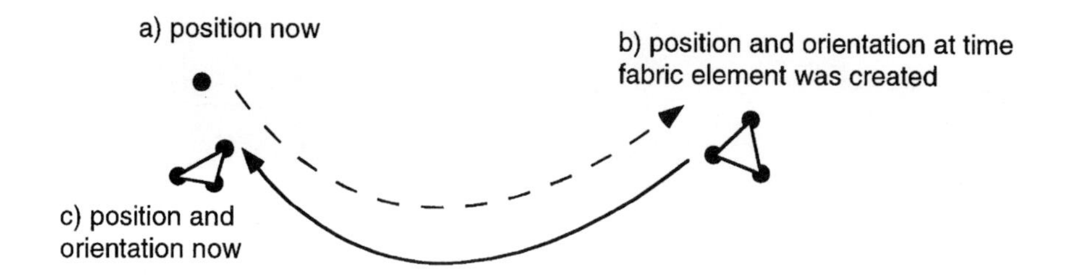

Fig. 6. Calculating orientation data. See text.

range, taking into account the difference in height between the layer and the sensor. For magnetic modelling, the geometric coefficients are also a function of the inclination and the intensity of the Earth's magnetic field.

2. At each measurement locality the field contributed by a particular layer is calculated by convolving the geometric coefficients with the local density or susceptibility values for the layer.

3. Steps 1 and 2 are repeated for each layer and the results from each layer are summed vertically to give the total field strength.

Applications of the Technique

We have investigated two main uses for this integrated modelling package: as a training tool in the teaching of the relationship between structural geology and geophysics, and as a method of interpreting natural structures.

Educational Use

One of the difficulties in teaching structural geology has always been the leap of imagination required to go from the description of individual structures to the interaction of these structures in three-dimensions. The structural forward modelling demonstrated in this paper was originally developed with this goal in mind and, by itself, provides a useful teaching aid. The addition of geophysical modelling allows the structural geologist to make useful predictions as to the expected appearance of particular structures in gravity and magnetic data sets. Similarly, the introduction of geological constraints into the teaching of geophysical interpretation ensures the student never loses sight of the geological origin of the data.

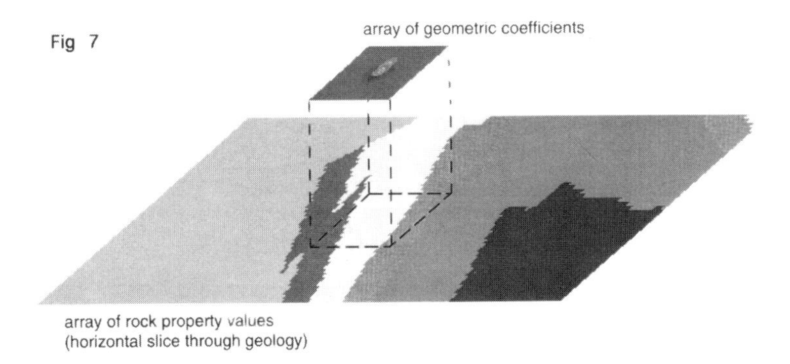

Fig. 7. Calculation of total field strength from voxel models. Array of geometric coefficients is calculated for each horizontal layer (coloured according to value), based on inclination and intensity of Earth's field, voxel size, and distance between layer and potential field measuring device. Array is convolved with density/susceptibility values to provide contribution to total field calculated by summing results vertically.

Modelling Regional Structure

The first constraint on the plausibility of a given geological history is its ability to account for outcrop patterns, orientation data, and age relationships seen in the field. However these generally represent an essentially two dimensional data set. The additional constraint of gravity and/or magnetic data allows for a significantly improved test of any proposed interpretation. If the geological model appears accurate, but fails to account for a significant part of the geophysical anomaly, then the model needs to be refined. One advantage of the system presented here is that modification of the history can be as simple as changing the throw on a fault and the final representation of the model is not a static description of the current geometry but, rather, a time history of structural events with geological significance. The development of three-dimensional structural models to account for the observed outcrop patterns is always difficult since, even if the surface outcrop is relatively unambiguous, the downward continuity of structures can only be guessed at. Gravity and magnetic data sets can confirm this continuity or suggest otherwise.

In Fig. 8, we compare geological and magnetic data for the Golden Dyke region of the Proterozoic Pine Creek Geosyncline, Northern Territory, Australia. The area (Fig. 8a,b) has undergone four main stages of deformation: NNW- to NNE-trending folding; a subsequent gentle refolding of these structures about an EW axis; NW-trending faulting (dominated by the Hayes Creek Fault); and, finally, emplacement of the Cullen Granite in the very south of the area. The aeromagnetic data (Fig. 8d) suggests that the resulting domal structures are probably uniformly developed across the map area, although it is difficult to confirm this since outcrop is sparse in the northwest. Poles to bedding (Fig. 8c)

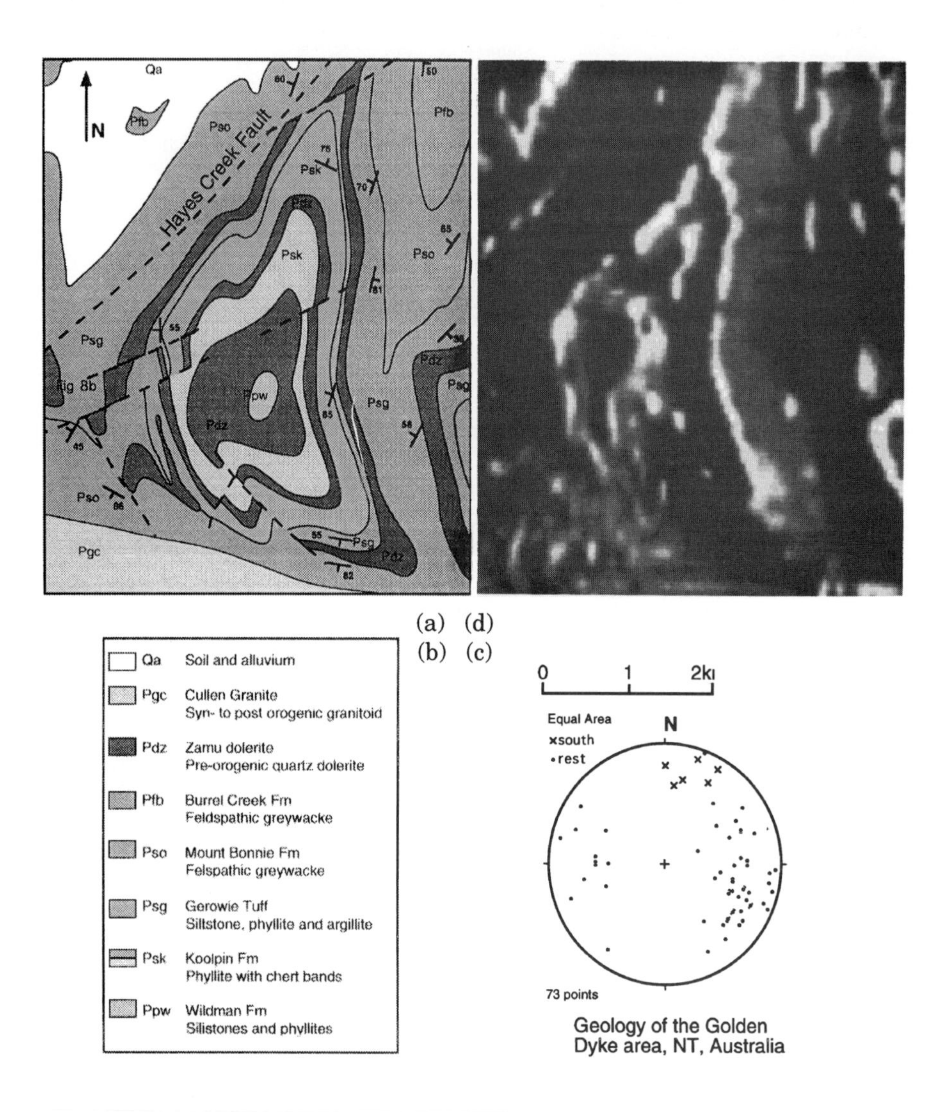

Fig. 8. Geology of Golden Dyke area, Pine Creek Geosyncline, Northern Territory, Australia. a) Geological Map, based on Northern Territory Geol. Surv. & Bureau of Mineral Resources Pine Creek Sheet 1:100,000. Coordinates of south west corner of map area: 131'30'50"E 13'35'11"S. b)Legend for (a). c) Lower hemisphere equal area projection of poles to bedding. Circles: main area. Crosses: area affected by Cullen Granite. d) Airborne total magnetic intensity image of same area. East-looking gradient filter applied to highlight north-south trending structures. (with permission, World Geoscience Corp.).

reflect the NNW-NNE folding with the recumbent nature showing up in the dominance of west-dipping layering. Based on comparisons between published maps and available aeromagnetic data, we can deduce that there are horizons within the Burrell Creek Formation, the Mount Bonnie Formation, and the Koolpin Formation which produce magnetic anomalies and it has been shown that pyrrhotite is the major magnetic mineral. Immediately adjacent to the Cullen Granite the bedding is uniformly steep and dipping to the south, and there is a marked decrease in the magnetic signature of the Koolpin Formation. We attribute the weakening of the magnetic signal to the alteration of the magnetic minerals during the emplacement of the Cullen Granite, and have modelled this by adding a 2000 metre alteration halo around the granite, with systematically reduced susceptibility values.

As a test of our structural understanding, we have modelled this area using *Noddy*. The model history we have developed reflects the history interpreted from the geological information, and is based on the same four deformation stages outlined above (Fig. 9), although in each case one general deformation event is modelled as a succession of small events so we can more accurately match the present-day structures. The resulting synthetic geological map (Fig. 10a) shows the main structural relationships in the area and poles to bedding also match, including the influence of the Cullen Granite, although there is a greater concentration of NW dipping poles in the real data. Compared to the real aeromagnetic image, the modelled response is much "cleaner" with a lower high frequency content due to the assumption of homogenous rock properties for each unit. Although there are some obvious differences in the detail of the geological and geophysical data sets that we have modelled, the modelling has confirmed the basic structural interpretation and suggested a local contact metamorphic event which could be confirmed by returning to the area. We believe the integrated modelling approach we have taken here will become an increasingly important tool in the interpretation of structures on the scale of 1 metre to 10 km (Valenta *et al.* 1992).

Discussion

In this paper we have outlined the use of a forward modelling approach to the reconciliation of gravity, magnetic, and structural data sets. The structures which may be modelled are based on simplistic displacement equations and take no account of the varying mechanical properties of different rock types. However, their combination still allows complex overprinting relationships to be investigated. The incorporation of more sophisticated deformation events is feasible and could be built onto the current framework. A major strength of this approach is the similarity between the types of geological data collected in the field and the predictions which result from the combination of structural events. The alteration associated with the Cullen Granite also points to the value of incorporating alteration or metamorphic events into the modelling scheme.

The scale limitations on modelling are those enforced by the range of geological processes that are approximated by the displacement equations. Thus, a 1 metre to 10 kilometre range is feasible but neither plate tectonic nor microstructural scale processes can be modelled.

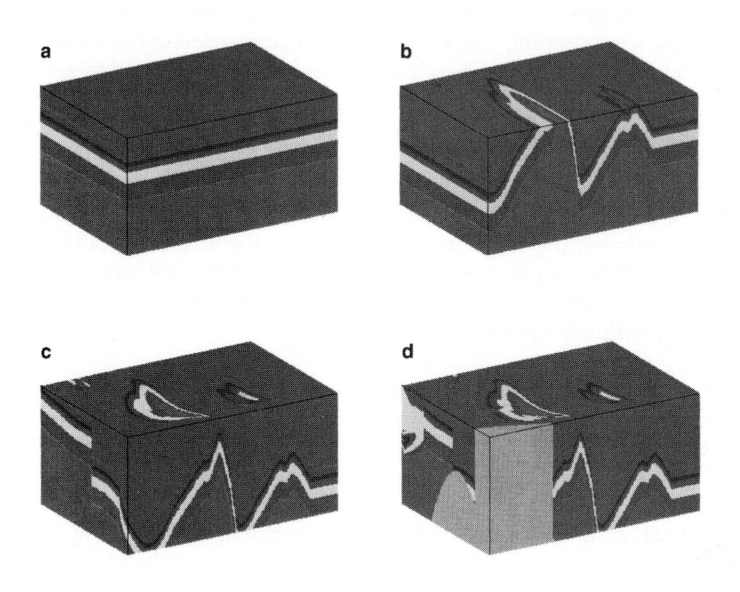

Fig. 9. Structural evolution of Golden Dyke area as modelled by *Noddy*. Block dimensions – 6500:4550:3250 m (East:North:Down). a) Initial stratigraphy. b) Major NS folding. c) EW refolding. d) Late faulting & emplacement of Cullen Granite.

The magnetic modelling is already quite sophisticated. It considers induced magnetism, remanence, and anisotropic susceptibility, with arbitrarily deformed remanently magnetised units, so that the remanent vectors track the deformation (Borradaile 1993). The addition of a further geophysical or geological property using the same structural base would allow new parallel sets of predictions, which would provide a new set of constraints to the overall geological interpretation, and would further improve the geologist's ability to discriminate between various hypotheses.

We believe that by forcing the constraints on geophysical modelling to be geological and kinematic, rather than merely geometric, we have developed a framework for the integration of not only gravity, magnetic, and structural data sets, but also other geological and geophysical rock properties and that this approach provides a pathway for the reconstruction of the evolution of the three-dimensional geology of a region using all available data sets.

Acknowledgements

We would like to thank Vic Wall from Mt. Isa Mines and Peter Williams from Western Mining Corporation for their encouragement and financial support of the project, Ernie Rutter and Gordon Lister for their enthusiasm at critical but temporally distinct stages of this project, and Carl Jacobsen for his thoughtful review. We would also like to thank World Geoscience Corporation for allowing us

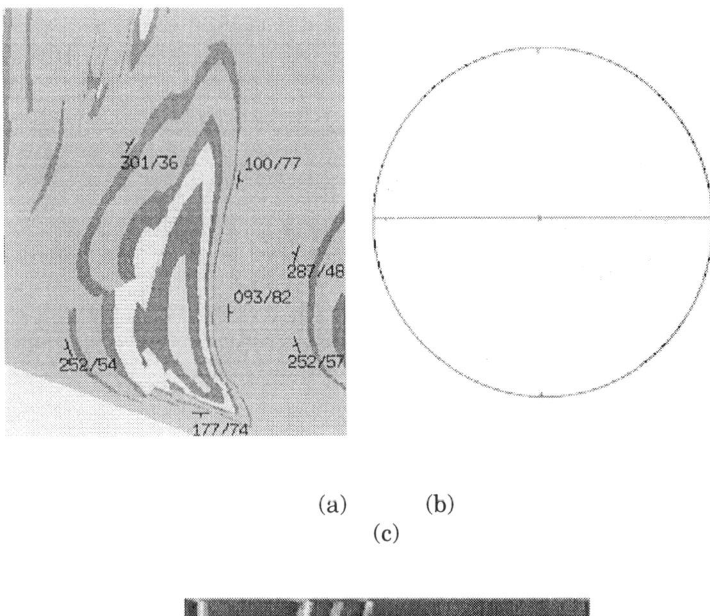

(a) (b)

(c)

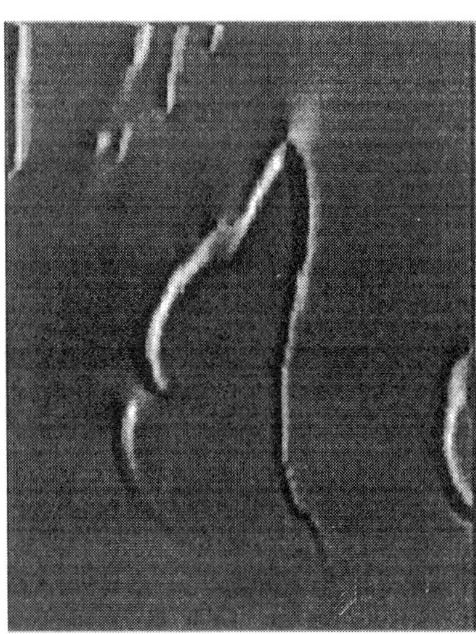

Fig. 10. Modelled geology and magnetics of the Golden Dyke area. a) Simulated geological map created using *Noddy* (legend as in Fig. 8). b) Lower hemisphere equal area projection of poles to bedding. c) Total magnetic intensity image calculated for this model geology, with east-looking gradient filter applied.

to reproduce some of their data. This project has been funded by the Australian Research Council, the Australian Geodynamics Cooperative Research Centre, and a consortium of companies *via* an Australian Mineral Industries Research Association project.

References

Barchi, M., Federico, C., Guzzetti, F. & Minelli, G. 1992. Computer modelling of surfaces: structural geology applications. In: Pflug, R. & W. Harbaugh, J. (eds.) *Computer Graphics in Geology*. Freiburg, Germany. pp. 89-97.

Barret, R. A. & Bailey, J. 1992. Three-dimensional modeling techniques in the analysis of a mature stream drive. In: D. E. Hamilton, D. E. & Jones, T. A. (eds.). *Computer Modeling of Geologic Surfaces and Volumes.* Amer. Assoc. Petrol. Geol., Tulsa OK. pp. 251-259.

Borradaile, C. J. 1993. Strain and magnetic remanence. *J. Struct. Geol.* **15**: 383-390.

Bonham-Carter, G. F. 1994. *Geographic Information Systems for Geoscientists: Modelling with GIS*. Pergamon Press. 398pp.

Casey, M. & Huggenberger, P. 1993. Numerical modelling of finite-amplitude similar folds developing under general deformation histories. *J. Struct. Geol.* **7**: 103-114.

Charlesworth, H. A. K. & McLellan, G. C. 1986. *REFOLD*; a Fortran 77 program to construct model block diagrams of multiply folded rocks. *Comp. & Geosci.* **12**: 349-360 .

Contreras, J. 1991. Kinematic modeling of cross-sectional deformation sequences by computer simulation; coding and implementation of the algorithm. *Comp. & Geosci.* **17**: 1197-1217 .

Dieterich, J. H. 1969. Origin of cleavage in folded rocks. *Amer. Journ. Sci.* **267**: 155-165.

Flewelling, D. M., Frank, A. U. & Egenhofer, M. J. 1992. Constructing geologic cross sections with a chronology of geologic events. *Proc. 5th Int. Symp. on Spatial Data Handling.* Charleston SC, USA. pp. 544-553.

Gratier, J.-P., Guillier, B., Delorme, A. & Odonne, F. 1991. Restoration and balance of a folded and faulted surface by best-fitting of finite elements; principle and applications. *J. Struct. Geol.* **13**: 111-115.

Guglielmo, G. J. 1993. Interference between pluton expansion and non-coaxial tectonic deformation; three-dimensional computer model and field implications. *J. Struct. Geol.* **15**: 593-608.

Hjelt, S. E. 1972. Magnetostatic anomalies of dipping prisms. *Geoexploration* **10:** 239-246.

Hjelt, S. E. 1974. The gravity anomaly of a dipping prism. *Geoexploration* **12:** 29-39.

Isles, D. J., Harman, G. & Cuneen, J. 1988. Aeromagnetics as part of the Archaean gold exploration in the Kalgoorlie region Western Australia. *Econ. Geol.* Monograph **6**: 389-397.

Jessell, M. W. 1981. *Noddy - An Interactive Map Creation Package.* Unpublished M.Sc. Thesis. University of London.

Leaman, D. E. 1991. Geophysical constraints on structure and alteration of the Eastern Creek Volcanics Mt. Isa Queensland. *Austr. Journ. Earth Sci.* **38**: 457-472.

Luthi, S. M. 1992. Solid computational models of geological structures in boreholes. In: Pflug, R. & W. Harbaugh, J. (eds.) *Computer Graphics in Geology*. Freiburg, Germany. pp. 51-61.

Mallet, J-L. 1992. *GOCAD*; a computer aided design program for geological applications. In: Turner, A. K. (ed.). *Three-dimensional Modeling with Geoscientific Information Systems*. NATO Adv. Study Inst. Ser. C. Col. Sch. Mines, Golden CO, pp. 123-141.

Mayoraz, R., Mann, C. E. & Parriaux, A. 1992. Three-dimensional modeling of complex geological structures: new development tools for creating 3-D volumes. In: D. E. Hamilton, D. E. & Jones, T.A. (eds.). *Computer Modeling of Geologic Surfaces and Volumes.* Amer. Assoc. Petrol. Geol., Tulsa, OK. pp. 261-271.

Oldenburg, D. W. and Li, Y. 1993 (abs). Inversion of induced polarization data: 63rd Ann. Int. Mtg., Sec. Expl. Geophys. 396-399.

Perrin, M., Oltra, P. H., & Coquillart. 1988. Progress in the study and modelling of similar fold interferences. *J. Struct. Geol.* **10**: 593-605 .

Pflug, R. 1992. Solid modelling of geological objects with 3D rasters. In: Pflug, R. & W. Harbaugh, J. (eds.) *Computer Graphics in Geology*. Freiburg, Germany. pp. 213-219.

Pflug, R., Klein, H., Ramshorn, C., Genter, M. & Stark, A. 1992. 3-D visualization of geologic structures and processes. In: Pflug, R. & W. Harbaugh, J. (eds.) *Computer Graphics in Geology*. Freiburg, Germany. pp. 29-39.

Prissang, R. 1992. Three-dimensional predictive deposit modelling based on the linear octree data structure. In: Pflug, R. & W. Harbaugh, J. (eds.) *Computer Graphics in Geology*. Freiburg, Germany. pp. 229-238.

Rouby, D., Cobbold, P. R., Szatmari, P., Demercian, S., Coehlo, D. & Rici, J. A. 1993. Least-squares palinspastic restoration of regions of normal faulting – applications to the Campos basin (Brazil). *Tectonophysics* **221**: 439-452.

Simmons, R. G., 1992. The roles of associational and causal reasoning in problem solving. *Artificial Intelligence* **53**: 159-207.

Theissen, R. L. & Means, W. D. 1980. Classification of fold interference patterns: a re-examination. *J. Struct. Geol.* **2**: 311-316.

Tipper, J. C. 1977. A method and FORTRAN program for the computerized reconstruction of three-dimensional objects from serial sections. *Comp. & Geosci.* **3**: 579-599.

Tipper, J. C. 1992. Reconstructing three-dimensional geological processes. *Proc. 29th Int. Geol. Congr.*, Kyoto, p. 958.

Valenta, R. K., Jessell, M. W., Jung, G. & Bartlett, J. 1992. Geophysical interpretation and modelling of three-dimensional structure in the Duchess area, Mt. Isa, Australia. *Expl. Geophys.* **23**: 393-400.

Verschuren, M. 1992. 3-D modeling of a complex fault pattern on an entry level 2-D workstation. In: Pflug, R. & W. Harbaugh, J. (eds.) *Computer Graphics in Geology*. Freiburg, Germany. pp. 83-88.

Walsh, J. J. & Watterson, J. 1987. Distributions of cumulative displacement and seismic slip on a single normal fault surface. *J. Struct. Geol.* **9**: 1039-1046.

Whiting, T. H. 1986. Aeromagnetics as a guide to geological mapping - a case history from the Aurunta Inlier, Northern Territory. *Austr. Journ. Earth Sci.* **33**: 34-41.

Zhang, Y., Scheibner, E., Ord, A. & Hobbs, B. E. 1994. Numerical modelling of crustal deformation of the eastern Australian passive margin. *Proc. Austr. Geol. Conv.*, Geological Society of Australia, Perth. **37:** 485.

Appendix I: Programming Notes

Noddy is written in *C*, and makes use of a commercial multi-platform programming environment available from XVT Software Inc., Box 18750, Boulder, CO, USA. This allows us to use essentially one set of source code for Macintosh, MS Windows, Sun OpenLook, and SGI Motif systems. The 3D visualization makes use of 3D Graphic Tools source code (Micro System Options, PO Box 95167, Seattle, WA 98145, USA) which we have converted to run under XVT.

Table 1. XVT Performance

Hardware	System	Time
PowerMac 8100/80 (native mode)	*Mac OS* 7.x	210 secs
Sun Sparc Classic	*UNIX / OL*	81 secs
Pentium PC 75Mz	*Windows* 3.1.1	50 secs
Silicon Graphics (Indigo II)	*UNIX*/GL	27 secs

A summary of the performance characteristics for each platform is provided in Table 1 for calculation of a gravity-magnetics model with dimensions 10k.7k.5k at 200 m voxel resolution. These are typical results, as the actual complexity of the model, the model size, and the resolution can be varied arbitrarily.

Appendix II: Displacement Equations used in Modelling

In order to simplify the formulation of the displacement equations used in this system we perform a transformation between the global (real world) reference frame and a local reference frame which is used only for the displacement equations. This transformation consists of a translation followed by a rotation prior to enactment of each displacement equation, followed by the inverse rotation and translation after its enactment. In the following sections the displacement equations are given for the local reference frame, and only in the Lagrangian form. In most cases the Eulerian form of the equations may be easily derived from the Lagrangian form. However, for curved and elliptical faults a Newton-Raphson iterative algorithm is used to calculate the Eulerian displacements.

Faults

Five different fault geometries are currently supported, ranging from simple planar faults with uniform translation or rotation of one block relative to the other to arbitrarily curved surfaces defined by a profile where the translation vector is a function of position relative to the surface and dies off to zero outside the area of influence of the fault. In each case the equations show the example of hanging wall only slip.

Translation Faults

In the local reference frame, the fault plane is the YZ plane, with slip in the Y direction. The kinematics are defined by:

$$\begin{pmatrix} X' \\ Y' \\ Z' \end{pmatrix} = \begin{pmatrix} X \\ Y+s \\ Z \end{pmatrix} \quad for\ X > 0$$
$$\begin{pmatrix} X' \\ Y' \\ Z' \end{pmatrix} = \begin{pmatrix} X \\ Y \\ Z \end{pmatrix} \quad for\ X \leq 0 \tag{1}$$

where s is the slip vector.

Rotational Faults

In the local reference frame the fault plane is the YZ plane, and the X axis is the rotation axis. The kinematics are defined by:

$$\begin{pmatrix} X' \\ Y' \\ Z' \end{pmatrix} = \begin{bmatrix} 1 & 0 & 0 \\ 0 & \cos\theta & -\sin\theta \\ 0 & \sin\theta & \cos\theta \end{bmatrix} \begin{pmatrix} X \\ Y \\ Z \end{pmatrix} \quad for\ X > 0$$
$$\begin{pmatrix} X' \\ Y' \\ Z' \end{pmatrix} = \begin{pmatrix} X \\ Y \\ Z \end{pmatrix} \quad for\ X \leq 0 \tag{2}$$

where θ is the rotation angle.

Ring Faults

In the local reference frame, the tangent to the fault plane and the slip vector are always parallel to the Y axis. The kinematics are defined by:

$$\begin{pmatrix} X' \\ Y' \\ Z' \end{pmatrix} = \begin{pmatrix} X \\ Y+s \\ Z \end{pmatrix} \quad for\ X^2+Z^2 > R^2 \tag{3a}$$

and

$$\begin{pmatrix} X' \\ Y' \\ Z' \end{pmatrix} = \begin{pmatrix} X \\ Y \\ Z \end{pmatrix} \quad for\ X^2+Z^2 \le R^2 \tag{3b}$$

where s is the slip vector and R is the radius of the ring fault.

Elliptical Faults

The displacement equations for elliptical faults, where slip decays away from the centroid of the fault, are based on the empirical equations developed in Walsh & Watterson (1987). In the local reference frame, the fault plane is the YZ plane, with slip in Y direction. The kinematics are defined by:

$$\begin{pmatrix} X' \\ Y' \\ Z' \end{pmatrix} = \begin{pmatrix} X \\ Y+s.p \\ Z \end{pmatrix} \quad for\ X > 0$$

$$\begin{pmatrix} X' \\ Y' \\ Z' \end{pmatrix} = \begin{pmatrix} X \\ Y \\ Z \end{pmatrix} \quad for\ X \le 0 \tag{4}$$

where s is the maximum slip and p is a function of position within the ellipsoid of deformed rock defined by:

$$p = \left(1 - \frac{X}{Ex}\right)^2 + \left(1 - \frac{Y}{Ey}\right)^2 + \left(1 - \frac{Z}{Ez}\right)^2 \tag{5}$$

where Ex, Ey, and Ez are the principal axes of the ellipsoid.

Curved Faults

The displacement equations for curved faults are similar to those used for elliptical faults where slip decays away from the centroid of the fault but when combined with curved fault surfaces they require a term for the displacement of points perpendicular the fault surface (parallel to local X), so that all points maintain a constant distance from the fault. The kinematics are defined by:

$$\begin{pmatrix} X' \\ Y' \\ Z' \end{pmatrix} = \begin{pmatrix} X+d \\ Y+s.p \\ Z \end{pmatrix} \text{ for } X > 0$$

$$\begin{pmatrix} X' \\ Y' \\ Z' \end{pmatrix} = \begin{pmatrix} X \\ Y \\ Z \end{pmatrix} \text{ for } X \leq 0 \qquad (6)$$

where s is the maximum slip, d is the distance between a point and the fault surface before and after deformation, and p is a function of position within the ellipsoid of deformed rock defined by eqn. (5).

Unconformities

Unconformities do not imply any displacement, as they merely define a planar discontinuity.

Folds

The similar fold model currently supported is probably the weakest representation of its true counterpart and does not allow upper crustal flexural slip geometries. The kinematics are defined by:

$$\begin{pmatrix} X' \\ Y' \\ Z' \end{pmatrix} = \begin{pmatrix} X \\ Y \\ a.e^{\frac{-\gamma^2}{c}}.f(w.Z) \end{pmatrix} \qquad (7)$$

where $f()$ is either a simple sinusoid function or a Fourier series which allows arbitrary fold profiles to be generated, w is the fold wavelength, a is the fold amplitude, and c is a parameter that controls fold cylindricity.

Shear Zones

Shear zone kinematics are the same as those for faults, except that the slip is distributed continuously within a parallel sided shear zone, rather than being concentrated on a discrete fault plane.

Dykes

Dykes are infinite parallel-sided bodies which can either be stope-like, with the country rock replaced by dyke material, or dilatational, which implies the creation of a void space which is filled by the dyke material.

Dilatational Dyke

The kinematics are defined by:

$$\begin{pmatrix} X' \\ Y' \\ Z' \end{pmatrix} = \begin{pmatrix} X+w \\ Y \\ Z \end{pmatrix} \text{ for } X > w$$
$$\begin{pmatrix} X' \\ Y' \\ Z' \end{pmatrix} = \begin{pmatrix} X \\ Y \\ Z \end{pmatrix} \text{ for } X \leq 0 \tag{8}$$

where w is the dyke width.

Plugs

Plugs can only be stope-like, with a replacement model for plug emplacement. Simple geometric forms can be defined (cylinders, paraboloids, cones, and ellipsoids) or complex voxel geometries can be imported. In the future, the displacement equations described by Guglielmo (1993) could become the basis for a forced emplacement model.

Homogenous Strains

A strain tensor can be defined to homogeneously deform the geology. The kinematics are defined by:

$$\begin{pmatrix} X' \\ Y' \\ Z' \end{pmatrix} = \begin{bmatrix} \varepsilon_{11} & \varepsilon_{12} & \varepsilon_{13} \\ \varepsilon_{21} & \varepsilon_{22} & \varepsilon_{23} \\ \varepsilon_{31} & \varepsilon_{32} & \varepsilon_{33} \end{bmatrix} \begin{pmatrix} X \\ Y \\ Z \end{pmatrix} \tag{9}$$

Tilts

A uniform tilt can be defined to rotate the geology. The kinematics are defined by:

$$\begin{pmatrix} X' \\ Y' \\ Z' \end{pmatrix} = \begin{bmatrix} j^2+(1-j^2)\cos\theta & jk(1-\cos\theta)+l\sin\theta & jl(1-\cos\theta)-k\sin\theta \\ jk(1-\cos\theta)-l\sin\theta & k^2+(1-k^2)\cos\theta & kl(1-\cos\theta)+j\sin\theta \\ jl(1-\cos\theta)+k\sin\theta & kl(1-\cos\theta)-j\sin\theta & l^2+(1-l^2)\cos\theta \end{bmatrix} \begin{pmatrix} X \\ Y \\ Z \end{pmatrix} \tag{10}$$

where j,k,l is the unit vector defining the axis of rotation, and θ is the angle of rotation.

Penetrative Foliations and Lineations

These deformation events are not explicitly associated with a volume change or strain. However, they can be combined with homogeneous strains if needed. Some of the deformation events implicitly produce cleavages and lineations, for example folds produce an implicit axial plane cleavage and a fold axis lineation, whose orientation can be displayed symbolically on map views.

Principal Stress Orientations from Faults: a C^{++} program

Bruno Ciscato
Dipartimento di Scienze della Terra,
Via La Pira 4, 50127 Firenze, Italy.

Abstract– The technique developed by Lisle is useful for estimating the orientation of principal stresses from measured orientations of fault planes and the slip direction within the fault planes. Lisle's technique, although very simple in concept, is practically impossible to implement by hand, even with few faults, as the stereonet plot becomes cluttered very rapidly. The program discussed in this paper is an improved version of Lisle's original. The new program evaluates about 500,000 tensors to find the most likely orientation of the principal stresses and produces an easy to read output in three different formats. Some programming techniques used in this program could also be used in other programs that deal with directional data.

Introduction

Techniques for estimating the orientations of principal stresses commonly employ fault plane orientations and sense of slip data. Several methods have been devised to reconstruct the orientation of the paleostress field by combining data from individual faults. Some methods solve the inverse problem once the mathematical relationship between the stress field, the orientation of the fault, and the resulting slip direction on the fault plane has been established (see Angelier 1984 for a review). Other methods use a graphical search procedure that allows limits to be placed on the possible directions of the principal stresses (Angelier & Mechler 1977, Lisle 1987). The program presented in this paper is an implementation of Lisle's graphical method and is an improved version of a BASIC program originally written by Lisle (1988).

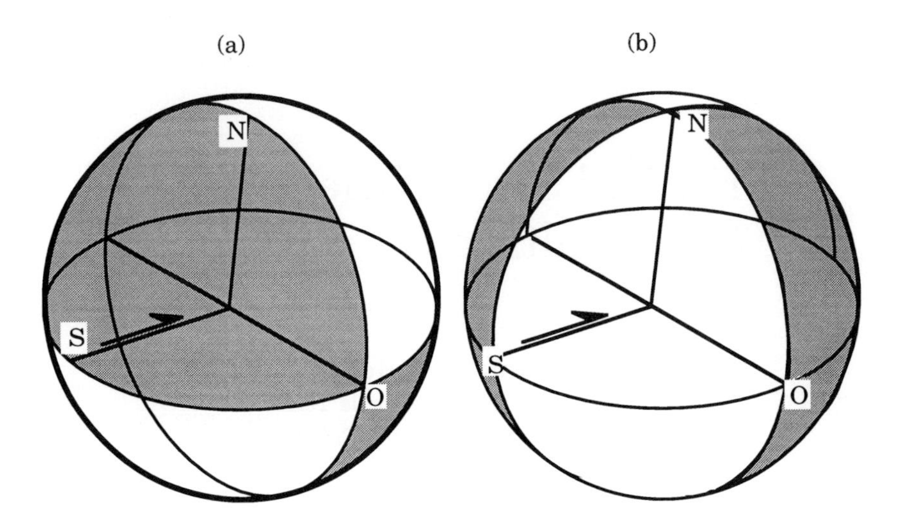

Fig. 1. a) σ_1 (unshaded) and σ_3 (shaded) dihedra defined by planes *OS* (fault) and *ON* (auxiliary plane). b) Shaded and unshaded dihedra bounded by planes *SN* and *ON*. One pair contains σ_1, the other σ_3. Modified after Lisle (1987).

Lisle's Stress Inversion Method

If slickenlines or striations on a fault surface are taken to be parallel to the direction of maximum shear stress on the fault, σ_1 and σ_3 are confined within different pairs of opposing right dihedra, bounded by the fault plane and a plane perpendicular to the direction of the striations (McKenzie 1969). If we know the sense of slip on the fault, we can deduce which pair of dihedra contains σ_1 and which ones contain σ_3 (Fig. 1).

Based on these assumptions, Angelier and Mechler's method (1977) superimposes data from several faults on a single stereogram. If we shade the σ_1 dihedron and leave the σ_3 dihedron unshaded for each fault, the most likely orientations for σ_1 are those which lie in the darker shaded area of the stereogram. These are the orientations which lie in the σ_1 dihedron with respect to the largest number of fault planes (Fig. 2).

Lisle's technique introduces an additional constraint, by taking into account that σ_1 and σ_3 lie in different pairs of dihedra bounded by two planes perpendicular to the fault: one containing the slip direction, the other perpendicular to it (Lisle 1987). These additional pairs of dihedra are labeled *A* and *B*. With this new constraint we can further reduce the σ_1 field eliminating those directions in the σ_1 field that lie in an *A* dihedron and do not have a corresponding σ_3 direction in a *B* dihedron, and *vice versa*.

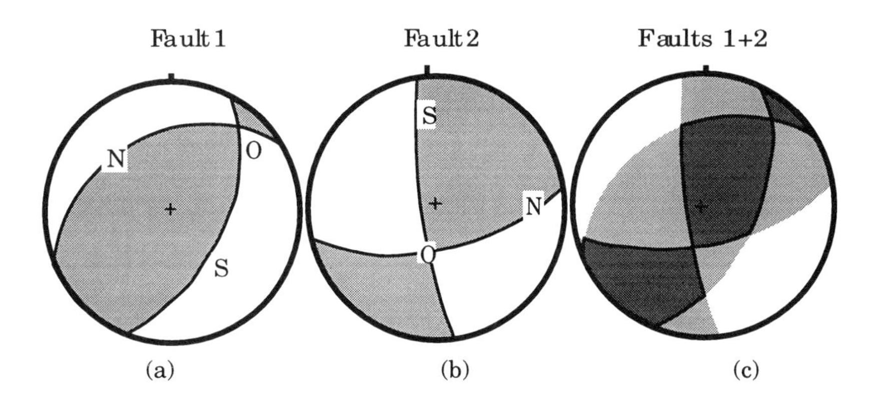

Fig. 2. Principle of Angelier and Mechler method. In stereographic projection, σ_1 (shaded) and σ_3 (unshaded) dihedra from several fault planes are superimposed. Most likely σ_1 directions are those which lie in darker shaded area of stereogram on right and correspond to directions in σ_1 dihedra of each individual fault. After Lisle (1988).

Assumptions

The application of stress inversion techniques is limited to situations in which a number of assumptions can adequately be satisfied. Lisle's method has the advantage that it employs the least restrictive assumptions of any of the inversion techniques: 1) all the faults within a data set are assumed to have been reactivated at the same time, 2) each fault moved independently of the other faults, and 3) the rock has not undergone subsequent deformation.

Implementation

The new technique developed by Lisle, although very simple in concept, is practically impossible to implement by hand even with few faults, as the stereonet plot becomes cluttered very rapidly. For this reason Lisle wrote a program that calculates the probability of any given direction being σ_1 (Lisle 1988). The program is rather slow, only a few points are calculated and the exact shape of the fields of the most likely σ_1 directions cannot be reconstructed. In order to increase the efficiency of Lisle's program, a new version has been written which allows the calculation of a large number of points in a short time.

The Program

The program reads data from a text file of the same format used in Lisle's original program, thus existing data files can be readily analyzed with the new program. The data file for each fault contains: fault dip, fault dip direction, striae plunge,

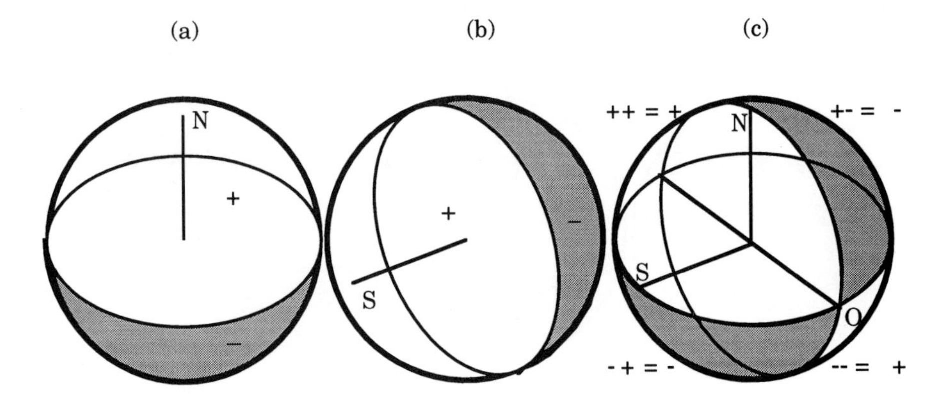

Fig. 3. a,b) If we divide a sphere into two hemispheres with a plane normal to the direction from which we calculate the angles, (a), one hemisphere is characterized by positive values of cosines, the other by negative values (b). c) If we superimpose two such spheres, one for the normal to the fault and the other for the striae, (a) and (b), the product of the signs of the cosines is positive for σ_1 dihedra and negative for σ_3.

striae plunge direction, vertical sense of slip, and horizontal sense of slip (see 'Using the program' for an exact description of the format).

For each fault, the directional cosines of the normal to the fault plane, the striae, and the *O* direction (perpendicular to the normal to the fault and to the striae) are calculated. These directions are stored in an array of objects of type `Fault`.

Forty thousand potential σ_1 orientations about one degree apart are evaluated. This resolution produces a very clear graphical output, but can be easily increased or decreased by changing the loop increments in the two main loops (x and y).

For each potential σ_1 orientation evaluated, the number of faults for which the given orientation lies in the σ_1 dihedra is calculated first. From simple trigonometry we can see that cosines of angles <90° from a given direction have positive values. Therefore we can divide a sphere into two hemispheres with a plane normal to the direction from which we calculate the angles (Fig. 3a). One hemisphere is characterized by positive values of the cosines, the other by the negative values. If we superimpose two such spheres, one for the normal to the fault and the other for the striae, (Fig. 3a and 3b), we can see that the product of the signs of the cosines is positive for the σ_1 dihedra and negative for the σ_3 dihedra (Fig. 3c).

For each fault the program calculates the sign of the product of the cosine of the angle from the fault normal to the given direction and the cosine of the angle from the striae to the given direction. If the product is positive the direction lies in the σ_1 dihedron and the variable `NumFaultP1`, representing the number of faults for which the given direction lies in the σ_1 dihedra, is increased by one.

Orientations normal to the σ_1 orientation (`GivDir`) being evaluated are then considered, to determine which one is the most likely σ_3 direction. Considering 180 potential σ_1 directions gives approximately the same resolution as that used for scanning the grid. However this value can also be modified easily by changing the *SCAN_ANGLE* and *SCAN_ANGLE_RAD* macros.

A loop of 180 steps pre-multiplies a rotation matrix by the S_3 vector. This rotates the S_3 direction 1 degree around the potential σ_1 orientation under investigation (`GivDir`). The initial S_3 direction must be perpendicular to `GivDir`. The easiest one to calculate is the S_3 direction of dip = 0 and dip direction = given direction dip + 90°. At each step, the proportion of faults for which the S_3 direction lies in the σ_3 dihedra is calculated. The same technique that was used for the σ_1 dihedra is used here, the only difference is that the S_3 direction lies in a σ_3 dihedron if the sign of the product of the cosines is negative.

Using the same technique, the program also calculates the proportion of faults for which the σ_1 (`GivDir`) and the σ_3 (S_3) orientations under investigation lie in different pair of *A* and *B* dihedra. The cosines of the angles to the striae and to the *O* direction of both directions are considered, and the signs of the products are compared. If they agree, the `GivDir` and the S_3 directions lie in the same pair of *A* and *B* dihedra. If they do not agree, the S_3 direction satisfies Lisle's constraint.

The S_3 direction that lies in the σ_3 dihedra and satisfies Lisle's constraint for more faults than any other is taken as the most likely σ_3 direction associated with the given direction (`GivDir`). The product *P2*P3* represents the proportion of faults for which σ_3 lies in the σ_3 dihedra times the proportion of faults for which S_3 satisfies Lisle's constraint. The product *P2*P3* for this direction is multiplied by *P1* (proportion of faults for which `GivDir` lies in σ_1 dihedra) and the result is taken as the overall probability that the given direction is the paleo σ_1 direction. This probability is written to a file along with the coordinates of the point that represent the given direction on the equal area net.

The output of the program is a series of values that represent x and y coordinates and the probability of any one node being the paleo σ_3 direction. These data can be handled by any data visualization program. Plotting the points with different shades, corresponding to different values of the probability, produces an equal area net divided into sectors with different probabilities. More complex representations are also possible. For example, shading can be added to a color representation, so that the areas with the highest value also stand out visually.

Directions that lie in sectors with the highest probability are the more likely directions for the paleo σ_1. The associated paleo σ_3 directions do not show up in the plot though. So, while the most likely directions for paleo σ_1 can be read on the plot, the most likely associated σ_2 and σ_3 directions are tabulated and presented in a separate file.

Table 1. Example text file with data of 11 faults. Note the number of faults in the first line. See the text for an explanation of the format. Umpublished data of G.D. Harper and A.J. Coulton.

```
11
10   52   4   116   1  0
54  290  53   272   1  0
55  302  35   242   1  0
40  280  28   332   1  0
85  240   6   151  -1  0
82   56  14   144   1  0
58  278   4     5   1  0
37  232   4   315   1  0
69  200  57   147   1  0
32  214  32   224   1  0
50  231  50   220  -1  0
```

Dealing with Directions

This program deals mainly with directional data. Such data are very common in geology, and therefore, some techniques used in this program could be used in many other applications. While a description of the basic functions and data structures that are sufficient to handle directions can be found in many textbooks (*e.g.*, Angell & Tsoubelis 1992), reading this program may help in the application of the general theory to a different geological problem.

Using the Program

The program reads the input from a plain text file and produces three output files. The input file `extension.dat` begins with the number of faults on the first line followed by the fault data, one line per fault. See Table 1 for an example. Starting on line 2, the format of each line is: fault dip, fault dip direction, striae plunge, striae plunge direction, vertical sense of slip (1 = normal, -1 = reverse, 0 = strike slip), horizontal sense of slip (1= dextral, -1 = sinistral, 0 = normal or reverse). The three output files are:

1) the grid file `extension.grd`. This file contains the entire list of results representing points on the stereo plot and the probability of each point being σ_1.

2) the tensor file `extension.ten`. This file contains the most likely paleo stress tensors (σ_1, σ_2, σ_3 directions) for the given set of faults. See Table 2 for an example.

3) the *xyz* file `extension.xyz` containing the grid nodes with the associated

Table 2. Most likely principal stress orientations for data in Table 1.

81.82% Probability Stress Tensors:

S1	S2	S3
S1: 231/43	S2: 51/47	S3: 321/0
S1: 231/41	S2: 51/49	S3: 321/0
S1: 230/42	S2: 50/48	S3: 320/0
S1: 229/43	S2: 49/47	S3: 319/0
S1: 229/43	S2: 49/47	S3: 319/0
S1: 228/44	S2: 48/46	S3: 318/0
S1: 230/40	S2: 50/50	S3: 320/0
S1: 230/41	S2: 50/49	S3: 320/0
S1: 229/41	S2: 49/49	S3: 319/0
S1: 229/42	S2: 49/48	S3: 319/0
S1: 228/43	S2: 48/47	S3: 318/0
S1: 227/43	S2: 47/47	S3: 317/0
S1: 226/44	S2: 48/46	S3: 317/1
S1: 231/38	S2: 51/52	S3: 321/0
S1: 230/39	S2: 50/51	S3: 320/0
S1: 230/40	S2: 50/50	S3: 320/0
S1: 229/40	S2: 49/50	S3: 319/0
S1: 228/41	S2: 48/49	S3: 318/0
S1: 228/42	S2: 48/48	S3: 318/0
S1: 227/42	S2: 47/48	S3: 317/0
S1: 226/43	S2: 48/47	S3: 317/1
S1: 226/43	S2: 50/46	S3: 318/2
S1: 230/38	S2: 50/52	S3: 320/0
S1: 230/38	S2: 50/52	S3: 320/0
S1: 229/39	S2: 49/51	S3: 319/0
S1: 228/40	S2: 48/50	S3: 318/0
S1: 228/40	S2: 48/50	S3: 318/0
S1: 227/41	S2: 47/49	S3: 317/0
S1: 226/42	S2: 48/48	S3: 317/1
S1: 226/42	S2: 50/48	S3: 318/2
S1: 229/37	S2: 49/53	S3: 319/0
S1: 229/38	S2: 49/52	S3: 319/0
S1: 228/38	S2: 48/52	S3: 318/0
S1: 228/39	S2: 48/51	S3: 318/0
S1: 227/40	S2: 47/50	S3: 317/0
S1: 226/40	S2: 48/50	S3: 317/1
S1: 228/37	S2: 48/53	S3: 318/0
S1: 228/38	S2: 48/52	S3: 318/0
S1: 227/38	S2: 47/52	S3: 317/0
S1: 226/39	S2: 50/51	S3: 318/2
S1: 227/37	S2: 47/53	S3: 317/0
S1: 226/38	S2: 50/52	S3: 318/2

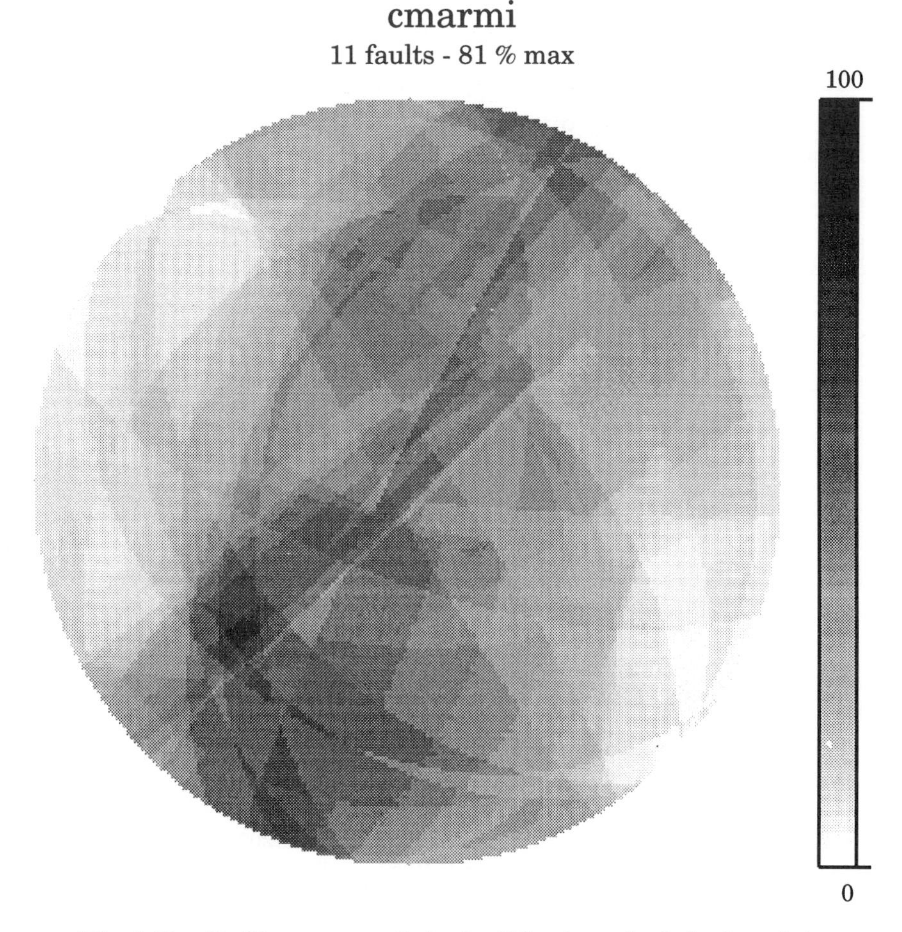

Fig. 4 . Result of the program. Output grid has been shaded using a data visualization program. Unpublished data of Harper.

probability of being σ_1 in the format x, y, z. x and y represent the position on the net in an arbitrary scale, the vertical axis being at $x = 0$ and the horizontal one at $y = 0$. z represents the probability. The *xyz* file can be used with any data representation program in order to obtain a graphic output similar to the one shown in Fig. 4.

Using the program is a two step process: 1) the data are written in a plain text file with `extension.dat`, *e.g.*, `mydata.dat`, 2) the program is executed inserting the name of the data file (without extension) on the command line, *e.g.*, romsa mydata. When the program has finished processing the data the results can be read in the grid file and in the tensor file, *e.g.*, `mydata.grd` and `mydata.ten` respectively. If a data visualization program is available, the *xyz* file can then be processed to obtain a graphical output.

The program is available for FTP at ftp.geo.unifi.it, in the directory /pub/romsa.

Conclusions

The program allows the use of a technique that, although simple in principle, is impossible to implement in practice without the aid of a computer. In addition, it demonstrates some good techniques for handling directional data, probably the most common type of data in geology. Parts of this program may also be useful to geologists who are not directly interested in stress inversion problems.

For people who are not familiar with *C* and *C*++ programming it should not be too difficult to translate the code into their language of choice, for the code has been kept as simple as possible, avoiding those cryptic expressions so common in *C* programs. Overloaded operators and classes can be replaced by functions and structures of data.

Acknowledgements

Credit goes to Lisle for writing the original program (*ROMSA*) whose general structure is the same used in this new version. However the algorithms used to implement the operations on directions are very different: matrix algebra and object oriented language constructs. This results in a more efficient program that is easier to understand and modify, and whose functions can be readily reused in other programs. The program was written in 1994, while I was studying at the Department of Geological Sciences at the State University of New York at Albany.

References

Angelier, J. 1984. Tectonic analysis of fault slip data sets. *J. Geophys. Research* **89**: 5835-5848.

Angelier, J. & Mechler, P. 1977. Sur une methode graphique de reserche des constraintes principales Ègalement utisable en tectonique et en seismologie: la methode des di Ëdres droits. *Bull. Soc. Geol. France* **19**: 1309-1318

Angell, I. O. & Tsoubelis, D. 1992. *Advanced graphics on VGA and XGA cards using Borland C*++. Wiley, New York.

Lisle, R. J. 1987. Principal stress orientations from faults: an additional constraint. *Annales Tectonicae* **1**: 155-158.

Lisle, R. J. 1988. *ROMSA*: a basic program for paleostress analysis using fault-striation data. *Comp. & Geosci.* **14**: 255-259.

McKenzie, D. P. 1969. The relation between fault plane solutions and the directions of the principal stresses. *Bull. Seismolog. Soc. America* **59**: 591-601.

Appendix: Code

```
#include <iostream.h>
#include <stdio.h>
#include <math.h>
#include <string.h>

#define SCAN_ANGLE 1
#define SCAN_ANGLE_RAD 0.017453292

const double pi=3.1415926535;
const double pi_2=pi/2;
const double pi2=pi*2;
const double epsilon=0.000001;

// CLASS DECLARATIONS

class vector3
{
	public:
	float L,M,N;

	friend vector3 operator * (vector3 V, float f);
	friend float operator * (vector3 A, vector3 B);

};

class Matrix
{
	float M[3][3];

	public:
	Matrix();
	void Set(float m[3][3]);
	void rotate(int m, float theta);
	vector3 dirTransf(vector3 d);

	friend Matrix operator* (Matrix A, Matrix B);
	friend vector3 operator* (Matrix A, vector3 V);
};

class Fault
{
	public:
	vector3 Norm,Stri,O;
};

class Table
{
	public:
	float Probab;
	vector3 S1, S3;
};

//

// FUNCTION DECLARATIONS

Matrix genrot(float phi, vector3 d);
vector3 dirCos(float,float);
vector3 cross(vector3, vector3);
float dire(float x,float y);
float rad(float);
float deg(float);
```

```
//

int main(const int aC, char *aV[])

{

   // VARIABLE DEFINITION

        FILE    *in,*xyz,*ten,*grd;

        int     i,j,n=0,NData,S3TPointer,
                NormalReverse,DextralSinistral,
                NumFaultP1,NumFaultP2,NumFaultP3;

        long int k;

        double NormL, NormM, NormN,
                NormDip, NormDipDir,
                NormAng, NormAngX, NormAngZ,
                StriaeAng, StriaeAngX, StriaeAngZ,
                OAng, OAngX, OAngZ;

        float  P1,P2,P3,P123,
                P123Max,P23Max,
                GivDirPlunge, GivDirPlungeDir,
                Dip,DipDir,Plunge,PlungeDir,
                R,x,y,perc,
                S1MaxDipDir,S1MaxDip,S2MaxDipDir,S2MaxDip,
                S3MaxDipDir,S3MaxDip;

        vector3        GivDir,S3,S2Max,S3Maxtemp;

        Table  S3Table[40401];

        Fault  Faul[100];

        Matrix Rotation;

        char   filein[30]="", filexyz[30]="", fileten[30]="", filegrd[30]="";

   //

   // OPENS FILES

   strcat(filein,aV[1]);
   strcat(filexyz,aV[1]);
   strcat(fileten,aV[1]);
   strcat(filegrd,aV[1]);
   strcat(filein,".dat");
   strcat(filexyz,".xyz");
   strcat(fileten,".ten");
   strcat(filegrd,".grd");


   if ((in = fopen(filein, "r")) == NULL)
   {
         fprintf(stderr, "Cannot open input file %s.\n", filein);
         return 1;
   }

   if ((xyz = fopen(filexyz, "w")) == NULL)
   {
         fprintf(stderr, "Cannot open output file %s.\n", filexyz);
         return 1;
   }
```

```
if ((ten = fopen(fileten, "w")) == NULL)
{
      fprintf(stderr, "Cannot open output file %s.\n", fileten);
      return 1;
}
if ((grd = fopen(filegrd, "w")) == NULL)
{
      fprintf(stderr, "Cannot open output file %s.\n",filegrd);
      return 1;
}

printf("\nReading file : %s\n",filein);

//


// LOADS DATA FROM FILE

fscanf(in,"%d",&NData);

for(n=1;n<=NData;n++)
      {
      /* loads data */

      fscanf(in, "%f %f %f %f %d %d", &Dip, &DipDir, &Plunge, &PlungeDir,
&NormalReverse, &DextralSinistral);


      /* checks if the fault is pure strike slip */

      if (NormalReverse==0)

            {
            PlungeDir=DipDir+90*DextralSinistral;
            NormalReverse=1;
            }


      /* calc dir cos of normal to fault (down end) */

      NormDip=rad(90-Dip);
      NormDipDir=rad(DipDir+180);

      Faul[n].Norm=dirCos(NormDip,NormDipDir);


      /* calculates dir cos of striae */

      Faul[n].Stri=dirCos(rad(Plunge),rad(PlungeDir));

      Faul[n].Stri=Faul[n].Stri*NormalReverse;


      /* calculates dir cos of O */

      Faul[n].O=cross(Faul[n].Stri,Faul[n].Norm);

      }

 fclose(in);

 printf("\nNumber of data: %d\n\n",NData);


 //
```

```
// MAIN LOOP

perc=0;
P123Max=P123=0.0;
S3TPointer=0;
    for (y=1;y>=-1;y-=0.01)
            {
            for(x=-1;x<=1;x+=0.01)
                    {

                    perc+=1;

                    printf("\r%.2f %% done.",perc/40401.0F*100.0f);

                    R=sqrt(x*x+y*y);



                    if(R<=1.0)
                            {
                            /* calc plunge & plunge direction of point of
coords x and y on a Schmidt net of radius 1 */

                            GivDirPlunge=pi_2-2.0F*asin(R*0.707106781F);
                            GivDirPlungeDir=dire(x,y);

                            GivDir=dirCos(GivDirPlunge,GivDirPlungeDir);

                            /* calc % faults with given direction in s1
dihedra */

                            NumFaultP1=0;

                            for (n=1;n<=NData;n++)
                                    {
                                    NormAng=Faul[n].Norm*GivDir;
                                    StriaeAng=Faul[n].Stri*GivDir;

                                    if(NormAng*StriaeAng>=0)
                                    NumFaultP1++;
                                    }
                            P1=float(NumFaultP1)/float(NData);

                            /* calc horz vector _|_ to GivDir */

                            S3=dirCos(0.0F,GivDirPlungeDir+pi_2);

                            /* calculates the matrix that performs the rota-
tion of a vector around an axis GivDir*/

                            Rotation=genrot(SCAN_ANGLE_RAD,GivDir);


                            P23Max=0.0;

                    /* considers possible sigma 3 directions */

                            for (i=0;i<=179;i+=SCAN_ANGLE)
                                    {
                                    NumFaultP2=0;
                                    NumFaultP3=0;

                                    for (j=1;j<=NData;j++)
                                            {
                                            /* calculates the number of faults
```

```
for which the S3 direction lies in the sigma 3 dihedron */

                                        NormAngZ=Faul[j].Norm*S3;
                                        StriaeAngZ=Faul[j].Stri*S3;
                                        if(NormAngZ*StriaeAngZ<0)
                                              NumFaultP2++;

                                        /* calculates the number of faults
for which the given direction and the S3 direction lie in opposite A and B
dihedra */

                                        OAngX=Faul[j].O*GivDir;
                                        StriaeAngX=Faul[j].Stri*GivDir;
                                        OAngZ=Faul[j].O*S3;
                                        StriaeAngZ=Faul[j].Stri*S3;


if((OAngX*StriaeAngX>=0)!=(OAngZ*StriaeAngZ>=0))
                                              NumFaultP3++;

                                        }

                                  P2=float(NumFaultP2)/float(NData);
                                  P3=float(NumFaultP3)/float(NData);


                                  if(P2*P3>P23Max)
                                        {
                                        P23Max=P2*P3;
                                        S3Maxtemp=S3;
                                        }
                                  S3=Rotation*S3;
                                  }
                                /* calculates the total probability of GivDir
being the paleo sigma 1 */

                          P123=P23Max*P1;

                          S3TPointer++;
                          S3Table[S3TPointer].Probab=P123;
                          S3Table[S3TPointer].S1=GivDir;
                          S3Table[S3TPointer].S3=S3Maxtemp;


                          if(P123>P123Max)
                                {
                                P123Max=P123;
                                }

                          fprintf(xyz,"%3.3f %3.3f %3d
\n",x,y,int(P123*100.0));
                          fprintf(grd,"%2d ",int(P123*100.0));
                          }
                    else
                          {

                          fprintf(xyz,"%3.3f %3.3f %3d \n",x,y,-1);
                          fprintf(grd,"   ");
                          }
                    }

              fprintf(grd,"\n\n");
              }

  fclose(xyz);
```

```
  fclose(grd);

  printf("\nP123Max : %.0f%%\n\n",P123Max*100);
  printf("The table of the most likely principal stress orientations\nis
saved in the file: %s\n\n",fileten);

// WRITES THE TABLE OF TENSORS TO A FILE

  fprintf(ten,"%.2f%% Probability Stress Tensors:\n\n",P123Max*100);

  for(k=0;k<=40401;k++)
       {
       if(fabs(S3Table[k].Probab-P123Max)<epsilon)
               {
               S2Max=cross(S3Table[k].S1,S3Table[k].S3);

               S1MaxDipDir=dire(S3Table[k].S1.M,S3Table[k].S1.L);

S1MaxDip=dire(S3Table[k].S1.N,sqrt(S3Table[k].S1.L*S3Table[k].S1.L+S3Table[k].S1.M*S3Table[k].S1.M));

               S2MaxDipDir=dire(S2Max.M,S2Max.L);
               S2MaxDip=dire(S2Max.N,sqrt(S2Max.L*S2Max.L+S2Max.M*S2Max.M));

               S3MaxDipDir=dire(S3Table[k].S3.M,S3Table[k].S3.L);

S3MaxDip=dire(S3Table[k].S3.N,sqrt(S3Table[k].S3.L*S3Table[k].S3.L+S3Table[k].S3.M*S3Table[k].S3.M));

               fprintf(ten,"S1: %3.0f/
%2.0f\t",deg(S1MaxDipDir),deg(S1MaxDip));
               fprintf(ten,"S2: %3.0f/
%2.0f\t",deg(S2MaxDipDir),deg(S2MaxDip));
               fprintf(ten,"S3: %3.0f/
%2.0f\n",deg(S3MaxDipDir),deg(S3MaxDip));
               }
       }

  fclose(ten);
//


//

return 1;
}






/*****************

 FUNCTIONS

*****************/


// MATRIX CLASS FUNCTIONS

/* Initialization of the matrix */

Matrix::Matrix()
{
       for(int i=0;i<=2;i++)
              for(int j=0;j<=2;j++)
```

```
                M[i][j]=0.0;
}

/* Assignement of initial values to the matrix */

void Matrix::Set(float m[3][3])
{
        for(int i=0;i<=2;i++)
                for(int j=0;j<=2;j++)
                        M[i][j]=m[i][j];
}

/* Product of a matrix and a vector */

vector3 Matrix::dirTransf(vector3 d)
{
        vector3 w;
        w.L=M[0][0]*d.L+M[0][1]*d.M+M[0][2]*d.N;
        w.M=M[1][0]*d.L+M[1][1]*d.M+M[1][2]*d.N;
        w.N=M[2][0]*d.L+M[2][1]*d.M+M[2][2]*d.N;
        return(w);
}


/* Rotation of the matrix around the axis m (1=x, 2=y, 3=z) of an angle theta
*/

void Matrix::rotate(int m, float theta)
{
        int m1=(m%3),
                m2=((m1+1)%3);
        float c=cos(theta),
                s=sin(theta);

        for(int i=0;i<=2;i++)
                for(int j=0;j<=2;j++)
                        M[i][j]=0.0;

        M[m-1][m-1]=1.0;
        M[m1][m1]=c;
        M[m2][m2]=c;
        M[m1][m2]=s;
        M[m2][m1]= -s;
}


//

// GENERAL FUNCTIONS

/* Return the matrix that can be used to rotate any vector around a direction
   d of an angle phi */

Matrix genrot(float phi, vector3 d)
{
        Matrix G,H,W,H1,G1,A;
        float beta,theta,v;
        theta=dire(d.M,d.L);
        G.rotate(3,theta);
        G1.rotate(3,-theta);
        v=sqrt(d.L*d.L+d.M*d.M);
        beta=dire(v,d.N);
        H.rotate(2,beta);
        H1.rotate(2,-beta);
        W.rotate(3,-phi);
```

```
        A=G1*(H1*(W*(H*(G))));
        return(A);
}

/* Cross product of two vectors */

vector3 cross(vector3 A, vector3 B)
{
        vector3 V;

        V.L=A.M*B.N-A.N*B.M;
        V.M=A.N*B.L-A.L*B.N;
        V.N=A.L*B.M-A.M*B.L;

        return (V);
}

/* Returns a vector of plunge Plunge and direction of plunge PlungeDir */

vector3 dirCos(float Plunge, float PlungeDir)
{
        vector3 V;

        V.L=cos(Plunge)*cos(PlungeDir);
        V.M=cos(Plunge)*sin(PlungeDir);
        V.N=sin(Plunge);

        return(V);
}


/* Cartesian coordinates to polar coordinates */

float dire(float x,float y)
{
        if(fabs(y)<epsilon)
                {
                if(fabs(x)<epsilon) return (0.0);
                if(x>=0) return(pi_2);
                else return(pi_2*3);
                }
        if(y<0.0) return(pi+atan(x/y));
        else {
                if(x>=0) return(atan(x/y));
                else return(pi2+atan(x/y));
                }
}

float rad(float n)
{
        return(n/180.0F*pi);
}

float deg(float n)
{
        return(n/pi*180.0F);
}

//



// OPERATOR OVERLOADING
```

```
/* Product of two matrices */

Matrix operator * (Matrix A, Matrix B)
{
       int i,j;
       Matrix C;

       for(i=0;i<=2;i++)
              {

              for(j=0;j<=2;j++)
                     {


C.M[i][j]=A.M[i][0]*B.M[0][j]+A.M[i][1]*B.M[1][j]+A.M[i][2]*B.M[2][j];

                     }

              }

       return(C);
}


/* Product of a matrix and a vector */

vector3 operator * (Matrix A, vector3 B)
{
       vector3 V;

       V.L=A.M[0][0]*B.L+A.M[0][1]*B.M+A.M[0][2]*B.N;
       V.M=A.M[1][0]*B.L+A.M[1][1]*B.M+A.M[1][2]*B.N;
       V.N=A.M[2][0]*B.L+A.M[2][1]*B.M+A.M[2][2]*B.N;

       return(V);
}

/* Dot product of two vectors */

float operator * (vector3 A, vector3 B)
{
       return (A.L*B.L+A.M*B.M+A.N*B.N);
}

/* Product of a vector and a scalar */

vector3 operator * (vector3 V, float f)
{
       V.L*=f;
       V.M*=f;
       V.N*=f;

       return V;
}

//
```

A Spring-Network Model of Fault-System Evolution

Norihiro Nakamura, Kenshiro Otsuki,
and Hiroyuki Nagahama
Institute of Geology and Paleontology, Graduate School of Science,
Tohoku University, Sendai 980-77, Japan.
nakamura@dges.tohoku.ac.jp

Abstract– Fault systems are restricted by the elasticity equation under the mechanical constraints of both the weakening stiffness under cyclic rupturing and constant fault rupture strength. In addition, according to existing experimental data, the ratio ϕ of stiffness after rupture to stiffness before rupture under the cyclic rupturing of rocks is constant and is not always one. So, based on these constraints on rock fracture mechanics, we propose a new spring network model which has various ϕ values to simulate the evolution of fault systems. Our model is supported not only by fracture mechanics but also by natural observations of fault systems (*e.g.*, "time-predictable recurrence" for large earthquakes). The result of our simulation shows clearly that varying ϕ affects patterns of spatial and temporal evolution of fault systems. An original ANSI-*C* program is presented for the simulation algorithm of our model. This simulation is a valuable tool for investigating fault system evolution in tectonically active regions of the upper lithosphere.

Introduction

Several researchers have studied the evolutionary process of brittle fracturing by computer simulation (Takayasu 1985, 1986, Louis *et al.* 1986, Louis & Guina 1987, 1989, Li & Duxbury 1988, Meakin 1988, 1991, Ausloos & Kowalski 1992, Reuschlé 1992, Andersen *et al.* 1994). In addition, there are several studies on the evolution of fault systems (Segall & Pollard 1980, Pollard & Segall 1987, Cox & Paterson 1989, Chelidze 1986, 1993; Sornette *et al.* 1992, Cowie *et al.* 1993, Sornette *et al.* 1994).

Fracturing during earthquakes or faulting can be regarded as cyclic rupturing of rocks. In general, the stiffness of brittle rocks decreases as rupturing progresses under cyclic rupturing in uniaxial compression or beam tests (Cook 1965, Wawersik & Fairhurst 1970, Bieniawski 1971, Peng & Podnieks 1972, Brady *et al.* 1973, Hardy *et al.* 1973, Hudson *et al.* 1973). Moreover, Shimazaki & Nakata (1980) have shown that the occurrence of repeated earthquakes, called "time-predictable recurrence" for large earthquakes, can be explained by a time-predictable model based on a constant rupture strength for faults. Therefore, it is necessary to assume weakening stiffness under cyclic rupturing and a constant rupture strength of faults in modeling the evolution of fault systems; there have been no simulation models incorporating these assumptions.

The mechanical property of rocks in the upper lithosphere is approximately elastic at low temperature (Ranalli 1987). Thus brittle crust with fault systems has been considered analogous to a network of Hookean springs. In this paper, we first present constraints of fracture mechanics on fault systems and then introduce a new spring network model of fault systems. Furthermore, we derive the displacement field of this model from the discretized elasticity equation by the relaxation method (Schmid *et al.* 1988; pp. 88-99). Next, we propose an algorithm for the evolution of fault systems, which contains the ratio ϕ of stiffness after rupture to stiffness before rupture, and our original program written in *C* language according to ANSI standards is presented for the simulation algorithm of our model. Moreover, we show that the difference of fault patterns for the spatial and temporal evolution of the fault system depends on the ratio, ϕ. Finally, we discuss how our spring network model is supported by fracture mechanics for cyclic rupturing of rocks and natural observations, and is a good tool for researching true fault system evolution in tectonically active regions of the upper lithosphere.

Constraints of Fracture Mechanics of the Fault System

Three constraints on the fault system are as follows:

1. We consider brittle crust with fault systems to be an elastic body which obeys Hooke's law. Thus, the mechanical condition of a fault system is always in elastic equilibrium, which obeys the elasticity equation.

2. In a real fault system, Shimazaki & Nakata (1980) have shown that the occurrence of repeated earthquakes can be explained by the time-predictable model based on a constant rupture strength of faults. Thus, we consider that rupture phenomena such as earthquakes occur when the elastic stress reaches a constant rupture strength.

3. In the cyclic rupturing process of brittle rocks, the stiffness of the rocks decreases as rupturing progresses under uniaxial compression or beam tests (Cook 1965, Wawersik & Fairhurst 1970, Bieniawski 1971, Peng & Podnieks 1972, Brady *et al.* 1973, Hardy *et al.* 1973, Hudson *et al.* 1973).

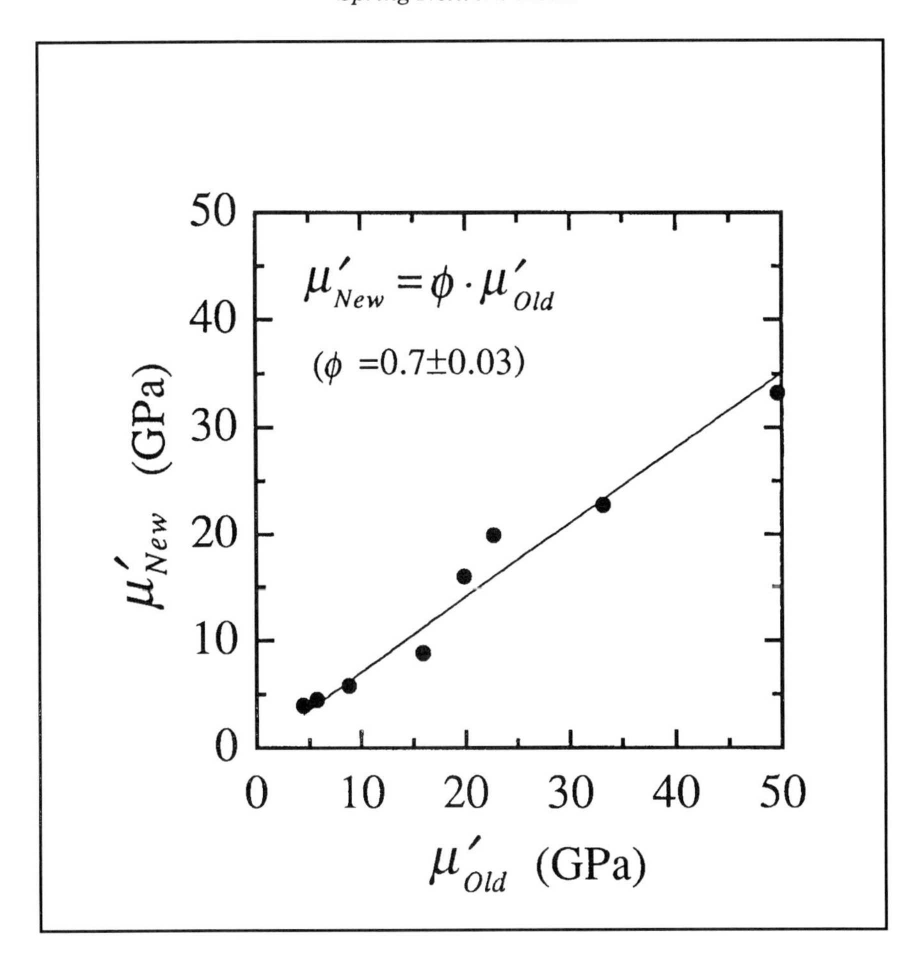

Fig. 1 Stiffness μ'_{New} after rupture as a linear function of stiffness μ'_{Old} before rupture under cyclic rupturing in a uniaxial compression test. Data from Bieniawski (1971). Solid line is fitted to data by least squares regression.

From the experimental result of Bieniawski (1971), we can point out that the stiffness after rupture is a linear function of the stiffness before rupture under the cyclic rupturing process (Fig. 1). Therefore, using the ratio ϕ, we can estimate the stiffness after rupture, which is weaker than the stiffness before rupture. Next, we introduce a spring network model and a simulation algorithm of fault systems undergoing cyclic rupturing based upon the three constraints stated above.

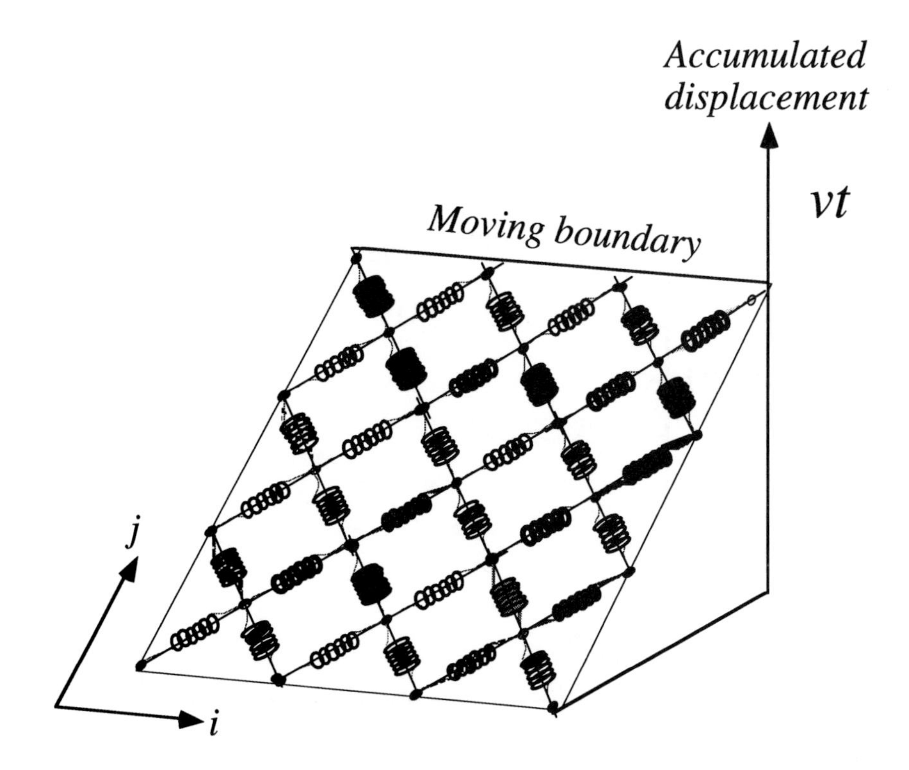

Fig. 2. Lattice network of a spring network model. i and j are coordinates. Height of each lattice node is accumulated displacement vt. Each spring element is characterized by two material properties; heterogeneous stiffness and constant rupture strength σ_c. Top end of lattice network (moving boundary) moves at constant velocity. A periodic boundary condition is applied to the other two sides of the lattice network as if it were periodic in the i direction.

Spring Network: Fault System Simulation Algorithm

The Model

We present a new spring network model of fault systems in the brittle crust. This model consists of a two-dimensional square lattice network that is a mesh of spring elements oriented at 45° with respect to the boundary (Fig. 2). A boundary condition with a constant velocity is applied to the moving boundary (the top end of the lattice network) while the bottom end of the lattice network is kept fixed. A periodic boundary condition is applied to the other two sides of this lattice network as if the lattice network is periodic in the i direction.

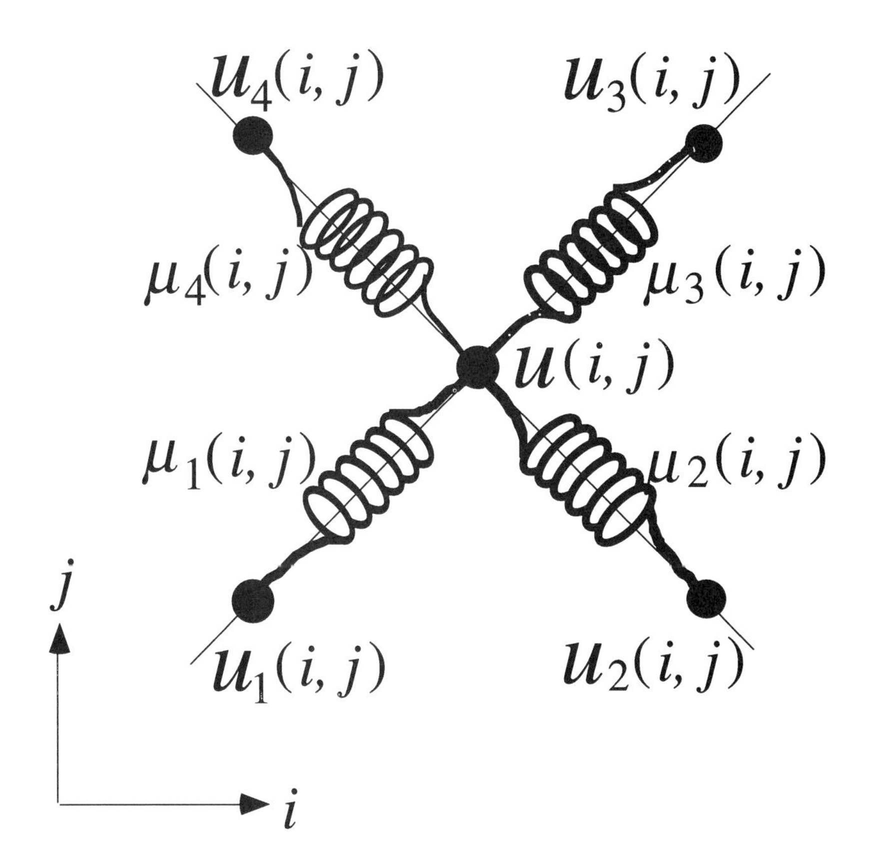

Fig. 3. Local field on lattice network. $u(i, j)$ denotes displacement of (i, j)-th lattice node, $u_k(i, j)$ $(k = 1, 2, 3, 4)$ is displacement at one of four nearest neighbors of (i, j)-th lattice node. $\mu_k(i, j)$ indicates stiffness of corresponding spring element.

Our spring network model is restricted by the elasticity equation (Laplace equation), which represents a balance of forces:

$$\mu \nabla^2 u = 0 \tag{1}$$

where u is the displacement of each lattice node and μ is the stiffness (Lamé coefficient). This displacement is given to be perpendicular to the plane (anti-plane shear problem). In this model, each spring element obeys Hooke's law :

$$\sigma = \mu \bullet \varepsilon \tag{2}$$

where σ and ε are the stress and strain of a spring element, respectively.

Moreover, each element is characterized by two material properties: a heterogeneous stiffness and a constant rupture strength σ_c.

The displacement field in this model can be given by the discretized version of eqn. (1) in the square lattice network:

$$\sum_{k=1}^{4} \mu_k(i,j)\{\mu_k(i,j) - u(i,j)\} = 0 \quad (3)$$

where $u(i, j)$ is the displacement of the (i, j)-th lattice node. Here, $u_k(i, j)$ ($k = 1, 2, 3, 4$) is the displacement at one of the four nearest neighbors of the (i, j)-th lattice node and $\mu_k(i, j)$ ($k = 1, 2, 3, 4$) indicates the stiffness of the corresponding element (Fig. 3). Equation (3) can be solved iteratively for each set of four elements meeting at each lattice node, using the relaxation method (Schmid *et al.* 1988, pp. 88-99). The relaxation method enables solving of the displacement $u(i, j)$ at each lattice node iteratively. When the displacement $u(i, j)$ at each lattice node no longer changes between iterations, $u(i, j)$ at all lattice nodes satisfies the discretized eqn. (3). In this model, the error criteria of displacement by iteration is 10^{-3} %. Rupture phenomena occur when the stress of a spring element reaches a constant rupture strength. The displacement distribution throughout the lattice network is recalculated instantaneously after all rupturing, using the relaxation method.

Algorithm

The temporal evolution of fault systems is modelled by the following procedure. Routines (A) through (E) are accomplished during one time step t.

(A) Assign a stiffness $\mu_k(i, j)$ to each element. $\mu_k(i, j)$ is defined by

$$\mu_k(i,j) = \overline{\mu}_k + \mu_{k0} \times w \quad (4)$$

where w is a random number distributed uniformly over the range from 0 to 1 and μ_k and μ_{k0} are constants ($\mu_k = 1$, $\mu_{k0} = 2$), respectively. The size of the lattice network is 70 × 70 spring elements (see Fig. 2) and the normalized constant rupture strength is $\sigma_c = 1$.

(B) Add a constant velocity as follows: The velocity of a moving boundary v (the top end of the lattice network: see Fig. 2) is given by 0.2 per time step t, while the bottom end of the lattice network boundary is fixed.

(C) Solve the discretized elasticity equation (3) for the displacement iteratively for each set of four elements meeting at each lattice node, using the relaxation method (Schmid *et al.* 1988; pp. 88-99). Then calculate the stress of each spring element throughout the lattice network.

(D) Compare the stress of a spring element with a constant rupture strength. If the stress of a spring element reaches the constant rupture strength σ_c, then the stiffness decreases after rupture to a new stiffness given by

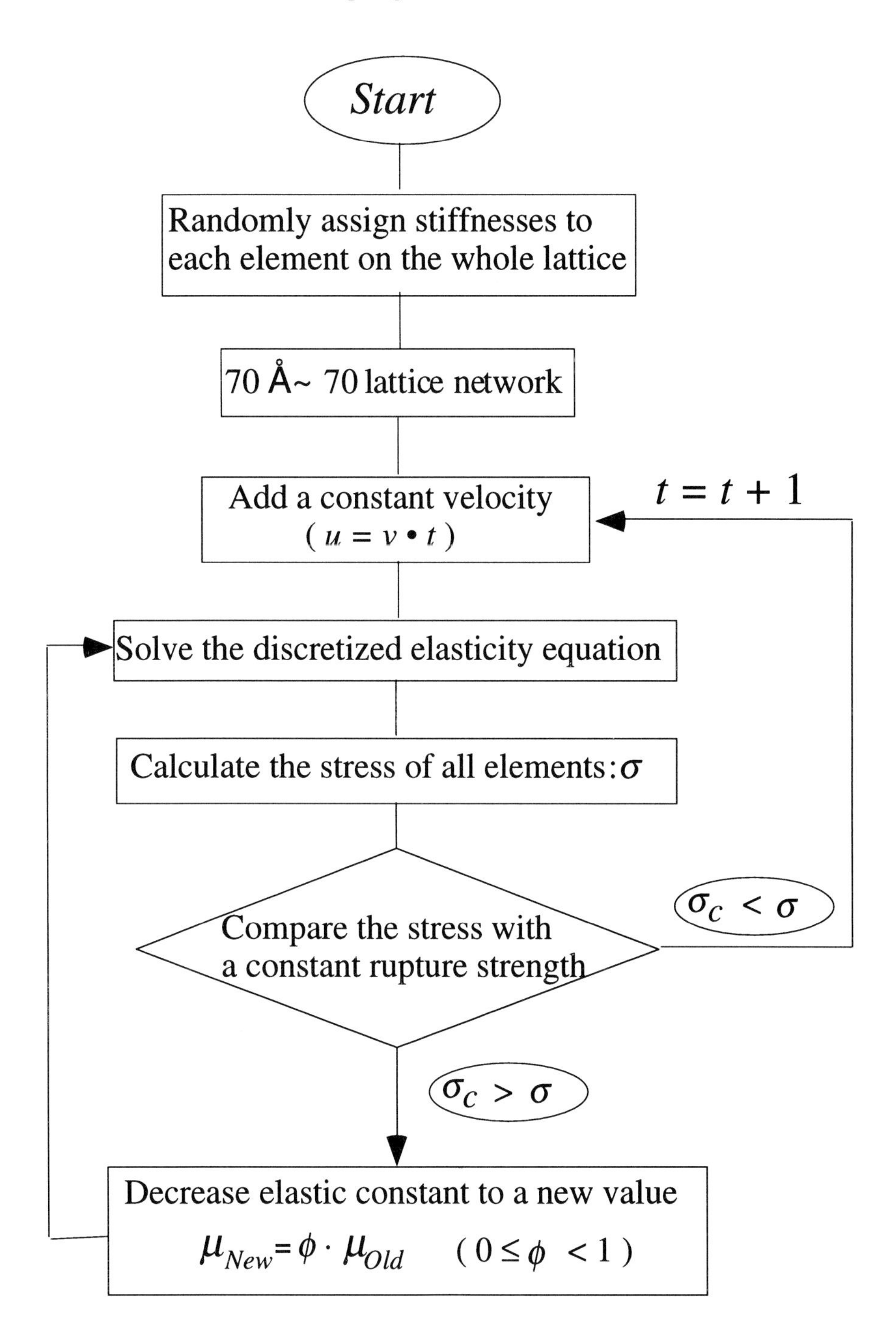

Fig. 4. The flow chart of the simulation algorithm.

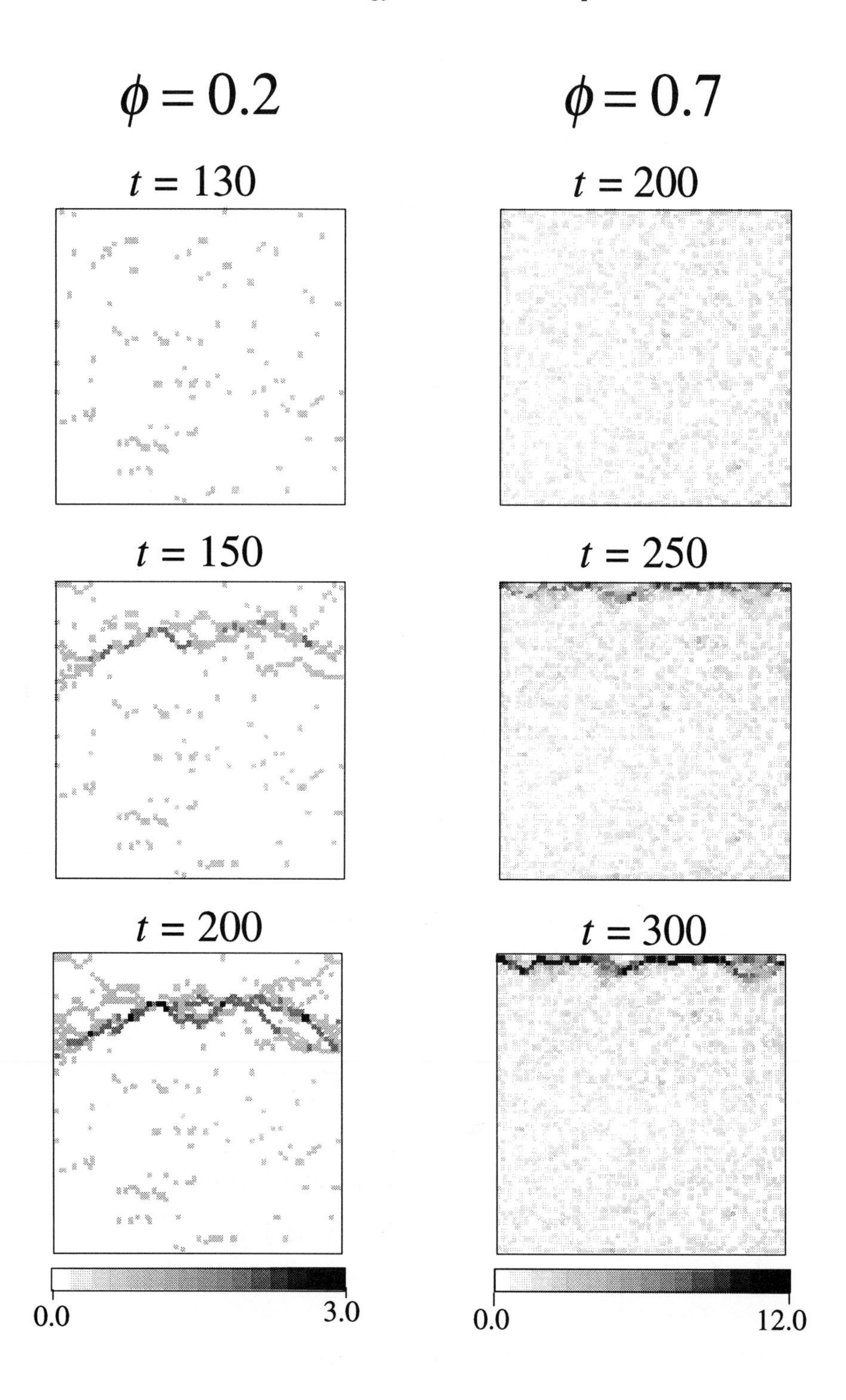
$\phi = 0.2$
$\phi = 0.7$
t = 130
t = 200
t = 150
t = 250
t = 200
t = 300
0.0
3.0
0.0
12.0

$$\phi \bullet \mu_{old} = \mu_{New} \tag{5}$$

where μ_{Old} and μ_{New} are the stiffnesses before and after rupture, respectively. The parameter ϕ takes various values in the range from 0 to 1. Then, return to procedure (C) until no element has newly ruptured during the preceding procedure.

(E) Go back to procedure (B). Then continue the routine.

The flow chart of this algorithm is shown in Fig. 4. Here, we present the original simulation program written in the *C*-language, ANSI standard (Appendix). We carried out the whole computation on a SPARC station 5, a desktop workstation manufactured by SUN Microsystems Inc.

Results

Fig. 5 shows the spatial and temporal distributions of ruptured areas in the fault system for $\phi = 0.2$ and 0.7, which are illustrated using the graphical software *Spyglass Transform*™ from Spyglass Inc. Open areas of each lattice network shown in Fig. 5 represent unruptured areas on the elements. The gray color (varying from white to black) indicates the accumulated number of ruptures on a spring element. The gray shaded scale is normalized by the color black, which is assigned to the element with the maximum number of ruptures.

Though the initial distribution of stiffnesses in each element remains the same for both $\phi = 0.2$ and 0.7, the fault pattern for $\phi = 0.7$ is more penetrative than that for $\phi = 0.2$ at early stage of fracturing. After the step time, the fault patterns for both $\phi = 0.2$ and 0.7 show the anisotropic distribution of ruptured elements across the lattice network in a direction approximately parallel to the moving boundary. Also the fault patterns for the different parameters ($\phi = 0.2$, $\phi = 0.7$) show different patterns in the spatial and temporal evolution of the fault system. Therefore, this result shows that the spatial and temporal evolution of the fault system depends on the ratio ϕ of the stiffness after rupture to the stiffness before rupture during cyclic rupturing of faults.

Discussion

We have proposed an evolutionary model for fault systems that considers various ratios ϕ in the range from 0 to 1. In the ratio $\phi = 0$, the strength after rupturing is always equal to 0. This case, such as the fracture of a glass, can be modeled by brittle fracture models (Takayasu 1985, 1986, Louis *et al.* 1986, Louis & Guina 1987, 1989, Li & Duxbury 1988, Meakin 1988, 1991, Ausloos & Kowalski 1992,

Fig. 5 (opposite). Spatial and temporal distribution of ruptured areas in fault system for $\phi = 0.2$ and 0.7. Gray color (varying from white to black) indicates accumulated number of ruptures on a spring element. Size of the lattice network is 70×70 spring elements.

Reuschlé 1992, Andersen *et al.* 1994). On the other hand, where $\phi = 1$, the stiffness after rupturing is always equal to the stiffness before rupturing. This case has been modeled by Cowie *et al.* (1993) and Sornette *et al.* (1994). However, in this paper, we show that the stiffness of rocks after rupturing weakens as rupturing progresses (weakening stiffness), using the experimental results of Bieniawski (1971). This indicates that ϕ is not always equal to one. There has been no computer simulation for faulting which has satisfied this mechanical constraints for fault systems. Therefore, we propose a spring network model for fault systems, which contains varying ϕ in the range from 0 to 1. Moreover, we discover that the fault patterns for different values of ϕ (ϕ = 0.2, 0.7) show different patterns of spatial and temporal evolution of the fault system.

Mohr Coulomb failure theory (Jaeger & Cook 1976) suggests that faulting occurs at lower stresses in pre-fractured rocks than in pristine material. On the other hand, rupturing phenomena for the fault system in our spring network model occur when the elastic stress of the spring element reaches a constant rupture strength. Understanding the reason for this striking contradiction is critical to understanding fault system processes. We must consider the relationship between our model and empirical data. Shimazaki & Nakata (1980) discovered that the time-interval between two successive earthquakes is approximately proportional to the amount of seismic displacement of the preceding earthquake, which is called "time-predictable recurrence" for large earthquakes, at three different sites (Nankaido, Boso Peninsula, and Kikai Island in Japan) in tectonically active regions. They presented the time-predictable model to explain this discovery and concluded that the rupture strength of fault systems is constant. Sykes & Quittmeyer (1981) also supported this conclusion. According to these discussions, the rupture strength of a fault is nearly constant and that is one of the constraints in real fault systems. Consequently, our spring network model for fault systems is supported by fracture mechanics and natural observations and is a valuable tool for researching the evolution of real fault systems in tectonically active regions of the upper lithosphere.

Conclusions

In this paper, from the experimental results of Bieniawski (1971), we first found that the stiffness after rupture is a linear function of the stiffness before rupture during cyclic rupturing. Next, three constraints of fracture mechanics on fault systems were given. Then, based on these three constraints, we presented a new spring network model of fault systems which contains the ratio ϕ of stiffness after rupture to stiffness before rupture during cyclic rupturing of faults. Furthermore, we described a computer simulation program written in ANSI *C*. We derived the displacement field of this model from the discretized elasticity equation by the relaxation method. The results of our simulation showed that the fault pattern for the larger ratio ϕ is more penetrative than that for the smaller ratio ϕ. Thus we conclude that the fault patterns over the spatial and temporal evolution of the fault system depend on ϕ. Finally, we discussed how our spring network model for the fault system is supported by fracture mechanics and natural observations and is a valuable tool for the research of the evolution of real fault systems in tectonically active regions of the upper lithosphere. We plan to report the detailed relationship between ϕ and seismological parameters (*e.g.*, fractal dimension and *b*-values) in a future paper.

Acknowledgements

We wish to thank Declan De Paor and two anonymous reviewers for valuable and critical comments to an earlier version of this manuscript. We would like to thank Robert M. Ross for improving the English style of this manuscript and Ken Kurokawa and Takuma Sasaki for valuable suggestions regarding the computer program. The authors would also like to thank K. Michibayashi for helpful reviews of this manuscript. This work was supported by a Grant-in-Aid from the Fukada Geological Institute.

References

Andersen, J. V., Brechet, Y. & Jensen, H. J. 1994. Fracturing described by a spring-block model. *Europhys. Lett.* **26**: 13-18.

Ausloos, M. & Kowalski, J. M. 1992. Stochastic models of two-dimensional fracture. *Phys. Rev.* **B45**: 12,830-12,833.

Bieniawski, Z. T. 1971. Deformational behavior of fractured rock under multi-axial compression. In: *Structure, Solid Mechanics and Engineering Design* (ed. Te'eni, M.). Proc. Southampton 1969 Civil Engineering Materials Conference. London: Wiley- Interscience, pp. 589-598.

Brady, B. T., Duvall, W. I., & Horino, F. G. 1973. An experimental determination of the true uniaxial stress-strain behavior of brittle rock. *Rock Mechanics* **5**: 107-120.

Chelidze, T. L. 1986. Percolation theory as a tool for imitation of fracture process in rocks. *Pure and Appl. Geophys.* **124:** 731-748.

Chelidze, T. 1993. Fractal damage mechanics of geomaterials. *Terra Nova* **5**: 421-437.

Cook, N. G. W. 1965. The failure of rock. *Int. J. Rock Mech. Min. Sci.* **2**: 389-403.

Cowie, P. A., Vanneste, C. & Sornette, D. 1993. Statistical physics model for the spatiotemporal evolution of faults. *Jour. Geophy. Res.* **98**: 21,809-21,821.

Cox, S. J. D. & Paterson, L. 1989. Tensile fracture of heterogeneous solids with distributed breaking strength. *Phys. Rev.* **B40**: 4,690-4,695.

Hardy, M. P., Hudson, J. A. & Fairhurst, C. 1973. The failure of rock beams. Part I: Theoretical studies. *Int. J. Rock Mech. Min. Sci.* **10**: 53-67.

Hudson, J. A., Hardy, M. P. & Fairhurst, C. 1973. The failure of rock beams. Part II: Experimental studies. *Int. J. Rock Mech. Min. Sci.* **10**: 69-82.

Jaeger, J. C. & Cook, N. G. W. 1976. *Fundamentals of Rock Mechanics*. Chapman and Hall, London.

Li, Y. S. & Duxbury, P. M. 1988. Crack arrest by residual bonding in resistor and spring networks. *Phys. Rev.* **B38**: 9,257-9,260.

Louis, E. & Guinea, F. 1987. The fractal nature of fracture. *Europhys. Lett.* **3**: 871-877.

Louis, E. & Guinea, F. 1989. Fracture as a growth process. *Physica* **D 38**: 235-241.

Louis, E., Guinea, F. & Flores, F. 1986. The fractal nature of fracture. In: *Fractals in Physics*. (eds. Pietronero, L. & Tosatti, E.). Elsevier Sci. Pub., Amsterdam, 177-180.

Meakin, P. 1988. Simple models for crack growth. *Crystal Properties & Preparation* **17&18**: 1-54.

Meakin, P. 1991. Models for material failure and deformation. *Science* **252**: 226-234.

Peng, S. & Podnieks, E. R. 1972. Relaxation and the behavior of failed rock. *Int. J. Rock Mech. Min. Sci.* **9**: 699-712.

Pollard, D. D. & Segall, P. 1987. Theoretical displacements and stresses near fractures in rock: with applications to faults, joints, veins, dikes, and solution surfaces. In: *Fracture Mechanics of Rocks*. Academic Press, London, 277-349.

Ranalli, G. 1987. *Rheology of the earth: Deformation and flow processes in geophysics and geodynamics*. Allen & Unwin, Boston.

Reushlé, T. 1992. Fractal in a heterogeneous medium: a network approach. *Terra Nova* **4**: 591-597.

Schmid, E. W., Spitz, G. & Lösch, W. 1988. *Theoretical Physics on the Personal Computer*. Springer-Verlag, Berlin.

Shimazaki, K. & Nakata, T. 1980. Time-predictable recurrence model for large earthquakes. *Geophys. Res. Lett.* **7**: 279-282.

Segall, P. & Pollard, D. D. 1980. Mechanics of discontinuous faults. *Jour. Geophys. Res.* **85**: 4,337-4,350.

Sornette, D., Vanneste, C. & Knopoff, L. 1992. Statistical model of earthquake foreshocks. *Phys. Rev.* **A45**: 8,351-8,357.

Sornette, D., Miltenberger, P. & Vanneste, C. 1994. Statistical physics of fault patterns self-organized by repeated earthquakes. *Pure & Appl. Geophys.* **142**: 491-527.

Sykes, L. R. & Quittmeyer, R. C. 1981. Repeat times of great earthquakes along simple plate boundaries. In: *Earthquake Prediction, An International Review*. (eds. Simpson, D. W. & Richards, P. G.) . AGU, Washington D.C., pp. 217-247.

Takayasu, H. 1985. A deterministic model of fracture. *Prog. Theor. Phys.* **74**: 1,343-1,345.

Takayasu, H. 1986. Pattern formation of dendritic fractals in fracture and electric breakdown. In: *Fractals in Physics* (eds. Pietronero, L. & Tosatti, E.), Elsevier Sci. Pub., Amsterdam, 181-184.

Wawersik, W. R. & Fairhurst, C. 1970. A study of brittle rock fracture in laboratory compression experiments. *Int. J. Rock Mech. Min. Sci.* **7**: 561-575.

Appendix : A Program for the Evolution of a Fault System Using a Spring Network Model

```
#include <stdio.h>
#include <math.h>
#include <stdlib.h>

float u[150][150], m[150][150];
int Nr[100][100];
int i, j, L, w, t, it, k;
float ratio, v;

main()
{
        L = 71; w = 5; it = 200; v = 0.2; ratio = 0.2;

        init();

        dynamics();

                for (j = 0; j < L - 1; j++)  {
                        for (i = 0; i < L - 1; i++)  {
                                printf (“%d/n”, Nr[ i ][ j ]);
                        }
                }


}
```

```
init()
{
        srandom(w);

                for (j = 0; j < L - 1; j++)  {
                        for (i = 0; i < L - 1; i++)  {
                                m[ i ][ j ] = 1.0 + 2.0 * random() / 2147483649.0;
                        }
                }

}

dynamics()
{
        t = 1;
                while (t <= it )  {
                        for (i = 0; i < L; i+= 2)  {
                                u[ i ][ 0 ] = v * t;
                                u[ i ][ L - 1 ] = 0.0;
                        }

                                do  {

                                        relaxation();

                                }  while ( compare() = = 1 );

                        t++;

                }

}

relaxation()
{
                float err[150][150];
                float cmax = 1;

    while ( cmax > 0.00001)  {
                cmax = 0;
        for (j = 1; j < L - 1; j++)  {   k = j % 2;
                if (k = = 0)  {
                        err[ 0 ][ j ] = (m[ L-2 ][ j ]*u[ L-2 ][ j+1 ] + m[ 0 ][ j ]*u[ 1 ][ j+1
]
                                + m[ L-2 ][ j-1 ]*u[ L-2 ][ j-1 ] + m[ 0 ][ j-1 ]*u[ 0 ][ j-1 ])
                              / (m[ L-2 ][ j ] + m[ 0 ][ j ] + m[ L-2 ][ j-1 ] + m[ 0 ][ j-1 ]);

                                        if ( fabs(err[ 0 ][ j ] - u[ 0 ][ j ]) >= cmax)
                                                cmax = fabs(err[ 0 ][ j ] - u[ 0 ][ j ]);
                                                err[ L-1 ][ j ] = err[ 0 ][ j ];
```

```
                for (i = 2; i < L - 2; i+=2)  {
                    err[ i ][ j ] = (m[ i-1 ][ j ]*u[ i-1 ][ j+1 ] + m[ i ][ j ]*u[ i+1 ][ j+1
]
                              + m[ i-1 ][ j-1 ]*u[ i-1 ][ j-1 ] + m[ i ][ j-1 ]*u[ i+1 ][ j-
1 ])
                              / (m[ i-1 ][ j ] + m[ i ][ j ] + m[ i-1 ][ j-1 ] + m[ i ][ j-1 ]);

                                   if ( fabs(err[ i ][ j ] - u[ i ][ j ]) >= cmax)
                                         cmax = fabs(err[ i ][ j ] - u[ i ][ j ]);

                }
            }
            else  {
                for (i = 1; i < L - 1; i+=2)  {
                    err[ i ][ j ] = (m[ i-1 ][ j ]*u[ i-1 ][ j+1 ] + m[ i ][ j ]*u[ i+1 ][ j+1
]
                              + m[ i-1 ][ j-1 ]*u[ i-1 ][ j-1 ] + m[ i ][ j-1 ]*u[ i+1 ][ j-
1 ])
                              / (m[ i-1 ][ j ] + m[ i ][ j ] + m[ i-1 ][ j-1 ] + m[ i ][ j-1 ]);

                                   if ( fabs(err[ i ][ j ] - u[ i ][ j ]) >= cmax)
                                         cmax = fabs(err[ i ][ j ] - u[ i ][ j ]);
                }
            }
        }

            for (j = 1; j < L - 1; j++)  {
                    for (i = 0; i < L; i++)  {
                            u[ i ][ j ] = err[ i ][ j ];
                    }
            }
    }
}

int compare()
{
        int b = 0, s = 1;

        for (j = 1; j < L - 1; j+=2)  {
                for (i = 1; i < L - 1; i+=2)  {
                        if (fabs(m[ i-1 ][ j-1 ]*(u[ i-1 ][ j-1 ] - u[ i ][ j ])) > s)  {

                                Nr[ i-1 ][ j-1 ]++;
                                m[ i-1 ][ j-1 ] = ratio*m[ i-1 ][ j-1 ];
                                b = 1;

                        }

                        if (fabs(m[ i ][ j-1 ]*(u[ i+1 ][ j-1 ] - u[ i ][ j ])) > s)  {

                                Nr[ i ][ j-1 ]++;
```

```
                m[ i ][ j-1 ] = ratio*m[ i ][ j-1 ];
                b = 1;

            }

            if (fabs(m[ i ][ j ]*(u[ i+1 ][ j+1 ] - u[ i ][ j ])) > s)  {

                Nr[ i ][ j ]++;
                m[ i ][ j ] = ratio*m[ i ][ j ];
                b = 1;

            }

            if (fabs(m[ i-1 ][ j ]*(u[ i-1 ][ j+1 ] - u[ i ][ j ])) > s)  {

                Nr[ i-1 ][ j ]++;
                m[ i-1 ][ j ] = ratio*m[ i-1 ][ j ];
                b = 1;

            }
        }
    }

    return (b);

}
```

Linear-Elastic Crack Models of Jointing and Faulting

Juliet G. Crider, Michele L. Cooke, Emanuel J. M. Willemse, and J. Ramón Arrowsmith
Department of Geological and Environmental Sciences, Stanford University, Stanford CA 94305-2115, U.S.A.
crider@pangea.stanford.edu

Abstract– To gain insight into the formation and distribution of joints and faults and to illuminate the origin of associated secondary structures, we have developed methods for the interactive calculation and visualization of the stresses and displacements around these features. We assume that joints and faults can be idealized as cracks in homogeneous elastic bodies. Linear elastic fracture mechanics (LEFM) provides analytical expressions for the stress and displacement fields around cracks of various geometries for various loading conditions. We use a widely available symbolic math program to solve the general equations for a crack under uniform remote load and to display the results in two and three dimensions.

Fracture mechanics has wide application in structural geology. The linked graphics and mathematics capabilities of symbolic math packages make LEFM accessible to a wide audience for both analytical and heuristic applications. We offer several geologic examples to demonstrate the power of visualizing the analytical solutions to linear elastic fracture problems. By examining the stress pattern around a modeled strike-slip fault we can predict the expression of secondary structures around the fault, including basins and ridges, or splay cracks and pressure solution surfaces. An analysis of variation in tensile stress away from an existing joint can be used to predict where the next parallel joint might initiate, defining the relationship between bed thickness and joint spacing. The crack model of a simple strike-slip fault also yields displacement vectors comparable in direction and relative magnitude to measured displacements produced by the 1989 Loma Prieta (California) earthquake.

Introduction

Joints, faults, and fractures are present throughout the world at all scales and in every tectonic province. At the outcrop scale, we observe joints and veins associated with folds (*e.g.,* Price 1966, Stearns & Friedman 1972, Hancock 1985, Ramsay & Huber 1987) and faults (*e.g.,* Rispoli 1981, Martel *et al.* 1988, Steward & Hancock 1990). On a broader field scale, local and regional faults and joint sets have been documented by many geologists (*e.g.,* Kelley & Clinton 1960, Engelder & Geiser 1980, Sani 1990, Lorenz & Finley 1991). Earthquakes within the upper crust and volcanic fissure eruptions are examples of processes that produce brittle deformation (*e.g.,* Sammis & Julian 1987, Scholz 1990, Gudmundsson 1987). Equipped with a mechanical understanding of fracture growth, we can use outcrop observations to infer the mechanical development of geologic structures. The importance of understanding such features is not limited to earthquakes and volcanic hazards. For example, joints and faults can control the flow of subsurface water (*e.g.,* Hsieh 1991), hydrocarbons (*e.g.,* Nelson 1985) and geothermal fluids (*e.g.,* Takahshi & Abé 1989). The presence of these structures also impacts landscape evolution (*e.g.* Selby 1982), and they hold important tectonic information (*e.g.,* Engelder 1984, Stauffer *et al.* 1987, Evans 1994, Tobisch & Cruden 1995).

Linear-elastic fracture mechanics (LEFM) has emerged in the last three decades as a framework for understanding the development of fractures and faults in rock (*e.g.*, Chinnery 1961, Lachenbruch 1961, Jaeger & Cook 1969, Pollard & Segall 1980, Atkinson 1987, Pollard & Aydin 1988). LEFM permits a mathematical description of the stress around an elongate flaw in an elastic medium. Rock within the upper part of the earth's crust is idealized as a linear elastic medium and planar brittle structures are idealized as cracks. Various graphical methods are used to visualize the mathematical fields in such a way that we can compare them to geologic structures. The goal of such comparison is to predict and explain the geometry of observed structures by understanding the conditions under which they are produced.

Neotectonic structures such as active faults and propagating dikes have been understood in the context of fracture mechanics (*e.g.*, Rudnicki 1980, Scholz 1990). Other studies concern ancient structures such as veins (*e.g.,* Olson & Pollard 1991), joints (*e.g.*, Lachenbruch 1961, Engelder & Geiser 1980, Pollard & Aydin 1988), dikes (*e.g.,* Delaney *et al.* 1986, Rubin 1993), and faults (*e.g.,* Pollard & Segall 1980, Rudnicki 1980, Lin & Parmentier 1988, Cowie & Scholz 1992, Bürgmann *et al.* 1994a, Willemse *et al.* 1996). The work of these researchers and others attests to the utility and instructiveness of this approach, but the sometimes complicated mathematics of fracture mechanics models may have inhibited wider use of this approach by geologists in research and teaching. We have developed methods for the interactive visualization of the stress and displacement fields around such structures using symbolic math software available for microcomputers. In this paper we explain our approach and introduce the general features of symbolic programs. We begin by outlining elementary fracture mechanics and then present several geologic examples of the application of these principles and the utility of the microcomputer in this approach. The Appendices detail the mathematics and illustrate examples of computer input and output.

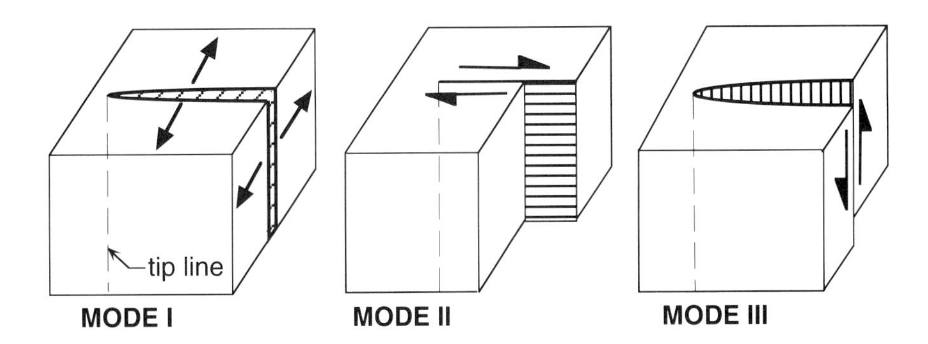

Fig. 1. The model crack may be a combination of any of three modes of displacement discontinuity. Modes of crack deformation are defined relative to the *tip line*, where crack surfaces meet. Mode I displaces the walls of the crack away from each other (*e.g.,* a vein or dike). The crack surfaces slide past each other perpendicular to the tip line in mode II displacement discontinuity (*e.g.,* map view of strike-slip fault). Mode III displacement slides the crack surfaces past each other in a direction parallel to the tip line (*e.g.,* map view of dip-slip fault).

Theoretical Background

In order to apply LEFM to fractures in the crust, we make the following assumption: planar faults, dikes, joints, veins, and pressure solution surfaces are idealized as two planar surfaces. These surfaces are contacts of the dike, margins of the fault zone, or opposing walls of the joint. The surfaces are *bounded in extent*; that is, there is a line (called the tip line) at which the fault or fracture ends, and at this line the two surfaces come together (Fig. 1). This is an important distinction from the block-faulting conceptualization of faults and joints from which the ends of the features are missing. We assume that the structures we model have small offsets relative to their lengths. Dikes and joints, for example, are generally between 100 and 1000 times longer than they are wide (Delaney *et al.* 1986). The rock is idealized as a homogeneous, isotropic, linear-elastic medium. We consider only two-dimensional models in plane strain; therefore, the geometry of the structure and the boundary conditions must not change significantly along the third dimension.

For the purposes of LEFM analysis, a fault or joint is a discontinuity in the displacement field where adjacent rocks move in opposite directions. The two surfaces of the model crack may displace in three different modes (Fig. 1; see Broek 1991). Mode I occurs when the crack opens perpendicular to its walls, like a joint or a dike. Mode II occurs when the crack surfaces slide past one another, in a direction perpendicular to the crack tip-line. Examples of mode II sliding are a map view of a strike slip fault and a normal fault viewed in cross section. Finally, mode III occurs when the crack surfaces slide past one another in a direction

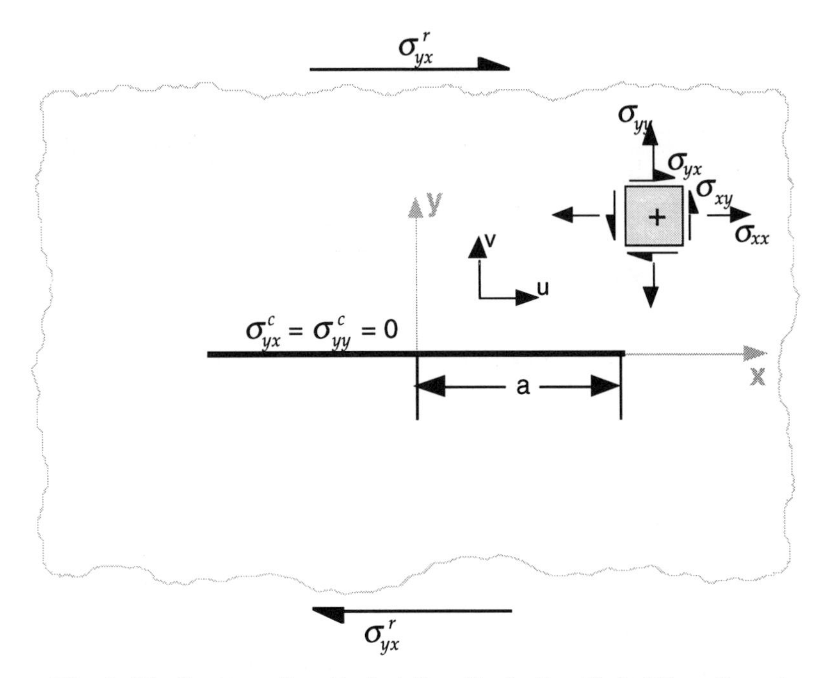

Fig. 2. Idealization of vertical strike-slip fault with half-length, a, in map view (a mode II crack). x-axis lies along fault, y-axis is perpendicular to fault, and origin of coordinate system is at center of fault trace. Remote shear stress, σ^r_{yx}, drives fault motion. Displacement vectors are expressed in terms of displacement in x-direction (u-component) and y-direction (v-component). Representative element (upper right) illustrates stress-subscript and sign convention used in this paper (see also Appendix 1).

parallel to the crack tip-line. Examples of mode III tearing are a map view of a normal fault and a strike-slip fault viewed in cross section. Relative motion near the tip-line of a crack may be any combination of modes I, II, and III, and this varies with position along the tip line.

The following sections describe the general background for the analytical solutions of this plane-elastostatic problem. The framework of such analyses is discussed in general terms only, to emphasize the basic concepts and the flow of logic. The corresponding equations, which are not essential to follow the overall approach, are given in Appendix 1.

The Concept of Stress Function

Elasticity theory in two dimensions is governed by three equations: two stress equilibrium equations that relate forces and body accelerations (eqns. A1 & A2,

Appendix 1), and a strain compatibility equation to prevent holes or overlaps in the body (eqn. A4). The strain compatibility equation can be rewritten in terms of the stress components using the constitutive properties of rock. We thus have three equations in three unknowns: two normal stresses, σ_{xx} and σ_{yy}, and a shear stress, $\sigma_{xy} = \sigma_{yx}$ (see Fig. 2 and Appendix 1 for notation). Given information about the geometry of the body and the tractions on its boundaries, we can solve this system of equations to determine the components of stress.

A glance at the equations in Appendix 1 suggests that the solution for the three unknown stress components can be arduous; however, the solution is made considerably easier by using a stress function which *automatically* satisfies the compatibility and equilibrium conditions (eqns. A6 - A9). The Airy stress functions do this by exploiting the mathematical interrelationships among the three governing equations. For a two-dimensional, continuous elastic medium, the stress function is bi-harmonic. The three unknown stress components are found by differentiating the stress function with respect to the spatial coordinates. There exist many different Airy stress functions to solve a whole range of problems with different body geometries and loading conditions (*e.g.,* Chou & Pagano 1967, Malvern 1969, Timoshenko & Goodier 1970).

The Stress Function for a Body with Cracks Such As Joints, Dikes, and Faults

Solutions to the equations describing the stress and strain around cracks such as joints and faults are commonly found using a sub-class of Airy stress functions that reflect the symmetry of the crack problem and the displacement discontinuity at the crack. The simplest functions involve complex variables and are known as Westergaard stress functions (Anderson 1991, Broek 1991). Stress functions for a variety of crack problems can be found in fracture mechanics handbooks (*e.g.*, Sih 1973, Tada *et al.* 1985, Anderson 1991).

The overall approach follows a fixed recipe. The desired solutions, such as the three stress components, displacement components, and slip or opening distribution, are obtained by differentiating and integrating the stress function, and taking the appropriate real and imaginary parts (eqns. A11-A13) This often requires cumbersome mathematics and it is very easy to make mistakes in the multiple pages of equations required to obtain the desired stress or displacement fields. Indeed, it is not uncommon to find mistakes in the standard textbooks. Symbolic math software for the computer greatly facilitates the mathematics, and reduces the risk of errors.

Symbolic Math Packages

Symbolic math packages such as *Mathematica*™, *MATLAB*™, *MAPLE V*™, and *Theorist*™ (see Appendix 3) are high-level programming languages designed to solve intricate mathematical expressions analytically or approximately numerically. High-level languages allow the user to communicate with the program using a syntax which resembles mathematical expressions rather than a syntax

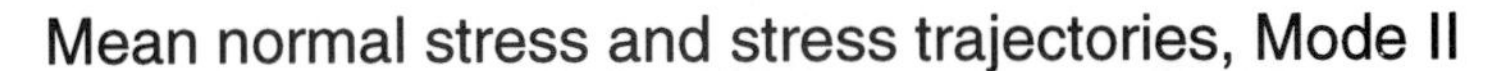

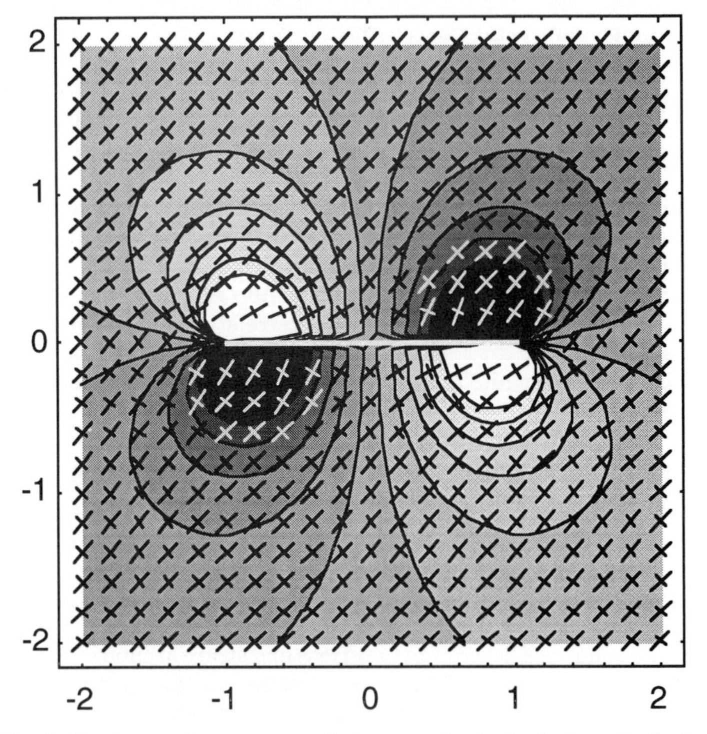

Fig. 3. Contours of mean normal stress and principal stress trajectories around an idealized vertical strike-slip fault in map view (mode II crack). Brightness indicates increasing tensile mean stress. Tick marks show maximum (long ticks) and minimum (short ticks) principal stress directions. Basins are predicted to form in regions of overall tension (upper left and lower right quadrants) while ridges may form in compressive regions (upper right and lower left quadrants) around the idealized fault.

required by lower-level languages (such as *C* or *FORTRAN*). They are also typified by more advanced user interfaces and large libraries of analytic, numerical and graphics functions. Equations are typed almost exactly as they would appear on the printed page, following the few, relatively simple rules required by the software. Example input is given in Appendix 2. Output may be in the form of another mathematical expression (*e.g.,* the derivative of $\sin(x)$ is $\cos(x)$), a numerical value (*e.g.,* $\sin(\pi/2) = 1$), or in graphical form (*e.g.,* a sinusoidal curve, see also Figs. 3 & 4). Because symbolic math packages are programming languages, it is possible to write executable routines which perform a series of mathematical tasks or plotting procedures repeatedly. This feature allows the user to organize text, computations, and graphics in such a manner that the sequence of research processes is preserved. Furthermore, preservation of the investigation sequence facilitates the development, exchange, and use of pedagogically innovative courseware.

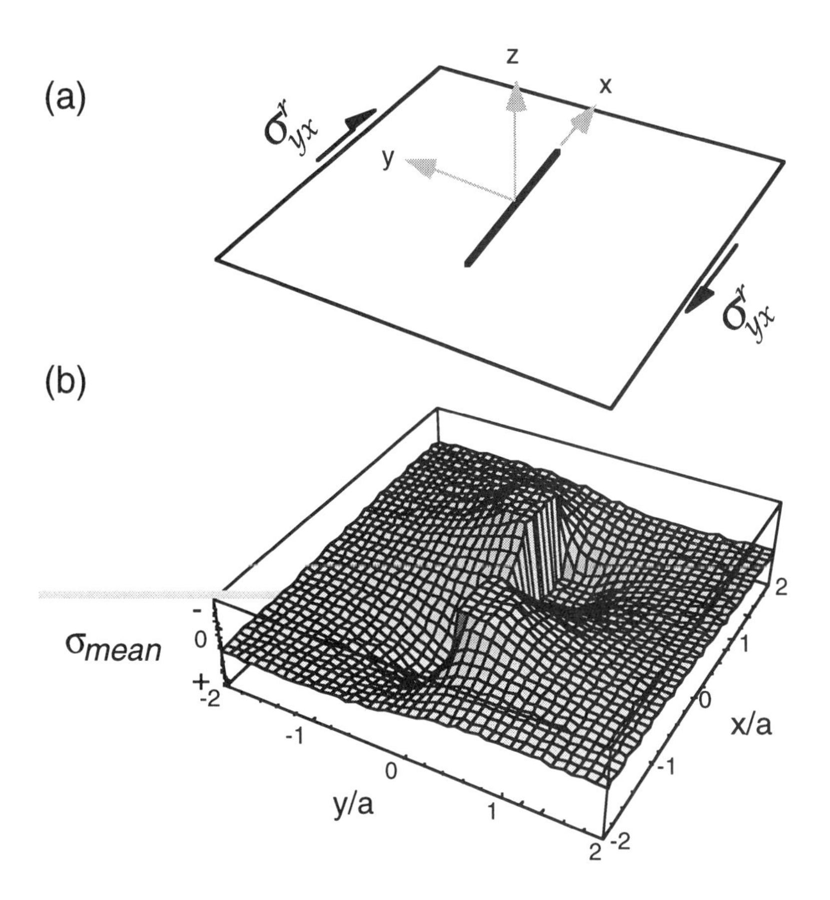

Fig. 4. a) Map view of strike-slip fault showing remote loading and position of fault in (b). b) Mean normal stress (σ_{mean}) around strike slip fault in map view (mode II crack). Height of three-dimensional surface shows magnitude of σ_{mean} to represent the relative warping of the earth's surface around a strike-slip fault. Subsidence is expected in regions of tensile mean stress (positive σ_{mean}) while uplift is expected in areas of compressional mean stress (negative σ_{mean}). Predicted uplift and subsidence are greatest at the fault tips and diminish with distance from the fault.

These programs provide two important innovations in desktop computer graphics. The first is the close integration of high-quality graphics with computation. Instead of calculating in one program and plotting in another, it is possible to directly visualize calculations within these programs. Secondly, the software treats the resultant graphics in a symbolic manner that allows the graphics routines to be customized for a given problem. The strengths of graphics in symbolic math packages are threefold; one can improve upon a program by testing the output graphically, interactively visualize relationships to develop intuition, and easily develop high quality presentations of the results.

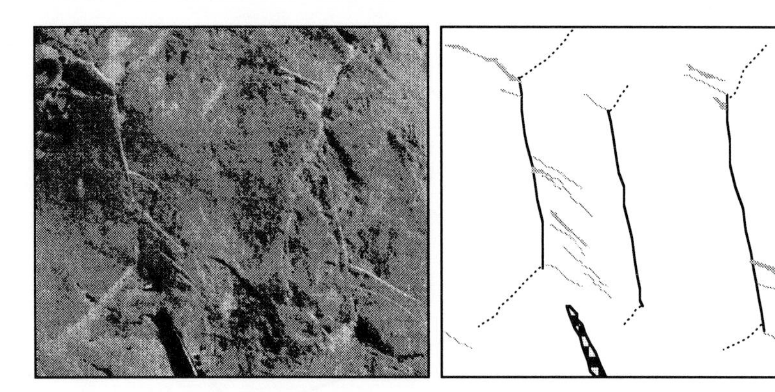

Fig. 5. Example of secondary features associated with left-lateral strike-slip faults within limestone of the Matelles exposure, Languedoc, France. Fault length is approximately 15 cm. Veins (shown grey in sketch) have developed in the upper left and lower right quadrants of the faults. Solution surfaces (shown dotted in sketch) are present in opposing quadrants. Checkered region in sketch corresponds to pen in the photograph.

Computer Implementation: How Intimidating Mathematics Are Made Simple

Here we discuss how complicated equations such as the Westergaard functions have been solved with symbolic math software and summarize some other important benefits of this approach. The symbolic nature of these computer packages is ideal for our problem because the programs can perform laborious mathematical and graphical tasks. Although many software packages produce similar results, we use *Mathematica* in the following examples.

As an example, consider a map view of a vertical strike-slip fault. We idealize this fault in two dimensions as a mode II crack embedded in an infinite linear-elastic medium (Fig. 2). In addition, we specify that there is a remotely-applied uniform stress that drives the crack and we require that there are no tractions along the crack surfaces. Strike-slip motion along the fault perturbs the stress field and displaces the material surrounding it.

Although the boundary conditions are very simple and do not reproduce the complicated loading and resistance to slip expected along a natural fault, this basic model illustrates the major changes in stresses and displacements in vicinity of a fault (Figs. 3, 4, 6, and 8). The stress function appropriate for this particular geometry and boundary conditions is:

$$Z(z) = \frac{\sigma_{yx}^{r}}{\sqrt{\left(1 - \left(a/z\right)^{2}\right.}}, \tag{1}$$

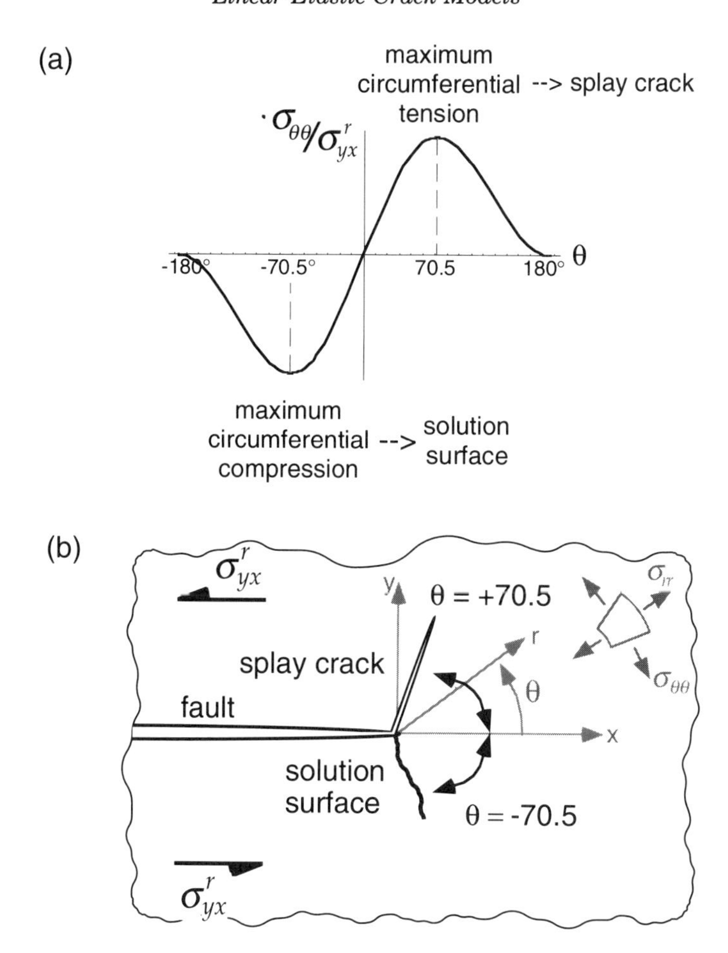

Fig. 6. a) Circumferential stress around tip of mode II crack. Maximum tension and maximum compression are located at $\theta = 70.5$ and $\theta = -70.5$ respectively. Polar coordinate system is illustrated in (b). b) Based on laboratory and numerical evidence, we expect tensile features (veins or splay cracks) to form at the location of maximum circumferential stress, $\theta = 70.5$, and compressional features (solution surfaces) to form at the location of minimum circumferential stress, $\theta = -70.5$. The computed orientations agree with field observations in Fig. 5.

Z(z) is the Westergaard stress function, σ^r_{yx} is the uniform remote stress that drives the fault motion, and a is the half-length of the fault (Tada *et al.* 1985). *Z(z)* is a function of the complex variable $z = x + iy$, where $i = \sqrt{-1}$. Differentiating and integrating this stress function and separating the resulting complex numbers into real and imaginary parts may not be not straightforward tasks and would require a few pages of analysis by hand; however, the computer implementation is simple and short, as shown in Appendix 2. Once the stress function and constants (such as fault length and material properties) are fully defined, the variables of interest can be printed or graphed immediately. Analyzing functions for stress or displacement requires fewer than 15 lines of code. This is one example of the tremendous power of symbolic math packages that makes a whole range of complicated problems accessible to a large user group.

Application to Geologic Problems

Many researchers document geologic fractures (small faults and joints) that can appropriately be described or investigated in the context of fracture mechanics (*e.g.,* Engelder & Geiser 1980, Pollard & Segall 1980, Pollard & Aydin 1988, Lin & Parmentier 1988, Cruikshank 1991, Petit & Mattauer 1995). Delaney and others (1986) document the growth of dikes. Fletcher and Pollard (1981) model small faults in limestone of the Matelles exposure at Languedoc, France, and used secondary structures to interpret the stress field around the faults as described by Rispoli (1981). Cruikshank and Aydin (1995) unravel the history of jointing in sandstone in Arches National Park, Utah. In each of these cases, the model of a crack in an elastic solid was successfully applied to interpretation of the structure. To illustrate the broad range of geologic problems to which these concepts may be applied and the utility of a symbolic math package in the analysis, we offer the following additional examples inspired by Pollard & Segall (1987):

Example A: Secondary Features Associated with Strike-Slip Faults

Strike slip fault deformation is observed on scales varying from 100-kilometer-long tectonic faults to centimeter-long shear fractures. Secondary fault deformation on any scale can be investigated by approximating faults as simple fractures loaded in any combination of modes I, II, and III. As discussed above, we use a mode II fracture to approximate the plan view of a vertical strike-slip fault.

Map-Scale Structures

Kilometer-scale secondary fault features include regions of uplift (ridges) and depressions (basins) that develop around strike-slip faults (Chinnery 1961, Pollard & Segall 1980, Rodgers 1980). We can predict the location and relative magnitudes of these features by investigating the mean normal stress field: *compressive* mean normal stress can promote uplift whereas *tensile* mean normal stresses can promote subsidence. In two dimensional analyses, the mean normal stress, σ_{mean}, is given by:

$$\sigma_{mean} = (\sigma_{xx} + \sigma_{yy})/2. \quad (2)$$

For a mode II fracture, five simple lines of *Mathematica* code calculate σ_{mean} (eqns. B1, B2, B4, B6, & B7, Appendix 2) and the results can be displayed in a variety of ways. Figure 3 combines information about σ_{mean} and the maximum and minimum principal stresses around the idealized strike-slip fault. The contours illustrate the relative magnitude of σ_{mean}. The tic marks are trajectories which indicate the direction in which the principal stresses act. Figure 3 illustrates an increase in mean stress within the upper left and lower right quadrants around the fault. Because tensile stresses are taken positive in our analysis, this corresponds to an overall increase in the tensile stress in these regions which promotes basin development. An increase in compressional stresses within the upper right and lower right quadrants may result in uplift within those regions. The effect is magnified when two faults overlap; for example, a right-stepping, right-lateral fault pair will show a large tensile stress in the overlap region. This may be manifest as a pull-apart basin (*e.g.*, Pollard & Segall 1980, Rodgers 1980).

An alternative to contour plots is a three-dimensional surface plot where the height of the surface represents the parameter of interest. Figure 4 shows the relative magnitude of the mean stress around our idealized strike slip fault. The height of the surface is the negative mean stress corresponding to relative amounts of surface subsidence (positive σ_{mean}) and uplift (negative σ_{mean}). This graphical representation vividly displays regions of relative uplift and subsidence.

Near-tip Structures

Determination of the stress field near the tip of a mode II fracture can be use to investigate the development of secondary features around fault tips. The presence of joints at the tips of faults has motivated investigation into the fault-tip stress field as a causative agent for fracturing (*e.g.,* Segall & Pollard 1983, Cruikshank 1991, Cruikshank & Aydin 1994). In limestone, opening fractures (splay cracks and tension gashes) and pressure solution features (stylolites) form in opposing quadrants around small strike slip faults (Fig. 5; see Rispoli 1981). As secondary fault features, splay cracks are useful indicators of the sense of relative motion on the fault in cases where offset markers are absent, such as slip along bedding planes (Cruikshank 1991). While the relative location of these features can be explained by the mean normal stress distribution described above, we investigate the orientation of splay cracks by investigating the circumferential stresses around the tip of an idealized fault.

Using LEFM, researchers have postulated that, within brittle materials such as rock, the concentration of stresses ahead of faults promotes the initiation of obliquely oriented opening-mode fractures (Lawn 1993). Laboratory and numerical experiments have been used to investigate the conditions that control the propagation path of single fractures initiating at mode II crack tips (Brace & Bombolakis 1963, Erdogan & Sih 1963, Nemat-Nasser & Horii 1982, Thomas & Pollard 1993). They indicate that opening mode cracks commonly grow from a fracture tip in the direction of maximum circumferential stress. For mode II loading, the circumferential stress is approximated by:

$$\sigma_{\theta\theta} = \sigma_{xy}^{r}\sqrt{\frac{a}{2r}}\left[-\frac{3}{2}\cos\left(\frac{\theta}{2}\right)\sin\theta\right]. \tag{3}$$

The distribution is shown in Figure 6. Although eqn. (3) can be found in textbooks (*e.g.*, Lawn 1993), the circumferential stress can also be obtained with the computer by transforming the solution from crack-centered Cartesian coordinates, σ_{xy}, used in the previous equations, to crack-*tip*-centered polar coordinates, $\sigma_{\theta\theta}$, used here (Fig. 6). In polar coordinates, r indicates the radius measured outward from the crack tip and q marks an angle, measured anticlockwise from directly in front of the crack. The notation $\sigma_{\theta\theta}$ designates a stress acting *on* a plane normal to a line at position θ and in the direction of that normal. For a uniformly loaded crack, the maximum tensile circumferential stress occurs at $\theta = 70.5°$ from the crack plane and the maximum compressive circumferential stress at $\theta = -70.5°$ (Fig. 6a). Equation (3) is easily modified to find the maximum circumferential stress for any combination of mode II and mode I loading. The ±70.5° angles match tail crack and stylolite angles commonly observed in limestone (Fig. 5; see Rispoli 1981, Petit & Mattauer 1995). This hypothesis for joint propagation from fault tips does not explain the presence of multiple fractures (sometimes called horsetail cracks). Recent research has begun to investigate the conditions under which multiple splay cracks may form (Martel *et al.* 1988, Cruikshank and Aydin 1994).

Example B: Joint Spacing Versus Bed Thickness

Joints are ubiquitous in rock. In layered sediments, the joints in thick beds are commonly more widely spaced than those in thin beds (Price 1966, Gross 1993, Wu & Pollard 1995a). Figure 7(a), for example, shows the relationship between bed thickness and joint spacing measured in the Monterey Formation,

Fig. 7 (opposite). a) Relationship between sedimentary bed thickness and bed-perpendicular joint spacing in Monterey Formation, California (from Wu & Pollard 1995). b) Sketch of idealized joint within sedimentary layer of thickness T. Bed thickness is equivalent to height, H, of crack and twice crack half-length, a. Crack is loaded in uniform remote tension. c) Variation of layer-parallel tension with distance from crack on line through crack center (x=0). Values normalized to applied remote stress. Shorter cracks (upper curve, H=0.5) have smaller stress shadows: stresses less than 0.8 of remote value confined to distance y<0.625. Longer cracks (dashed lower curve, H=2) have more extensive stress shadows: stresses less than 0.8 of remote value occur as far away as y=2.5. Since new joints are most likely to develop outside stress shadow, short joints in thinner beds may have closer neighbors than tall joints in thicker beds. Thus thinner sedimentary layers have closer bed-perpendicular joint spacing as in (a).

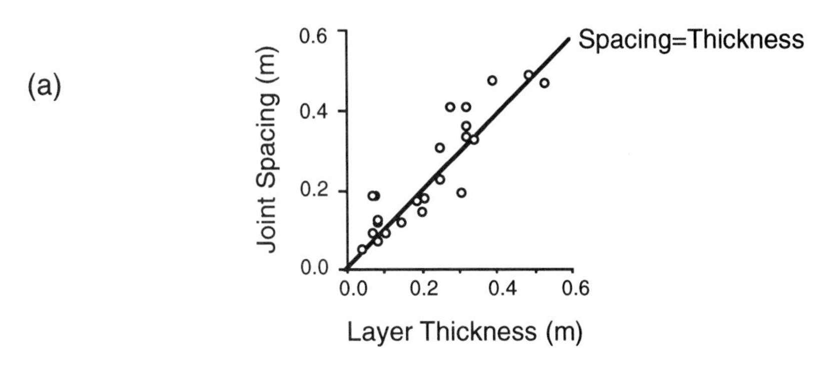
(a)
Spacing=Thickness
Joint Spacing (m)
Layer Thickness (m)
0.0
0.2
0.4
0.6

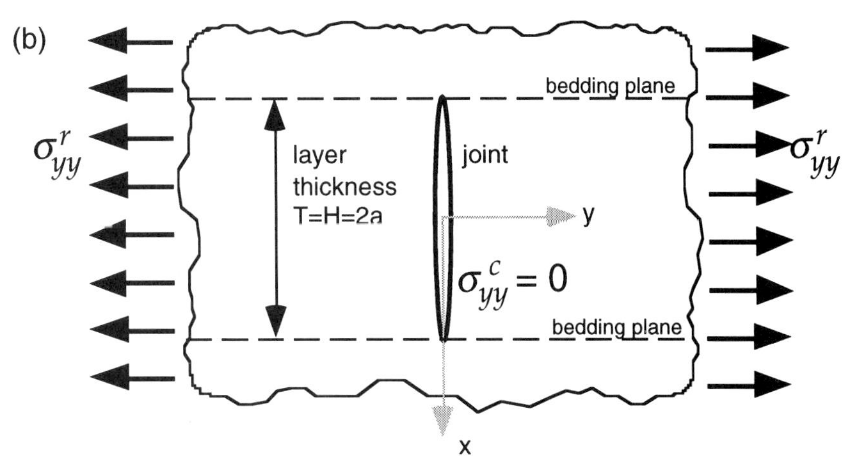
(b)
bedding plane
σ_{yy}^{r}
layer thickness T=H=2a
joint
y
$\sigma_{yy}^{c} = 0$
bedding plane
x

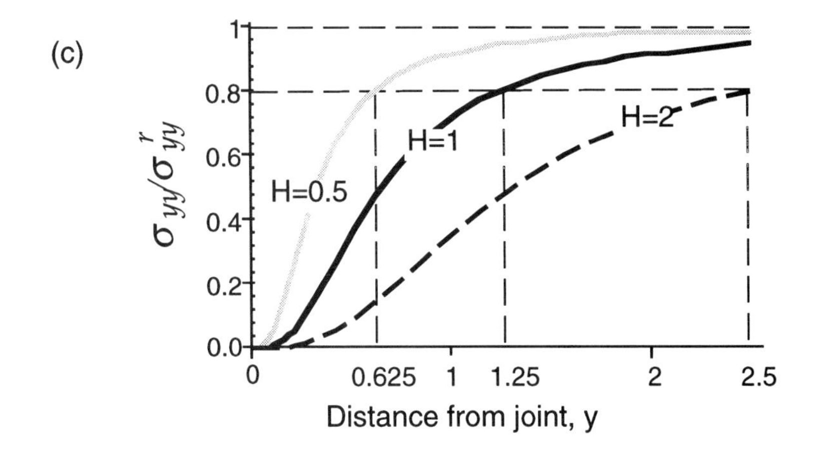
(c)
$\sigma_{yy}/\sigma_{yy}^{r}$
H=0.5
H=1
H=2
0
0.625
1
1.25
2
2.5
Distance from joint, y

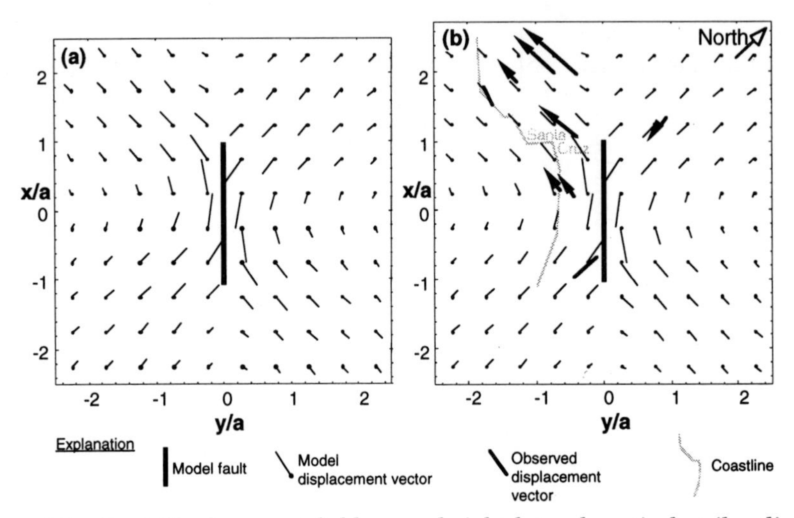

Fig. 8: a) Displacement field around right-lateral vertical strike-slip fault in map view (mode II crack) loaded in uniform remote shear. Location of crack is represented by thick vertical line in center of plot. Model displacement vector at each point is illustrated by tadpoles, and relative magnitude by length of tail. Tail of each vector is on its grid-point (dot) and vector direction is away from grid-point. Displacement is greatest at crack walls and becomes small within twice crack half-length. Material is displaced away from crack in upper-left and lower-right quadrants and toward crack in opposing quadrants. b) Model results from (a) compared with displacements observed during 1989 Loma Prieta earthquake, shown by bold arrows. Maximum observed displacement was approx. 45 cm (Williams *et al.* 1993). Horizontal dimensions (x and y) scaled by half-length of fault (approx. 10 km).

California. Although this observation is typically interpreted in terms of joint spacing versus layer thickness, it can equally well be interpreted and quantified as joint spacing versus joint height if the joints are perpendicular to bedding. Because most bed-perpendicular joints terminate at bedding plane boundaries or lithologic contacts, the joint height generally equals the layer thickness.

One possible explanation for the observed relationship between bed thickness and joint spacing was first discussed by Lachenbruch (1961). The formation of a joint perturbs the ambient stress, causing a reduction in the driving stress to either side of the joint. This stress shadow around one joint may deter or even prevent the growth of other parallel joints nearby. The hypothesis that stress reduction around a joint explains some commonly observed spacings can be investigated using the stress function approach (Pollard & Segall 1987).

We presume that a joint is a mode I fracture; in other words, it forms by opening. If there is no opening or slip along the upper and lower boundaries of the bed containing the joint, we can apply analytical solutions for a crack in a homogenous elastic medium to calculate the stress field around the idealized joint. The simplest model is one with a uniform remote driving stress, σ^r_{yy}, and zero shear and normal tractions on the crack walls (Fig. 7b). This is the mode I equivalent of the strike-slip model discussed above, with eqn. (1) as the appropriate stress function.

To test the above hypothesis, it is appropriate to quantify how the normal stress that drives the joint opening, σ_{yy}, is perturbed near the joint. Three lines of computer code calculate the values of σ_{yy} (eqns. B1, B2, and B6, Appendix 2). Figure 7(c) shows how the driving stress, σ_{yy}, changes along a line that is perpendicular to the joint and goes through the joint center. Three different joint heights are modeled. For all cases, the driving stress decays from the unit remote value far away from the joint to zero at the joint wall. The decay towards the joint illustrates the stress shadow where the stress magnitude is decreased significantly.

For short joints (upper, gray curve, Fig. 7c), significant reduction in driving stress occurs only in the immediate joint vicinity. However, for tall joints (lower, dashed curve, Fig. 7c) the stress shadow extends across a much larger distance. If new parallel joints cannot form in the stress shadow because the driving stress is too low, then the spacing of joints should correlate loosely with joint height. We suggest that new neighboring parallel joints cannot form at distances less than about 1.0 to 1.5 times the joint height from the original joint, because in this interval the driving stress is less than between 70 and 85 percent of the remote value (Pollard & Segall 1987). In cases where bedding planes control the joint height, joint spacing is expected to correlate with the bedding thickness.

This analysis provides some understanding of the commonly observed relationship between bed thickness and bed-perpendicular joint spacing (*e.g.*, Price 1966, Gross 1993). In other instances, however, the joints may be clustered, or be spaced at distances that are considerably less than the layer thickness (Olson & Pollard 1991, Cruikshank & Aydin 1995). Ongoing research focuses on these observations as well as on the temporal evolution of joint spacing (*e.g.*, Wu & Pollard 1995).

Example C: Visualizing Earthquake Displacement Fields

Although slip along a fault may be measured by the separation of marker horizons, the displacement field that characterizes the deformation of the wall rock is not commonly preserved in the geologic record (Fig. 8). Displacement fields from large earthquakes (*i.e.*, greater than magnitude 5) are measured using geodetic methods such as radar interferometry, Global Positioning System (GPS) triangulation, and surveying and are important in neotectonics research (Thatcher 1979, Stein & Barrientos 1985, Savage & Plafker 1991, Feigl *et al.* 1993,

Bürgmann *et al.* 1994b, Freymuller *et al.* 1994). The magnitude and geometry of the observed displacement fields relate directly to the fault geometry and fault-slip magnitude — key topics in earthquake studies (Thatcher and Bonilla 1989, Arnadottir & Segall 1994, Du *et al.* 1994).

Return to the idealized strike-slip fault (Fig. 2). Such an idealization permits us to model the fault geometry and stress drop due to an earthquake, and gain intuition about the resulting displacement field around the fault. For the first-order solution, we do not consider dynamic effects, such as radiation of seismic energy and rupture propagation. In our model, we assume that far-field plate tectonic motion generates stresses that are uniform and remote compared to fault length. The driving stress for the fault slip is the difference between the remote stress component and the shear traction acting on the fault surface. A slip event produces a displacement field that is distinctive: it is maximal along the fault and decreases to zero at an infinite distance. The distance at which the displacement of the wall rock is significant scales with the length of the fault, in a manner similar to the stresses near joints described above. Figure 8(a) illustrates the displacement field around an idealized right-lateral strike-slip fault. The components of displacement were calculated using only four lines of computer code (eqns. B1, B3, B8, and B9, Appendix 2). Right-lateral sliding along the model fault produces displacements toward the crack in the upper left and lower right quadrants and away from it in the upper right and lower left quadrants.

The displacement field calculated using the crack model described above can be compared to the apparently complex displacement field caused by the 1989 Loma Prieta earthquake. The earthquake-induced displacements of benchmarks were determined by repeated geodetic measurements using the GPS (Williams *et al.* 1993). Those displacements are shown with arrows in Fig. 8(b). Note how they, like the model results (tadpoles), are pulled northward in the lower portion of the plot and pushed westward in the upper portion.

The relative displacement magnitude distribution and direction are explained fairly well by our simple model. However, the magnitudes in the northwest are generally too large relative to those closer to the source. The discrepancies between the results from our simple two dimensional model and the observations may be explained by the three dimensional nature of the rupture. The event apparently occurred along a steeply-dipping (rather than vertical) rupture plane with oblique-slip (rather than pure strike-slip) motion (Arnadottir & Segall 1994). The discrepancy between our two-dimensional analysis and the observations is not a failure of the model, but rather can be used to illuminate the important three dimensional aspects of the rupture not incorporated in our first-order model.

Finally, the graphical comparison between model results and observations shown in Fig. 8 was done entirely within *Mathematica* and illustrates the flexibility of high-level graphics packages. The sophisticated graphical tools available within most symbolic math packages can be used to document the displacement field and compare it to observed displacements and geographic features, such as the coastline and the labeled city location, by combining graphical building blocks (points and lines).

Conclusion

The foregoing discussion demonstrates how complicated problems can be tackled relatively easily using a symbolic math package. Our approach gives a wide audience access to LEFM analyses of structural geologic problems, greatly reduces the potential for errors in calculation, saves time and aids interpretation with interactive graphics. By using a symbolic math package to solve stress functions, we circumvent the effort of solving the analytical equations using paper and pencil. Neither is there a need to subsequently code the resulting solutions in a separate program to further manipulate the functions or to visualize the results. Analysis of the stress and displacement fields is also made simpler with this approach and the mathematical details of the computations are hidden to the user. We have also used this approach to study the radial and angular variations in the stress distribution around the crack tip in both Cartesian and polar coordinates without requiring complicated transformations.

The examples of simple, uniformly-loaded opening and sliding cracks as analogs for joints and faults give an introduction to the application of elastic fracture mechanics to geologic structures. At this level, it is possible to study first-order behavior of fractures in rock. Simple models, such as these, are most appropriate for developing intuition based on a quantitative and mechanical understanding of brittle deformation. They are exceptional teaching tools, and we have used them in student exercises. The models teach the basic relationships between the geometry and loading states of fractures in the earth, as well as the concept of boundary value problems. They also provide an illustration of the use of models as tests of hypotheses, and may be employed to evaluate plans for field investigations or the consequences of certain assumptions. Because input parameters (geometry, applied stresses, and material properties) are easily changed and plotting the results takes only a few seconds, the program can be used interactively and is an excellent heuristic tool.

This approach is by no means limited to the classroom. The simple model of a single crack under uniform load closely approximates the first order behavior of many structures, such as the joints and faults described above. More realistic models might be obtained by adding complexity to the LEFM simulation. For example, the geometry of a fault may be better approximated by several planar segments than by a single crack. Superposition of various analytic solutions, intricate or virtually impossible using the pencil and paper approach, is made relatively simple with these symbolic math packages. Linked graphical output gives instantaneous feed back on the chosen model. Combined, these features speed interpretation and understanding of the mechanics of brittle structures in rock.

Acknowledgements

The authors acknowledge the financial support of the Stanford Rock Fracture Project. This manuscript benefited from discussions with David Pollard, Donal M. Ragan, Simon Kattenhorn, Lisa Koenig, and an anonymous reviewer.

Notebooks Available On-Line

We have written several *Mathematica* Notebooks that solve the stress and displacement equations for a variety of loading conditions and others that plot the results. These notebooks are available by anonymous ftp at bishop.stanford.edu or through the World Wide Web from URL http://geo.stanford.edu/~geomech.

References

Anderson, T. L. 1991. *Fracture Mechanics Fundamentals and Applications*. CRC Press, Boca Raton.

Arnadottir, T. & Segall, P. 1994. The 1989 Loma Prieta earthquake imaged from inversion of geodetic data. *J. Geophys. Research* **99**: 21,835-21,855.

Atkinson, B. K. 1987. *Fracture Mechanics of Rock*. Academic Press, London.

Brace, W. F. & Bombolakis, E. G. 1963. A note on brittle crack growth in compression. *J. Geophys. Research* **68**: 3709-3713.

Broek, D. 1991. *Elementary Engineering Fracture Mechanics*. Kluwer Academic Publishers, Dordrecht, The Netherlands.

Bürgmann, R., Pollard, D. D. & Martel, S. 1994a. Slip distributions on faults: effects of stress gradients, inelastic deformation, heterogeneous host-rock stiffness and fault interaction. *J. Struct. Geol.* **16**: 1675 - 1690.

Bürgmann, R., Segall, P., Lisowski, M. & Svarc, J. P. 1994b. Post-seismic strain following the 1989 Loma Prieta earthquake from repeated GPS measurements. *U. S. Geol. Surv. Prof. Pap.* 1,550.

Chinnery, M. A. 1961. The deformation of the ground around surface faults. *Bull. Seismol. Soc. Amer.* **51**: 355-372.

Chou, P. C. & Pagano, N. J. 1967. *Elasticity*. D. Van Nostrand Company, Inc., Princeton, New Jersey.

Cowie, P. A. & Scholz, C. H. 1992. Physical explanation for the displacement-length relationship of faults, using a post-yield fracture mechanics model. *J. Struct. Geol.* **14**: 1,133 - 1,148.

Cruikshank, K. M. 1991. Analysis of minor fractures associated with joints and faulted joints. *J. Struct. Geol.* **13**: 865-886.

Cruikshank, K. M. & Aydin, A. 1994. Role of fracture localization in arch formation, Arches National Park, Utah. *Geol. Soc. Amer. Bull.* **106**: 879-891.

Cruikshank, K. M. & Aydin, A. 1995. Unweaving the joints in Entrada Sandstone, Arches National Park, Utah, U.S.A. *J. Struct. Geol.* **17**: 409-421.

Delaney, P. T., Pollard, D. D., Ziony, J. I. & Mckee, E. H. 1986. Field relations between dikes and joints, emplacement processes and paleostress analysis. *J. Geophys. Research* **91**: 4,920-4,938.

Du, Y., Segall, P. & Gao, H. 1994. Dislocations in inhomogeneous media via a moduli-perturbation approach: general formulation and 2-D solutions. *J. Geophys. Research* **99**: 12,767-12,779.

Engelder, T. 1984. Loading paths to joint propagation during a tectonic cycle: an example from the Appalachian Plateau, U.S.A. *Int. Conf. on Multiple Def. & Foliation Dev.* **7**: 459-476.

Engelder, T. & Geiser, P. 1980. On the use of regional joint sets as trajectories of paleostress fields during the development of the Appalachian Plateau, New York. *J. Geophys. Research* **85**: 6,319-6,341.

Erdogan, F. & Sih, G. C. 1963. On the crack extension in plates under plane loading and transverse shear. *ASME J. Basic Eng.* 51-52.

Evans, M. A. 1994. Joints and decollement zones in Middle Devonian shales; evidence for multiple deformation events in the central Appalachian Plateau. *Geol. Soc. Amer. Bull.* **106**: 447-460.

Feigl, K. L., Agnew, D. C., Bock, Y., Dong, D., Donnellan, A., Hager, B. H., Herring, T. A., Jackson, D. D., Jordan, T. H., King, R. W., Larsen, S., Larson, K. M., Murray, M. H., Shen, Z. & Webb, F. H. 1993. Space geodetic measurement of crustal deformation in Central and Southern California, 1984-1992. *J. Geophys. Research*: 21,677-21,712.

Fletcher, R. C. & Pollard, D. D. 1981. Anticrack model for pressure solution surfaces. *Geology* **9**: 419-424.

Freymuller, J., King, N. E. & Segall, P. 1994. The co-seismic slip distribution of the Landers earthquake. *Bull. Seismol. Soc. Amer.* **84**: 646-659.

Gay, N.C. & Ortlepp, W.D. 1979. Anatomy of a mining induced fault zone. *Geol. Soc. Amer. Bull.* **90**: 47-58.

Gross, M. R. 1993. The origin and spacing of cross joints: examples from the Monterey Formation, Santa Barbara coastline, California. *J. Struct. Geol.* **15**: 737-751.

Gudmundsson, A. 1987. Tectonics of the Thingvellir fissure swarm, SW Iceland. *J. Struct. Geol.* **9**: 61-69.

Hancock, P.L. 1985. Brittle microtectonics: principles and practice. *J. Struct. Geol.* **7**: 437-457.

Hsieh, P. A. 1991. Fluid flow in fractured rocks. *U.S. Geol. Surv. Water Res. Inv.* **91-4084:** 80-85.

Irwin, G. R. 1958. Fracture. In: *Encyclopedia of Physics* (ed. S. Flugge). Springer-Verlag, Berlin, pp. 551-590.

Jaeger, J. C. & Cook, N. G. W. 1969. *Fundamentals of Rock Mechanics*. Methuen, London.

Kelley, V. C. & Clinton, N. J. 1960. Fracture systems and tectonic elements of the Colorado Plateau. *Univ. New Mexico Pub. Geol.* **6**: 104 pp.

Lachenbruch, A. H. 1961. Depth and spacing of tension cracks. *J. Geophys. Research* **66**: 4,273-4,292.

Lawn, B. 1993. *Fracture of Brittle Solids, Second Edition*. Cambridge University Press, New York.

Lin, J. & Parmentier, E. M. 1988. Quasistatic propagation of a normal fault: a fracture mechanics model. *J. Struct. Geol.* **10**: 149 - 162.

Lorenz, J.C. & Finley, S.J. 1991. Regional fractures II: fracturing of the Mesa Verde reservoirs in the Pineance Basin, Colorado. *Bull. Amer. Assoc. Petrol. Geol.* **75**: 1,738 - 1,757.

Malvern, L. E. 1969. *Introduction to the Mechanics of a Continuous Medium*. Prentice-Hall, Englewood Cliffs, New Jersey.

Martel, S., Pollard, D. D. & Segall, P. 1988. Development of simple strike-slip fault zones, Mount Abbot Quadrangle, Sierra Nevada, California. *Geol. Soc. Amer. Bull.* **100**: 1451-1465.

Nelson, R. A. 1985. *Geological Analysis of Naturally Fractured Reservoirs*, Gulf Publishing Company, Houston Texas.

Nemat-Nasser, S. & Horii, H. 1982. Compression-induced nonplanar crack extension with application to splitting, exfoliation, and rockburst. *J. Geophys. Research* **87**: 6,805-6,821.

Olson, J. E. & Pollard, D. D. 1991. The initiation and growth of en echelon veins. *J. Struct. Geol.* **13**: 595-608.

Petit, J.-P. & Mattauer, M. 1995. Paleostress superposition deduced from mesoscale structures in limestone: the Matalles exposure, Languedoc, France. *J. Struct. Geol.* **17**: 245-256.

Pollard, D. D. & Segall, P. 1980. Mechanics of discontinuous faults. *U.S. Geol. Surv. Prof. Pap.* **1,175**: 245-255.

Pollard, D. D. & Segall, P. 1987. Theoretical displacements and stresses near fractures in rock: with applications to faults, joints, veins, dikes, and solution surfaces. In: *Fracture Mechanics of Rock* (ed. Atkinson, B. K.). Academic Press, London, pp. 227-349.

Pollard, D. D. & Aydin, A 1988. Progress in understanding jointing over the past century. *Geol. Soc. Amer. Bull.* **100**: 1,181-1,204.

Price, N. J. 1966. *Fault and Joint Development in Brittle and Semi-Brittle Rock*. Pergamon Press, Oxford.

Ramsay, J. G. & Huber, M. I. 1987. *The Techniques of Modern Structural Geology. II: Folds and Fractures.* Academic Press, London.

Rispoli, R. 1981. Stress fields about strike-slip faults inferred from stylolites and tension gashes. *Tectonophysics* **75**: T29-T36.

Rodgers, D. A. 1980. Analysis of pull-apart basin development produced by en echelon strike-slip faults. *Spec. Pub. Int. Assoc. Sed.* **4**: 27-41.

Rubin, A. M. 1993 Tensile fracture of rock at high confining pressure: implications for dike propagation. *J. Geophys. Research* **98**: 15,919-15,935.

Rudnicki, J. W. 1980. Fracture mechanics applied to the Earth's crust. *Ann. Rev. Earth & Planet. Sci.* **8**: 489-525.

Sammis, C. G. & Julian, B. R. 1987. Fracture instabilities accompanying dike intrusion. *J. Geophys. Research* **92**: 2,597-2,605.

Sani, F. 1990. Extensional veins and shear joint developments in a thrust-fold zone (Northern Apennines, Italy). In: *Deformation Mechanisms, Rheology and Tectonics* (eds. Knipe, R. J. & Rutter, E. H.). Geol. Soc. Lond. Spec. Pub. **54**: 475-490.

Savage, J. C. & Plafker, G. 1991. Tide gauge measurements of uplift along the south coast of Alaska. *J. Geophys. Research* **96**: 4,325-4,335.

Scholz, C. H. 1990. *The Mechanics of Earthquakes and Faulting*. Cambridge Univ. Press, Cambridge.

Segall, P. & Du, Y. 1993. How similar were the 1934 and 1966 Parkfield earthquakes? *J. Geophys. Research* **98**: 4,527-4,538.

Segall, P. & Pollard, D. D. 1983. Nucleation and growth of strike-slip faults in granite. *J. Geophys.Res.* **88**: 555-568.

Selby, M. J. 1982 Controls on the stability and inclinations of hillslopes formed on hard rock. *Earth Surfaces Processes and Landforms* **7**: 449-467.

Sih, G.C. 1966. On the Westergaard method of crack analysis. *J. Fracture Mech.* **2**: 628-630.

Sih, G. C. 1973. *Handbook of Stress-Intensity Factors*. Inst. Fract. and Solid Mech., Lehigh University, Bethlehem, Pennsylvania.

Stauffer, M. R. & Gendzwill, D. J. 1987. Fractures in the Northern Plains, stream patterns, and the Midcontinent stress field. *Can. J. Earth Sci.* **24**: 1086-1097.

Stearns, D.W. & Friedman, M. 1972. Reservoirs in fractured rock. *Amer. Assoc. Petrol. Geol. Mem.* **16**: 82-106.

Stein, R. S. & Barrientos, S. 1985. The Borah Peak, Idaho earthquake: geodetic evidence for deep rupture on a planar fault. In: *Proc. Workshop XXVIII on the Borah Peak earthquake, U. S. Geol. Surv. Open File Rep. 85-290* (eds. Stein, R. S. & Bucknam, R. C.), 459-485.

Steward, I. S. & Hancock, P. L. 1990. Brecciation and fracturing within neotectonic normal fault zones in the Aegean region. In: *Deformation Mechanisms, Rheology and Tectonics* (eds. Knipe, R. J. & Rutter, E. H.). Geol. Soc. Lond. Spec. Pub. **54**: 105-110.

Tada, H., Paris, P. C. & Irwin, G. R. 1985. *The Stress Analysis of Cracks Handbook*. Paris Productions Inc. (& Del Research Corp.), St. Louis.

Takahashi, H. & Abé, H. 1989. Fracture mechanics applied to hot, dry rock geothermal energy. In: *Fracture Mechanics of Rock*, (ed. B.K. Atkinson), Academic Press, London, 241-176.

Thatcher, W. 1979. Systematic inversion of geodetic data in central California. *J. Geophys. Research* **84**: 2,283-2,295.

Thatcher, W. & Bonilla, M. 1989. Earthquake fault slip estimation from geologic, geodetic and seismologic observations: implications for earthquake mechanics and fault segmentation. In: *Fault Segmentation and Controls of Rupture Initiation* (eds. Schwartz, D. P. & Sibson, R. H.), USGS Open File Report **89-135**, pp. 386-399.

Thomas, A. L. & Pollard, D. D. 1993. The geometry of echelon fractures in rock: implications from laboratory and numerical experiments. *J. Struct. Geol.* **15**: 323-334.

Timoshenko, S. P. & Goodier, J. N. 1970. *Theory of Elasticity*. McGraw-Hill Publishing Company.

Tobisch, O. T. & Cruden, A. R. 1995. Fracture-controlled magma conduits in an obliquely convergent continental magmatic arc. *Geology* **23**: 941-944.

Williams, C. R., Arnadottir, T. & Segall, P. 1993. Coseismic deformation and dislocation models of the 1989 Loma Prieta earthquake derived from global positioning system measurements. *J. Geophys. Research* **98**: 4,567-4,578.

Willemse, E. J. M., Pollard, D. D., & Aydin, A. 1996. Three dimensional analysis of slip distributions on normal fault arrays with consequences for fault scaling. *J. Struct. Geol.* **18**: 295-309.

Wu, H. & Pollard, D. D. 1995. An experimental study of the relationship between joint spacing and layer thickness. *J. Struct. Geol.* **17**: 887-905.

Appendix 1:
The Stress Function Approach for Linear-Elastic Materials

The first part of this appendix presents the concept of stress functions for continuous bodies to illustrate how they relate to equilibrium and compatibility considerations. This introductory material is covered in many standard elasticity textbooks (*e.g.,* Chou & Pagano 1967, Jaeger & Cook 1969, Malvern 1969, Timoshenko & Goodier 1970). The second section explains how to use a specific type of stress function that is suitable for to bodies with holes or cracks. A comprehensive treatment of this class of problems, including equations, is given in some fracture mechanics books (*e.g.*, Tada *et al.* 1985, Anderson 1991).

The nomenclature and sign conventions used here are summarized in Fig. 2. We define a coordinate system, centered on the fault, with the x-axis parallel to the fault and the y-axis perpendicular. Stresses are designated by σ with two subscripts. The first subscript indicates the normal to the plane on which the stress is acting; the second subscript indicates the direction in which the stress is acting. Superscripts indicate where the stress is applied: r indicates a remote stress, and c indicates a stress applied to the crack surface. A remote shear stress is thus designated σ^r_{yx}. Shear stress components are positive if they act on a surface with a positive normal in a positive direction or on a surface with a negative normal in a negative direction. Tensile normal stress components are positive. Components of the stress tensor are indicated by σ_{ij}; ε_{ij} denotes components of the strain tensor.

Stress Functions in a Continuous Medium

For a linearly-elastic material, stress is proportional to strain as described by Hooke's Law. For a homogeneous and isotropic body, the material behavior is completely defined by two elastic moduli such as Poisson's ratio and the shear modulus. Equilibrium is established by Newton's law which equates the net force on a body to the product of mass and acceleration. For a two-dimensional static body (one with zero acceleration), or for a quasistatic body (one with negligible acceleration), and neglecting body forces, the two equilibrium equations are:

$$\frac{\partial \sigma_{xx}}{\partial x} + \frac{\partial \sigma_{xy}}{\partial y} = 0 \qquad \text{(A1)}$$

and

$$\frac{\partial \sigma_{xy}}{\partial x} + \frac{\partial \sigma_{yy}}{\partial y} = 0. \qquad \text{(A2)}$$

To assure no (or negligible) angular accelerations, the shear stresses on adjacent faces of any element must be equal (Fig. 2). That is,

$$\sigma_{xy} = \sigma_{yx}; \qquad \text{(A3)}$$

therefore, there are only three independent stress components (σ_{xx}, σ_{yy}, and σ_{zz}). Finally, the three strain components are directly related to the displacements and cannot vary arbitrarily with respect to the spatial coordinates. In fact, the strain components must be related to each other by the compatibility equation:

$$\frac{\partial^2 \varepsilon_{xx}}{\partial^2 y} - 2\frac{\partial^2 \varepsilon_{xy}}{\partial x \partial y} + \frac{\partial^2 \varepsilon_{yy}}{\partial^2 x} = 0. \tag{A4}$$

This constrains the deformation of the body in order to prevent the creation of holes or overlaps. The compatibility equation can be expressed in terms of the stress components by using the stress-strain relationships for a linear elastic material (Hooke's Law). The result can then be simplified further using the equilibrium equations, to yield (for plane strain):

$$\left[\frac{\partial^2}{\partial x^2} + \frac{\partial^2}{\partial y^2}\right]\left(\sigma_{xx} + \sigma_{yy}\right) = 0$$

or

$$\nabla^2\left(\sigma_{xx} + \sigma_{yy}\right) = 0. \tag{A5}$$

The plane strain condition is defined such that the out-of-plane displacement component is zero everywhere, and the two in-plane components vary as functions of x and y only.

The stress function is based on the mathematical interrelationships between this particular harmonic equation (A5) and the two equilibrium equations (A1 & A2). All three equations are automatically satisfied if there exists a function $Y(x,y)$ that satisfies the bi-harmonic equation, *i.e.*:

$$\frac{\partial^4 \Psi}{\partial x^4} + 2\frac{\partial^4 \Psi}{\partial x^2 \partial y^2} + \frac{\partial^4 \Psi}{\partial y^4} = 0$$

or

$$\nabla^2\left(\nabla^2 \Psi\right) = 0 \tag{A6}$$

which is related to the stress components by:

$$\sigma_{xx} = \frac{\partial^2 \Psi}{\partial y^2}, \tag{A7}$$

$$\sigma_{yy} = \frac{\partial^2 \Psi}{\partial x^2}, \tag{A8}$$

and

$$\sigma_{xy} = -\frac{\partial^2 \Psi}{\partial x \partial y}. \tag{A9}$$

The function *Y(x,y)* is called the Airy stress function and it has many forms. Airy stress functions cover a wide variety of problem geometries and loading conditions.

In summary, the solutions of many two-dimensional linear elastic boundary value problems are obtained by following a fixed recipe. The first step is to obtain the appropriate stress function. The known functions can found in elasticity textbooks. The second step is to find the desired stress components by differentiating twice according to A7 through A9. The third step is to check that the results match the boundary conditions. If desired, one may determine strains using the constitutive stress-strain relations (such as Hooke's Law) and calculate displacements with the relations between strain and displacement.

Stress Functions for Bodies with Holes or Cracks

The solution technique for two-dimensional linear-elastic problems involving bodies with holes or cracks employs stress functions *Z(z)* that are functions of the complex variable $z = x + iy$, where $i = \sqrt{-1}$ and x and y are as shown in Fig. 2. The simplest stress functions are the Westergaard stress functions. The Westergaard stress function for opening mode (mode I) cracks is related to the Airy stress function by:

$$\Psi = \mathrm{Re}\left[\bar{\bar{Z}}\right] + y\,\mathrm{Im}\left[\bar{Z}\right], \tag{A10}$$

where *Re* and *Im* indicate the real and imaginary parts of the complex variable function, respectively (Tada *et al.* 1985). The bar(s) over Z represent one (or two) integration(s) of Z with respect to z. The crack is a discontinuity in the displacement field which coincides with a discontinuity in the corresponding complex functions.

Stress and displacement fields can be computed from the Westergaard function as follows:

For Mode I –

$$\sigma_{xx} = \mathrm{Re}[Z] - y\,\mathrm{Im}[Z'] \tag{A11a}$$

$$\sigma_{yy} = \mathrm{Re}[Z] + y\,\mathrm{Im}[Z'] \tag{A11b}$$

$$\sigma_{xy} = -y\,\mathrm{Re}[Z'] \tag{A11c}$$

$$\sigma_{zz} = \nu\left(\sigma_{xx} + \sigma_{yy}\right) \tag{A11d}$$

$$u = \frac{1}{2\mu}\left((1-2\nu)\mathrm{Re}\left[\bar{Z}\right] - y\,\mathrm{Im}[Z]\right) \tag{A11e}$$

$$v = \frac{1}{2\mu}\left(2(1-\nu)\,\mathrm{Im}\left[\bar{Z}\right] - y\,\mathrm{Re}[Z]\right) \tag{A11f}$$

where Z' is the derivative of Z with respect to z and u and v are the displacements in the x and y directions, respectively (Fig. 2). Poisson's ratio and the elastic shear modulus are indicated by n and m. The displacement discontinuity along the opening crack itself is given by:

$$\Delta v = \frac{1}{\mu}\left(2(1-\nu)\,\mathrm{Im}\left[\bar{Z}\right]\right) \quad \text{for } y = 0; |x| \le a, \tag{A11g}$$

where Δv is the amount of opening along the crack.

Irwin (1958) and Sih (1966) extended Westergaard's approach to Mode II and III cracks. They found:

For Mode II –

$$\sigma_{xx} = 2\,\text{Im}[Z] + y\,\text{Re}[Z'] \tag{A12a}$$

$$\sigma_{yy} = -y\,\text{Re}[Z'] \tag{A12b}$$

$$\sigma_{xy} = \text{Re}[Z] - y\,\text{Im}[Z'] \tag{A12c}$$

$$\sigma_{zz} = \nu\left(\sigma_{xx} + \sigma_{yy}\right) \tag{A12d}$$

$$u = \frac{1}{2\mu}\left(2(1-\nu)\,\text{Im}\left[\bar{Z}\right] + y\,\text{Re}[Z]\right) \tag{A12e}$$

$$v = -\frac{1}{2\mu}\left((1-2\nu)\,\text{Re}\left[\bar{Z}\right] - y\,\text{Im}[Z]\right) \tag{A12f}$$

$$\Delta u = \frac{1}{\mu}\left(2(1-\nu)\,\text{Im}\left[\bar{Z}\right]\right) \text{ for } y = 0; |x| \leq a, \tag{A12g}$$

where Δu is the slip along the crack in a direction perpendicular to the crack tip-line.

For Mode III –

$$\sigma_{yz} = \text{Re}[Z] \tag{A13a}$$

$$\sigma_{xz} = \text{Im}[Z] \tag{A13b}$$

$$w = \frac{1}{\mu}\left(\mathrm{Im}\left[\bar{Z}\right]\right) \tag{A13c}$$

$$\Delta w = \frac{1}{\mu}\left(2\,\mathrm{Im}\left[\bar{Z}\right]\right) \text{ for } y = 0; |x| \leq a, \tag{A13d}$$

where Δw is the distribution of out-of-plane slip parallel to the crack tip line. Where stress or displacement components are not explicitly defined in the equations above, they are zero.

Many alternative approaches and formulations can be found in the literature (see Jaeger & Cook 1969). The formulations presented above are frequently found in the engineering literature. Pollard and Segall (1987) present a geologically oriented formulation of complex stress functions that more clearly separates remote loads and tractions on the crack. As such it is somewhat easier to visualize the significance of the various parameters, although that comes at the expense of slightly more complicated mathematics during the analysis.

Using the above formulas, other attributes of the stress or displacement fields can be computed. For example, for in-plane problems, the mean normal stress is given by:

$$\sigma_{mean} = \frac{\left(\sigma_{xx} + \sigma_{yy}\right)}{2}. \tag{A14}$$

Principal stress magnitudes are computed by:

$$\sigma_1 = \frac{\left(\sigma_{xx} + \sigma_{yy}\right)}{2} + \sqrt{\frac{\left(\sigma_{xx} - \sigma_{yy}\right)^2}{4} + \sigma_{xy}^2} \tag{A15a}$$

and

$$\sigma_2 = \frac{\left(\sigma_{xx} + \sigma_{yy}\right)}{2} - \sqrt{\frac{\left(\sigma_{xx} - \sigma_{yy}\right)^2}{4} + \sigma_{xy}^2}\ , \tag{A15b}$$

and the angle between the maximum principal stress and the positive x-axis is given by:

$$\alpha = 0.5\,\mathrm{Arctan}\left(\frac{2\sigma_{xy}}{\left(\sigma_{xx}-\sigma_{yy}\right)}\right). \tag{A15c}$$

The computed components can also be expressed in other coordinate systems (for example, polar coordinates, Fig. 6) by applying the appropriate transformations.

Appendix 2:
Computer Implementation of the Stress Function

This appendix reproduces *Mathematica* code used to calculate the three Cartesian stress components, the mean normal stress, and displacement fields around a uniformly loaded mode II crack that serves as an idealization for a strike-slip fault. Each line of code corresponds to an equation in the main text or Appendix 1.

```
zee[z_] = remoteLoad / Sqrt[1-(a/z)^2] (stress function)                 (B1)
zeePrime[z_] = D[zee[z],z] (derivative of stress function)              (B2)
zeeBar[z_] = Integrate[zee[z],z] (integrated stress function)           (B3)
```

The first line is the stress function appropriate for mode II loading configuration (eqn. 1, main text) whereas the second and third lines define the derivative and integrated stress function needed to calculate the stress components. The **[z_]** indicates that the named function will carry the argument **z**. The Sqrt[ ], D[,z], and Integrate[,z] are *Mathematica* commands for taking the square root, and for differentiating and integrating with respect to z.

Note that some care must be taken when coding the stress function. Many crack stress functions employ $\sqrt{(z^2-a^2)}$ (*e.g.*, Tada *et al.* 1985). Coding this directly into *Mathematica* results in asymmetric solutions, with sign error in the two quadrants where x < 0, and correct solutions where $x > 0$. The correct symmetry can be forced by rewriting $\sqrt{(z^2-a^2)}$ into $z\sqrt{(1-(a/z)^2)}$.

For mode II problems, the stress components are given by:

```
sxx[x_,y_]    = 2*Im[zee[z]]+y*Re[zeePrime[z]] /.z->x+I*y              (B4)
sxy[x_,y_]    = Re[zee[z]]-y*Im[zeePrime[z]] /.z->x+I*y                (B5)
syy[x_,y_]    = -y*Re[zeePrime[z]] /.z->x+I*y                          (B6)
```

This is exactly equations A12a-A12c. The additional /. z->x+I*y tells the program to use the complex variable $z = x + iy$ when evaluating the equation. The mean normal stress is calculated by eqn. (3) as:

```
meanNormalStress[x_,y_] = (sxx[x,y]+syy[x,y])/2
```
(B7)

Similarly, the displacements of the elastic medium in x and y directions are defined, respectively, as:

```
u[x_,y_] = (2*(1-nu)*Im[zeeBar[z]]+y*Re[zee[z]])/(2*mu) /.z->x+I*y
```
(B8)

```
v[x_,y_] = (-(1-(2*nu))*Re[zeeBar[z]]-y*Im[zee[z]])/(2*mu)/.z->x+I*y
```
(B9)

where nu represents Poisson's ratio, n, and mu stands for the elastic shear modulus, m. Just eight lines are required to provide all the functions for this particular problem.

Quantitative answers can be obtained and graphed as soon as the four elastic and geometrical constants are defined. All units must be in a consistent unit system such as the SI system. For the strike-slip fault example, we might chose:

```
nu = 0.25 (Poisson's ratio, dimensionless)
mu = 30000 (Shear modulus in units of stress, e.g., MPa)
a = 1000 (Fault half length in units of length, e.g., meters)
remoteLoad = 10 (Remote uniform driving stress in units of stress, e.g., MPa)
```
(B10)

Because of the speed of the calculations and ease of graphical output, interactive exercises can be created to illustrate to effect of each of these parameters on the stress or displacement fields.

For other types of mode II boundary conditions or geometry, only the first line defining the stress function is changed, and all other lines are the same. For mode I or III problems, the first three lines remain the same, but stresses and displacements are found by replacing the last six lines by the appropriate equations given in Appendix 1.

Appendix 3:
Symbolic Math Packages: *MATHEMATICA*

Widely available symbolic math packages such as *MATLAB*, *MAPLE V*, *Theorist*, and *Mathematica* share similar analytical, computational and graphical features. Each varies in price, cross-platform compatibility and ease of use. *Mathematica* is a high-level programming language and symbolic manipulation package developed by Wolfram Research. We use *Mathematica* because it runs on both *MacOS* and *UNIX* platforms and has excellent graphics capabilities. We encourage careful evaluation of system requirements, features of interest and potential applications before any one of the packages is acquired. Many colleges and universities have distributed computing resources and software servers so that higher-end hardware and software tools are commonly available.

The *Mathematica* Notebook interface has a high level of formatting capability and allows the user to organize text, computations, and graphics so that the Notebook is easy to read, use, and understand. Running a Notebook that is already written is quite simple. Writing new Notebooks does not require an immense effort. A week might be spent learning commands and keystrokes and becoming familiar with programming syntax and function definitions. Some familiarity with the *C* programming language is helpful but not essential. Steven Wolfram (Wolfram 1991) has written an extensive textbook describing in detail the functions and features of the package. Others have written useful introductory guides (see below). To learn to use *Mathematica*, we suggest beginning with a Notebook that has been written by someone else, learn how to modify it, and eventually create original routines.

Annotated Bibliography for Mathematica

Blachman, N. 1992. *Mathematica: A Practical Approach*. Englewood Cliffs, New Jersey, Prentice Hall, 365 pp. This is an excellent beginners book with many practical exercises to get users going with *Mathematica*.

Blachman, N. 1992. *Mathematica* Quick Reference (Version 2). Menlo Park, CA, Addison-Wesley Publishing Company, 304 pp. Blachman furnishes a compact guide that provides a complete listing of *Mathematica* commands.

Wickham-Jones, T. 1994. *Mathematica Graphics: Techniques and Applications*. Santa Clara, CA, Springer-Verlag Publishers, 721 pp. Wickham-Jones provides a thorough review of *Mathematica* graphics, as well as details on customizing pictures using *Mathematica* and his own programming packages that extend the standard graphics, notably in three dimensions.

Wolfram, S. 1991. *Mathematica: A System For Doing Mathematics By Computer*. Menlo Park, CA, Addison-Wesley Publishing Company, 961 pp. This is the official and most comprehensive guide to *Mathematica*. Wolfram provides a practical introduction, presentation of the principles of *Mathematica*, discussion of advanced mathematics in the program, and a complete reference guide.

Relevant Internet Sites

Mathematica (Wolfram Research): http://www.wri.com/
MATLAB (The Mathworks, Inc.): http://www.mathworks.com/
MAPLE V (Waterloo Maple Software): http://www.maplesoft.com/
Theorist (Waterloo Maple Software): http://www.maplesoft.com/

Bézier Curves and Geological Design

Declan G. De Paor
Department of Earth and Planetary Sciences,
Harvard University, 20 Oxford Street, Cambridge
MA 02138, U.S.A. depaor@eps.harvard.edu

Abstract– Most structural geologists probably don't think of themselves as designers and yet in practice, their work is directly analogous to that of designers in other fields ranging from high fashion to aircraft manufacture. Structural geology can greatly benefit from design tools that have been developed for microcomputer applications in a variety of other disciplines. Among the most important of these tools are Bézier curves which may be used in forward modelling or reconstruction. The underlying theory of Bézier curves has the potential to revolutionize our treatment of heterogeneous deformation and kinematics in the future, much as tensor algebra radically improved our understanding of homogeneous strain during the past two decades.

Introduction

Structural geologists frequently need to manipulate complex curves and surfaces. For example, when compiling a geological map, you might wish to interpolate contacts among known field locations and then edit those contacts when extra data becomes available at a later time. Or you may want to create a numerical model in which the smoothly curved boundaries of a structure change with time. Personal computer programs, including drawing, GIS, and CAD applications, provide simple graphical tools for creating and editing curves but they can be frustrating because the user generally does not have access to the governing mathematical equations and so cannot invent new, unanticipated ways of using curves. It can be difficult, for example, to determine a curve's arc length, enclosed area, or radius of curvature. This paper examines the theoretical basis of the most popular class of curves – Bézier polynomials – and presents new methods for modeling geological structures using Bézier curves and surfaces.

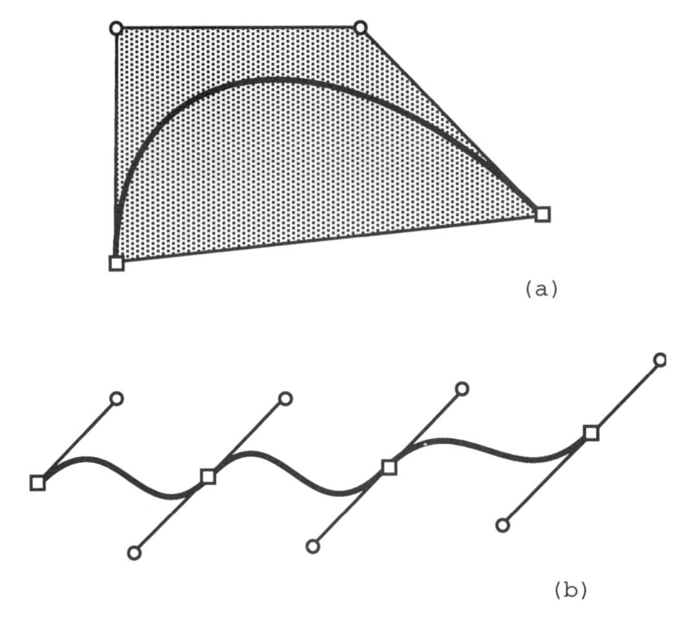

Fig. 1. a) Bézier cubic (thick line), tangents (thin lines), control points (squares and circles), and convex hull (shaded). b) Curve with three cubic segments; note extra leading tangent.

The Design Philosophy

Designers are neither artists nor scientists but occupy the spectrum between these end members. They must create functional objects that meet preset criteria (clothes that fit, aircraft that fly) but, subject to these constraints, they may employ creative talent to achieve an acceptable style. Geologists act as designers whenever they construct cross sections or three dimensional models of structures in areas of incomplete data. Their sections are constrained by observations and must work (*i.e.*, must be physically realistic) but within these constraints there is room for choice based on current knowledge about geological processes or on currently popular styles of cross section construction or model evolution.

The construction of kink-fold cross sections (*e.g.*, Suppe 1983) is a typical example of design philosophy in use in modern structural geology. As recently as fifteen years ago it was quite common to find published examples of ‘stylistic’ cross sections which today we recognise do not balance and could never be restored to their undeformed condition by a plausible sequence of events. Whilst some people may justifiably question the suitability of kink-fold constructions for cross sections that involve rocks deformed at high confining pressures and /or moderate metamorphic grade, no one cannot doubt the improved predictive performance of kink-fold constructions in hydrocarbon exploration. With the aid of Bézier curves, structural geologists may be able to apply the same design criteria and constraints to the construction of cross sections with a less "kinky" style.

Forward modeling, the opposite of restoration, is the mathematical description of phenomena as they change with the passage of time. In a well-designed geological model, the history of development of structures, as represented in sequential sections or movies, must be compatible with the evidence preserved in the final deformed rocks. The model must also conform to current theory regarding how the structures formed. To facilitate the design approach to geological modeling, interactive tools such as Bézier curves and surfaces are of great value.

By calling structural geologists designers, I do not mean to imply that they are inferior to pure scientists; quite the opposite. Suppose we are given the positions of a feature in a cross section at two instances. We know that natural motion involves periods of acceleration and deceleration but these are not measurable from the geological record and so a pure scientist will argue for the simplest inferred path – a straight line from the earlier position to the later one. However, a designer will place bounds upon reasonable acceleration rates and will use a curved trajectory. The designer gets closer to the truth by incorporating implicit information (limits to the rates of natural processes) with the explicitly measured data.

What are Bézier Curves?

All designs begin with sketches or rough plans but these need not be drawn freehand. Personal computer drafting software includes design tools that allow more freedom than a template yet less than the unguided hand. Foremost among these tools are Bézier curves which were invented by design engineers in the French auto industry (de Casteljau 1959, 1963, Bézier 1966, 1967). de Casteljau's work was never published and so the curves became known after the latter author. The recent growth in popularity of personal computer graphics programs, along with CAD/CAM and GIS, has led to the widespread use of Bézier curves and to an explosion of interest in the general subject of interactive graphics.

As those of you who have used the Bézier drawing tool in a graphics software package will be aware, the shape of a Bézier curve is determined by a set of control points here labelled $\mathbf{P}_j$, $j = 0,1,2,3,....$ The number of control points per polynomial is determined by the degree of the polynomial. One control point defines a point, two a straight line, three a quadratic, four a cubic, five a quartic, and so on. Just as a tensor is assumed to mean a rank-2 tensor unless otherwise stated, the default Bézier polynomial is usually taken to be a cubic, thus requiring four control points.

A Bézier curve consists of one or more polynomial segments concatenated in daisy-chain fashion. The start of a curve is defined by clicking and dragging the computer mouse to define two ends of a leading tangent (usually marked by a small square and circle connected by a thin line; Fig. 1). Thereafter, each click and drag of the mouse adds a new cubic segment to the curve. The circle marking the point where the mouse was released is reflected through the square where the mouse was depressed to define a trailing tangent which, together with the previous leading tangent, controls the curve (like knitting, the procedure is easier

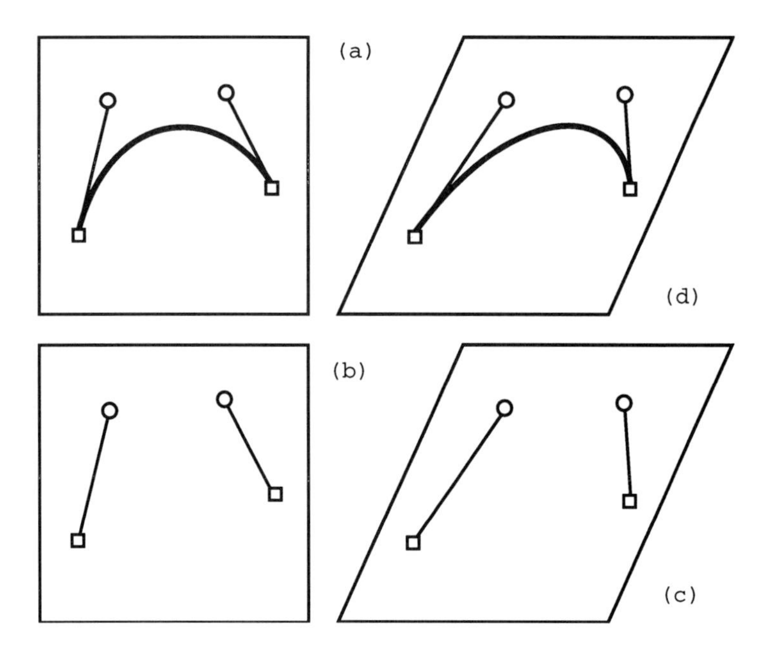

Fig. 2 Affine invariance property. a) Bézier curve, b) control tangents, c) sheared control tangents, d) curve fitted to sheared tangents which is identical to sheared equivalent of curve in (a).

to demonstrate than to describe in words). The trailing tangent of one cubic and leading tangent of the next cubic are usually colinear, ensuring continuity and differentiability at the square control points where each pair of cubic polynomials meet. A Bézier curve always passes through its leading and trailing square control points; its shape is modified by the intervening control points (Fig. 1a).

Why Use Cubic Bézier Curves?

Designers in other fields use a variety of more or less sophisticated curves such as interpolating splines (curves that pass through all of their control points), Hermite splines (curves controlled by the orientations of tangents through their endpoints but not by the lengths of these tangents), NuSplines, Catmulls, NURBs, *etc.* (for details, see for example Foley *et al.* 1990). All of these are available in different degrees – linear, quadratic, cubic, quartic, *etc.* However, the broad availability of cubic Bézier drawing tools makes them a natural choice for structural geologists. Higher degree polynomials require sophisticated and expensive CAD applications usually available only on sophisticated and expensive graphics workstations. Also, you run the risk of introducing extraneous turning points in the model structure. If you do wish to generate multiple turning points, as when modeling a parasitic fold train for example (Fig. 1b), a chain of cubic curves is easier to control than a single high degree polynomial. Most importantly, Bézier curves are the simplest curves that possess a number of attributes considered essential for structural applications. The following is a brief description of these attributes (for details and derivations, see Farin 1988):

1) *Symmetry*: The cubic defined by four control points $\mathbf{P}_0$, $\mathbf{P}_1$, $\mathbf{P}_2$, and $\mathbf{P}_3$ is identical to that controlled by points $\mathbf{P}_3$, $\mathbf{P}_2$, $\mathbf{P}_1$, and $\mathbf{P}_0$. It is simply drawn in the opposite direction.

2) *Local effect*: The effect of a change in any one of the four control point is concentrated near the chosen point.

3) *Differentiation*: The first, second, and third derivatives of a cubic Bézier curve are a quadratic Bézier curve, a straight line Bézier, and a point Bézier, respectively. These derivatives may be used to describe curvature.

4) *Convex hull*: If a polygon called a "convex hull" is drawn by imagining a rubber band stretched around the perimeter of the set of four Bézier control points, then every point on the cubic will lie within this polygon. If the vertices of the convex hull are in the order $\mathbf{P}_0$, $\mathbf{P}_1$, $\mathbf{P}_2$, and $\mathbf{P}_3$, counting either clockwise or counterclockwise, then the curve forms a box fold, otherwise it forms a syncline-anticline pair separated by an inflection point. Bézier cubics never become infinite in amplitude or otherwise unstable, even if their terminal tangents are parallel, as in an isoclinal fold.

5) *Affine invariance*: Bézier curves deform homogeneously with their convex hulls. Instead of transforming all points on the curves (of which there may be hundreds or thousands depending on resolution), it is sufficient to deform the vertices of the convex hulls and refit the curves (Fig. 2).

6) *Subdivision*: Every subdivision of a Bézier cubic is itself a Bézier cubic (Fig. 3). This means, for example, that a Bézier curve representing a bedding trace may be cut and displaced across a Bézier curve representing a fault.

7) *Continuity*: Adjacent cubics in a Bézier curve are continuous and differentiable if adjacent control tangents are colinear. They are continuous but are not differentiable if the tangents form a kink at a control point.

8) *Tractability*: It is possible to calculate arc-lengths, curvatures, derivatives, forward differences, enclosed areas, *etc.*, either exactly or by approximation.

9) *Expansion to surfaces*: Arrays of Bézier curves can be combined to define Bézier surfaces, either using triangulation or quadrilateral interpolation.

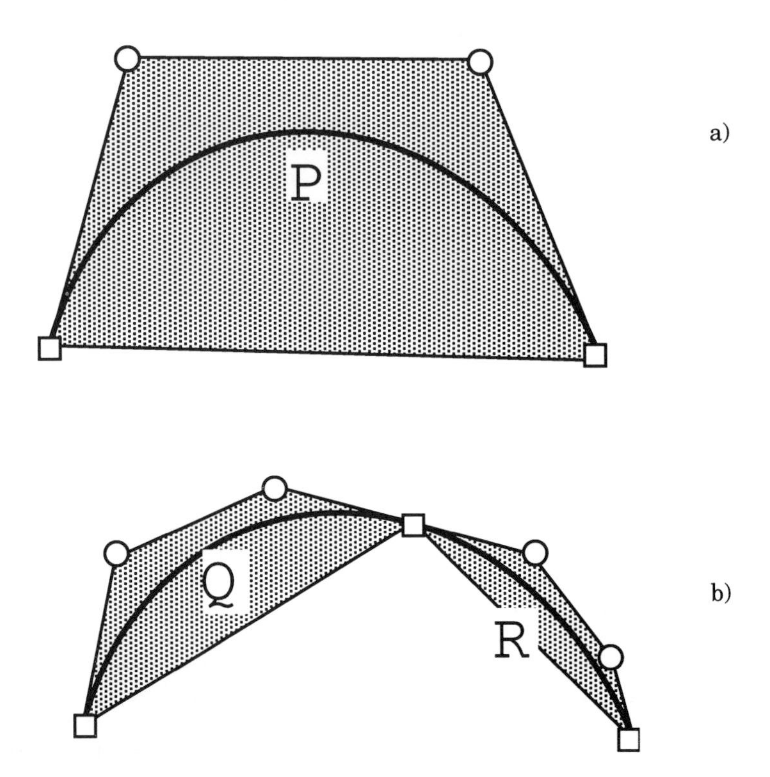

Fig. 3. Subdivision of a Bézier curve **P** in (a) to form two smaller Bézier curves, **Q** and **R** in (b). Convex hulls shaded.

Mathematical Representation of Bézier Curves

The traditional mathematical representation of a curve employs the functional format $y = f(x)$. For example, the equation of a straight line of slope m and intercept c may be written

$$y = mx + c. \tag{1}$$

However, such functional descriptions are unsuitable for use in modeling because m and c are reference frame dependent parameters that are not intrinsic attributes of the curve and also because there is no solution for vertical slopes or for looped curves that form a knot.

The designer's approach to the mathematical definition of curves employs sets of points $\{\mathbf{P}_j\}$, $j = 1,2,3,\ldots,n$. **P** stands for each coordinate in a Cartesian reference frame, so for example

$$\mathbf{P}_1 = 2\mathbf{P}_0 \tag{2}$$

can be expanded to

$$x_1 = 2x_0 \tag{3a}$$
$$y_1 = 2y_0 \tag{3b}$$
$$z_1 = 2z_0 \tag{3c}$$

A great advantage of the parametric form is that equations are similar in two and three dimensions; eqn. (2) may represent the two eqns. (3a,b) or the three eqns. (3a, b, c) as desired. To overcome the problems associated with function descriptions, designers use simultaneous equations for the coordinates (x, y, z) in terms of an independent parameter t

$$x = x(t) \tag{4a}$$
$$y = y(t) \tag{4b}$$
$$z = z(t) \tag{4c}$$

or in short

$$\mathbf{P} = \mathbf{P}(t) \tag{5}$$

One can think of t as time. Even if the slope is locally vertical or if the curve crosses over itself, every point on the curve is drawn at a unique instant in time. This analogy may also help one to understand that parametric 'distance' along a curve need not be correlated with arc-length. The curve may be drawn fast at first and then more slowly, so that the parametric halfway point, $\mathbf{P}(0{\cdot}5)$, is not necessarily the mid-point in terms of arc-length. Note that parametric curves can be defined in reference frames where one of the axes is itself time, so it is best to think of t as an abstract parameter rather than physical time.

Because the fundamental physical properties of an object must be independent of reference frame, we need mathematical operations on the point set $\{\mathbf{P}_j\}$ that are reference frame indifferent. One such operation, which we shall employ later, is the difference, Δ, of any two points,

$$\Delta\mathbf{P}_{ij} = \mathbf{P}_j - \mathbf{P}_i \tag{6}$$

This defines a vector directed from point $\mathbf{P}_i$ towards point $\mathbf{P}_j$ (Fig. 4a). Points are not themselves vectors because the sum of two points, $\mathbf{P}_i + \mathbf{P}_j$ is not reference frame independent (Fig 4b). However, we can define a weighted sum, here denoted by the caret operator ^

$$\mathbf{P}^\wedge = s\mathbf{P}_i + t\mathbf{P}_j \tag{7}$$

where we have introduced the dependent parameter s,

$$s = 1 - t \tag{8}$$

simply because $s\mathbf{P}_i$ looks neater than $(1\text{-}t)\mathbf{P}_i$. Such a weighted sum is reference frame indifferent and is called a 'barycentric' combination because it would give the center of gravity if s and t were physical weights located at the end points. As

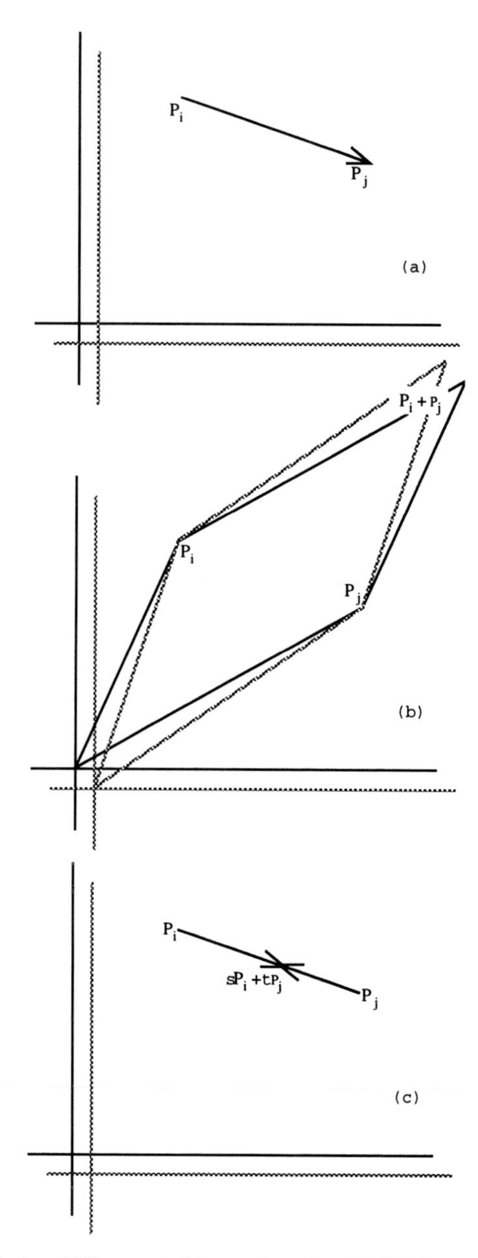

Fig. 4. a) Vector difference of two points *i* and *j*. b) Simple sum showing reference frame dependence. c) Barycentric combination is reference frame indifferent. Solid, dashed axes are alternative reference frames.

the barycentric coordinates vary, eqn. (7) yields the set of points on the straight line through $\mathbf{P}_i$ and $\mathbf{P}_j$. One may say that a point $s = 0{\cdot}3$; $t = 0{\cdot}7$ on the line is composed of 30% $\mathbf{P}_i$ and 70% $\mathbf{P}_j$. However, the point $s = 0{\cdot}3$; $t = 0{\cdot}4$ does not lie on the line because s and t do not sum to 1. We express s and t as decimals rather than percentages because they are free to adopt negative values subject to eqn. (8). Where s and t are both positive, the interpolated point lies between $\mathbf{P}_i$ and $\mathbf{P}_j$; otherwise it is externally colinear.

A member of a set of geometric objects such as points, lines, or curves may be denoted $\mathbf{P}_i^{\wedge j}$, spoken "p-i-hat-j", where the superscript denotes polynomial degree (0 for a point, 1 for a line, 2 for a quadratic, 3 for a cubic, *etc.*) and the subscript is the object number. Thus $\mathbf{P}_2^{\wedge 0}$ means point number 2 whilst $\mathbf{P}_0^{\wedge 1}$ means line number 0. This hierarchy of objects is the basis of the de Casteljeau algorithm outlined below (de Casteljeau 1959).

Section Construction Example

To draw a cross section through a geological structure, we need data from specific locations and a methodology for interpolating between data points. If nothing is known except for the orientation of a stratum at one location, the only option is to extrapolate along dip to the boundaries of the section. Similarly, a trivial solution ensues if two locations of the same lithology are known but the dips are not; the best estimate of bedding is a straight line join defined by eqn. (7),

$$\mathbf{P}_0^{\wedge 1}(t) = s\mathbf{P}_0^{\wedge 0} + t\mathbf{P}_1^{\wedge 0} \tag{9}$$

(Fig. 5a). To proceed further, we need to know the two bedding dips at the two locations; let us say 30°W at $\mathbf{P}_0^{\wedge 0}$ and 20°E at $\mathbf{P}_1^{\wedge 0}$. Correcting for the relative height in the stratigraphic column if necessary, (see point labelled Q) we extrapolate the dip lines to an intersection point (Fig. 5b). Relabeling the three points $\mathbf{P}_0^{\wedge 0}$, $\mathbf{P}_1^{\wedge 0}$, and $\mathbf{P}_2^{\wedge 0}$ as in Fig. 5b so that $\mathbf{P}_1^{\wedge 0}$ is in the middle, the linear interpolation of eqn. (7) may be applied twice,

$$\mathbf{P}_0^{\wedge 1}(t) = s\mathbf{P}_0^{\wedge 0} + t\mathbf{P}_1^{\wedge 0} \tag{10}$$

$$\mathbf{P}_1^{\wedge 1}(t) = s\mathbf{P}_1^{\wedge 0} + t\mathbf{P}_2^{\wedge 0} \tag{11}$$

This gives the well-known kink fold model for the cross section (Suppe 1983). A more smoothly curved interpolation is obtained, however, if eqn. (9) is now applied to the outcome of eqns. (10) and (11),

$$\mathbf{P}_0^{\wedge 2}(t) = s\mathbf{P}_0^{\wedge 1} + t\mathbf{P}_1^{\wedge 1} \tag{12}$$

$$= s^2\mathbf{P}_0^{\wedge 0} + st\mathbf{P}_1^{\wedge 0} + t^2\mathbf{P}_2^{\wedge 0} \tag{13}$$

(Fig. 5c). When eqn. (13) is solved for various ratios $s{:}t$ it gives a quadratic curve $\mathbf{P}_0^{\wedge 2}$ which passes through the control points $\mathbf{P}_0^{\wedge 0}$ and $\mathbf{P}_2^{\wedge 0}$ with tangency determined by $\Delta\mathbf{P}_{01}$ and $\Delta\mathbf{P}_{21}$ (Fig. 5d). In practice, a personal computer would be used to draw the curve but a manual solution is not difficult; for $t = 0{\cdot}75$, for

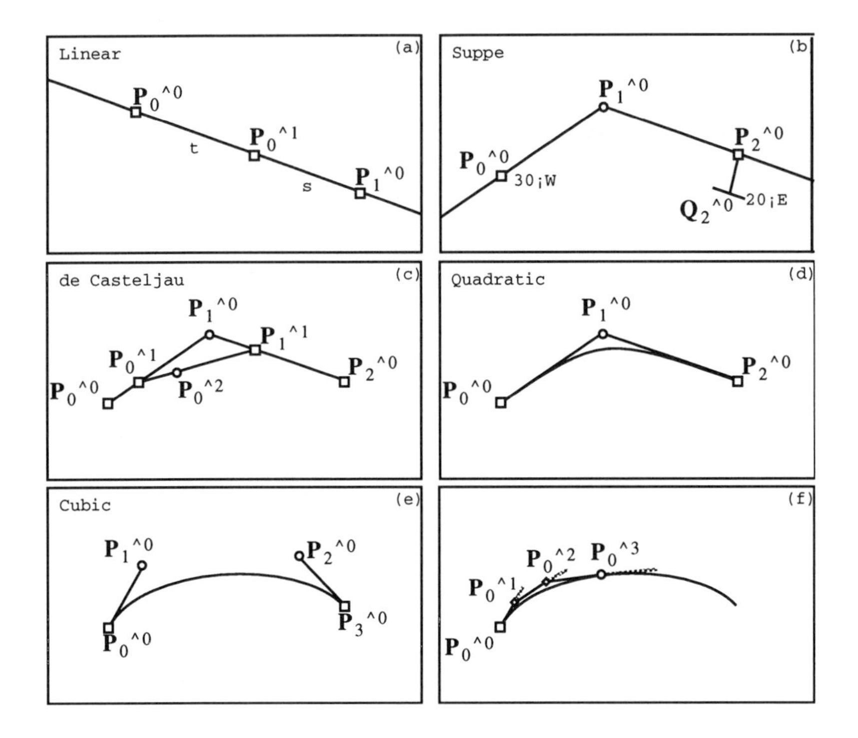

Fig. 5. a) Interpolation of bedding at locations $\mathbf{P}_0^{\wedge 0}$ and $\mathbf{P}_1^{\wedge 0}$. b) Kink interpolation of dips at $\mathbf{P}_0^{\wedge 0}$ and $\mathbf{Q}_2^{\wedge 0}$. $\mathbf{P}_2^{\wedge 0}$ is on same stratum as $\mathbf{P}_0^{\wedge 0}$. $\mathbf{P}_1^{\wedge 0}$ is intersection of dip lines. c) Interpolation of $\mathbf{P}_1^{\wedge 2}$ between $\mathbf{P}_0^{\wedge 1}$ and $\mathbf{P}_1^{\wedge 1}$. d) Quadratic Bézier. e) Cubic Bézier. f) Recursive application of (c).

example, one must set off three quarters of the distance from $\mathbf{P}_0^{\wedge 0}$ to $\mathbf{P}_1^{\wedge 0}$ and from $\mathbf{P}_1^{\wedge 0}$ to $\mathbf{P}_2^{\wedge 0}$, then divide the distance between the points thus obtained in the same ratio. Repeating for a range of t values other than 0·75 defines the full quadratic curve.

Quadratic interpolation is an advance on the linear but there are cases where no real solution is possible (isoclinal folds, for example). To interpolate a curve through any two dips, a box-fold or fold pair is required, thus necessitating a cubic expression. The two end points are relabelled $\mathbf{P}_0^{\wedge 0}$ and $\mathbf{P}_3^{\wedge 0}$ and now two intermediate points $\mathbf{P}_1^{\wedge 0}$ and $\mathbf{P}_2^{\wedge 0}$ define independent tangent vectors $\Delta\mathbf{P}_{01}$ and $\Delta\mathbf{P}_{32}$ (Fig. 5e). Any point $\mathbf{P}_0^{\wedge 3}(t)$ lies on the cubic curve if it satisfies

$$\mathbf{P}_0^{\wedge 3}(t) = s\mathbf{P}_0^{\wedge 2} + t\mathbf{P}_1^{\wedge 2} \tag{14a}$$

$$= s(s^2\mathbf{P}_0^{\wedge 1}+st\mathbf{P}_1^{\wedge 1}+t^2\mathbf{P}_2^{\wedge 1}) + t(s^2\mathbf{P}_1^{\wedge 1}+st\mathbf{P}_2^{\wedge 1}+t^2\mathbf{P}_3^{\wedge 1}) \tag{14b}$$

$$= s^3\mathbf{P}_0^{\wedge 0}+3s^2t\mathbf{P}_1^{\wedge 0}+3st^2\mathbf{P}_2^{\wedge 0}+t^3\mathbf{P}_3^{\wedge 0} \tag{14c}$$

or in matrix format,

$$\mathbf{P}_0^{\wedge 3}(t) = \left(s^3 \; 3s^2t \; 3st^2 \; t^3\right) \begin{bmatrix} \mathbf{P}_0^{\wedge 0} \\ \mathbf{P}_1^{\wedge 0} \\ \mathbf{P}_2^{\wedge 0} \\ \mathbf{P}_3^{\wedge 0} \end{bmatrix} \tag{15}$$

A point on the cubic curve is thus obtained by recursive application of the linear interpolation procedure (Fig. 5f).

In effect, we are attributing significance to the lengths and not just the orientations of the dip lines at the control locations even though directions and not lengths are given by dip measurements in the field. The fact that tangent lengths are unconstrained by data should be viewed, not as a barrier, but rather as a design advantage. Tangents may be varied interactively in an effort to discover a design that accommodates mechanical aspects of the structure in addition to raw field data. In contrast, a Busk or Suppe-style construction leaves no room for modification and may become mechanically unreasonable or internally inconsistent. This is most pronounced where small surface dip variations are extrapolated to depth to form large synclinal structures for which there is no mechanical argument.

Higher degree Bézier curves require additional intermediate control points. If $\mathbf{P}_0^{\wedge 0}$ and $\mathbf{P}_n^{\wedge 0}$ are the end points, an nth degree Bézier curve is written

$$P_i^{\wedge r} = P_i^{\wedge r-1} + P_{i+1}^{\wedge r-1} \tag{16}$$

where $r = [1,\ldots,n]$ and $i = [0,\ldots,n-r]$. This is best understood using a Pascal Triangle of coefficients (Fig. 6). Equation (16) simply states that each coefficient in the triangle is a barycentric combination of the coefficients under it in the triangle.

Fault Displacement Example

As illustrated in Fig. 3, a Bézier curve may be subdivided at any point. If $\mathbf{P}_0^{\wedge 3}(t)$ is the point of division, then the equations of the two newly created Bézier curves are given by the coefficients on the sides of the Pascal triangle in Fig. 6 (Farin 1990). Thus one sub-curve has control points $\mathbf{P}_0^{\wedge 0}, \mathbf{P}_0^{\wedge 1}, \mathbf{P}_0^{\wedge 2}, \mathbf{P}_0^{\wedge 3}$, and the other has control points $\mathbf{P}_0^{\wedge 3}, \mathbf{P}_1^{\wedge 2}, \mathbf{P}_2^{\wedge 1}, \mathbf{P}_3^{\wedge 0}$. We can now apply this theory to the modeling of beds and faults in cross section.

In Fig. 7a, a horizontal stratum is represented by the Bézier curve $\mathbf{B}_0^{\wedge 3}$. Its control points are colinear and equally spaced in this special case. A listric normal fault (Fig. 7b) is represented by the Bézier curve $\mathbf{F}_0^{\wedge 3}$. Its control points are arranged orthogonally so that fault dips range from 90° to 0° (other fault shapes may be chosen provided they are sufficiently simple to be represented by a single cubic polynomial). The point of intersection of the bed and fault simultaneously

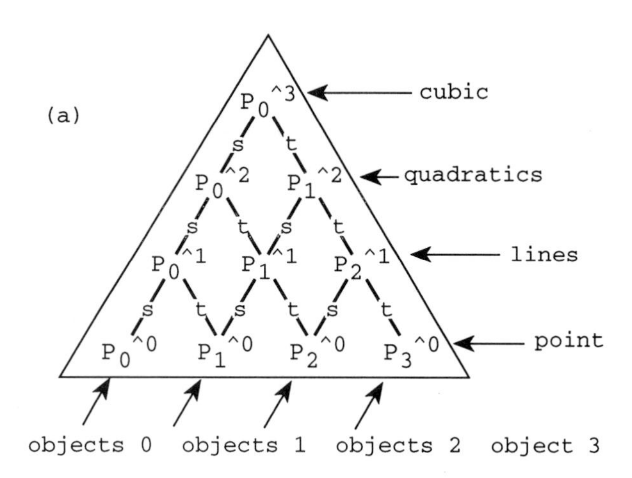

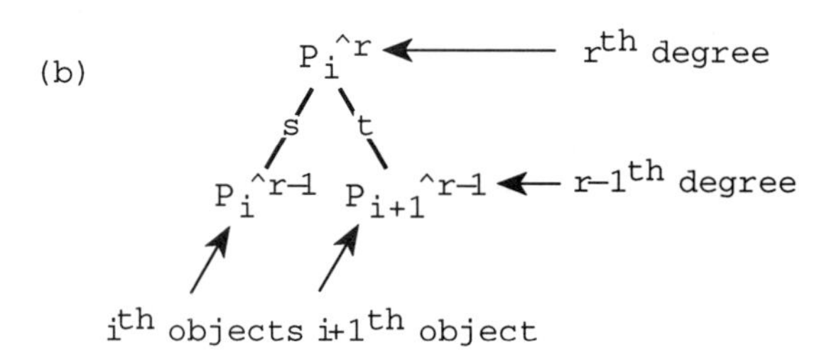

Fig. 6. Pascal triangle of Bézier coefficients. Each is s times element below left plus t times element below right. a) Cubic, b) Arbitrary degree.

satisfies the two Bézier curves. If we denote the fault's barycentric coordinates (u,v) where $u + v = 1$, to distinguish them from the barycentric coordinates of the bed, (s,t), then the intersection point is given by

$$\mathbf{B}_0{}^{\wedge 3}(t) = \mathbf{F}_0{}^{\wedge 3}(v). \tag{17}$$

We may divide the bed into two sub-Bézier curves at $\mathbf{B}_0{}^{\wedge 3}$, as in Fig. 7c. and displace the hanging wall section along the fault using

$$\mathbf{B}_0{}^{\wedge 3}(t) = \mathbf{F}_0{}^{\wedge 3}(v+\delta). \tag{18}$$

Note that δ is a pure number such as 0·15, representing the change in the fault's barycentric coordinates from, say, (0·8,0·2) in Fig. 7b to, say, (0·75,0·25) in Fig. 7c. δ is not a linear measure of displacement. To account for the hanging wall deformation which must accompany displacement if material gaps and

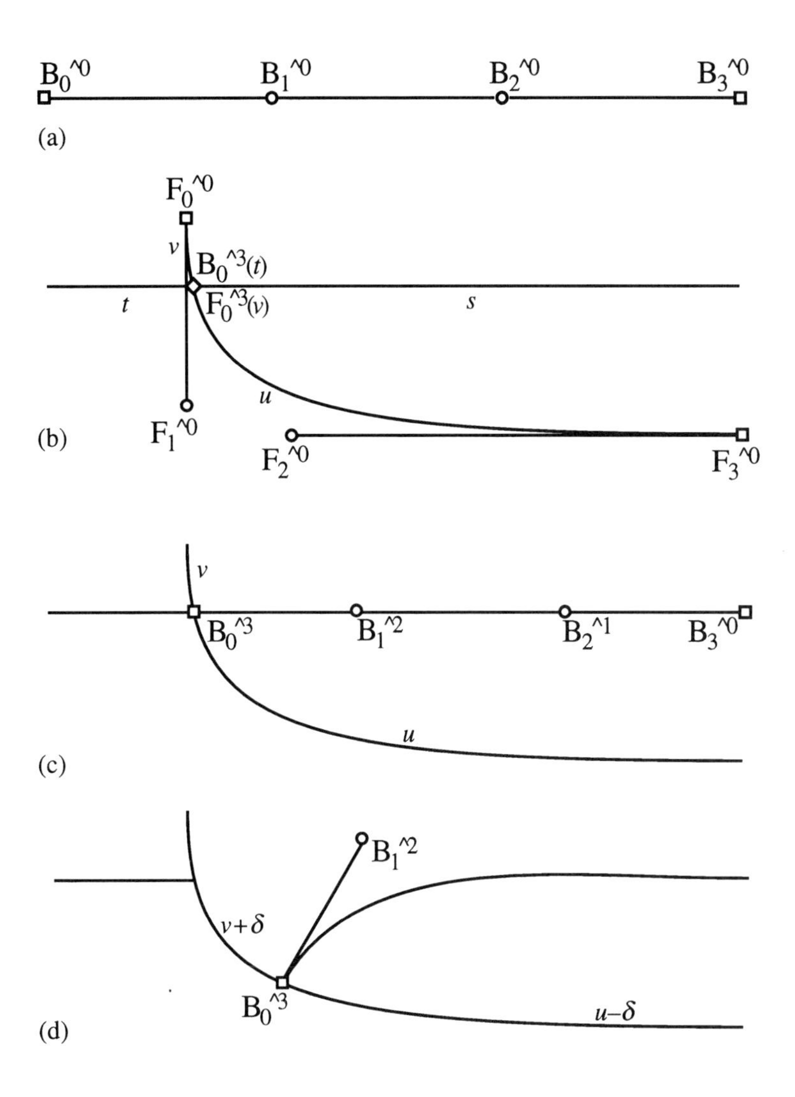

Fig. 7. Use of Bézier curves to model a half graben structure. a) Points $\mathbf{B}_j^{\wedge 0}$ are control points of bedding curve. b) Points $\mathbf{F}_j^{\wedge 0}$ control fault trace which intersects bedding at barycentric coordinates (u,v) on fault and (s,t) on bed. c) Sub-division of bedding trace into two Bézier curves. d) Displacement of $\mathbf{B}_0^{\wedge 3}$ by parametric amount δ along fault. See text.

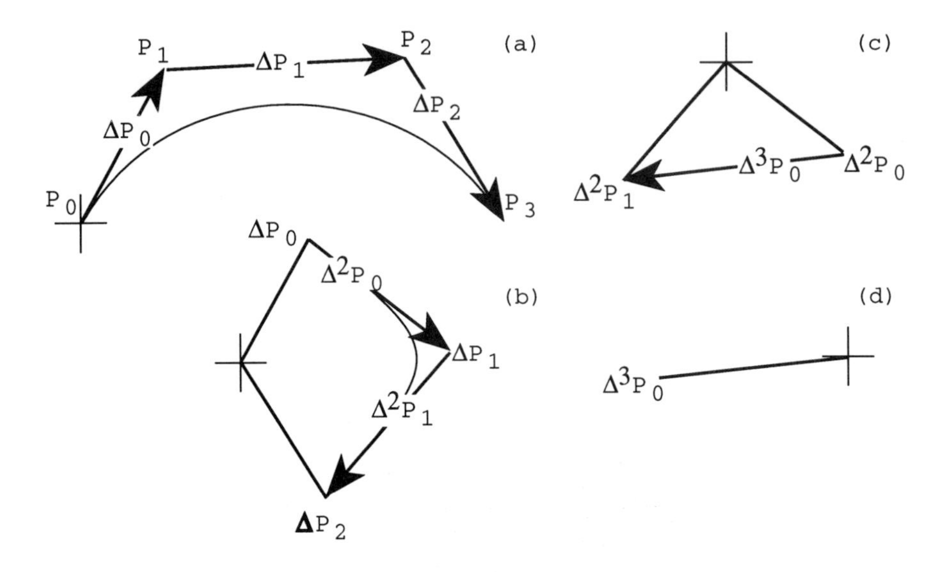

Fig. 8. Derivatives of a Bézier curve. a) Cubic curve with vector differences of control points. b) Vectors from (a) translated to common origin (cross) define quadratic derivative curve. c) Vector differences from (b) translated to a common origin define linear second derivative. d) Third derivative is a point.

overlaps are to be avoided it is necessary to decide on a displacement algorithm for the other control points. In Fig. 7c, I rigidly rotated the left control tangent and rigidly displaced the right control tangent horizontally so as to conserve heave. The deformed hanging wall bed segment is regenerated from its displaced control points. Alternative approaches would be to apply vertical or inclined simple shear to the control tangent or to conserve bedlength.

Length Measurement and Section Balancing

To control the amount of fault displacement in the above example, or to use Bézier curves in other structural applications where lengths need to be controlled, we need a formula for the arclength of a Bézier curve. The arclength L of an arc of the curve $\mathbf{P}(t)$ between $t = a$ and $t = b$ is given by

$$L(a,b) = \int_a^b \left|\dot{\mathrm{P}}(t)\right| dt. \tag{19}$$

Solving in two dimensions gives

$$L(a,b) = \frac{1}{3}\left[\frac{\left[\dot{x}^2(t)+\dot{y}^2(t)\right]^{3/2}}{\dot{x}(t)\ddot{x}(t)+\dot{y}(t)\ddot{y}(t)} \right]_a^b \tag{20}$$

where the primes denote derivatives. Substituting $(v, v+\delta)$ for (a,b) gives the fault displacement in Fig. 7d, for example. To use eqns. (19) and (20), we need expressions for the derivatives of a Bézier curve. To avoid clutter, we omit the ^0 notation and simply label the cubic control points $\mathbf{P}_0$, $\mathbf{P}_1$, $\mathbf{P}_2$, and $\mathbf{P}_3$ (Fig. 8a). Then the first derivative is a quadratic curve with control points $\Delta\mathbf{P}_j$ given by

$$\Delta\mathbf{P}_0 = \mathbf{P}_1 - \mathbf{P}_0 \tag{21}$$
$$\Delta\mathbf{P}_1 = \mathbf{P}_2 - \mathbf{P}_1 \tag{22}$$
$$\Delta\mathbf{P}_2 = \mathbf{P}_3 - \mathbf{P}_2. \tag{23}$$

(Fig. 8b). Similarly, the second derivative is a line with control points $\Delta^2\mathbf{P}_j$,

$$\Delta^2\mathbf{P}_0 = \Delta\mathbf{P}_1 - \Delta\mathbf{P}_0 \tag{24}$$
$$\Delta^2\mathbf{P}_1 = \Delta\mathbf{P}_2 - \Delta\mathbf{P}_1 \tag{25}$$

(Fig. 8c). Although it is not needed in the length calculation, the third derivative is a point with one control point (Fig. 8d):

$$\Delta^3\mathbf{P}_0 = \Delta^2\mathbf{P}_1 - \Delta^2\mathbf{P}_0. \tag{26}$$

Thus, to calculate the full length of a Bézier curve, set $a = 0$ and $b = 1$ in eqn. (20):

$$L(0,1) = \sqrt{\left(\Delta x_0 + 2t\Delta^2 x_0 + t^2\Delta^3 x_0\right)^2 + \left(\Delta y_0 + 2t\Delta^2 y_0 + t^2\Delta^3 y_0\right)^2} \tag{27}$$

An obvious application of the length formula is in line-length balancing of cross sections. Figure 9a is an illustration of a bed **B** cut by a fault **F** in the deformed state (primed) and the corresponding undeformed state (unprimed). For simplicity, the footwall bed segment is fixed (**B1'** = **B1**), however the hanging wall segment **B2** has been folded and displaced to **B2'**. The interpretation illustrated is purposefully unbalanced as the shaded section **B3'** represents excess bed length in the deformed state compared with the undeformed. A microcomputer can be programmed to calculate the Bézier curve's length in real time. The user must manipulate the curve's handles until the shaded section disappears (Fig. 9b is closer to this goal). However, if the Bézier Curve is shortened too much, a segment of excess bed length **B3** appears in the undeformed state (Fig. 9c). Thus by successive over- and underestimations, a balanced interpretation may be achieved.

Given routines to subdivide a curve and to measure curve length, other applications become possible. For example, one could allow a short Bézier curve to "slip" along a longer curve by recursive subdivision and length calculation.

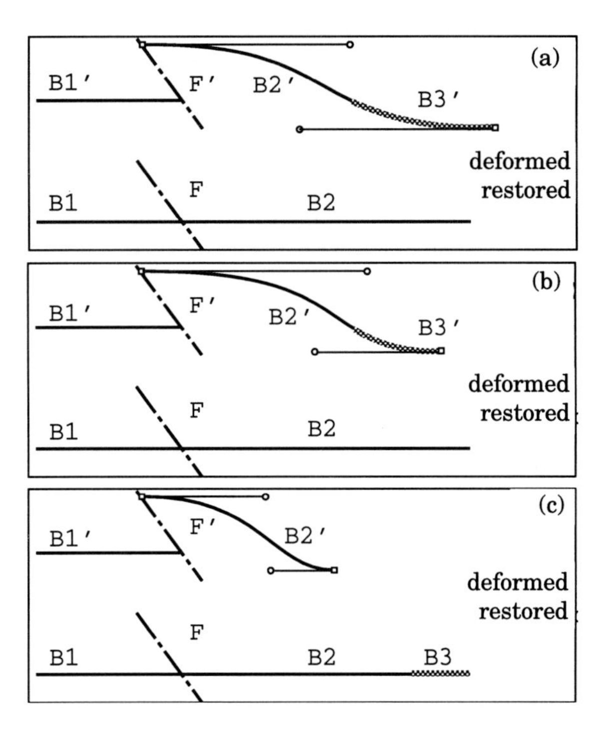

Fig. 9. Use of Bézier length calculation in section balancing. Deformed bed segments primed, restored unprimed. a) Large excess bedlength **B3'** in deformed state. b) Reduced excess **B3'**. c) **B2'** is now too short so excess bedlength occurs in restored section, **B3**.

Bézier Functions and Card Deck Simulations

Thus far, we have used the parametric form of Bézier curve which can handle non-functional curves, *i.e.*, ones that are not single-valued with respect to either coordinate axis. However, we can constrain a Bézier curve to the functional form by modifying eqns. (4),

$$x = x(t) \qquad (4a)$$
$$y = t \qquad (4d)$$

This restriction is illustrated by a Macintosh application called *CarDec* (a play on the names of the authors, Carol Simpson and Declan De Paor). As illustrated in Fig. 10, *CarDec* simulates a deck of filing cards or old computer cards undergoing simple shear. Standard Macintosh PICT images, such as the oolites of Fig. 11, may be imported and drawn on the deck. A pair of Bézier control tangents determines the relative horizontal offsets of individual cards but no vertical offset

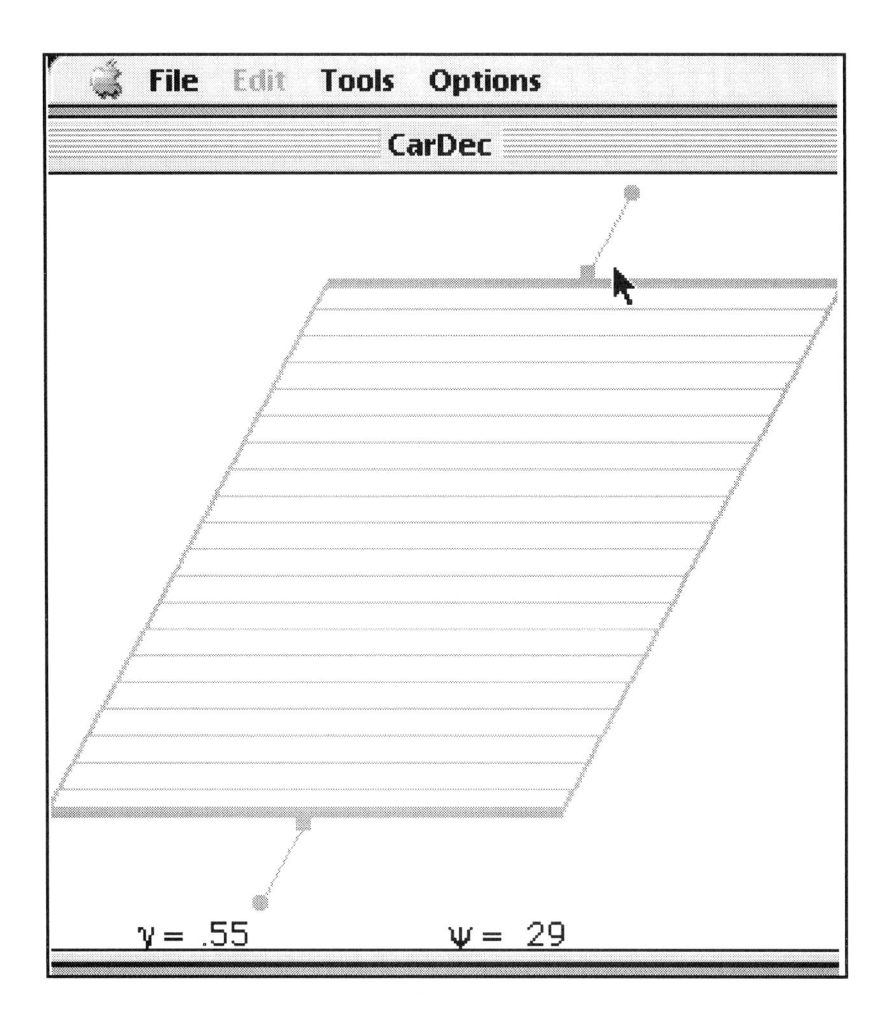

Fig. 10. User interface of the *CarDec* application. Bézier handles control shear. γ is shear strain, the tangent of the angular shear ψ.

is permitted because of eqn. (4d). In Fig. 11, the circular control point at the top of the deck has just been dragged, resulting in a heterogeneous strain state in the upper part of the card deck.

Bézier Curves and Surface Patches

A Bézier patch is a two dimensional array of Bézier curves which define a surface in two or three dimensions. A patch is described by two pairs of barycentric coordinates,

$$s + t = 1 \tag{8}$$

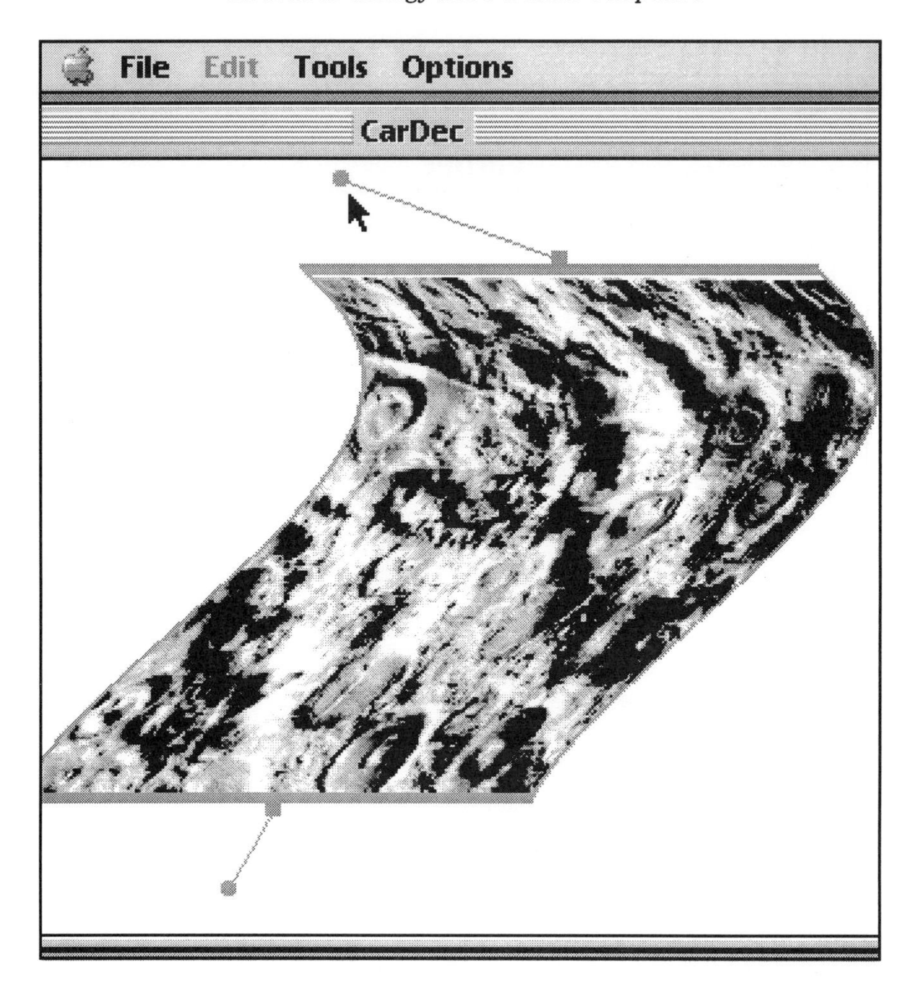

Fig. 11. Heterogeneous simple shear of a PICT image (oolites photographed by the late Ernst Cloos). Note that the top circular control point has been displaced to the left relative to Fig. 10.

$$u + v = 1, \tag{27}$$

and a double caret symbol ^^ implies interpolation to the same degree in both dimensions. The trivial case of a zeroth degree patch, $\mathbf{P}_{00}{}^{\wedge\wedge 0}$, defines a single point. A first degree patch is controlled by four points as in Fig. 12, while nine and sixteen control points are required to control bi-quadratic and bi-cubic patches, respectively (Fig. 13 a,b). More general approaches are possible; for example, a cylindrical fold might be modeled using one linear and one cubic degee $\mathbf{P}_{ij}{}^{\wedge 1 \wedge 3}$.

A point lies on the surface patch if it satisfies a two dimensional extrapolation of eqn. (16),

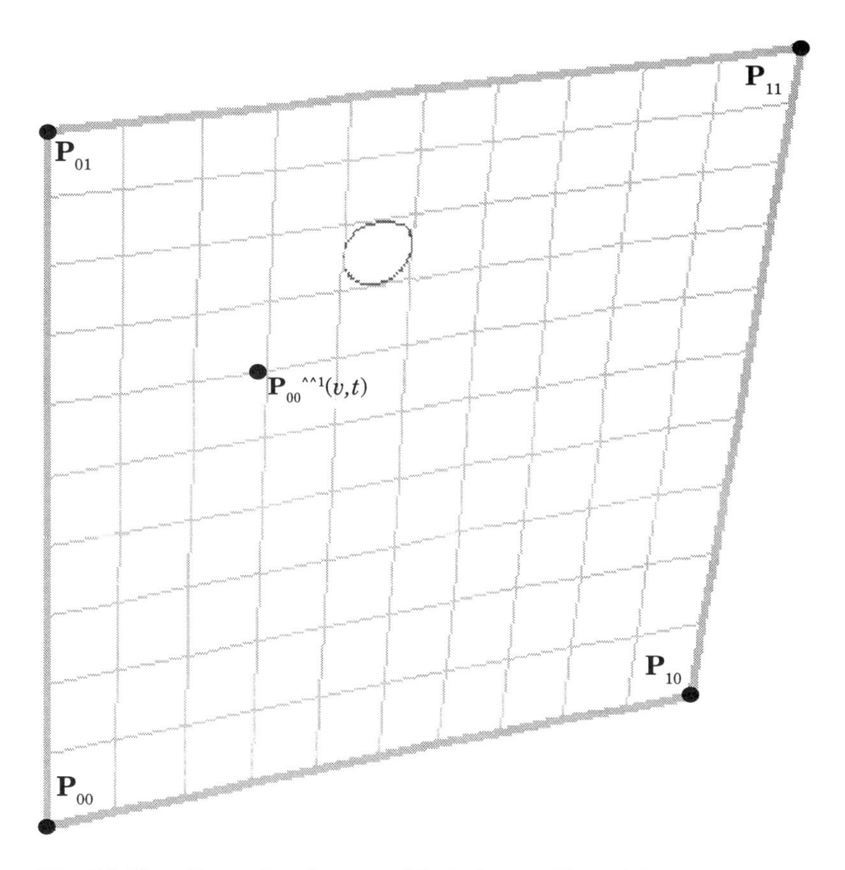

Fig. 12. Two dimensional array of 2nd. degree (linear) Bézier curves controlled by four points $\mathbf{P}_{00}$, $\mathbf{P}_{01}$, $\mathbf{P}_{10}$, $\mathbf{P}_{11}$ ($^{\wedge\wedge 0}$ omitted for simplicity). $\mathbf{P}_{00}{}^{\wedge\wedge 1}(v,t)$ is an arbitrary point on the surface.

$$\mathbf{P}_{ij}^{\wedge\wedge r}(v,t) = \begin{pmatrix} u & v \end{pmatrix} \begin{bmatrix} \mathbf{P}_{i\,j}^{\wedge\wedge r-1} & \mathbf{P}_{i\,j+1}^{\wedge\wedge r-1} \\ \mathbf{P}_{i+1\,j}^{\wedge\wedge r-1} & \mathbf{P}_{i+1\,j+1}^{\wedge\wedge r-1} \end{bmatrix} \begin{bmatrix} s \\ t \end{bmatrix} \tag{28}$$

where $r = 1, ..., n$ and $i, j = 0, ..., n-r$. For the relatively simple case of a bi-linear patch, any point is defined by substituting values of its barycentric coordinates (s,t) and (u,v),

$$\mathbf{P}_{00}^{\wedge\wedge 1}(v,t) = \begin{pmatrix} u & v \end{pmatrix} \begin{bmatrix} \mathbf{P}_{00}^{\wedge\wedge 0} & \mathbf{P}_{10}^{\wedge\wedge 0} \\ \mathbf{P}_{01}^{\wedge\wedge 0} & \mathbf{P}_{11}^{\wedge\wedge 0} \end{bmatrix} \begin{bmatrix} s \\ t \end{bmatrix} \tag{29}$$

$$= su\mathbf{P}_{00}{}^{\wedge\wedge 0} + sv\mathbf{P}_{01}{}^{\wedge\wedge 0} + tu\mathbf{P}_{10}{}^{\wedge\wedge 0} + tv\mathbf{P}_{11}{}^{\wedge\wedge 0} \tag{30}$$

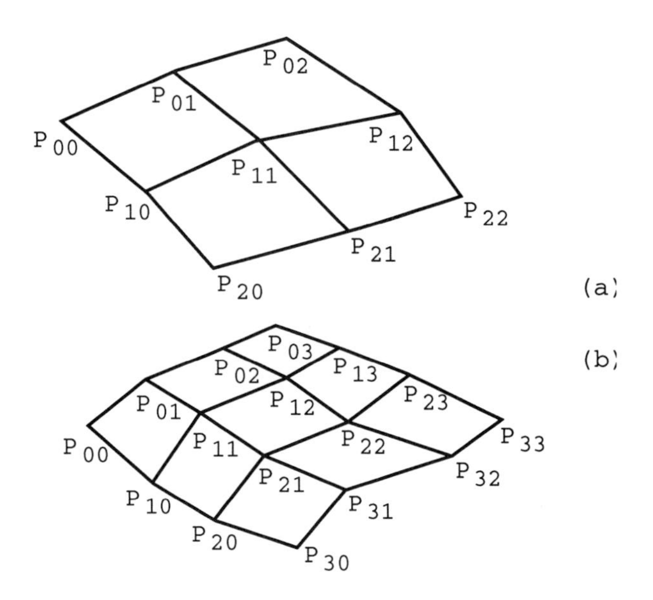

Fig. 13. Control points for a) bi-quadratic and b) bi-cubic patches. For simplicity, $^{\wedge\wedge 00}$ superscripts have been omitted from all points.

Area of Bézier Surface Patch

The area of an rth degee Bézier patch is determined by the double integral

$$\mathrm{A}(a,b,c,d) = \int_a^b \int_c^d \left| \frac{\partial}{\partial v}\left(\mathrm{P}_{ij}^{\wedge\wedge r}\right) \times \frac{\partial}{\partial t}\left(\mathrm{P}_{ij}^{\wedge\wedge r}\right) \right| dv\, dt \qquad (31)$$

where a,b and c,d are parameter ranges and the cross product gives the area of a small element. This equation must be solved by approximation.

The Strain Ovoid

In Fig. 12, a deformed unit circle has been inscribed in an initial square of the patch. Its deformed shape, here called the strain ovoid, is the heterogeneous equivalent of the strain ellipse. Its symmetry is triclinic, lacking the center and planes of symmetry of the homogeneous strain ellipse, but it does possess a centroid and semi-major axis and it may be irrotational or rotational depending on the initial orientation of that axis. The mean rotation rate for all of its radii defines the geological vorticity which may be coaxial or non-coaxial. Despite the triclinic shape of the strain ovoid, its mathematical representation in terms of barycentric coordinates is beautifully simple:

$$t = \cos\theta \tag{32a}$$

$$v = \sin\theta \tag{32b}$$

where θ is an independent parameter covering the range $[0,2\pi]$. The Cartesian coordinates of points on the ovoid are obtained by substituting (s,t) and (u,v) values from eqns (32) into eqn. (29). Equations (32) apply equally to undeformed and deformed states; they define a circle when the barycentric coordinates form a unit square grid, an ellipse when the bounding grid cell happens to form a parallelogram, and an ovoid in other cases.

Higher Order Patches

Just as linear equations fail to adequately represent structural traces, the bi-linear grid of Fig. 12 is inadequate for general surface modeling. To fit a surface to any four control tangent planes requires a third degree interpolation, *i.e.*, a bi-bi-cubic polynomial (Fig. 14). Furthermore, different surfaces may be fitted through the same boundary Bézier curves (Fig. 14a,b). Therefore four additional control tangents marked with diamond tips are added. Mathematically, they determine the twist vector of the surface at the control point, which is a function of the mixed second partial derivatives in barycentric coordinates, $d^2/dvdt$.

The *Strain Grid* Program

The bi-cubic Bézier patches in Fig. 14 are three-dimensional surfaces in general, however if all of a patch's handles are coplanar then the entire patch is confined to a plane. Internal Bézier curves then divide the surface into a grid of quadrilaterals. To illustrate the state of heterogeneous strain throughout the patch, a program called *Strain Grid* inscribes a circle in each grid element before deformation and the corresponding strain ovoid is shown after deformation (Fig. 14 c,d). Though the strain state is quite complex, it is fully defined by the handles. Thus the grid is constrained by a maximum of 16 independent points, arranged in four groups as follows,

$$\tilde{\mathbf{G}} = \left[\begin{array}{cc} \boxed{\begin{array}{cc} \mathbf{P}_{00} & \mathbf{P}_{10} \\ \mathbf{P}_{01} & \mathbf{P}_{11} \end{array}} & \boxed{\begin{array}{cc} \mathbf{P}_{20} & \mathbf{P}_{30} \\ \mathbf{P}_{21} & \mathbf{P}_{31} \end{array}} \\ \boxed{\begin{array}{cc} \mathbf{P}_{02} & \mathbf{P}_{12} \\ \mathbf{P}_{03} & \mathbf{P}_{13} \end{array}} & \boxed{\begin{array}{cc} \mathbf{P}_{22} & \mathbf{P}_{32} \\ \mathbf{P}_{23} & \mathbf{P}_{33} \end{array}} \end{array} \right] \tag{33}$$

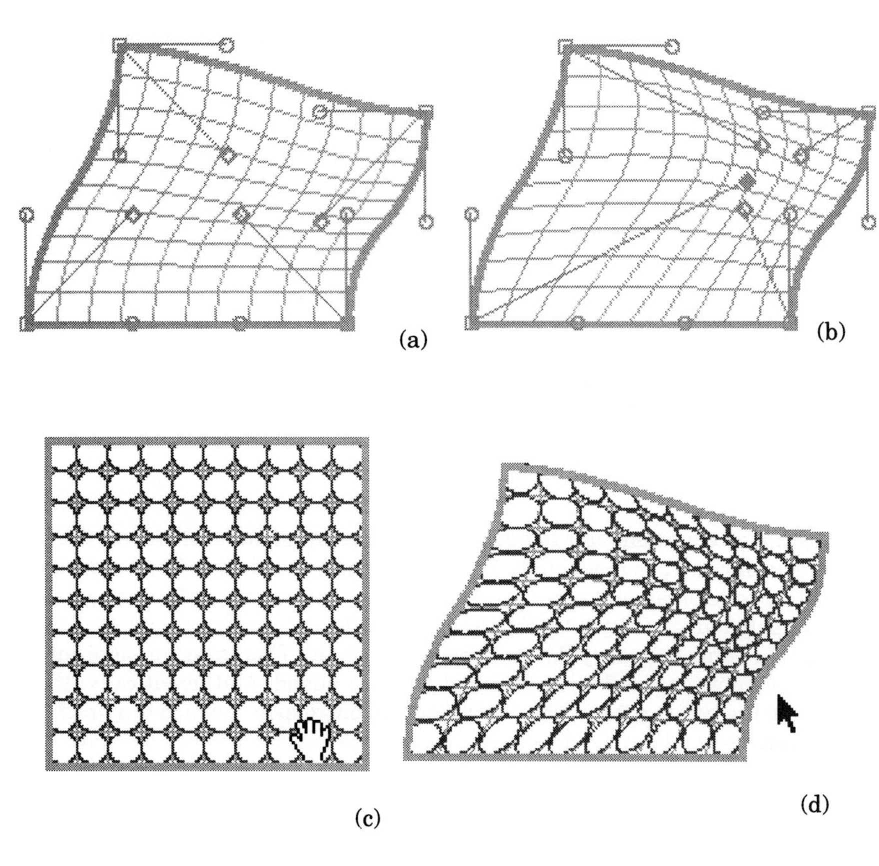

Fig. 14. a,b) Two bi-cubic patches produced by *Strain Grid*, (a) and (b), with identical outlines but different interiors controlled by diamond-headed control tangents. c) Undeformed state. d) Strain ovoids for case (b).

Assuming that the diamond-headed tangent vectors such as $\Delta\mathbf{P}_{0011}$ form the diagonals of parallelograms bounded by the circle-headed tangent vectors such as $\Delta\mathbf{P}_{0010}$ and $\Delta\mathbf{P}_{0001}$, the number of independent variables is reduced to twelve,

$$\tilde{\mathbf{G}} = \left[\begin{array}{cc} \boxed{\begin{array}{cc} \mathbf{P}_{00} & \mathbf{P}_{10} \\ \mathbf{P}_{01} & - \end{array}} & \boxed{\begin{array}{cc} \mathbf{P}_{20} & \mathbf{P}_{30} \\ - & \mathbf{P}_{31} \end{array}} \\ \\ \boxed{\begin{array}{cc} \mathbf{P}_{02} & - \\ \mathbf{P}_{03} & \mathbf{P}_{13} \end{array}} & \boxed{\begin{array}{cc} - & \mathbf{P}_{32} \\ \mathbf{P}_{23} & \mathbf{P}_{33} \end{array}} \end{array} \right] \qquad (34)$$

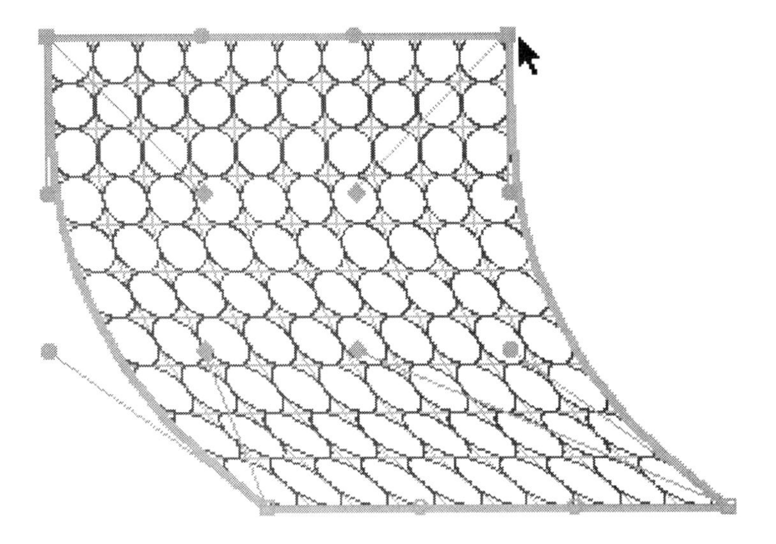

Fig. 15. Representation of strain in a thrust sheet. The bottom row of control handles were shifted to the right.

The undeformed and deformed states may be denoted by unprimed and primed control points, $\mathbf{P}_{00}$ --> $\mathbf{P'}_{00}$, *etc.*

Strain Grid and Model Structures

We can now demonstrate the power of Bézier patches in structural modeling using a number of simple examples. In each of these examples, we start with an undeformed square grid as illustrated in Fig. (14c). The circular control points divide the sides of the outer square into thirds and the diamond control points form an internal square one third the side length of the outer square. The state is represented by the unit strain grid,

$$\tilde{\mathbf{I}} = \begin{bmatrix} \boxed{\begin{matrix} \mathbf{P}_{00} & \mathbf{P}_{10} \\ \mathbf{P}_{01} & - \end{matrix}} & \boxed{\begin{matrix} \mathbf{P}_{20} & \mathbf{P}_{30} \\ - & \mathbf{P}_{31} \end{matrix}} \\ \boxed{\begin{matrix} \mathbf{P}_{02} & - \\ \mathbf{P}_{03} & \mathbf{P}_{13} \end{matrix}} & \boxed{\begin{matrix} - & \mathbf{P}_{32} \\ \mathbf{P}_{23} & \mathbf{P}_{33} \end{matrix}} \end{bmatrix} \tag{35}$$

where

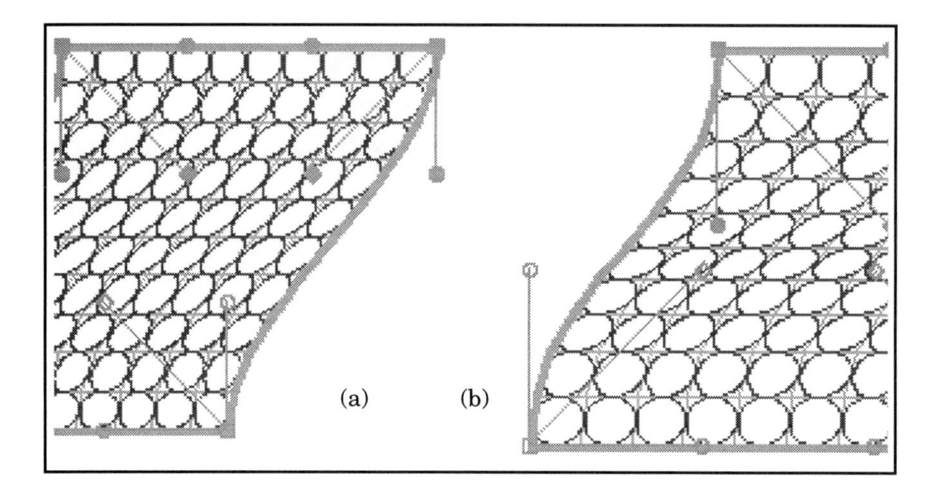

Fig. 16. Model of ductile simple shear zone. a) Two top rows of control handles were shifted to the right. b) Control points were also shifted vertically to change the shear zone's curvature.

$$\mathbf{P}_{10} = \frac{2}{3}\mathbf{P}_{00} + \frac{1}{3}\mathbf{P}_{30}$$
$$\mathbf{P}_{20} = \frac{1}{3}\mathbf{P}_{00} + \frac{2}{3}\mathbf{P}_{30} \quad (36a,b)$$

etc. In the following examples, the strain grid is described in terms of its departure from this initial state.

Thrust Sheet Example

The strain state at the base of a ductile thrust sheet is represented by starting with an undeformed square grid and shifting the bottom row of control points laterally by a fixed amount Δx

$$\tilde{\tilde{\mathbf{G}}}' = \tilde{\mathbf{I}} + \begin{bmatrix} \boxed{\begin{matrix} 0 & 0 \\ 0 & 0 \end{matrix}} & \boxed{\begin{matrix} 0 & 0 \\ 0 & 0 \end{matrix}} \\ \boxed{\begin{matrix} 0 & 0 \\ \Delta x & \Delta x \end{matrix}} & \boxed{\begin{matrix} 0 & 0 \\ \Delta x & \Delta x \end{matrix}} \end{bmatrix} \quad (37)$$

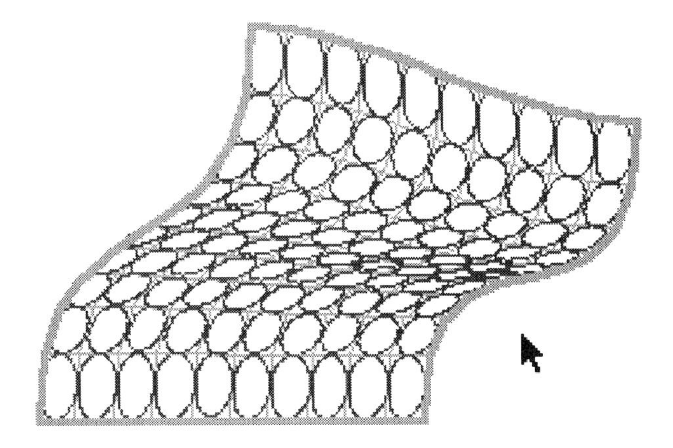

Fig. 17. General shear zone formed by independent displacement of top left and top right sets of control points.

(Fig. 15). This is equivalent to shifting all but the bottom row of control points in the opposite direction, which is closer to physical reality. The curvature of vertical Bézier curves can be modified by changing the vertical spacing of control points. This curvature is determined by the power law of the flow regime leading to the observed deformation state and by the relative proportions of pure (overburden) and simple (thrust) shear contributions.

Ductile Shear Zone Example

The strain in a zone of heterogeneous simple shear is modeled by displacing the top two rows of control points rigidly by a fixed amount Δx,

$$\tilde{\mathbf{G}}' = \tilde{\mathbf{I}} + \begin{bmatrix} \boxed{\begin{matrix} \Delta x & \Delta x \\ \Delta x & \Delta x \end{matrix}} & \boxed{\begin{matrix} \Delta x & \Delta x \\ \Delta x & \Delta x \end{matrix}} \\ \boxed{\begin{matrix} 0 & 0 \\ 0 & 0 \end{matrix}} & \boxed{\begin{matrix} 0 & 0 \\ 0 & 0 \end{matrix}} \end{bmatrix} \quad (37)$$

(Fig. 16a). The curvature of the shear zone can be adjusted by changing the vertical spacing of handles (Fig. 16b). If the top right and top left sets of handles are displaced independently, the result is a zone of convergent or divergent heterogeneous general shear (Fig. 17).

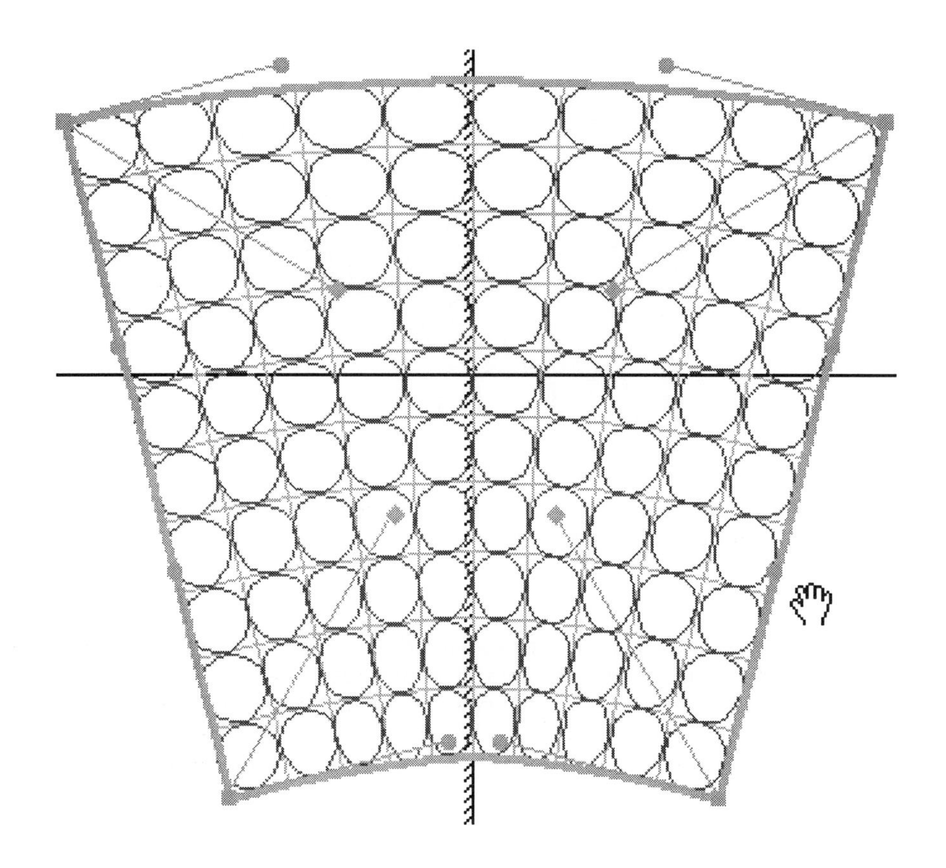

Fig. 18. Strain grid representation of neutral surface fold. Note vertical plane of symmetry.

Neutral Surface Fold

Perhaps the strongest demonstration of the strain grid's modeling power is provided by the example of a neutral surface fold (Fig. 17),

$$\widetilde{\mathbf{G}'} = \begin{bmatrix} \mathbf{R}_\theta & 0 & 0 & 0 \\ 0 & \mathbf{R}_\theta & 0 & 0 \\ 0 & 0 & -\mathbf{R}_\theta & 0 \\ 0 & 0 & 0 & -\mathbf{R}_\theta \end{bmatrix} \tilde{\mathbf{I}} \tag{38}$$

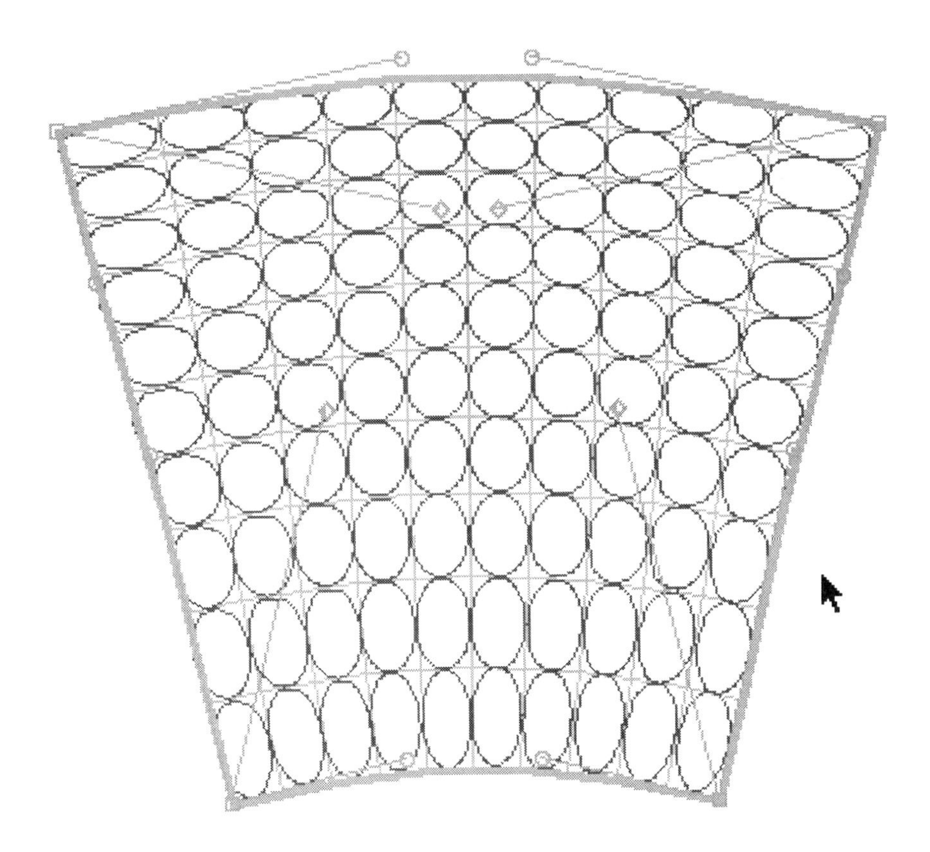

Fig. 19. Modified neutral surface. Circle-topped control tangents have been lengthed or shortened (but not rotated) to ensure that the neutral surface is in the center of the folded layer.

where

$$\mathbf{R}_\theta = \begin{bmatrix} \cos\theta & -\sin\theta \\ \sin\theta & \cos\theta \end{bmatrix} \tag{39}$$

The angle 2θ is the fold interlimb angle. In order to ensure that the neutral surface passes through the central layers of strain ovoids, it is necessary to lengthen control tangents $\Delta\mathbf{P}_{0010}$, $\Delta\mathbf{P}_{3020}$, $\Delta\mathbf{P}_{0302}$, and $\Delta\mathbf{P}_{3332}$, and to shorten $\Delta\mathbf{P}_{0001}$, $\Delta\mathbf{P}_{3031}$, $\Delta\mathbf{P}_{0313}$, and $\Delta\mathbf{P}_{3323}$ (Fig. 19).

A significant feature of the strain grid used to model these structures is its ability to accumulate successive increments of homogeneous and heterogeneous strain. For example, the grid in Fig. 20 was created by applying homogeneous simple shear to the control handles of an initial symmetric neutral surface fold in a oblique direction.

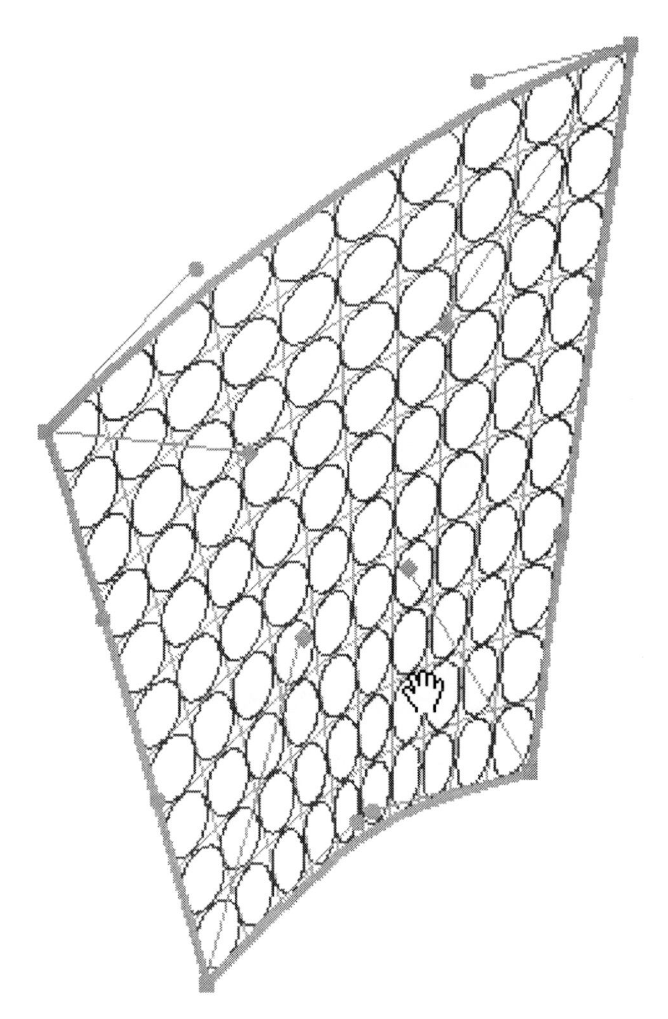

Fig. 20. Superposition of heterogeneous strain leading to neutral surface folding and subsequent homogeneous inclined simple shear.

Discussion

The computer programs illustrated in this paper, *CarDec* and *Strain Grid*, are intended as simple teaching aids. Nevertheless, they demonstrate the great potential of Bézier curves and patches for research applications in the field of structural modeling. Just as individual Bézier curves may be concatenated in daisy-chain fashion, so Bézier patches may be quilted together to model complex structures with complexly deformed geometries. It is also possible to extrapolate to three barycentric dimensions in order to model the deformation of a region and it is possible to use functional Bézier curves to control the time dimension. The *Strain Grid* application permits construction of movies by linear interpolation in the time dimension but a cubic interpolation in this dimension would result in a more realistic model of structural evolution, incorporating periods of acceleration and deceleration of displacement rates. Further research is planned in the development of Bézier modeling tools for map compilation and cross section construction.

Acknowledgements

This work was supported by NSF grant number EAR-9304879 and by a grant from the Mobil Corporation.

References

Bézier, P. 1966. Definition numérique des courbes et surfaces. *I. Automatisme* **11**: 625-632.

Bézier, P. 1967. Definition numérique des courbes et surfaces. *II. Automatisme* **12**: 17-21.

de Casteljau, P. 1959. Outillages méthodes calcul. *Technical Report A.*, Citroën, Paris.

de Casteljau, P. 1963. Courbes et surfaces à poles. *Technical Report A.*, Citroën, Paris.

Farin, G. 1988. *Curves and Surfaces for Computer Aided Geometric Design.* Academic Press.

Foley, J. D., van Dam, A., Feiner, S. K. & Hughes, J. F. 1990. *Computer Graphics: Principles and Practice* (2nd. ed). Addison Wesley Press.

Suppe, J. 1983. Geometry and kinematics of fault bend folding. *Amer. J. Sci.* **283**: 684-721.

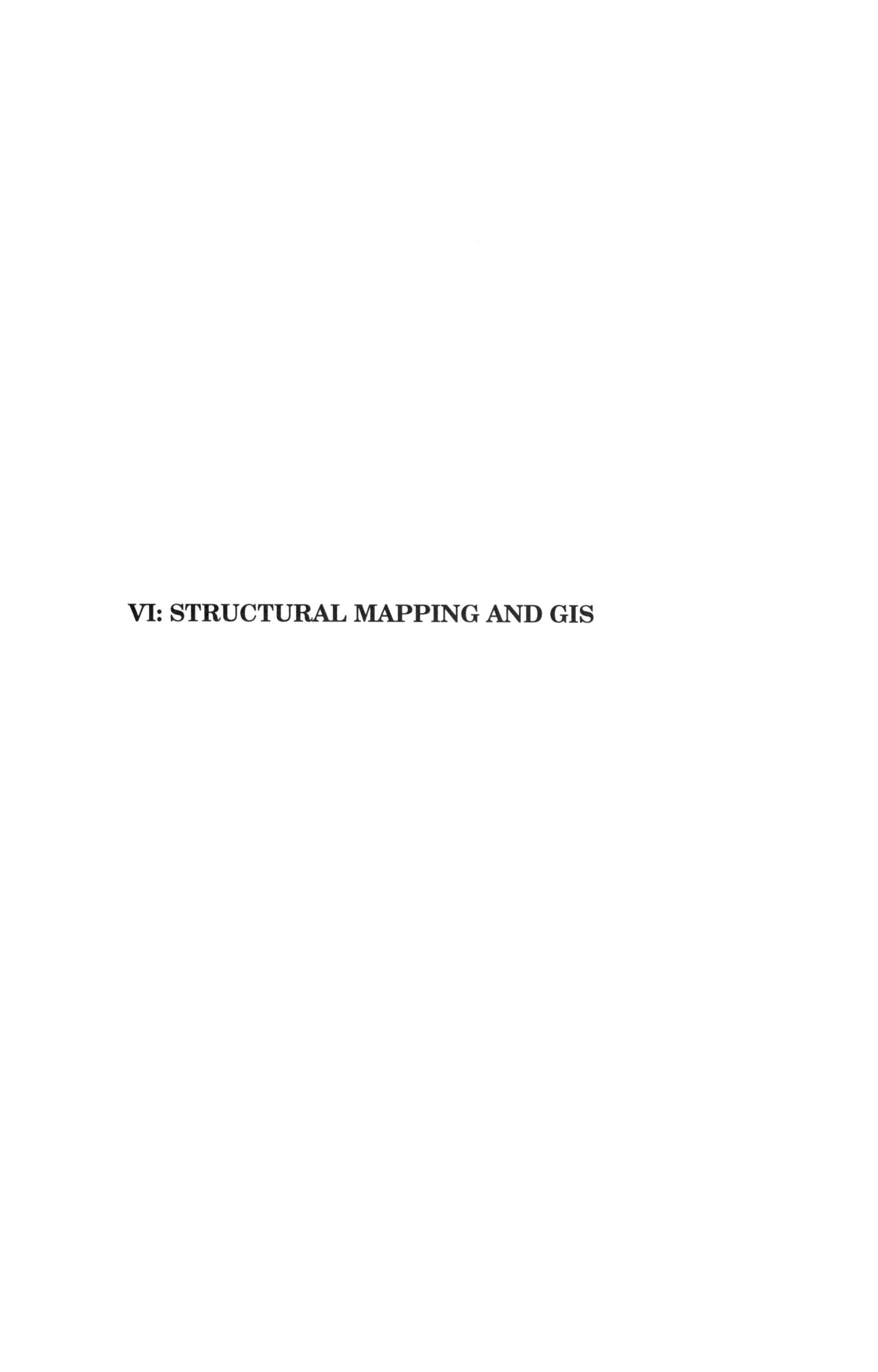

VI: STRUCTURAL MAPPING AND GIS

Digital Terrain Models and the Visualization of Structural Geology

Robert N. Spark* and Paul F. Williams
Department of Geology, University of New Brunswick,
Fredericton NB, Canada E3B 5A3. pfw@unb.ca

*Present address: Institute of Geological and Nuclear Sciences, Private Bag 1930, Dunedin, New Zealand.

Abstract– The visualization of three dimensional structure from two dimensional maps is greatly complicated by topographic relief. Digital Terrain Models (DTM), within a Geographic Information System (GIS) can significantly enhance the visual impact of a complexly deformed terrain. A DTM allows the viewing of landscape surfaces, from any vantage point, with geology, satellite imagery, or any two-dimensional map image draped over the surface. Data point representation as a Triangulated Irregular Network (TIN), involves the modelling of a surface as a sheet of triangular facets having data points as their vertices; thereby, honouring variable topography without data redundancy since the triangular mesh adapts to varying densities of data. All critical topographic features are thereby retained. The demonstrated data set is that of the Pingston Fold within the Monashee Mountains of southeastern British Columbia, Canada. The Pingston Fold is situated in the Thor-Odin structural culmination of the Monashee Complex, within the Omineca Belt of the Canadian Cordillera. The Pingston Fold is a major complex structural feature outcropping in the southern half of the Thor-Odin culmination. The structural interpretation is based on an orthographic projection of the Digital Terrain Model. It was produced by first dividing the map into fold domains defined on the basis of homogeneity of the orientations of fold axes where reasonably cylindrical (*i.e.,* where poles to the folded gneissosity plot as a great circle girdle). Once structural domains have been outlined on a map, all data can be projected down plunge and the structure joined at domain boundaries. This differs from conventional manual plotting in that all data points are projected from their position on the modelled topographic surface, rather than from a planar surface of a block. This makes a significant difference in such mountainous terrains. The present surface expression and the three-dimensional geometry of the Pingston Fold can be explained by fold interference and the structure is illustrated using a Digital Terrain Model.

Introduction

Three-dimensional visualization is crucial to the study of structural geology. From a two-dimensional map, the visualization of structural data can be difficult enough without the complication of topographic relief. By utilizing a Digital Terrain Model (DTM), one has the ability to significantly enhance the visual impact of a complexly deformed terrain. Landscape surfaces can be viewed from any vantage point and combined with geology, satellite imagery, or any other two-dimensional map image draped over the topographical image. The purpose of this paper is to introduce DTM theory and to demonstrate an application of DTM in structural geology using the *CARIS* Geographic Information System (GIS), developed by Universal Systems Limited of Fredericton, New Brunswick, Canada.

The structure chosen to demonstrate the technique is the Pingston Fold which occurs in the Thor-Odin culmination at the southern end of the Monashee Complex in the southeastern Cordillera of British Columbia. The Monashee Complex is a basement dome in the Omineca Belt. The latter is a structural and metamorphic zone separating oceanic/island-arc terranes from the western margin of the North American paleocontinent. The terranes were accreted during early Jurassic to mid-Cretaceous times. The culmination has been interpreted previously as a diapiric mantled gneiss dome (Reesor & Moore 1971), as a tectonically denuded antiformal crustal scale duplex (Brown & Journeay 1987) and as a northerly-directed thrust duplex and fold interference structure (Duncan 1982). Thor-Odin comprises early Proterozoic basement core gneisses of displaced North American cratonic origin, unconformably overlain by late Proterozoic to early Cambrian platformal metasedimentary cover gneisses. Both are complexly folded. The Pingston Fold, mapped as an antiformal syncline (Reesor 1970), is a complex regional structure occupying the southern half of the Thor-Odin culmination (Fig. 1). In this paper we present a three-dimensional representation of the Pingston Fold in the form of an orthographic block diagram derived from a DTM.

Digital Terrain Models

The development of digital terrain models chiefly occurred at the Massachusetts Institute of Technology during the 1950's (*e.g.*, Miller & Laflamme 1958). A continuous ground surface was statistically represented by a set of gridded $\{x,y,z\}$ coordinate datasets. A third order polynomial curve was calculated using cross-sectional height data. This allowed the interpolation of values at any x-y location along the cross-section.

Since the inception of DTMs, advances in computer automation have encouraged the rapid numerical representation of the Earth's surface. Numerous geoscience applications involving both gridded and non-gridded data sets have evolved during this time. The term digital elevation model (DEM) refers to a model of a featureless surface that contains only $\{x,y,z\}$ coordinates and no accompanying attributes. It can represent any continually varying surface (*e.g.*, topography, geophysical anomalies, *etc.*). A digital terrain model (DTM) is specifically a model of topography and usually has additional attributes (*e.g.*, drainage patterns, geological horizons, or symbols).

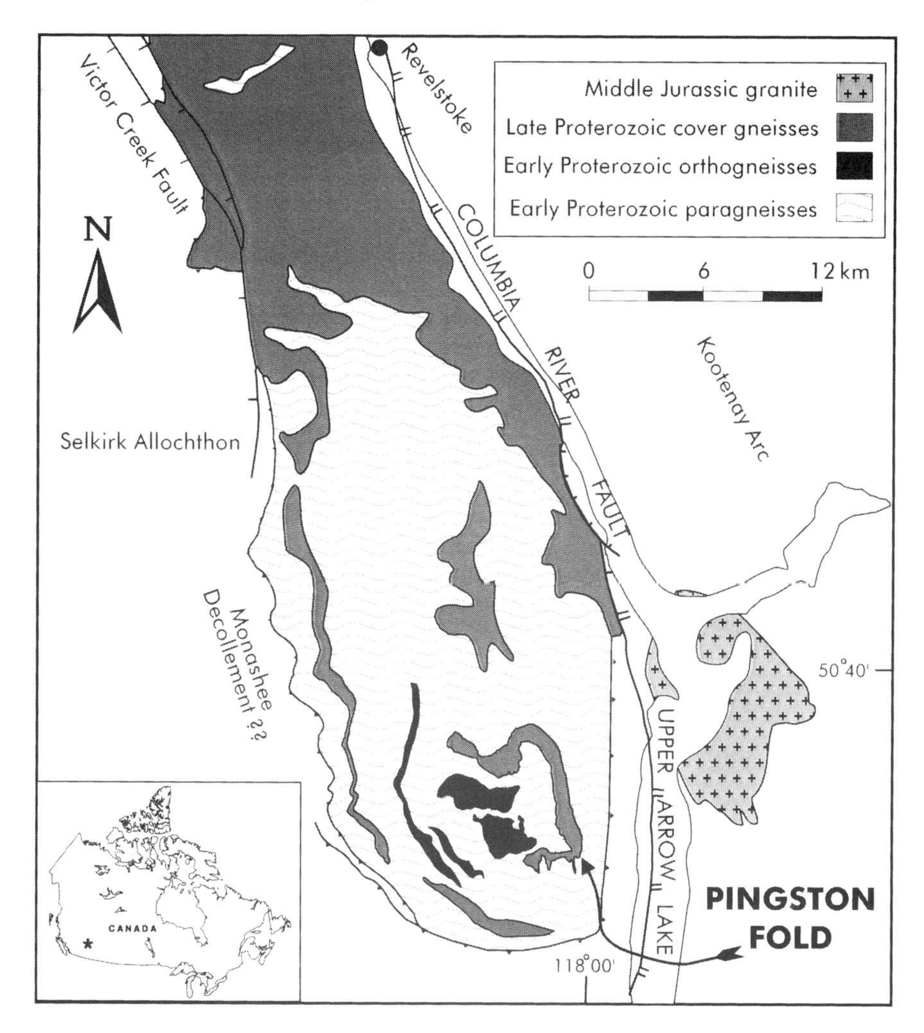

Fig. 1. Location of the Pingston Fold (outlined by folded cover gneiss contacts) within the southern Thor-Odin culmination.

Data Capture and Data Structure

The accuracy of the representation of topography on a map depends on the method of acquiring the elevation data (Kennie & Petrie 1990, Weibel & Heller 1991). The most accurate data collection originates from ground survey methods using electronic tacheometers and theodolites that feed data directly into computer files. Ground surveys allow the measurement of significant terrain locations but can be very time-consuming and expensive and are only practical in small map areas. For larger areas, photogrammetric methods involving the stereoplotting of

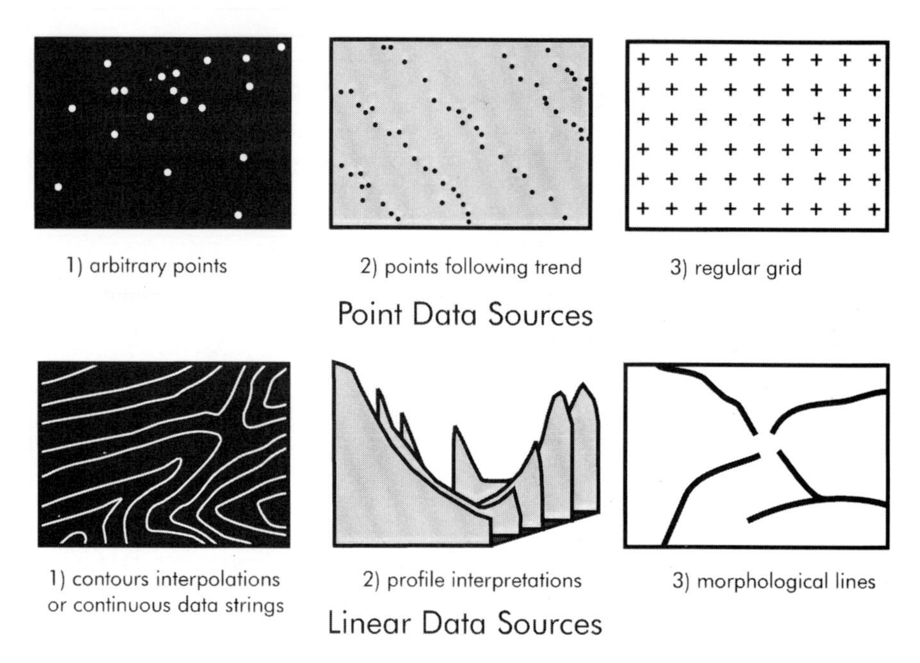

Fig. 2. Point and linear data sources for digital terrain models. (Universal Systems Ltd., 1992)

aerial and satellite images are preferable. Photogrammetry can provide data of similar precision and is cheaper than ground surveys. However, it also can become a very labour intensive and tedious process.

Coordinate data may be acquired from existing maps by graphical digitizing techniques. Data may be collected manually (using a digitizing tablet), semi-automatically (involving user interaction), or automatically (raster to vector conversion). If the accuracy of existing maps is sufficient for the purpose intended, then screen digitization is the most accurate and cost-effective method of obtaining digital data.

DTM data generally comprise point and linear data (Fig. 2). Ground survey measurements and spots heights are examples of point data sources. Any surface can be sampled as random arrays, as a regular geometric pattern of measurements, or as points along a trend (*e.g.*, ridge, road, valley). Point data may not adequately reflect all significant terrain features, especially if the sampling density is too low. Linear data may be obtained through photogrammetric stereoplots that produce continuous streams of data points. This data source possesses a sampling bias (*e.g.*, photogrammetric contours or profiles). As a surface is sampled through photogrammetry, an abundance of data is collected along the contour lines. However, between contours there are no data and that creates a less precise interpolation. However, linear data may also be digitized as contours obtained by the interpolation of point data.

Interpolation Methods

Surface sampling, especially as point data, usually creates sets of unordered, randomly located, and unrelated data points. Interpolation functions are used to estimate elevations between original data points so that a continuous surface can be approximated. The accuracy of surface fitting depends on the distribution, density and type of initial data and on which interpolation method is chosen. A variety of spatial interpolation methods (*e.g.*, kriging, spline, least squares, *etc.*), such as those reviewed and compared by Schut (1976), Peucker (1978), Lam (1983), and McCullagh (1988), are used to order data and to establish the topological relations between data elements. Topology describes any spatial relationships between map elements in 2D or 3D space. Topological data models record the geometrical properties of individual map objects and the connectivity between its components and with other objects. The characteristics of the objects (attributes) are not contained within a topological model but reside within a GIS relational database model.

The coordinates of any point on the topographical surface that are not directly measured must be interpolated. The correlation between actual topography and the DTM is influenced by the type of model used. A regularly gridded DTM generates data points by interpolation. This model joins only the regularly spaced interpolated points to create a surface. Actual measurement points are shown on the surface approximation only if they happen to coincide with a grid point. The triangulated irregular network (TIN) is more precise. It models the topographic surface as a sheet of triangular facets, with each measured data point used as a vertex. Therefore, data points are directly involved. Any other points on the DTM facets are interpolated.

The simplest and computationally fastest method of surface modelling uses a grid-based interpolation. Topological relationships can be implicitly defined within a regular rectangular mesh (Fig. 3). At each unique grid node, elevation values are then interpolated based on the weighted average (where the closest values have the greatest influence) of the surrounding randomly located measurement points. Regular grids, however, cannot adapt to the variability of data density. Depending on the grid spacing, various scales of complexity in relief may be lost. Resampling, using a finer grid over the entire area, to capture localized finer detail (such as a coulee on the plains), would result in redundant and excessively large data sets (especially for surfaces of uniform relief). The mesh width of the grid would then have to be manually altered in different parts of the map (*e.g.*, along the course of the coulee) to accommodate critical terrain features. The other drawback of this method is a sampling bias parallel to grid axes.

An alternative to pointwise interpolation is to globally fit lines or mathematical surfaces to the data points, rather than considering the grid points in isolation. Linear interpolation structures are similar to grids but are stored as straight line paths of points fitted to curves in arbitrary increments (Miller & Laflamme 1958, Peucker 1978). This technique is aesthetically more pleasing but still has an inefficient number of unwieldy mathematical functions to represent data adequately. Subtle yet critical features are always lost and the search time (*i.e.*, nearest neighbour searches) becomes too great. Areal interpolation structures

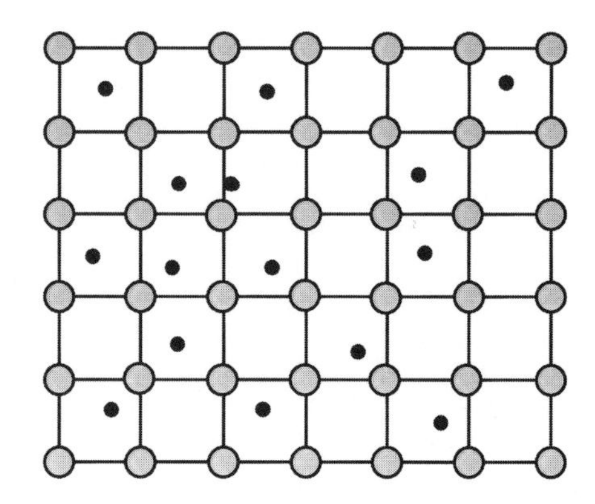

● Original Data Point

○ Interpolated Data Point

Fig. 3. Regular grid interpolation of randomly situated data.

implement global or patchwise methods to fit smooth, mathematically derived surfaces to data points (Peucker 1978, Kennie & Petrie 1990). A high-order polynomial expression can be used to derive *{x,y,z}* values for the entire surface. Computation time is greatly reduced by dividing the map into small sub-areas, and thus, smaller mathematical subsets. This patchwise method is inaccurate and needs substantial computer resources. Since most DTM data sets contain many thousands of points that cover a large and topographically complex area, the fitted polynomial expression would also be very complex and would therefore require a very powerful computer. The polynomial may not closely fit the terrain and may require the interactive removal of spikes and misfits, especially at boundaries. The use of overlapping patches avoids some of the boundary value problems. The most popular surface approximation involves the modelling of a surface as a sheet of triangular facets having data points as vertices; thereby, honouring each measured data point. The triangulated irregular network (TIN) developed by Peucker *et al.* (1978) produces triangles that are as equilateral as possible with minimal side lengths, thereby producing a unique solution for a set of data points. When a surface containing contour data is modelled as a TIN, every point on each line becomes a triangular vertex. *CARIS-DTM* utilizes Douglas-Peucker filter algorithms (Douglas & Peucker, 1973) to reduce the number of points required to characterize lines. Once the lines are thinned, the TIN is ideally suited for variable topography since the triangular mesh adapts to varying densities of data. All critical topographic features are retained as far as they are represented by the original data (*i.e.*, if they are not smoothed out of existence). The TIN has the advantage of variable resolution. The creation of a TIN involves the process of Delauney Triangulation (see McCullagh 1988) and creates the same unique solution in any of the many DTM packages that use it. The Delauney Triangulation algorithm creates Thiessen Polygons (also known as Dirichlet or Voronoi polygons) which define a vector topological structure that determines areas of influence about a point (Fig. 4a). Perpendicular bisectors of the triangle edges meet at Thiessen vertices, defining

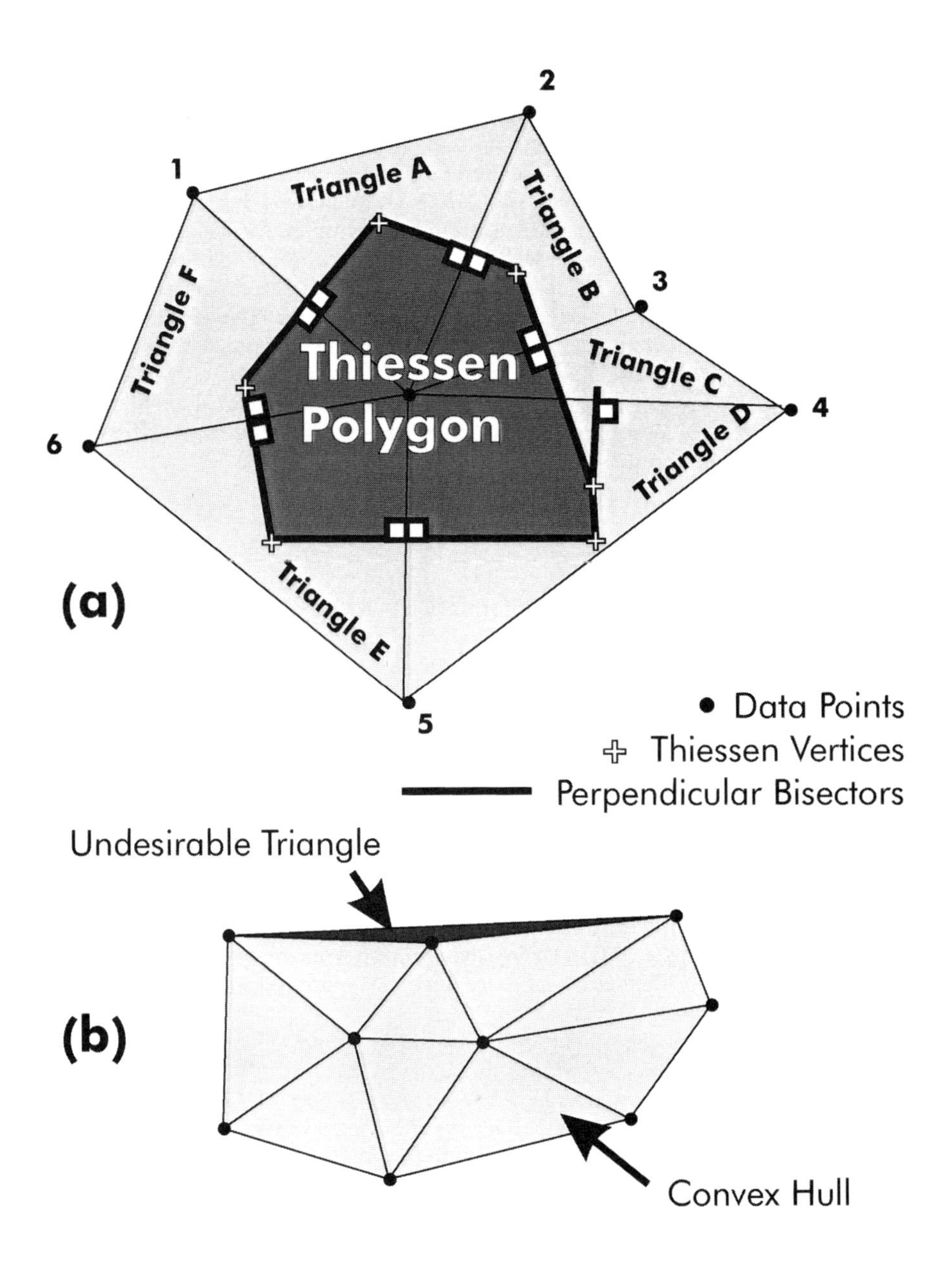

Fig. 4. a) Construction of a Thiessen polygon by Delauney triangulation (from Kennie & Petrie 1990). b) Undesirable triangle formed during Delauney triangulation as bounding shell is created around data set.

a Thiessen polygon around that point. These polygons and their Thiessen neighbours are used in search processes within the data set (Davis 1986, p. 358) during triangulation and also for subsequent data searches within a relational database. At any location, a z-axis value can be computed through a local search for vertices and a subsequent interpolation. The Delauney Triangulation may produce long, thin undesirable triangles as it creates shells around data groups (Fig. 4b). Software packages such as *CARIS-DTM* possess triangle trimming subroutines (Universal Systems Ltd. 1992). If a triangular facet has two angles that are less than 10° (default value), then the facet is automatically eliminated. Overall, the TIN models surface terrain most accurately. A comparison between gridded and triangulated DTMs (Kumler 1990) showed that TINs produced fewer discrepancies from the actual surface and that shaded relief maps were found to be more realistic. Although TIN is a more complex data structure than gridded DTMs, the vertices required for TIN are fewer, compared with gridded methods. Thus, TIN requires less computer storage capacity and is more efficiently organized resulting in much faster computation time.

General DTM Applications

The ability to display landscape information with a DTM makes numerous applications possible (Burrough 1986, Weibel & Heller 1991). Attributes associated with the surface and their spatial relationships are visibly apparent and can be quantified. The most obvious use of a DTM is for landscape visibility for planning purposes (*e.g.*, for recreational facilities or military purposes). One of the earliest DTM packages (*PREVIEW*) was developed exclusively for the visual management of proposed timber harvesting and regrowth areas from various vantage points (Myklestad & Wagar 1977). This module required six hours of key punch card input just to enter elevation data in a 100 × 80 array. More recently, the visual impact of clear-cut logging practices in British Columbia, Canada was portrayed using *CARIS-DTM* (Smart *et al.* 1990, Universal Systems Ltd. 1992). Using entered tree height and crown closure information, tree symbols appeared on the DTM surface to simulate forest cover viewed from vantage points outside or within the DTM. This facilitated more aesthetic planning.

The DTM can also depict relief by hill-shading. This technique, first developed by Yoéli (1965), simulates the effect of the sun's illumination on the terrain. Reflected light intensity on each DTM facet is expressed as an appropriate tint according to arbitrary light ray inclination and orientation. The intensity of the diffusely reflected light is proportional to the cosine of the angle between the illumination vector and the vector that is normal to the facet surface. The tint is determined by the vector dot product of the sun's vector and the vector normal to the facet (Yoéli 1965, Weibel & Heller 1991). The direction of the illumination can be changed in order to emphasize faults, lineaments, *etc*. Hill-shading is useful in agricultural and urban planning applications, to determine the number of hours that daylight strikes each DTM facet.

Z-axis values (*e.g.*, elevation, geophysical data, *etc.*) within the DTM can be used directly for automatic contour generation. Contoured data can then be used, for example, to delineate flood risk zones (Universal Systems Ltd. 1992) which proves beneficial for rapid emergency planning. Quantitative analyses of a DTM can also provide volume integrations for dredging or cut and fill applications. Two DTM surfaces (*e.g.*, before and after dredging) are created and the volume difference between the two can be calculated automatically.

Each facet can be given a colour code based on the calculation of geomorphometric parameters from elevation data (Burrough 1986, Weibel & Heller 1991). Elevation data allow the creation of maps indicating altitude ranges or maps showing degrees of steepness (the slope tangent to the DTM at any point). The first derivative of elevation data yields maps that indicate gradient (rate of change of altitude) and aspect (the direction normal to the facet, or slope direction). Convexity or curvature maps represent the second derivative of elevation data and indicate the rate of change of slope. This type of classification facilitates various hydraulic analyses such as: soil erosion, watershed modelling, and landslide potential (Band 1989). Auxiliary attribute keys can also be assigned to individual facets for the purpose of differentiating terrains such as lithology type, forest, clear cut, water (*e.g.*, Smart *et al.* 1990).

The application demonstrated in this paper involves three-dimensional block diagrams that portray the orientation of structural features. The computation of the block, such as that used in the *CARIS-V3D* module (Universal Systems Ltd. 1993), permits either a perspective or orthographic view from any vantage point. As stated above, two-dimensional map features, such as geological contacts, roads, rivers and symbols, can be draped over the DTM surface.

CARIS DTM - V3D Software Models

Using *CARIS-V3D*, a three-dimensional DTM surface is drawn by the sorting of facets and the removal of hidden surfaces. Using the painter's algorithm (Newman & Sproull 1979, p.239), a more realistic view is created whereby surfaces in the foreground overlap those behind them. Polygons are drawn starting with those farthest from the viewpoint and ending with the nearest. It is often necessary to place finite limits on the view volume to prevent any foreground objects from obscuring the rest of the image and to avoid excessive overplotting of distant images that will not be visible in the final model. Back and front clipping planes are used to define the view volume (Fig. 5).

Listed below are some of the various parameters used in the creation of the DTM models in this demonstration:

> *Projection*: Two modes of projection are possible: perspective and parallel/orthographic. The orthographic mode is used in this study in order to minimize the distortion inherent in perspective views.
>
> *User-defined Classification*: Classification definition files are created by means of any text editor (*e.g.*, UNIX vi-editor). These files list ranges, colours and the type of classification (*e.g.*, Fig. 6).

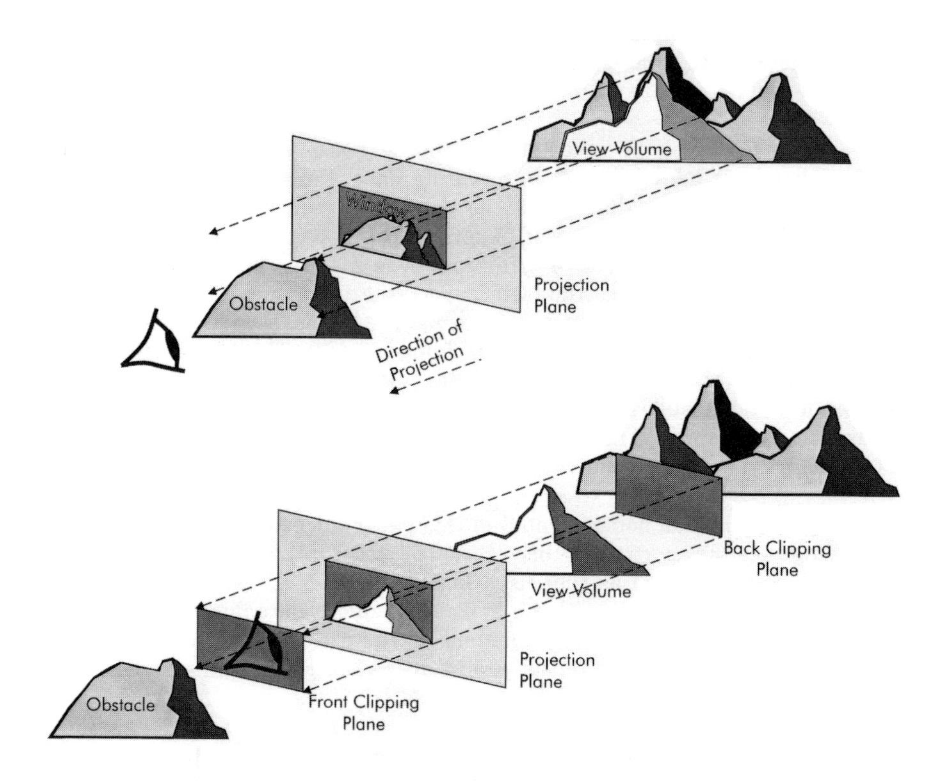

Fig. 5. Orthographic projection - parallel lines of sight (top) with front and back clipping planes (bottom) to limit an infinite field of view. Front clipping plane effectively moves vantage point forward beyond intervening obstacles. Back clipping plane eliminates unnecessary background features.

Vertical Exaggeration: Vertical exaggeration is user-defined and can enhance (or distort) the data presentation. In this study the exaggeration facility was used since the digitized map data consisted of elevations and ground coordinates in imperial units (feet), and the x, y coordinates of the DTM were in metres. Therefore a vertical exaggeration of 0.304 was used as a conversion. This of course resulted in a zero vertical exaggeration for this image.

Changing Viewpoint: The vantage point of any image can be modified by altering the bearing (bird's eye view or any oblique view defined by compass bearing) and inclination (where 0° = horizontal and 90° = vertical).

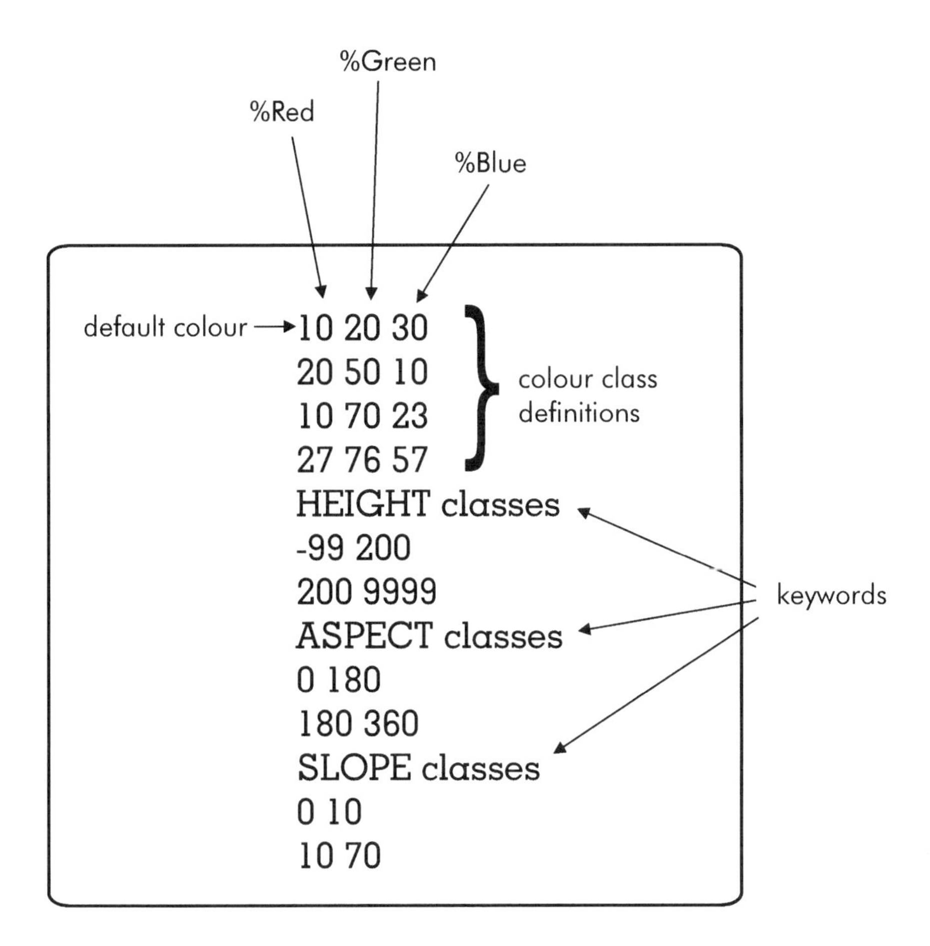

Fig. 6. *CARIS-DTM* class definition file format (Universal Systems Ltd., 1992).

Hillshading: The default illumination direction from a bearing of 325° with a declination of 45° is used in this study (*i.e.*, simulates the sun situated in the northwest quadrant of the image). Any other direction is also obtainable.

Pingston Fold

A bird's eye view DTM of the Thor-Odin culmination, comparable in quality to a satellite image, is presented in Fig. 7. The index contours (interval = 500 ft.) from two adjacent topographic map sheets (NTS 82L/9 and 82L/16) were merged for this image. Hillshading causes a shadow effect for the slopes facing away from the illumination "source". The shadows greatly amplify the NNW and ENE-striking fault sets. The Pingston Fold is located in the southeastern-most corner of Fig. 7.

Fig. 7. Thor-Odin culmination bird's eye view DTM (produced from index contour lines from NTS map sheets 82L/9 - Gates Creek and 82L/16 - Revelstoke).

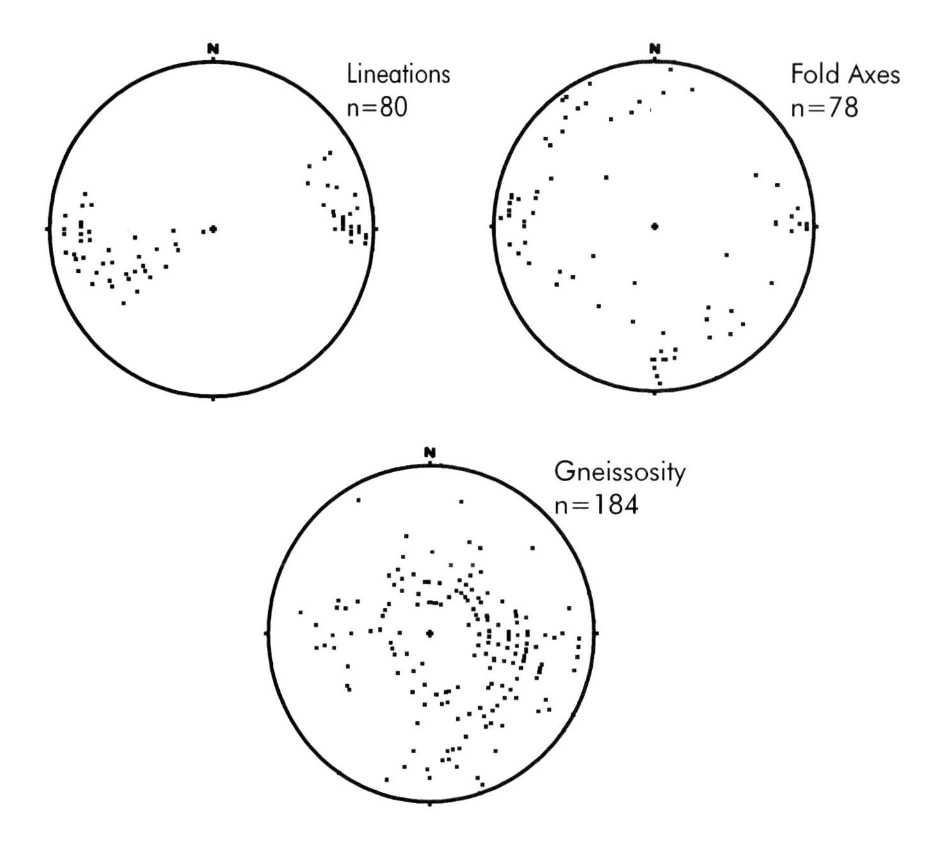

Fig. 8. Equal-area projections of structural data of Reesor & Moore (1971) from their mapping of southern Thor-Odin culmination (5° rings = their data approximation).

The Pingston Fold was originally mapped by Reesor & Moore (1971). Structural data was extracted from the map of Reesor & Moore (1971, figure 13) and is presented in Fig. 8 on equal area hemispherical plots. The data from the entire southern Thor-Odin region, indicates a pervasive E-W lineation, variable and undifferentiated fold axes and a domal distribution of dipping gneissosity.

Fold Data

In order to create a three-dimensional geological model from two-dimensional structural data, the map area must be divided into domains on the basis of the homogeneity of the orientations of fabric elements. If the fabric is cylindrically folded (*i.e.*, where poles to the folded surface plot on a great circle), the area of consistent fold axis orientation defines a domain. For these domains, the map data is projected down the plunge of fold axes and then the structural surfaces are merged at domain boundaries (Hobbs *et al.* 1976, p.373).

Reesor & Moore (1971) divided their map area into domains, generally with respect to the orientation of small fold axes but some were more arbitrary and coincided with lithological contacts. Their stereographic projections were extracted (figure 12 of Reesor & Moore 1971) and data maxima were read from them and are summarized in Table I. The foliation ($S_{0\text{-}1}$ of Reesor & Moore 1971) represents primary layering and parallel F_1 axial planar foliation. The statistical orientation for each domain is listed in Table I as *dip direction;dip*. Mineral lineations ($L_{2\text{-}3}$) generally trend east-west and are oblique to the domain-defining small F_2 fold axes. In some domains, a crenulation lineation or mineral lineation subparallel to F_2 fold axes was noted. The structural maxima from each domain of Reesor & Moore (1971) are presented on a portion of a 1:50,000 *CARIS* map in Fig. 9 (SE corner of NTS sheet 82L/9), displaying all structural domain boundaries.

Digital Terrain Model

A DTM was constructed using all contour data (interval = 100 feet) available on a portion of a 1:50,000 *CARIS* map in Fig. 9 (SE corner of NTS sheet 82L/9). The marker quartzite (unit 2) that commonly separates core from the overlying cover gneisses is highlighted, and is used to trace out larger scale structures. The DTM (Fig. 10) presents a bird's eye view of the map sheet including some of the geological data. Domain boundaries and marker quartzite contacts, compiled from the geological maps of Reesor & Moore (1971) are draped over the DTM.

The DTM offers the ability to view the three-dimensional relationship of map features that a two-dimensional map sheet cannot provide. Much of what appears to be late folding of the earlier tight folds in Fig. 10 is simply due to a cut effect. The influence of topographical relief can be demonstrated by selecting the appropriate oblique view of the DTM. A northeastern vantage point (Fig. 11) enhances the image of the Pingston Fold and avoids obscuring the image with intervening mountains. Additionally, this same technique allows the dip of any surface to be obtained by varying the line of sight. For example, the isoclinal folds that traverse across several ridges in the background of Fig. 11 have a more obvious steep axial plane. The optimal view direction of a structure would be down the plunge of folds or along strike of planar features.

Construction of Orthographic Block Diagram

To further elucidate three-dimensional structures, an orthographic block diagram can be created from the DTM. Previously, block diagrams were all painstakingly hand-drawn. Topography was represented by the raising of individual points to their correct vertical axis value above the surface of the block (*e.g.*, Lobeck 1924, p.151-155). The use of a DTM can save a great deal of time by rapidly providing the correct elevations for relevant points.

Many different types of projection can be used to depict a structural feature; however, most are distorted and are not to scale. Perspective projections are probably the most visually pleasing images (*e.g.*, Lobeck 1924) but are quite difficult to prepare and use. The perspective projection introduces distortion as

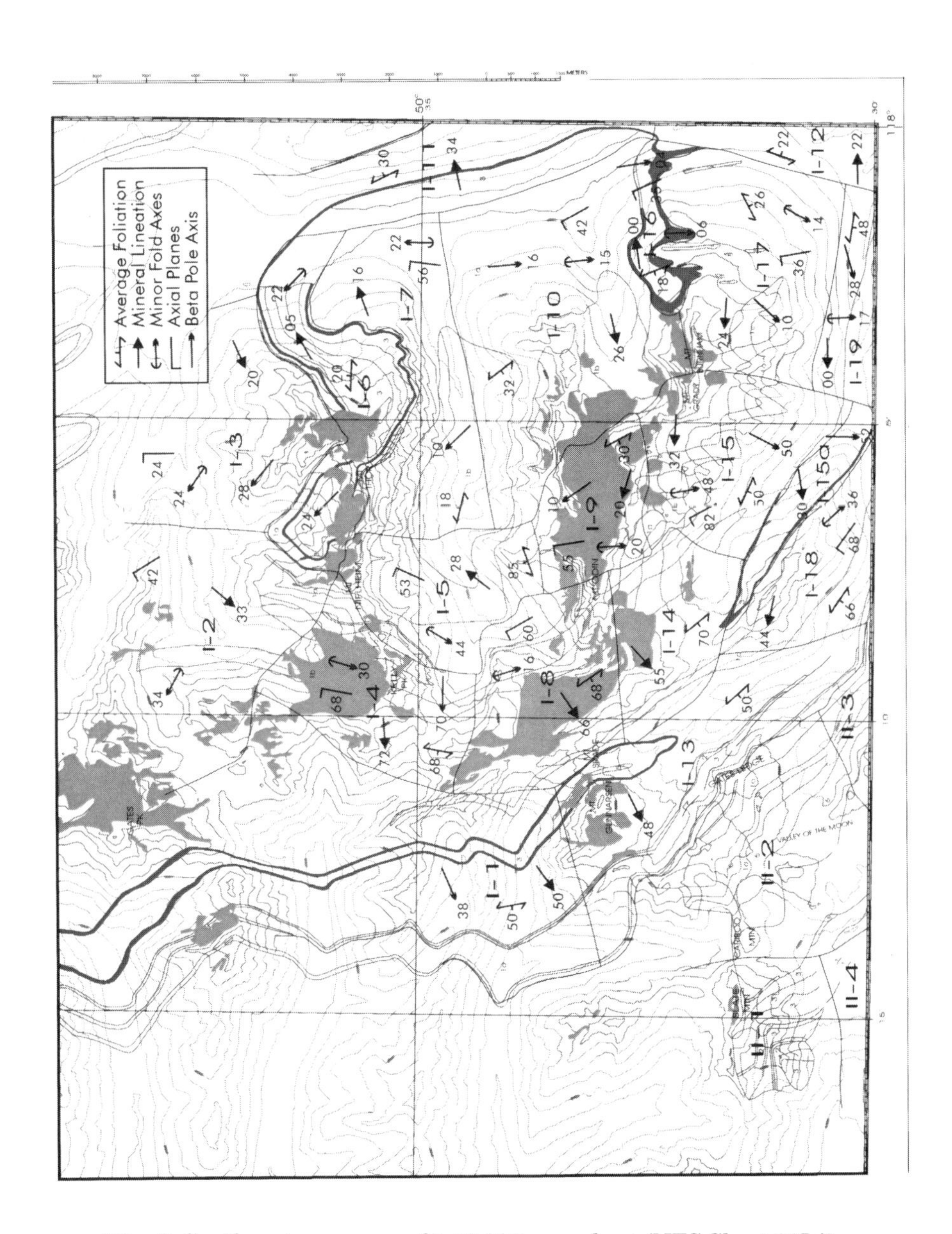

Fig. 9. Southeastern corner of 1:50,000 map sheet (NTS Sheet 82L/9 - Gates Creek) displaying structural domains and domainal data. Red lines are geological contacts. Marker quartzite unit highlighted. Index contours (interval = 500 feet) were digitized semi-automatically from the 82L/9 map. Blue bodies mark glaciers.

Table 1

Reesor & Moore (1971) Domainal Structure Data

Domain	Average Foliation	Mineral Lineation	Small Fold Axes	Fold Axial Planes	Crenulation Lineation	Beta Axis to Fol'n Poles	Domainal Axis	Type
I-1	260->50	235-50	AxPl pole girdle	261->59	------	246->38	246->38	beta
I-2	scatter	220-33	300-34	240->42	------	scatter	300->34	F.A.
I-3	F.A. girdle	246-20	306-24	270->24	------	312->28	312->28	beta
I-4	281->68	267-72	197-30	281->68	210-40	272->70	272->70	beta
I-5	340->85	038-28	211-44	294->53	------	scatter	340->85	S1
I-6	003->20	060-05	316-22	F.A. girdle	------	316->24	316->24	beta
I-7	015->18	070-16	002-22	278->56	000-12	318->10	318->10	beta
I-8	237->68	236-66	164-16	242->60	178-48	------	237->68	S1
I-9	243->30	288-20	178-20	262->55	168-24	326->10	326->10	beta
I-10	234->32	258-26	175-15	242->42	158-20	176->16	176->16	beta
I-11	065->30	078-34	------	-------	------	------	065->30	S1
I-12	109->22	090-22	scatter	153->10	------	------	109->22	S1
I-13	230->50	244-48	Foln pole girdle	220->68	------	------	230->50	S1
I-14	232->70	228-55	Foln pole girdle	240->70	Foln pole girdle	------	232->70	S1
I-15	206->50	274-32	170-48	243->82	178-38	207->50	207->50	beta
I-15a	F.A. girdle	262-30	202-64	F.A. girdle	202-54	182->52	182->52	beta
I-16	250->18	083-00	179-06	258->35	------	172->04	172->04	beta
I-17	152->26	275-24	210-14	261->36	------	222->10	222->10	beta
I-18	216->66	284-44	140-36	218->68	137-27	------	216->66	S1
I-19	190->48	270-00	178-17	-------	------	254->28	254->28	beta

Fig. 10. Bird's eye view of Pingston Fold Digital Terrain Model acquired from SE corner of 82L/9 1:50,000 digital map (red contacts indicate marker quartzite, white lines indicate domain boundaries each containing a unique minor fold axis symbol).

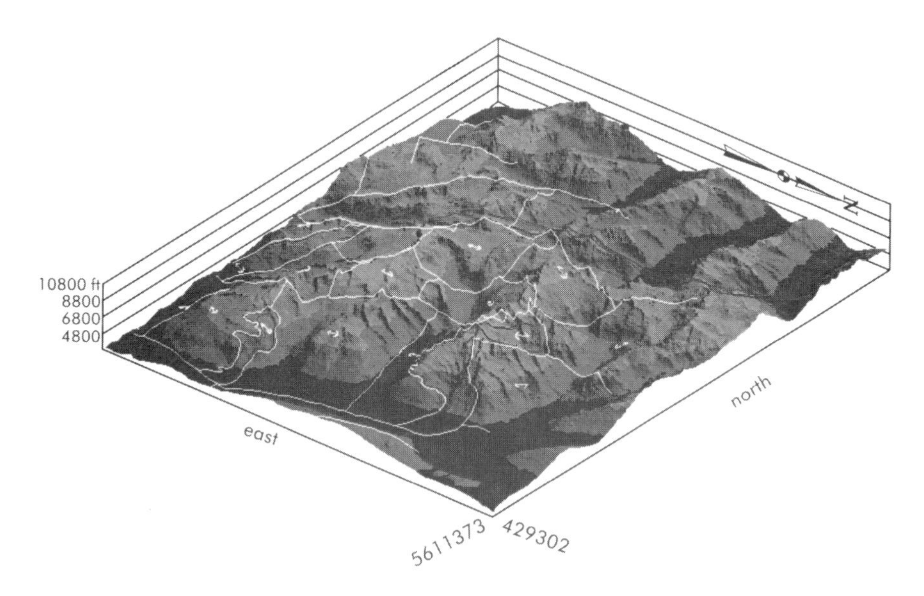

Fig. 11. Optimal viewpoint of Pingston Fold vector map data draped over DTM (orthographic projection viewed from northeast inclined down 30° without vertical exaggeration).

lines of sight converge towards a vanishing point and therefore alters the scale in any given direction with distance from the viewer and varies the angles between lines for each angle of sight. Distant objects appear smaller and thus possess smaller angles between lines. The perspective projection loses some important geometrical information. Distances cannot be measured, and angles retain their geometrical relationships only on the faces situated parallel to the projection plane.

The orthographic projection (McIntyre & Weiss 1956), however, projects all points in the object along parallel lines that are perpendicular to the plane of the projection (*i.e.*, the page). Essentially, it is a perspective drawing, but with the vanishing point located at infinity. Unlike perspective projections, the scale within an orthographic projection remains constant, for a given direction, throughout the visualized object, irrespective of the distance of the object from the viewpoint. Thus lines that are parallel on the object remain parallel upon projection.

Isometric projections (Lobeck 1924) are a unique subset of orthographic projections whereby the upper surface of the block has corners of 60° and 120°, with sides that are equally distorted. Lobeck illustrates how isometric projections accurately portray distances and angles, since objects appear of equal size regardless of their distance from the observer. All vertical or parallel lines retain their angular relationships on the block. Isometric diagrams are of limited value because of their restricted range of angles of sight that would keep block sides equally foreshortened and 120° apart. The original block diagram of the Pingston Fold (Fig. 12) was prepared as an isometric projection by Reesor (1970).

The orthographic projection net, or orthonet, can be used to plot structural orientation data as well as block edges (McIntyre & Weiss 1956). The orthographic block diagram is created from one block corner from which three mutually perpendicular axes emerge as would be visible from a designated line of sight. It is customary to use one axis to represent vertical and any mutually perpendicular horizontal orientations. The three axes are inclined to the plane of the projection in order to simultaneously display the top block face (map surface) and the two vertical block faces. This inclination is produced by first viewing one side of the block along one horizontal axis (*e.g.*, N-S or E-W) (Fig. 13(1)). An adjacent side is brought into view by the rotation of the axes through an angle α about the N-S diameter of the orthonet (Fig. 13(2)). A second rotation around the E-W diameter through an angle β (*e.g.*, the declination of intended line of sight) will bring the top surface into view (Fig. 13(3)).

The choice of α and β will depend upon the viewpoint, which is dependent on the relative importance of the data representation on each block face, or on the orientation of a particular structure (*e.g.*, views down fold axes or fault planes), or on intervening topography. The rotation of the axes moves their endpoints which define lines that have equal length in the projection sphere (they are radii) but appear unequal in projection. A block diagram can then be scaled from the ratio of the lengths of the projected axes.

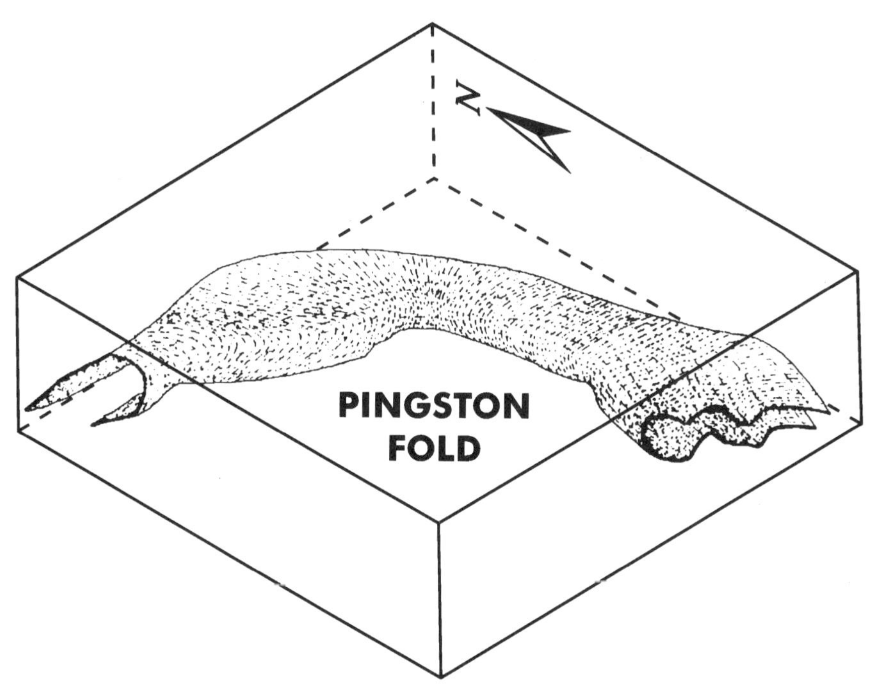

Fig. 12. Early version of a Pingston Fold block diagram (Reesor 1970). Projection is isometric with no vertical exaggeration.

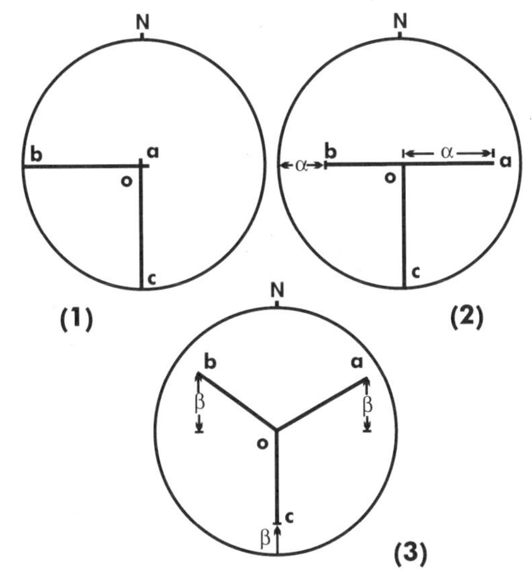

Fig. 13. Rotation of orthographic block diagram using lower hemisphere projection of three orthogonal axes *(a,b,c)* that intersect at *o*. (1) projection of *bc*-plane, (2) rotation of α around N-S diameter, (3) further rotation of β around E-W diameter resulting in three correctly positioned axes. (from McIntyre & Weiss 1956)

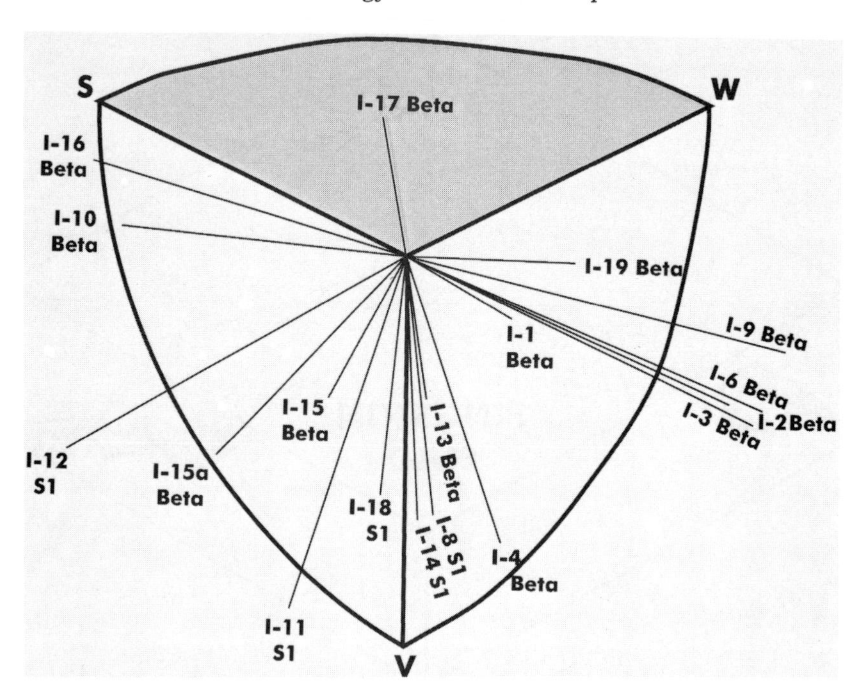

Fig. 14. Domainal fold axes in orthographic projection, view from NE, inclined down 30°.

Pingston Fold Three Dimensional Structure

The shape of the Pingston Fold becomes more apparent through the use of DTM orthographic visualization. Figure 14 shows a lower hemisphere orthographic projection of the Pingston domainal axes using a view from the northeast, looking down 30°. This orthonet figure represents one corner of the block diagram that intersects a lower hemisphere. All axes end where they impinge on the surface of the sphere as seen in projection, thus appearing unequal in length. The DTM that was created within *CARIS-DTM*, along with the draped geology and domainal boundaries, was exported to a graphics package (*Corel Draw*™) (Fig. 15). The orthonet projections of the domainal fold axes were scanned and imported into the same graphics image. In each domain, the quartzite contact was projected down plunge in a direction parallel to the corresponding projected axis, and was blended between domains.

The structural interpretation derived from this model is that of polyphase fold interference pattern. The initial folds were isoclinal similar to the long, shallowly plunging folds visible in the background of the DTM. From field mapping within the central Thor-Odin area to the north during the field seasons of 1991-1993, the authors were able to distinguish two overprinting fold generations. F_2 fold axes trend roughly NW and SSE and are folded about F_3 SW-trending fold axes (Fig. 16). F_2/F_3 distinctions were possible using rare overprinting relationships in outcrop as well as by using style and orientation criteria.

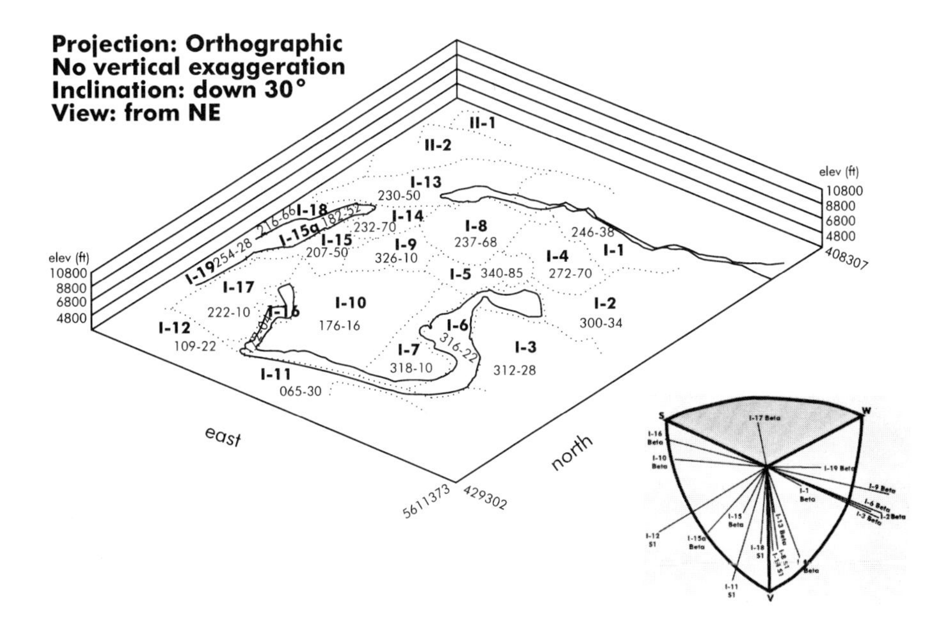

Fig. 15. Basal quartzite contacts and structural domain boundaries draped over topography of Pingston Fold DTM. Fold axes for each domain are listed on model with orientations shown on orthonet (lower right corner).

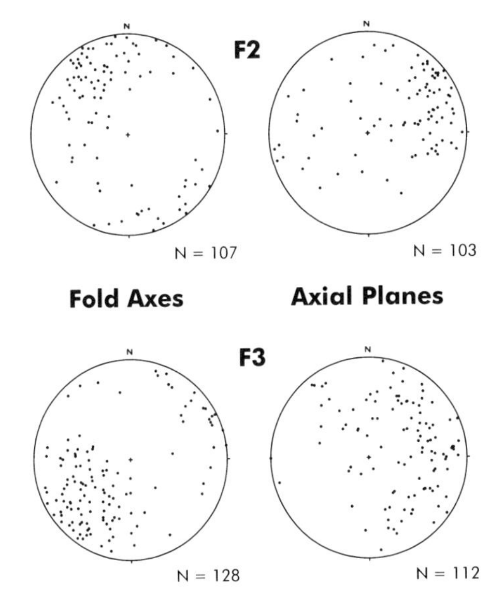

Fig. 16. F_2/F_3 overprinting fold orientation data, obtained from mapping throughout central Thor-Odin culmination.

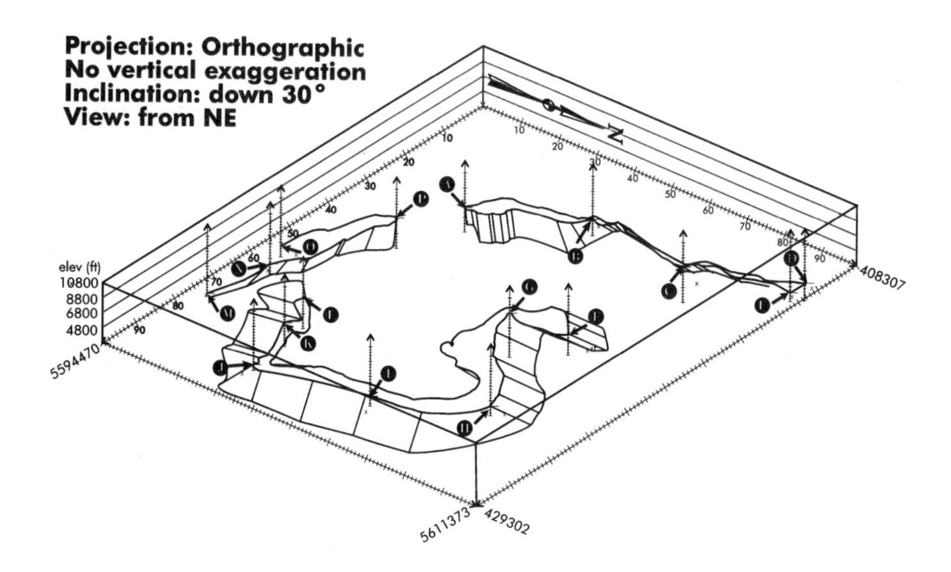

Fig. 17. Using projected orientations of each domainal fold axis, quartzite contact is projected down plunge. All projection lines are of equal length. Measurement of depth to block bottom for various critical localities on topographical surface (labelled A-P). 'x' symbols indicate intersection of domainal axes with bottom of block for these sites.

The block diagram portrayal of the Pingston Fold must be completed by projecting structural elements to the block faces. Elevation must be calculated in order to determine where the images lie vertically within the block. Figure 17 shows selected z-calculation sites (A-P) and corresponding scale bars, indicating the vertical distance to the bottom of the block. To calculate where a projected axis would intersect the block, a simple trigonometric calculation can be applied. Using an arbitrary grid coordinate system, one can read the $\{x_1,y_1\}$ location at the DTM topographic surface, as well as the elevation at that point $\{z_1\}$. It is then possible to determine the depth to the bottom of the block $\{x_2,y_2\}$. The vertical and horizontal components of the domainal axis projection can now be determined to indicate the intersection of the domain axis with the block face $\{z_2\}$ (indicated on Fig. 17 with an 'x' symbol).

The manner in which the structures intersect the sides of the block depends on the orientation of the axial plane where the fold axis emerges. Axial planes can be estimated by bisecting the distribution of S-attitudes on the fold limbs. The plotting of the trace of S-planes or axial planes is a simple exercise on any stereographic projection. Figure 18 displays a modified block diagram that contains structures projected to the bottom and sides of the block.

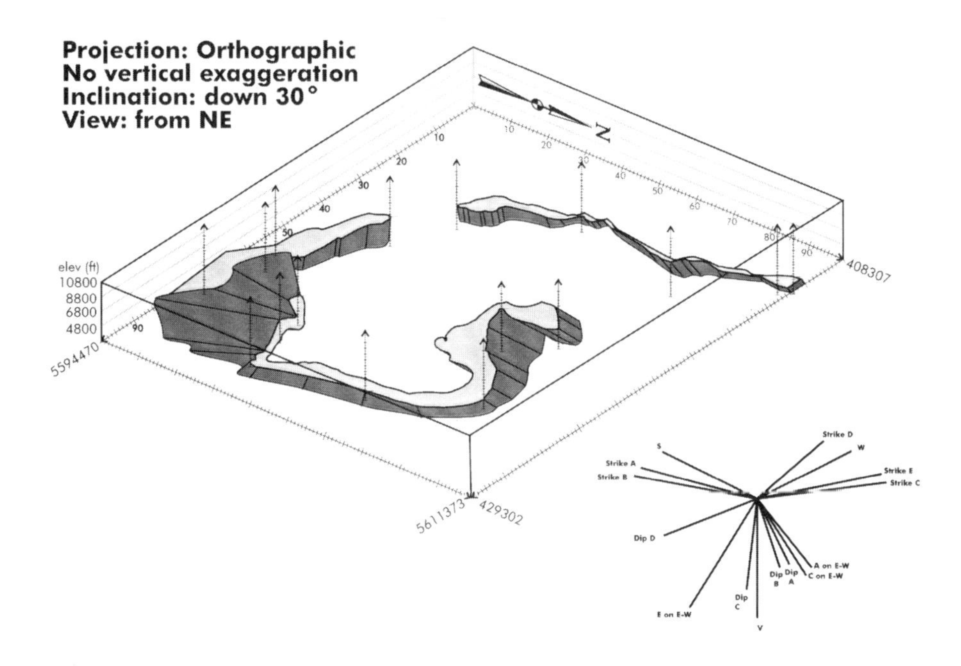

Fig. 18. Domainal axis projections extended to block faces. Traces of axial planes on block sides (projections shown in orthonet scan in bottom right) enabled drawing of fold limbs.

Towards the southeastern corner of the block, there is a problem in that the projection of the surface expression results in the intersection in space of the two rock bodies. This impossible situation indicates that we are projecting linear axes too far. It is necessary to interpret a change in plunge along the domainal axis. This is of course a reasonable interpretation in such an area of fold interference.

A wire-frame version of the DTM surface can be produced to see how structures relate to topography (Fig. 19). The resultant three-dimensional interpretation of the Pingston Fold consists of an early isoclinal trough (F_1) that was folded by F_2. These folds were then reoriented by a large F_3 WSW-plunging fold.

DTM visualization is an easily performed semi-automated procedure that rapidly provides views from any vantage point. Once the model is completed, it can be viewed from any orientation by simply changing the line of sight. This is in contrast to manual methods where each new view requires that all the steps for block construction be repeated. A block diagram can then be drawn using a computer graphics package and the imported images of structures superimposed

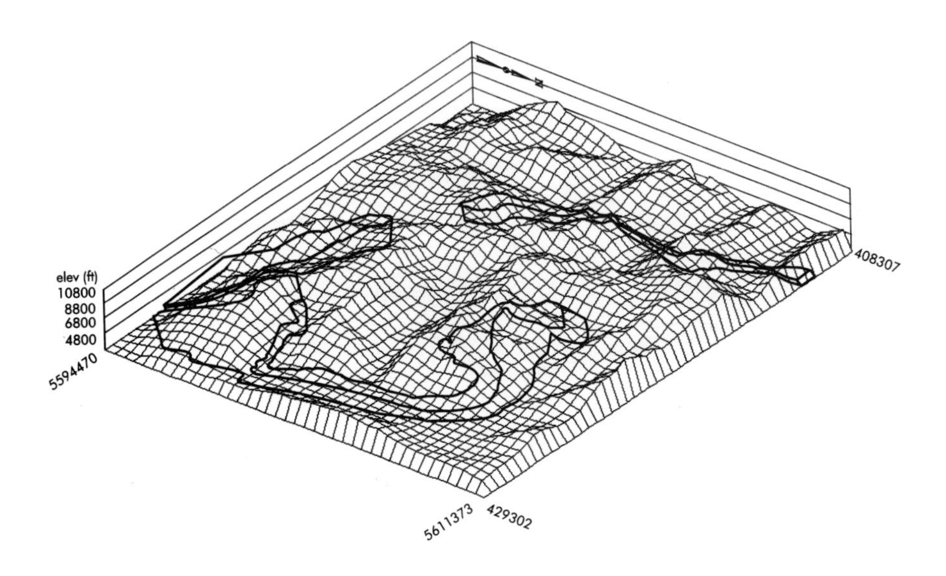

Fig. 19. Wire-frame DTM combines projected structures with topographic influences (sloped edges represent wire-frame profile averaging artifacts). The intersection of two bodies of rock in left-hand corner has been eliminated by blending plunge variation from one domain to another (see text).

on topography. Additionally, a DTM can be constructed for the structures themselves (*i.e.*, representations of axial planes given a series of *x,y,z* coordinates) which could then be viewed along with map data in any orientation. Therefore the DTM offers advantages of clear and rapid visualization of a complexly deformed terrain superimposed with any map data and can image any new surfaces constructed in the block diagram.

Acknowledgements

We thank Kevin Berry and Todd Jellett of the GIS laboratory at the Department of Geology, University of New Brunswick for helping create DTMs and topographic databases. *CARIS* GIS software was made available by Universal Systems Ltd. of Fredericton and was used in creating DTMs and compiling geological and topological. Lynne Cowan of UNB Department of Geodesy and Geomatics Engineering is thanked for critically reviewing early drafts of this manuscript.

References

Band, L. E. 1989. A terrain-based watershed information system. *Hydrological Processes* **3**: 151-162.

Brown, R. L. & Journeay, J. M. 1987. Tectonic denudation of the Shuswap metamorphic terrane of southeastern British Columbia. *Geology* **15**: 142-146.

Burrough, P. 1986. Digital elevation models. In: *Principles of Geographical Information Systems for Land Resources Assessment.* Clarendon Press, 39-56.

Davis, J. C. 1986. *Statistics and Data Analysis in Geology*. John Wiley & Sons, New York, 2nd. edition, 646 pp.

Douglas, D. H. & Peucker, T. K. 1973. Algorithms for the reduction of the number of points required to represent a digitized line or its caricature. *The Canadian Cartographer* **10**: 255-263.

Duncan, I. J. 1982. The evolution of the Thor-Odin gneiss dome and related geochronological studies. *Ph.D. thesis*, University of British Columbia, Vancouver, British Columbia.

Hobbs, B. E., Means, W. D., & Williams, P. F. 1976. *An Outline of Structural Geology*. John Wiley & Sons, Toronto, 571pp.

Kennie, T. J. M. & Petrie, G. 1990. Digital terrain modelling. In: *Engineering Surveying Technology*. T. J. M. Kennie & G. Petrie (eds.), John Wiley & Sons Inc., New York, pp. 391-426.

Kumler, M. P. 1990. A quantitative comparison of regular and irregular digital terrain models. *GIS / LIS '90 Proc.* **1**: 255-263.

Lam, N. S. 1983. Spatial interpolation methods: a review. *The American Cartographer* **10**: 129-149.

Lobeck, A. K. 1924. Block *Diagrams and Other Graphic Methods Used in Geology and Geography*. John Wiley & Sons, New York, 206pp.

McCullagh, M. J. 1988. Terrain and surface modelling systems: theory and practice. *Photogrammetric Record* **12**: 747-779.

McIntyre, D. B., & Weiss, L. E. 1956. Construction of block diagrams to scale in orthographic projection. *Proc. Geol. Assoc.* **67**: 142-155.

Miller, C.L. & Laflamme, R. A. 1958. The digital terrain model - theory and application. *Photogram. Eng.* **24**: 433-442.

Myklestad, E. & Wagar, J. A. 1977. *PREVIEW*: computer assistance for visual management of forested landscapes. *Landscape Planning* **4**: 313-331.

Newman, W. M., & Sproull, R. F. 1979. *Principles of Interactive Computer Graphics*. McGraw-Hill, Toronto, 541pp.

Peucker, T. K. 1978. Data structures for digital terrain models: discussion and comparison. In: *Harvard Papers on Geographic Information Systems*. G. Dutton (ed.), First Int. Adv. Study Symp. on Topological Data Structures for Geographic Information Systems **5**: 1-15.

Peucker, T. K., Fowler, R. J., Little, J. J., Mark, D. M. 1978. The triangulated irregular network. In: *Proceedings of the Digital Terrain Models Symposium*. Amer. Soc. Photogram., Falls Church, Virginia, pp. 516-540.

Reesor, J. E. 1970. Some aspects of structural evolution and regional setting in part of the Shuswap Metamorphic Complex. *Geol. Assoc. Canada, Spec.* Pap. **6**: 73-86.

Reesor, J. E., & Moore, J. M., -Jr. (1971). Petrology and Structure of Thor-Odin Gneiss Dome, Shuswap Metamorphic Complex, British Columbia. *Geol. Surv. Canada Bull.* **195**: 149pp.

Schut, G. H. 1976. Review of interpolation methods for Digital Terrain Models. *The Canadian Surveyor* **30**: 389-412.

Smart, J., Mason, M., & Corrie, G. (1990). Assessing the visual impact of development plans. *GIS'90 Symposium* 19-27.

Universal Systems Ltd. 1992. *CARIS DTM User's Guide*. Universal Systems Ltd., Fredericton NB 58 pp.

Universal Systems Ltd. 1993. *CARIS V3D User's Guide*. Universal Systems Ltd., Fredericton NB.

Weibel, R. & Heller, M. 1991. Digital terrain modelling. In: Maguire, D. J., Goodchild, M.F. & Rhind, D. W. (eds.). Geographical Information Systems: Principles and Applications, Longman Group UK Ltd., vol. 1, pp. 269-297.

Yoéli, P. 1965. Analytical hill shading. *Surveying and Mapping* **25**: 573-579.

Computation of Orientations for GIS – the ‘Roll’ of Quaternions

Declan G. De Paor
Department of Earth and Planetary Sciences,
Harvard University, 20 Oxford Street,
Cambridge MA 02138, U.S.A. depaor@eps.harvard.edu

Abstract– Geographical information systems are powerful research tools but they lack the ablility to handle basic operations involving the rotation and interpolation of orientation data and therefore are limited in their application to structural geology. This paper describes a solution to these limitations using mathematical operators called quaternions which were invented in the last century by Sir William Rohan Hamilton (who also developed vector theory). Quaternions combine the properties of complex numbers and vectors and incorporate the basic concept of Euler's Theorem.

Introduction

The specification of orientation and the process of rigid rotation of spatial data by a given amount about a given axis are important topics in strain and kinematic analyses of deformed rocks both from a theoretical and a practical viewpoint. In theory, we need to better understand the contributions of increments of vortical flow to rotational deformation states and in practice we often need to rotate large data sets – perhaps hundreds of points on a pole figure – so as to remove the effects of folding or tilting on earlier structures, for example. The need to manipulate spatial orientation data consistently and efficiently is especially pressing for users of geographic information systems. The classical method of defining spatial orientation employs a triad of Euler angles. Any orientation may be obtained by starting with a reference orientation and then applying three successive orthogonal tensors corresponding to rotations through Euler angles denoted α about reference axes denoted x, y, z:

$$\mathbf{R}_x = \begin{bmatrix} 1 & 0 & 0 \\ 0 & \cos\alpha_x & \sin\alpha_x \\ 0 & -\sin\alpha_x & \cos\alpha_x \end{bmatrix} \tag{1a}$$

$$\mathbf{R}_y = \begin{bmatrix} \cos\alpha_y & 0 & -\sin\alpha_y \\ 0 & 1 & 0 \\ \sin\alpha_y & 0 & \cos\alpha_y \end{bmatrix} \tag{1b}$$

$$\mathbf{R}_z = \begin{bmatrix} \cos\alpha_z & -\sin\alpha_z & 0 \\ \sin\alpha_z & \cos\alpha_z & 0 \\ 0 & 0 & 1 \end{bmatrix}. \tag{1c}$$

In this approach, specification of orientation is treated as a special type of deformation. The combined rotation may be written in shorthand using c for $\cos\alpha$ and s for $\sin\alpha$,

$$\mathbf{R} = \begin{bmatrix} c_y s_z & c_y s_z & -s_y \\ s_x s_y c_z - c_x s_z & s_x s_y s_z + c_x c_z & s_x c_y \\ c_x s_y c_z + s_x s_z & c_x s_y s_z - s_x c_z & c_x c_y \end{bmatrix}. \tag{2}$$

Instead of α_x, α_y, α_z, structural geologists use alternative equivalent triads of angles, strike / dip / pitch or plunge / trend / twist, in a well-defined format, *e.g.*,

Strike:	σ	=	040°	(3a)
Dip:	δ	=	45°SE	(3b)
Pitch:	ψ	=	12°	(3c)
Plunge:	ϕ	=	45°	(4a)
Trend:	θ	=	135°	(4b)
Twist:	τ	=	62°.	(4c)

(here pitch (*alias* rake) is a rotation of a line measured about the pole to a dipping plane and twist is a rotation of a plane about a plunging axis contained in the plane).

Limitations of Euler Angles

Unfortunately, the Euler angle approach involves several disadvantages. There is no defined value for strike when dip is zero, or trend when plunge is vertical, for example. Consequently, pitch becomes less and less accurate as dip decreases. Singularities lead to the phenomenon of gimbal lock (a term coined by nagivators) and make it difficult to handle orientation data numerically. Gimbal lock occurs when the $\mathbf{R}_y$ rotation of eqn. (1) is 90° so that the $\mathbf{R}_z$ rotation occurs about the same axis as $\mathbf{R}_x$ and not about an independent axis - effectively a loss of a degree of freedom. Euler angles are not actually independent in the same way that Cartesian axes are. Successive rotations defined by orthogonal matrices are non-commutative (that is why the puzzle called Rubik's Cube is so difficult to solve) so sequences of rotations expressed in terms of Euler angles, as in eqns. (1), are reference frame dependent (Shoemake 1994a). Most importantly, orientations defined by Euler angles cannot be uniquely interpolated. If one is given the

Fig 1. Interpolation of dip and strike readings on a map.

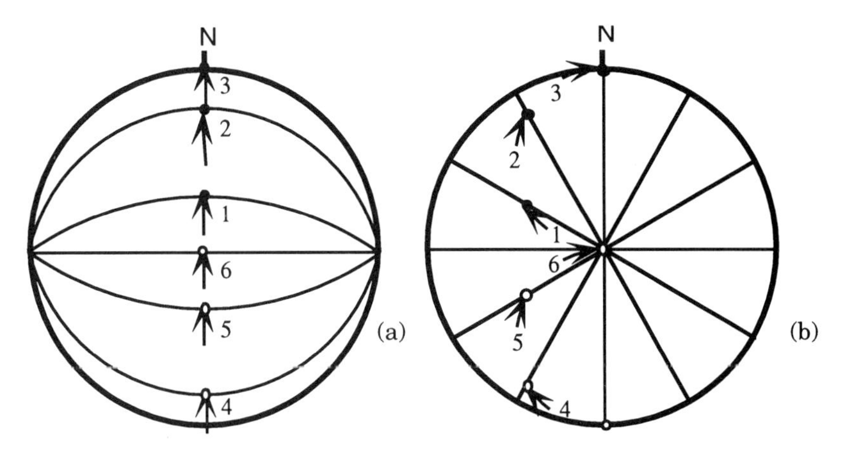

Fig 2. Alternative rotations (a), (b) between the same end orientations. Solid and open dots represent opposite ends of a lineation.

orientation of a plane at two locations on a map (Fig. 1), it is not possible to interpolate orientations at in-between locations in a simple linear fashion as one would interpolate scalar quantities such as density or age (note that the issue at this stage is not whether a linear interpolation is appropriate but rather whether it is possible). If the plane's orientation is 030/30SE at location A and 036/36SE at location B, then the correct linearly interpolated in-between orientations at locations one sixth, one third, one half, *etc.*, of the distance from A to B are certainly not 031/31SE, 032/32SE, 033/33SE, *etc.*! Furthermore, if a tensor **R** rotates the orientation at location A into that at location B, repeated application of the tensor **R**/6 (in which each matrix element of **R** is divided by six) will not generally result in a sensible array of intermediate orientations.

Equivalent limitations apply to 'morphing' in the temporal dimension. Suppose one wishes to construct a simulation in which the orientation 030/30SE gradually changes to 036/36SE over the period of six time frames. Morphing refers to the process of generating the intermediate frames and is subject to the same limitations as spatial interpolation. There is no linear solution (indeed, no there is no unique path) defined by in-between Euler angles.

To demonstrate the lack of a unique interpolation path, consider the problem of rotating a north-facing plane of strike and dip 090/90 with a lineation pitching straight down and transforming it into a south-facing plane 270/90 with its lineation directed vertically up (Fig. 2). One solution would be to rotate the initial plane about a horizontal east-west axis (Fig.2a), but it is also possible to rotate the plane about a vertical axis whilst simultaneously changing the pitch

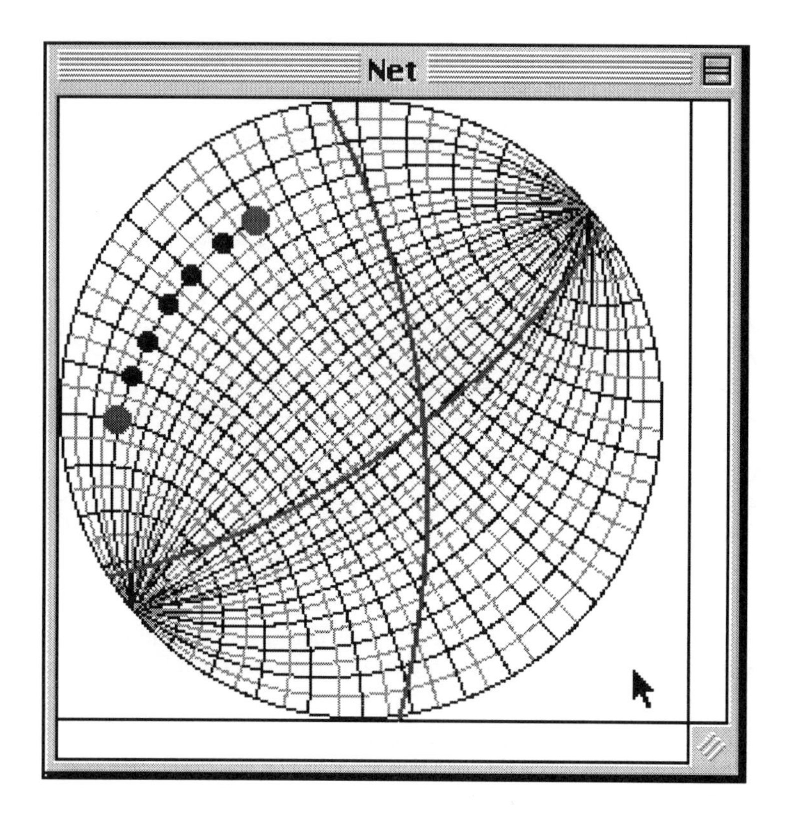

Fig. 3. Interpolation of orientations following Euler's theorem which states that any direction can be transformed into any other direction by a finite rotation about a unique axis. Five small dots divide arc between two poles (large dots) into six equal angles of pitch. Great circles represent planes perpendicular to end poles.

from +90° to -90° (Fig. 2b). This non-uniqueness is somewhat surprising because of Euler's theorem, which states that any two orientations can be connected by a single rotation about a unique axis. Ironically, Euler angles are not useful mathematical expressions for solving Euler's theorem.

Students of structural geology should know how to solve this interpolation or morphing problem graphically (Fig. 3). Poles to the planes at A and B are plotted on a stereographic net and a great circle is drawn through them. The arc of the great circle subtended between the poles is divided into six equal angles of pitch to yield the poles to the correctly interpolated planes for the five intermediate locations or time frames. The problem is to convey the above instructions to a computer; even if one could derive an expression for the required rotation, one would have to make special provision for singularities (gimbal lock) and for non-commutative operations.

Quaternion Theory

Fortunately, there is a relatively simple solution to the interpolation of spatial data. It was invented by the legendary Irish mathematician, Sir William Rohan Hamilton, in the mid-nineteenth century (Hamilton 1853) and has recently attracted the attention of computer graphics specialists (*e.g.*, Shoemake 1985, Foley *et al.* 1990, p.1063). The solution requires the definition of a novel group of numbers called quaternions which are a combination of vectors and complex numbers. Several years ago I began to investigate the problem of specifying fault slip movements using complex numbers in blissful ignorance of Hamilton's work. It was my intention to store the fault's orientation and amount of slip in a multi-dimensional complex number but I was unable to define a group in the mathematical sense. Luckily, I only wasted a couple of weeks on this project and later discovered that Hamilton had grappled with it for ten years (1843-1853) before arriving at his solution.

As readers may recall from their high school math education, a complex number $\mathbf{c} = (a,b)$ consists of a real part a and an imaginary part b, both scalars, which combine as follows,

$$\mathbf{c} = a + \mathrm{i}\, b \tag{5}$$

where i (termed iota) is equal to $\sqrt{-1}$. A quaternion $\mathbf{q} = (s,\mathbf{v})$ is a number consisting of two parts, a real scalar part and an imaginary vector part,

$$\begin{aligned} \mathbf{q} &= s + \mathrm{i}\,\mathbf{v} \\ \mathbf{q} &= s + \mathrm{i}_x v_x + \mathrm{i}_y v_y + \mathrm{i}_z v_z \end{aligned} \tag{6}$$

where $\mathrm{i}_x^2=\mathrm{i}_y^2=\mathrm{i}_z^2=-1$ and $\mathrm{i}_x\mathrm{i}_y=\mathrm{i}_z$; $\mathrm{i}_y\mathrm{i}_x=-\mathrm{i}_z$, *etc.* Quaternions may be added, just like complex numbers and vectors, by adding their respective components, but multiplication is a little trickier. The products of two complex numbers, $\mathbf{c}_1$ and $\mathbf{c}_2$, and two quaternions, $\mathbf{q}_1$ and $\mathbf{q}_2$, are given by

$$\mathbf{c}_1\mathbf{c}_2 = (a_1a_2 - b_1b_2,\ a_1b_2 + b_2a_1) \tag{7a}$$

$$\mathbf{q}_1\mathbf{q}_2 = (s_1s_2 - \mathbf{v}_1\bullet\mathbf{v}_2,\ s_1\mathbf{v}_2 + s_2\mathbf{v}_1 + \mathbf{v}_1\times\mathbf{v}_2). \tag{7b}$$

The extra cross product factor makes quaternion multiplication non-commutative.

Just as a complex number $\mathbf{c}$ has a complex conjugate $\overline{\mathbf{c}}$,

$$\mathbf{c} = (a,\ b);\ \overline{\mathbf{c}} = (a,-b) \tag{8a,b}$$

so a quaternion $\mathbf{q}$ has a conjugate $\overline{\mathbf{q}}$,

$$\mathbf{q} = (s,\ \mathbf{v});\ \overline{\mathbf{q}} = (s,-\mathbf{v}). \tag{8c,d}$$

Further properties of quaternions are briefly reviewed here (for details see, for example, Maillot 1990). By analogy with complex numbers, the (squared) magnitude of a quaternion is defined using the product of a quaternion and its conjugate,

$$\mathbf{q}\bar{\mathbf{q}} = \bar{\mathbf{q}}\mathbf{q} = |\mathbf{q}|^{2} = |s|^{2} + |\mathbf{v}|^{2} \tag{9}$$

A pure quaternion **p** is one with no scalar part,

$$\mathbf{p} = (0, \mathbf{v}) \tag{10}$$

whereas for a unit quaternion

$$s^2 + v_x^2 + v_y^2 + v_z^2 = 1. \tag{11}$$

Consequently, a unit quaternion **u** may always be written in the form

$$\mathbf{u} = \left(\cos\frac{\alpha}{2}, \sin\frac{\alpha}{2}\mathbf{n}\right) \tag{12}$$

where α is some angle and **n** is a unit vector, $|\mathbf{n}| = 1$. If we use the pure form **p** as a 'position quaternion' to specify a point in space, and the unit quaternion format of eqn. (12) to specify the axis of rotation **n**, and the amount of rotation α about **n**, then the equation

$$\mathbf{p}' = \mathbf{u}\,\mathbf{p}\,\bar{\mathbf{u}} \tag{13}$$

is a quaternion description of the rotation from **p** to **p'** (magnitude is unchanged, $|\mathbf{p}'| = |\mathbf{p}|$). The occurrence of both the unit quaternion and its conjugate in eqn. (13) explains why the angle of rotation in eqn. (12) is halved. This is a mathematical alternative to the pencil and tracing paper procedure for rotation about an inclined axis on a stereonet in which the first and third steps are also 'conjugate' (1– move the axis of rotation straight down to the center of the net and apply the corresponding rotation about a horizontal axis to the data, 2 – perform the required rotation α about the now-vertical rotation axis, 3 – restore the rotation axis to its initial inclination and apply the same rotation about a horizontal axis to the data).

Advantages and Disadvantages of Quaternions

Two advantages of the quaternion approach to rotation are the avoidance of gimbal lock-type singularities and automatic incorporation of Euler's theorem. If a quaternion $\mathbf{q}_1$ transforms **p** to **p'** along a great circle arc and then a quaternion $\mathbf{q}_2$ transforms **p'** to **p"** along a great circle arc, the quaternion product $\mathbf{q}_2\mathbf{q}_1$ transforms **p** to **p"** along the direct great circle arc. Unit quaternions can be thought of as defining points on the surface of a unit hypersphere in 4-dimensional space. The rotation described by eqn. (13) is a spherical linear interpolation (or SLERP) which finds the shortest path – a great circle arc – between two points on this hypersurface. In real space, this corresponds to the Euler rotation from **p** to **p'** or **p"**. Note however, that there are always two possible quaternion descriptions of a rotation - either a rotation of θ about axis **n** or of $-\theta$ about the antipodal direction $-\mathbf{n}$. Also note that a quaternion cannot be used to describe angles of rotation in excess of 360°. The equations we have used to define orientations and rotations cannot distinguish a 365° rotation from a 005° rotation (a solution to this limitation is discussed in the SLERP section).

Quaternion - Tensor Conversions

In order to implement an orientation rotation and interpolation scheme on a personal computer, it is necessary to convert a quaternion of the form in eqn. (6) into a rotation tensor

$$\mathbf{R} = \begin{bmatrix} R_{xx} & R_{xy} & R_{xz} \\ R_{yx} & R_{yy} & R_{yz} \\ R_{zx} & R_{xx} & R_{zz} \end{bmatrix} \tag{14}$$

and *vice versa*. To convert a quaternion into a rotation tensor, let q_w be the scalar part and (q_x,q_y,q_z) the components of the vector part so that the total quaternion is written

$$\mathbf{q} = \left(q_w, \begin{bmatrix} q_x \\ q_y \\ q_z \end{bmatrix} \right). \tag{15}$$

Then

$$\mathbf{R} = \begin{bmatrix} q_w^2+q_x^2-q_y^2-q_z^2 & 2q_xq_y-2q_zq_w & 2q_zq_x+2q_yq_w \\ 2q_xq_y+2q_zq_w & q_w^2-q_x^2+q_y^2-q_z^2 & 2q_yq_z-2q_wq_x \\ 2q_zq_x-2q_yq_w & 2q_yq_z+2q_wq_x & q_w^2-q_x^2-q_y^2+q_z^2 \end{bmatrix} \tag{16}$$

(for an explanation in terms of 4×4 homogeneous coordinates, see Shoemake 1991; also Watt & Watt 1992, p.362). Conversely, if one is given the components of a rotation tensor, **R**, the equivalent quaternion may be obtained. Since it is a unit quaternion, $q_w^2 + q_x^2 + q_y^2 + q_z^2 = 1$, we have

$$q_w^2 = 1 - q_x^2 - q_y^2 - q_z^2. \tag{17}$$

From the trace of **R** in eqn. (16),

$$q_w = \frac{\sqrt{R_{xx}+R_{yy}+R_{zz}+1}}{2} \tag{18a}$$

and the other quaternion elements follow by adding diagonally opposite elements of **R** in eqn. (16),

$$q_x = \frac{R_{zy}-R_{yz}}{4q_w};\ q_y = \frac{R_{xz}-R_{zx}}{4q_w};\ q_z = \frac{R_{yx}-R_{xy}}{4q_w}. \quad (18b,c,d)$$

SLERP

We now return to the problem presented graphically in Figs. 1 and 3. Given the orientations of two planes at locations A and B, how do we linearly interpolate plane orientations at five intervening locations? The first step is to represent the planes by pure unit quaternions $\mathbf{p}_a = (0,\mathbf{n}_a)$ at A and $\mathbf{p}_b = (0,\mathbf{n}_b)$ at B. Let θ and ϕ be the trend and plunge of the poles to the planes, so that

$$\mathbf{p_a} = \left(0, \begin{bmatrix} \cos\theta_a \cos\phi_a \\ \sin\theta_a \cos\phi_a \\ \sin\phi_a \end{bmatrix}\right);\ \mathbf{p_b} = \left(0, \begin{bmatrix} \cos\theta_b \cos\phi_b \\ \sin\theta_b \cos\phi_b \\ \sin\phi_b \end{bmatrix}\right) \quad (19a,b)$$

(if this seems odd, remember that the plunge of a line is the *supplement* of the angle it subtends with the vertical axis). The angle of pitch α subtended along the great circle arc between $\mathbf{p}_a$ and $\mathbf{p}_b$ is obtained from the 'quotient' of these two pure unit quaternions. Since their scalar parts are zero,

$$\mathbf{p_b}\mathbf{p_a}^{-1} = \mathbf{p_b}\overline{\mathbf{p}}_\mathbf{a} = \left(\mathbf{n_a}\bullet\mathbf{n_b},\ \mathbf{n_a}\times\mathbf{n_b}\right) \quad (20)$$

(Shoemake 1992, 1994b) and α is easily calculated from

$$\mathbf{n_a}\bullet\mathbf{n_b} = \cos\alpha. \quad (21)$$

A spherical linear interpolation is achieved using an independent parameter t = [0,1] and, for clarity, a dependent parameter $s = 1\text{-}t$ (see De Paor 1994),

$$\mathbf{p}(s,t) = \mathbf{p_a}\frac{\sin s\alpha}{\sin\alpha} + \mathbf{p_b}\frac{\sin t\alpha}{\sin\alpha}. \quad (22)$$

For example, if t = 1/6, 1/3, 1/2, *etc.* and s = 5/6, 2/3, 1/2, *etc.*, then eqn. (22) gives the interpolated poles shown in Fig. 3. There remain the problems of ensuring that the interpolated poles fall in the acute arc between the end members and of interpolating quaternions into arcs greater than one revolution. Comparing the magnitude of the sum versus the difference of the end member quaternions, if $|\mathbf{p}_a\text{-}\mathbf{p}_b| > |\mathbf{p}_a+\mathbf{p}_b|$, then we must replace $\mathbf{p}_b$ by $-\mathbf{p}_b$ in eqn. (22) in order to interpolate into the acute arc. Of course, we may wish to interpolate into the obtuse arc; indeed if we wish to model spiral trails in porphyroblasts or delta tails on porphyroclasts, we may need to interpolate into arcs of several revolutions. A solution to this problem is given by Morrison (1992),

$$\mathbf{p}(r,s,t) = \mathbf{p_a}\frac{\sin s\beta}{\sin\alpha} + \mathbf{p_b}\frac{\sin t\beta}{\sin\alpha} \quad (23)$$

where $\beta = \alpha + 2\pi r$ and the integer r determines the number of rotations. Note that eqns. (22) and (23) are equivalent when $r = 0$. When r is negative, the sense of rotation is reversed and so $r = -1$ yields the alternative arc of the great circle.

There are more sophisticated forms of interpolation than SLERPs – for a discussion of spherical cubic spline interpolation see, for example, Watt & Watt (1992, p. 365). The pole to a surface undergoing buckle folding might be expected to describe a non-spherical-linear path on an orientation net. However, spherical linear interpolation is adequate (indeed it is appropriate) for many structural applications, ranging from simple shearing of an inclined lineation to plate motions about tectonic Euler poles.

Discussion

Quaternions provide an algebraic method for performing the rotational operations on orientation data that have been traditionally done with stereonet, tracing paper, pencil, and thumb tack! They are thus of critical importance to those who wish to adapt GIS systems to cater for the needs of structural geology and tectonics. By studying quaternion operations, students may expect to obtain an understanding of the complexities of rotational transformations. In contrast, the traditional graphical techniques involve sets of steps and rules of thumb which yield little insight into the nature of spatial orientation data or the process of rotation.

Acknowledgements

This research was supported by NSF grant EAR-9304879 and by a grant from the Mobil Corporation.

References

De Paor, D. G. 1994. A parametric representation of ellipses and ellipsoids. *J. Struct. Geol.* **16**: 1331-1333.

De Paor, D. G. 1995. Quaternions, raster shears, and the modeling of rotations in structural and tectonic studies. *Geol Soc. Amer. Abs. with Prog.* **27**(6): A72.

Foley, J. D., van Dam, A., Feiner, S. K., & Hughes, J. F. 1990. *Computer Graphics: Principles and Practice - 2nd ed.* Addison-Wesley, Reading MA.

Hamilton, W. R. 1853. *Lectures on Quaternions.* Hodges & Smith, Dublin, Ireland.

Maillot, P.-G. 1990. Using quaternions for coding 3D transformations. In: *Graphics Gems* I (ed. A.S. Glassner). Academic Press, Boston, pp. 498-515.

Morrison, J. 1992. Quaternion interpolation with extra spins. In: *Graphics Gems* III (ed. D. Kirk). Academic Press, Boston, pp. 96-97.

Shoemake, K. 1985. Animating rotation with quaternion curves. *Siggraph* **85**: 245-254.

Shoemake, K. 1991. Quaternions and 4×4 matrices. In: *Graphics Gems* II (ed. J. Arvo). Academic Press, Boston, pp. 351-354.

Shoemake, K. 1992. Arcball: a user interface for specifying three-dimensional orientation using a mouse. In: *Proceedings of Graphics Interface,* pp. 151-156.

Shoemake, K. 1994a. Euler angle conversion. In: *Graphics Gems* IV (ed. P.S. Heckbert). Academic Press, Boston, pp. 222-229.

Shoemake, K. 1994b. Arcball rotation control. In: *Graphics Gems* IV (ed. P.S. Heckbert). Academic Press, Boston, pp. 175-192 .

Watt, A. H. & Watt, M. 1992. *Advanced Animation and Rendering Techniques: Theory and Practice.* ACM Press / Addison-Wesley, Reading MA.

Computerized Geologic Map Compilation

Mark G. Adams, Laura D. Mallard,
Charles H. Trupe, and Kevin G. Stewart
Department of Geology, University of North Carolina,
Chapel Hill, NC 27599-3315, U.S.A.
mga4470@email.unc.edu

Abstract – Compilation and presentation of geologic maps are greatly aided by utilization of basic computer equipment and software. Geologic data in computer format can easily be modified to accommodate new field data. Cross sections are also easily constructed on the computer. Paper or photographic slide copies of a map can be conveniently and economically produced at various scales for use in the field or for communicating data.

We scanned USGS 7.5 minute topographic base maps on a flatbed scanner. Each sheet was scanned in six 8.5 × 14 inch (~22 × 36 cm) sections with 2 to 4 inches (~5 to 10 cm) of overlap per section. The six scanned files were imported as PICT files into *Canvas*™, spliced, and edited to eliminate overlap. Structural data, lithologic contacts, text, *etc.*, were compiled into separate layers. The completed 7.5 minute quadrangle with all compiled structural and geologic data uses approximately 3.7 MB of disk space.

Introduction

Geologists often need to print and present geologic field data, either as figures in journal publications or poster presentations, as published geologic maps, or as projected illustrations for oral presentations. Traditional techniques for compiling and presenting geologic maps involve manual drafting using published topographic maps as base maps. Disadvantages of the traditional techniques include workspace needs, difficulties in correcting errors and updating, and problems in economically reproducing the finished product, either at original or modified scale. Basic computer equipment greatly aids in the compilation and presentation of geologic maps. However, adequate topographic base maps are not readily available in computerized format. Although digital elevation maps are available for many areas in the United States, cultural features (roads, buildings, *etc.*) are generally not included in these computer files and the topographic resolution is generally inadequate for 1:24,000 scale mapping. The need to present geologic field data inspired us to develop an inexpensive, convenient method for compiling data on computerized standard topographic base maps.

This paper provides a detailed description of the technique we used for creating geologic maps on the computer. Although the following technique is described in terms of Macintosh-compatible hardware and software, we have also successfully used the technique on the PC platform with equivalent software. Specific steps in the method can be modified or improved upon, depending on the needs of, and availability of resources to, the worker. The following description is intended to assist the user in avoiding time-consuming problems that we encountered during development of this technique.

Equipment Used

Computer System

The basic system used was a Macintosh Quadra 650™ computer (version 7.1 operating system) with 250 MB hard drive. The computer came equipped with 8 MB RAM. This minimum amount of RAM tended to be somewhat slow and occasionally presented memory problems when manipulating large graphics files. Addition of a 16 MB SIMM chip and installation of *RAM Doubler*™ (made by Connectix Inc.) effectively increased the amount of available RAM to 48 MB. Five megabytes of this available memory was assigned to a RAM DISK located in the memory file in the control panels folder. Saving the graphics files to this disk greatly increased the speed of the process. Certainly other computer models will suffice, but we recommend the minimum of 8 MB RAM (preferably more) and an operating system that provides the RAM DISK feature.

Scanner

We used an Apple Color OneScanner™ to scan the base maps. This is a flatbed page scanner with an image area of 8.5 × 14 inches (~22 × 36 cm). As a result, the complete 7.5 minute quadrangle (~ 18.5 × 22.5 inches or 47 × 57 cm) had to be scanned in several segments and later spliced to complete the quadrangle. Hand-held scanners or drum scanners may also suffice. Large drum scanners are capable of scanning an entire 7.5 minute quadrangle, thus eliminating the need for splicing segmented images; however, drum scanners are not as readily available to most workers as flatbed scanners.

Printers

A variety of printing devices are adequate for output of the final product. We used an Apple Color StyleWriter™ for viewing the product during initial stages of production. The page-size printer can also be used to output the final product, but it is then necessary to manually trim and splice the pages in order to present a complete 7.5 minute quadrangle. Many professional print shops possess larger ink-jet plotters capable of printing a full quadrangle or larger maps. For our final full-color product, a professional print shop printed the maps using a HP Design Jet 650™ for a charge of about $0.75 per linear inch.

Software

Many scanners include image processing software. Our Apple OneScanner came supplied with *Ofoto*™ version 2.02. We used a graphics application (*Canvas*™ version 3.5.2 by Deneba Software Inc.) for modifying the scanned base map and for compiling the geologic data. Deneba Software also produces a version of *Canvas* for *Windows*™ which includes virtually all of the features in the Macintosh version. Other graphics applications may also be used at the discretion of the worker, however some may not include all of the same features as *Canvas*.

Technique

Scanning the Base Map

The first step in this method is to scan the base map. Regardless of the scanning equipment used, it is important that the scanned image is oriented such that lines of latitude are horizontal. This is particularly important when scanning the map using a page scanner because of the importance of accurately aligning the separate segments during the splicing stage. *Ofoto* has an auto-aligning feature in the software, but was not always able to resolve the precise border to align. *Canvas* can rotate images in 1° increments and some other image-editing applications can rotate images in fractions of a degree; however, large rotated images greatly increase the memory requirements for the file. After trying several techniques for obtaining a precise alignment, we found the most consistent method was to accurately trim the maps along the northern and southern borders, which are defined by lines of latitude. The northern border was then aligned flush with the top edge of the scanner window. The width of the scanner window is 8.5 inches (~22 cm) and the northern border of the quadrangle is 18.5 inches (47 cm); therefore, we made three scans of the northern part of the map with 2 to 4 inches (5 to 10 cm) of overlap on adjacent scans. We then scanned the southern section of the map in the same manner, but had *Ofoto* rotate those images 180° so that they would have the same orientation as the northern segments.

An important consideration while scanning in *Ofoto* is the image type. The scanned image type depends upon how the user intends to use the image. Because most standard topographic maps are in color, the `Autodetect` setting would result in the mapped scanned as a `Color Photo`. We disengaged the `Autodetect` setting and scanned the maps as `Line Art` (Fig. 1). For our purposes, this setting had several benefits. For example, `Line Art` images consume much less memory than `Color Photo` images, and color on the scanned

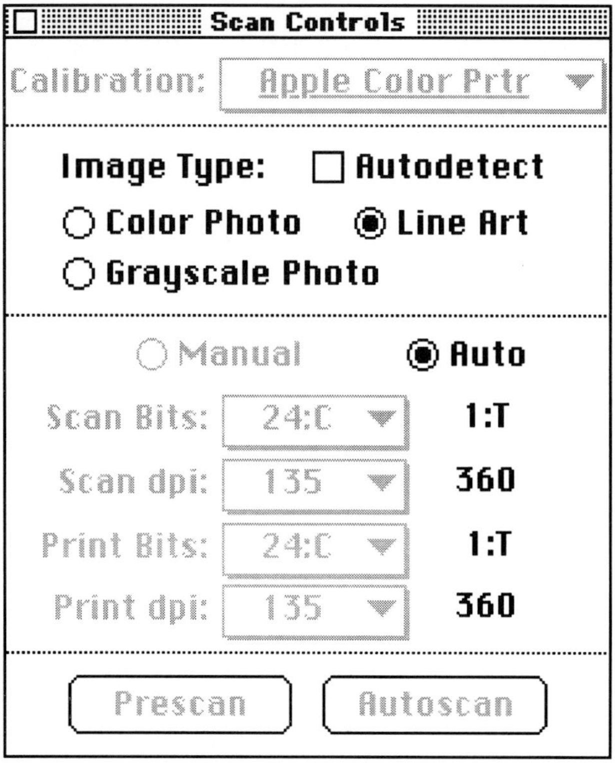

Fig. 1. Scan controls for *Ofoto*. `Autodetect` is turned off, image type is designated as `Line Art`, scanning and printing resolution are set on default settings (`Auto`).

image tends to interfere with the colored lithologic units on the compiled version.

After the image is prescanned, it is necessary to adjust the `Threshold` setting for the scanned image. The `Threshold` is adjusted within the Tone submenu in the `Options` menu. By adjusting this setting, one can eliminate unwanted shading (*e.g.*, the shaded forest area on standard topographic maps). However, relatively high threshold settings can result in a loss of parts of contour lines. We have found that `Threshold` settings in the range of 85 to 100 tend to provide satisfactory results (Fig. 2). After we were satisfied with the quality of the scanned image, we saved it as an *Ofoto* PICT format file to be exported to the graphics application.

Splicing the Scanned Images

Because of the large memory size of the PICT files, we allotted additional memory to the *Canvas* application. To change the memory setting, select the application

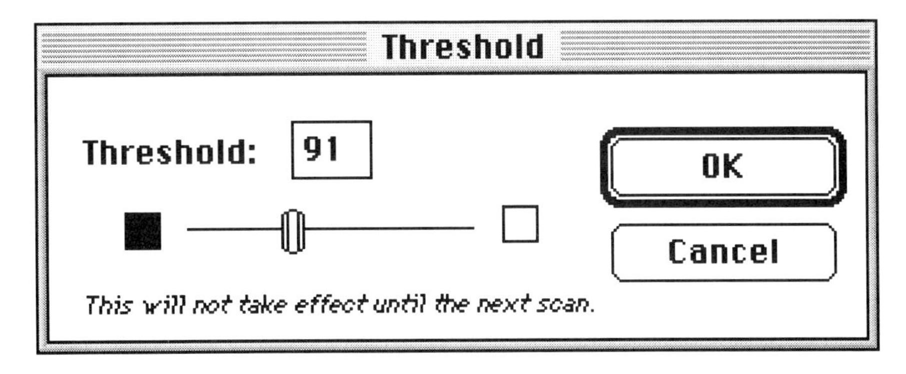

Fig. 2. Threshold window from *Ofoto*. The window is chosen from the Tone submenu under the Options menu.

from the desktop, choose Get Info... from the File menu, and enter the desired memory size in the appropriate boxes in the info window (Fig. 3). When *Canvas* is launched, the computer will now allocate the user-specified amount of RAM to the application. The PICT files are opened from *Canvas* and saved as *Canvas* format files. This reduces their memory size by 40 to 50%.

Using the Place command from the file menu, we then transported adjacent scan segments into a single file. By selecting the images, the adjacent scans can then be transported to a single layer via the Send to Layer command from the Arrange submenu under the Object menu. After the images were transferred to the same layer, the ink mode for each image was designated as OR (Fig. 4). The images were then aligned such that the overlap areas were superimposed (Fig. 5). Due to several possible factors (*e.g.*, imprecise alignment, possible aberration during scanning, *etc.*) certain zones within the region of overlap are more suitably aligned than other zones. Dragging the handles of each scanned object can then eliminate the overlap (Fig. 6). After images are aligned and overlap eliminated, the objects are selected and converted to a single paint object in the Group Specifications submenu under the Objects menu (Fig. 7). The resolution of the Paint object is set at 1 bit depth and 300 dpi. Stray dots, unwanted shading, or irregularities in the seams can then be modified using the eraser or pencil tools on the tool pallet. The entire quadrangle is now stored as a computer file occupying 2.7 to 3.2 MB of disk storage. Portions can be printed at virtually any scale for various uses (*e.g.,* field maps).

One problem we encountered with this technique involves the scale of the completed map. The completed maps are generally slightly larger (~ 0.3 to 0.6%) than the original topo map. We are unsure whether this distortion occurs as a result of aberration during scanning, or whether it is caused by the *Ofoto* scanning software or the *Canvas* graphics software. To correct for this the map can be scaled from the Object menu. Because bit mapped objects tend to become jagged when scaled, it is necessary to first convert the object type from a Paint object (bit mapped) to a Picture (vector) object. This can be accomplished from the Object Specs submenu in the Objects menu (Fig. 8). Then, scale the map by the

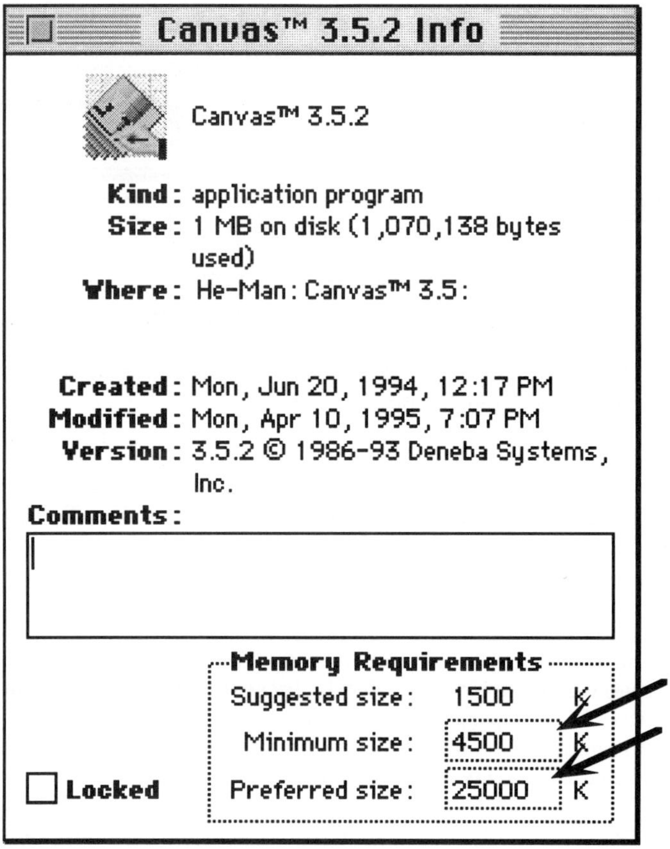

Fig. 3. *Canvas* `Info` window. Arrows point to modified memory requirements.

calculated percentage in the Scale submenu (Fig. 9). It is important to note that `Picture` objects cannot be edited (*e.g.*, colors changed, portions deleted, *etc.*); therefore, it may be necessary to convert the object back to a `Paint` object after scaling. We have been able to adjust the computerized maps in this manner such that they exactly match the scale of the original topo map.

Compiling Geologic Data

After the topographic base map has been computerized, geologic field data can be compiled onto the base map. For use as a base map with overlain geologic data, we selected the base map and set the fore color to a light gray so that the base map would appear subordinate to the geologic data. For convenience, several layers can be created so that different aspects can be compiled separately (Fig. 10). The

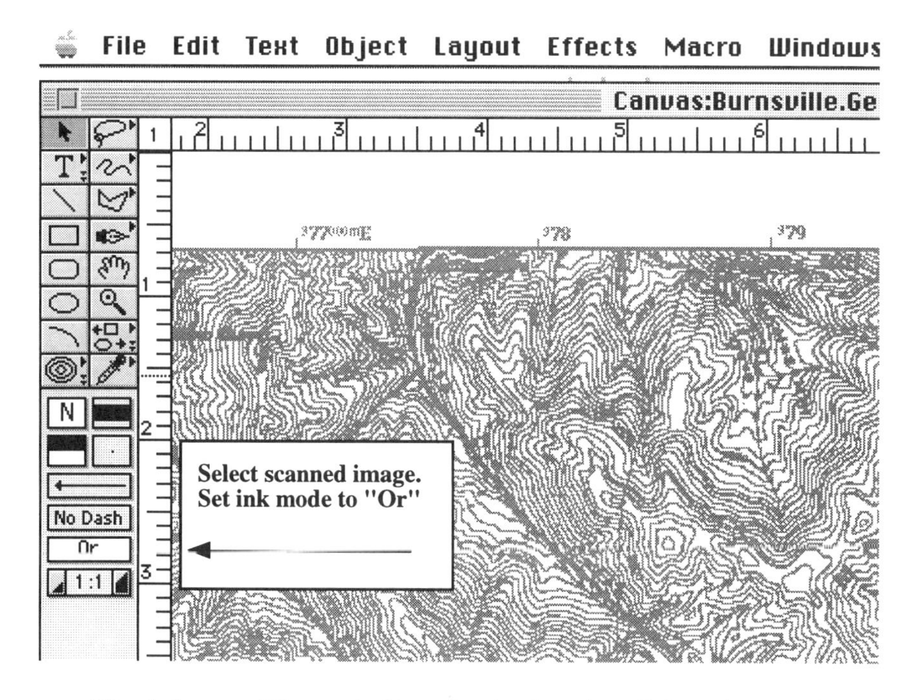

Fig. 4. Image of *Canvas* tool bar showing location of ink mode controls.

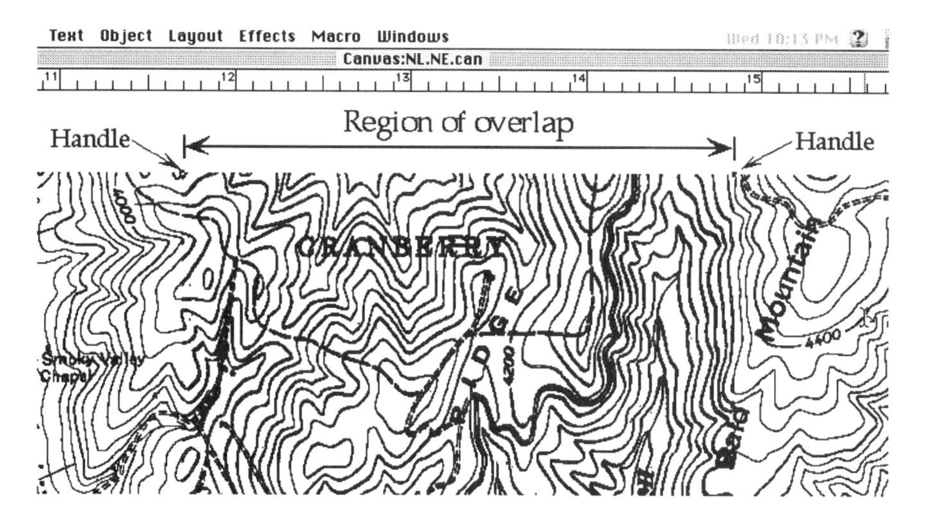

Fig. 5. Adjacent scans of topographic quadrangle with region of overlap. Heavier contour lines in region of overlap (most notable on right side) are due to slight misalignment of scans.

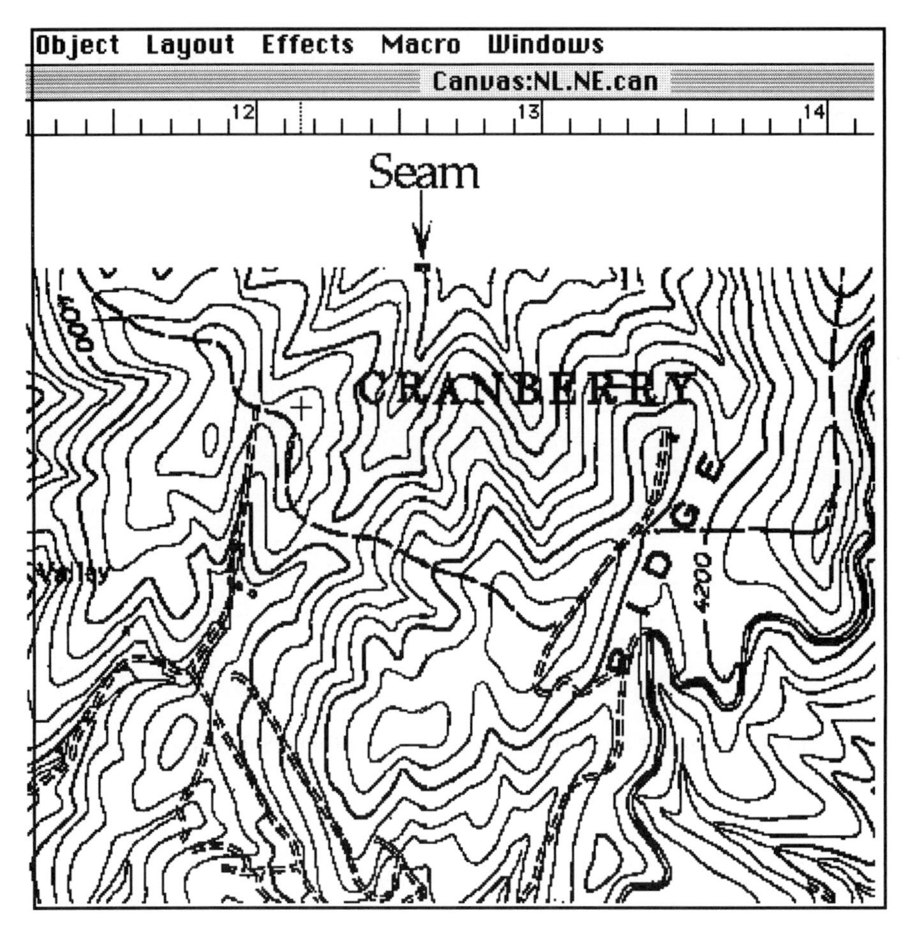

Fig. 6. Same images as Fig. 5, without overlap. Seam illustrates slight misalignment of adjacent scans.

layers are created in the `Layer Specs` submenu (`Layout` menu). During compilation, we created an extra `Working Layer`, to manipulate specific objects without inadvertently disturbing objects in other layers.

The first stage was to compile structural data. We created structure symbols (metamorphic foliation, mylonitic foliation, compositional layering, mineral lineations, fold hinges, *etc.*) as vertical symbols (*i.e.*, oriented north-south) and saved them as a `Macro` set. We then could place structure symbols in the structure layer by choosing them from the `Macro` menu. The symbols could then be rotated from the `Effects` menu to the appropriate orientation and plotted in the appropriate position relative to the base map. The values for the structural data were plotted in a separate layer so that symbols and values (*e.g.*, line density, font, style, *etc.*) could be grouped and modified separately if needed.

After compiling structural data, we created lithologic contacts as polygons in the contacts layer. For full color output, it is desirable for the lithologic units

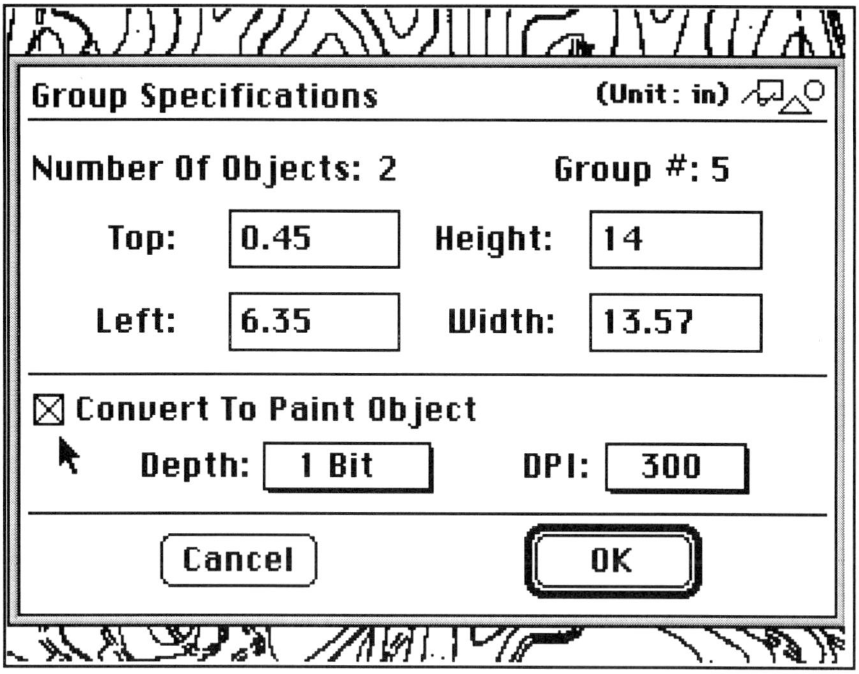

Fig. 7. Group Specifications window. This window is chosen from the `Objects` menu. The scanned images are selected, then converted to a single paint object at the resolution settings illustrated in the figure.

on the map to be in color. The most convenient method for this is to create the map units as filled polygons. Because filled polygons are opaque, we created the colored map units in a layer beneath the topographic base map layer (Fig. 10). When the ink mode of the base map is designated as `Or`, the colored map units underneath the base map can be viewed while retaining the topographic features. Copies of the contacts were pasted into the colored map units layer. These objects can be exactly overlaid on the original objects by pressing the `Tab` key and choosing the `Paste In Position` command from the `Edit` menu. The polygons in the map units layer were closed and filled with appropriate colors. The lines around the colored map units were then eliminated by designating no pen pattern. Labels for the map units and other text were then typed in the text layer. After this stage the geologic map is essentially complete.

Constructing Cross Sections from the Geologic Map

The completed geologic map can be used for the construction of cross sections on the computer. A duplicate of the geologic map should be made for construction of the cross section. In this completed form, the file consumes a relatively large amount of memory (3-4 MB); manipulation of such a large file thus requires abundant RAM. To reduce the memory requirements needed for constructing the cross section, it may be convenient to delete parts of the geologic map not needed.

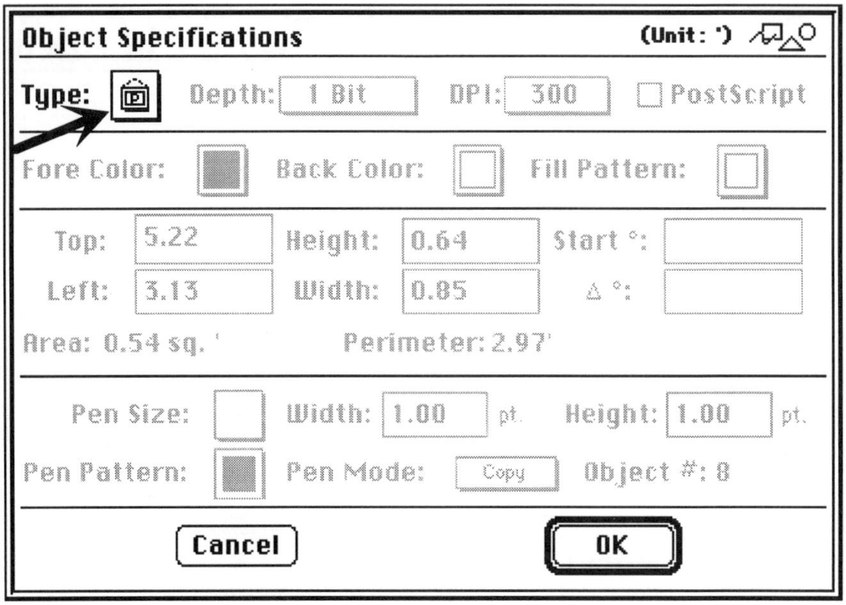

Fig. 8. `Object Specifications` submenu. Object is converted from `Paint` object to `Picture` object in the `Type` pop-up menu (see arrow).

The important parts of the map needed for the cross section are the base map, structural symbols, and lithologic contacts. Parts of the base map can be deleted by selecting the base map and dragging the handles as described in the "Splicing the Scanned Images" section. Structural symbols outside of the area of interest can be deleted by selecting the objects and pressing the `Delete` key. Because many of the geologic features (lithologic contacts, faults, *etc.*) are objects that cover large portions of the map, to work with just a band of the geologic map it is necessary to cut portions of these features outside of the area of interest. Using the `Scissors` tool on the tool bar, one can remove parts of the contacts not needed for the cross section by cutting the polygons (clicking on the polygon at the desired positions) and deleting the unwanted parts. After the map has been reduced to the area of interest, objects in each separate layer should be grouped via the `Objects` menu.

To avoid confusion during construction of the topographic profile, it may be necessary to make all layers, except the base map layer, invisible. This can be done by clicking the √ toggles under `V` in the `Layer Specifications` window (Fig. 10). The base map can then be rotated so that the chosen line of section is horizontal (Fig. 11). This is accomplished by selecting the base map and rotating it the desired amount using the `Rotate` command under the `Effects` menu. By creating a scaled grid using the `Grid Maker` command (in the `Managers` submenu, `Edit` menu), the program can be used in the same manner as graph paper. Horizontal lines can be ruled to represent different elevations on the cross section. Vertical lines can then be drawn from the intersections of contour lines with the line of section to the appropriate elevation on the section. This can be done by activating the `Line` tool on the `tool pallet`. The topographic profile is then constructed by connecting the end points (Fig. 11).

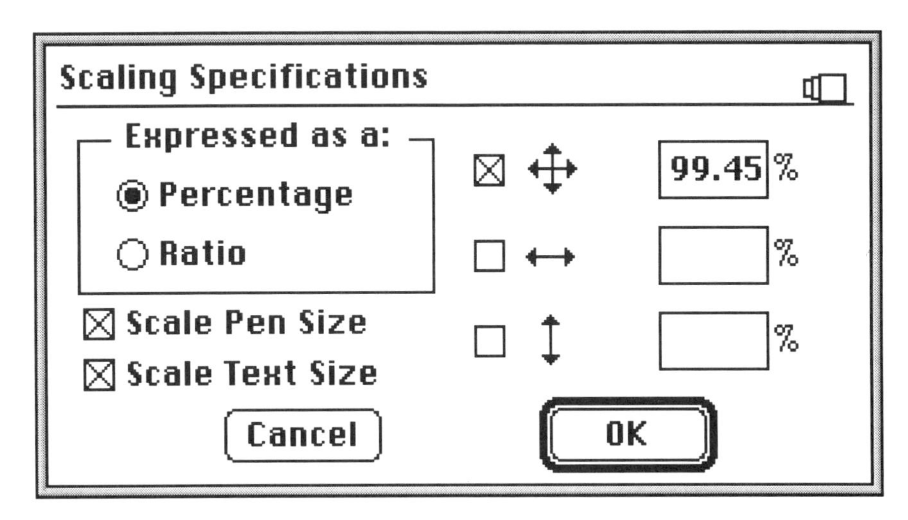

Fig. 9. `Scaling Specifications` window. This figure illustrates that a map is to be scaled by 99.45% in both the x– and y–directions in order to match the scale of the original topographic map.

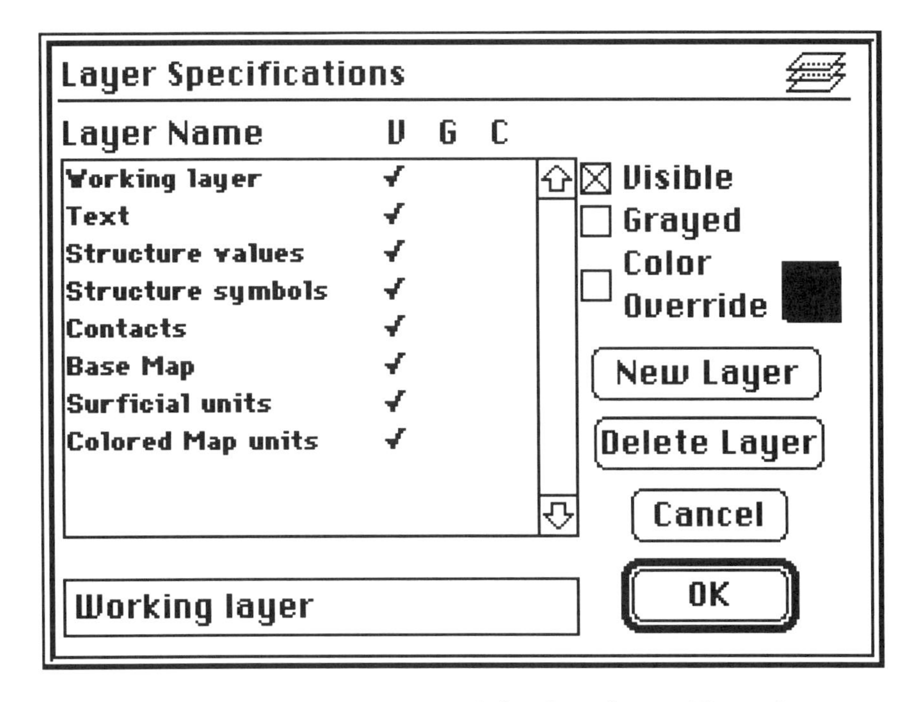

Fig. 10. `Layer Specifications` window from *Canvas*. Figure shows multiple layers used for compilation of geologic data onto scanned topographic base map.

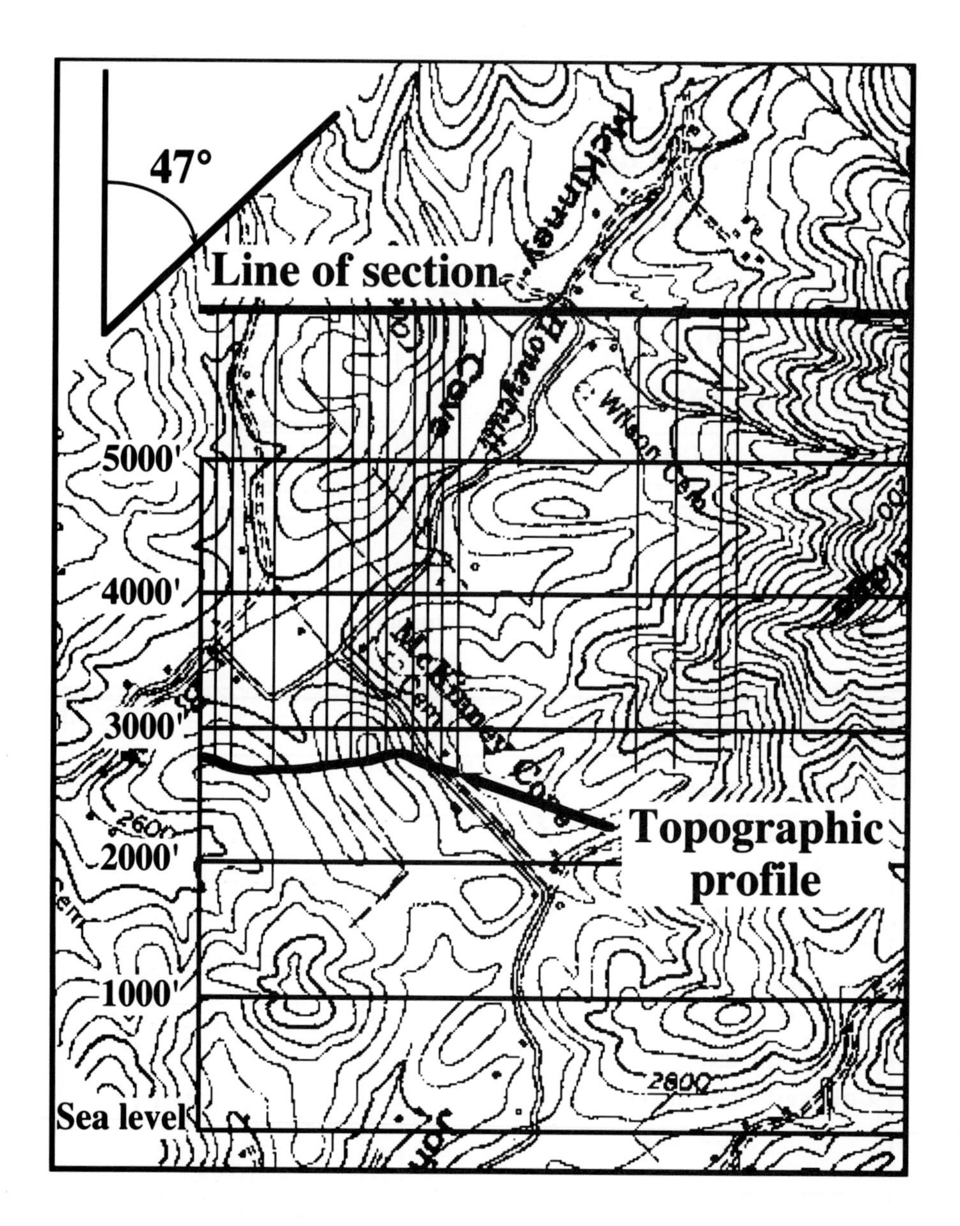

Fig. 11. Image showing construction of cross section. Base map was rotated so that chosen line of section (azimuth: N47E) is horizontal. Grid was created at appropriate scale. Vertical lines were drawn from the intersection of the line of section with contours and then extended to the appropriate elevation on the grid. Topographic profile was drawn by connecting end points of elevation lines.

Contacts and structural data can be plotted on the cross section in the same manner. It is necessary to rotate these features to the same orientation as the rotated base map. These data must be grouped in order to rotate them as a group. We have found that with such large memory files, occasionally all structural data points are not rotated to the appropriate orientation. A solution to this problem is to select all of the structural symbols and convert them to a single `paint` object, as in Fig. 7, before rotating. Lithologic units on the interpretive cross section can then be traced and filled with any desired pattern or color, similar to the technique for coloring the map units. The completed cross sections can be used to create restored or balanced cross sections following the method described by Groshong & Epard (1996).

Presenting the Map

In this computer format, the map and cross section can easily be modified to accommodate new data. Parts or all of the map can be output at any scale in a variety of media formats, such as printed maps or photographic slides. For a recent presentation (Adams *et al.* 1994), we had portions of five quadrangles (1:24,000 scale) printed by a professional print shop in full color on a HP Design Jet 650C ink jet plotter. One of the files was a full 7.5 minute quadrangle, using approximately 3.7 MB of memory space. In order to transfer the files to the printer, we compressed the files into self-expanding archives using *Compact Pro* version 1.34. The archives were then downloaded onto floppy disks and transported to the printer.

Summary

Basic Macintosh computer equipment can be useful for compiling geologic maps and constructing interpretive cross sections. Our technique for scanning topographic maps on the computer provides the geologist with a relatively inexpensive way of compiling field data on a computer format topographic base map. The geologic maps can be easily modified and output at a variety of scales and formats. The final products (maps and cross sections) can be output as printed maps, photographs, or photographic slides for written or oral presentations or for use in the field.

Acknowledgements

Financial assistance was provided by NSF Grant EAR-9316033. Critical reviews by Rick Groshong and an anonymous reviewer greatly improved the manuscript.

References

Adams, M. G., Trupe, C. H., Stewart, K. G., Butler, J. R., & Goldberg, S. A. 1994. Extent and field relations of eclogite in the Blue Ridge of western North Carolina. *Geol. Soc. Am. Abstr. with Prog.* **26**: 196.

Groshong, R. H., Jr. & Epard, J.-L., 1996. Computerized cross section balance and restoration. In: De Paor, D. G. (ed.). *Structural Geology and Personal Computers*. Elsevier Science Ltd., pp. 477-498.

Fieldlog: GIS Software as a Mapping Aid for Structural Geologists

Mohamed I. Matsah and Timothy Kusky
Department of Earth Sciences, Boston University, 675 Commonwealth Ave., Boston MA 02215, U.S.A. kusky@bu.edu

Abstract- We here describe the operation of *Fieldlog,* a Geographical Information System suited to the needs of structural geologists. This application will be of interest to PC users who wish to automate the management of field data and the generation of structural maps.

Introduction

Fieldlog version 2.8.3 is a PC-based limited Geographical Information System developed by Boyan Brodaric of the Geological Survey of Canada, and available as freeware from http://gold.gsc.ca. *Fieldlog* is used for recording and plotting field data, giving users tools to design and build their own database which can be imported into one of several Computer Aided Design (CAD) software packages such as *AutoCAD*™ by Autodesk Inc., and thereby represented in a map format. The *Fieldlog* database can be designed by the geologist to record geological directional data such as the orientation of planar and linear features, which can then be plotted on a map as symbols with their exact orientation. *Fieldlog* also permits the user to perform simple database operations on the recorded data. *Fieldlog* is a particularly useful tool for structural mapping because it helps in relieving the cumbersome chore of plotting structural data and symbols on a map in the correct geographic orientation. *Fieldlog* version 2.83 requires at least 540K free RAM memory, and DOS 3.0 or greater, and runs on most laptop computers. When run with *AutoCAD* and a small digitizing pad, *Fieldlog* becomes a powerful mapping tool that facilitates the nightly routine of plotting field data, and it increases the ability of the field geologist to analyze data while in the field, and adjust the mapping program accordingly.

The *Fieldlog* program is composed of two main modules including the *Fieldlog* database program module and the *Fieldlog-AutoCAD* interface module. The database program module is a simple database designing program in a table-like format while the *Fieldlog-AutoCAD* interface module is an add-on module for *AutoCAD* to utilize *Fieldlog* databases from within *AutoCAD* menus and commands to perform simple GIS-like operations.

The Database Program Module

The database program module is used to design the structure of a database project and for data entry. The data can then be plotted in a DXF file or printed in a table format. In general, the use of the *Fieldlog* database program module with geological field data involves four steps: 1) Designing and building the basic structure of the database, 2) Data entry and manipulation, 3) Map generation, and 4) Report generation. These steps are described below.

Designing and Building the Database

The *Fieldlog* database module is a database-designing module that gives the user some flexibility in designing their own database. The *Fieldlog* program can deal only with one database project at a time. A database project in *Fieldlog* is made of a set of tables and each table is composed of a set of related fields. *Fieldlog* can accommodate a maximum of 15 tables with 27 fields in each table. At this stage the structure of the database is built by the user by specifying the different tables and fields and the types of data that are to be recorded in each of them. Brodaric (1996) has designed a tutorial for the program, in which a data table might include information about stations, structures, samples, or photographs, and data fields would include information specific to those fields, such as location of stations, date visited, or the map sheet the station is located in. The structural table might include data fields such as strike and dip of planar features, or trend and plunge of linear features. For any project *Fieldlog* requires that at least one field must be common in all tables in any project. For instance, a project may consist of 4 tables, with table 1 including fields 1, 2, and 3, table 2 including fields 1, 4, and 5, table 3 including fields 1, 6, and 7, and table 4 consisting of fields 1, 8, and 9. This allows different data fields to be related to each other, giving *Fieldlog* the power of a relational database.

In the tutorial example, Brodaric (1996) uses four tables, including one for stations, one for structure, another for samples, and another for photographs. The station table is used for recording the field station information, the structure table is used for recording information about structural features, the sample table is used for recording sample information, and the photograph table is used for recording information about photographs. *Fieldlog* gives the user the ability to design each field according to the data to be recorded in that field. Several data types are available for designing fields. Fields can be designed to accommodate text, integer or real numbers, date, or location data, *etc*. One of the main features of *Fieldlog* is that the `location` type fields can accommodate coordinates of the location of a station in three formats (UTM, Lat-Long, or user grid) and at the data entry stage, data can be entered into location fields either by the keyboard, or by digitizing points using a digitizing tablet attached to the PC. The `symbol` field type records the symbol name of the feature to be plotted.

One of the advantages of *Fieldlog* is that a field can be assigned certain keywords, and the program will not accept any entry for that field except for one of these words. For example, for the feature field in the structure table of the tutorial project certain structural feature names are assigned to that field representing the known structural features of an area. At the data entry stage that field will accept only one of the preassigned keywords. This is a good feature that helps to make the data consistent for database operations and comparison as well as to eliminate typographic errors in the data entry.

Besides defining tables and fields, the database design includes setting some of the database controlling parameters such as selecting a common field for all the tables in a project (primary key), which is then used to associate data between tables (Brodaric, 1996). Secondary keys can also be set; they are a second level of association between tables.

The data base design of *Fieldlog* gives you the choice of setting the layer or layers onto which the different features or field data will be plotted in the plot generation stage. One or more features or fields can be plotted on one or more layers depending on the needs of the user. For example, station information can be plotted on one layer and structural features can each be plotted in a separate layer.

Data can be marked hidden or unhidden according to logical conditions set forth by the user in the designing stage. This eliminates plotting some of the data at the plotting stage but the data are still available in the database for statistical operations. This feature can be used to avoid plotting duplicate features at the same location. For example, if several foliation readings at the same station have nearly the same value then this feature can be used to plot only a single representative symbol.

Data Entry

After the database design is complete, then data are entered into the database according to the rules and constraints set forth in the designing stage. For fields that were assigned certain keywords, *Fieldlog* will pop up a window with keywords to choose from instead of making the user type them over and over again. Location fields permit the user to enter coordinates of a point according to the projection used but conversion is also available at the data entry level between UTM and Lat–Long projections, which allows correlation between the two projections. Furthermore, *Fieldlog* can enter location data from a digitizer attached to the PC. *Fieldlog* can also enter data into the database by importing it from a text file.

After entering the data into the database, several operations can be performed on the data, including sorting, hiding or unhiding, deleting or adding, grouping, and setting global conditions.

Map Generation

After entering the data into the database and manipulating it, map layers can be generated and plotted into a file in DXF format. The map generation command has several parameters that can be changed to control the appearance of the data on the map. For example, the symbol and text sizes can be changed as well as setting the rotation direction of the symbols. Also, all or parts of the data may be plotted according to some logical conditions that can be set by the user at this stage. For example, the user can ask the program to plot only data recorded after a certain date, or at certain locations, or plot only a certain data type or feature using simple menu-driven logical operations.

The *Fieldlog* map-generating command can generate maps in three different DXF formats that can be read and dealt with by the *AutoCAD* program. Two of these formats are general and can be read in several CAD programs other than *AutoCAD*. The third format is intended to be used with the *Fieldlog-AutoCAD* interface module to perform simple GIS operations from within *AutoCAD*.

Report Generation

Fieldlog has the capability to generate text files containing data from the tables of a project in a tabular format with user-specified delimiters for fields and text. Users can specify some conditions on the data to determine which part of the data to output to the report. These reports can be imported to other programs such as stereonet plotting programs or spreadsheets. This feature is tremendously useful for rapidly plotting stereographic projections of various structural data from user-definable domains, which change as mapping progresses.

The Fieldlog-AutoCAD Interface Module

This module is made as an add-on module to *AutoCAD* and it gives *AutoCAD* the ability to communicate interactively with *Fieldlog* databases (Brodaric, 1996). It also adds to *AutoCAD* some GIS-like operations and benefits. For example new data may be added, deleted, browsed, or modified to the map and the database at the same time and any changes made to either the database or the map will affect the other. This is to ensure the consistency of data sets for projects (Brodaric, 1996). The DXF files may be displayed by an *AutoCAD* program running on a PC or Macintosh personal computer.

Conclusion

Fieldlog version 2.83 is very useful to field geologists for recording and plotting geological point data and relieves one from the cumbersome work of plotting orientation data. When linked with *AutoCAD*, a sophisticated mapping tool emerges, as lines can be drawn representing geological contacts. The program can be linked to other tools, such as stereonet plotting programs. In this way, a map can be linked with structural data tables and plots, and updated and changed daily while in the field. *Fieldlog* has good and flexible capabilities for database design and operations. It can enter location coordinates from a digitizer and can record them in UTM, Lat-long, or a user-defined grid. It can generate maps in three DXF formats. It can import data from text files but it is limited to a specific format. A newer release that can produce plots to be used in *ArcView*™ is also available and is currently being tested by the structural geology research group at Boston University. All in all, it is hard to beat the features/cost relationships of *Fieldlog* (remember, this GIS is freeware!). However, some of the pitfalls of the program include:

> 1– The plot generating command does not have the capabilities to plot or at least show a rough map on the screen, which means that in order to see the plot, *Fieldlog* must be exited and a CAD program or at least a program that can read and show DXF files must be used to see the plot.
>
> 2– Hard copies of the map can not be produced directly from the *Fieldlog* database program, but must be done through the *AutoCAD* program.
>
> 3– The *Fieldlog-AutoCAD* interface module does not have the capabilities to design new *Fieldlog* databases and it can not redesign or restructure existing *Fieldlog* databases.

References

Brodaric, B. 1996. *Fieldlog version 2.8.3*. Geol. Surv. Can., Ottawa, Ontario, Canada, 89 pp.

Computerized Cross Section Balance and Restoration

Richard H. Groshong, Jr.
Department of Geology, University of Alabama,
Box 870338, Tuscaloosa AL 35487-0338, U.S.A.
rgroshon@wgs.geo.ua.edu

Jean-Luc Epard
Institut de Géologie, Université de Lausanne,
BFSH2, CH-1015 Lausanne, Switzerland.

Abstract– We describe how to make the necessary measurements and manipulations to balance and restore cross sections using the widely available programs *Canvas*™, *Photoshop*™, and *Excel*™. Data or cross sections are input to *Canvas* from a flatbed scanner or a video camera, or sections are constructed using program drafting tools. Scanned photos and captured video are significantly improved using *Photoshop* filters. Lengths and areas are automatically measured with line-drawing tools in *Canvas*. Automation of measurement makes area-depth-strain relationship a powerful and practical tool for creating and testing balanced cross sections independent of kinematic model. An *Excel* procedure is given for calculating layer-parallel strain, fitting least-squares lines to area-depth data to find depth to detachment, and plotting the results. Restorations can be made in *Canvas* by rigid-block translation, rotation, and simple-shear distortion of images or drawings. We outline techniques for constant bed-length and oblique-shear restorations.

Introduction

The balancing and restoring of cross sections refers to a group of techniques for developing, testing, and validating a structural interpretation. Restoring is returning the cross section to the geometry it had prior to deformation. A balanced cross section is restorable, area-constant or with explainable area changes, and has the correct structural style. Balancing and restoring techniques are based on measurements of bed lengths, thicknesses, and areas. The concepts are discussed in detail by Dahlstrom (1969, 1990), De Paor (1988), Geiser (1988), Woodward *et. al.* (1989), Mitra & Namson (1989), Epard & Groshong (1993), Groshong & Epard (1994), and Groshong (1994). The purpose of this paper is to describe how the necessary measurements and calculations can be made with the widely available, general-purpose software packages *Canvas* by Deneba Inc., *Photoshop* by Adobe Inc., and *Excel* by Microsoft Inc.

In the procedures to be outlined, a cross section is input directly into the computer (or constructed therein) and the required measurements are made on the screen. We use *Canvas* because it provides the necessary drawing tools, area measurements, and length measurements. *ArtWorks*™ by Deneba Inc. also provides length and area measurement capabilities. A variety of drawing and CAD programs should be suitable for this purpose although most low-cost drawing programs do not have the capability for measuring areas. *Canvas* also has `cut` and `paste` operations for bit-mapped images from a scanner. This discussion will be given in terms of *Canvas* operations, but any program where it is possible to measure line lengths, areas and angles, to work on multiple layers, and to manipulate scanned objects should be equally satisfactory. Courier font denotes program menu items or commands. *Canvas* is virtually the same on Macintosh and *DOS* based systems and files can be exchanged between the two. Our main experience is with *Canvas* 3.0 on a Macintosh and so the techniques will be described in this context. Program operations will be explained the first time they are used in the discussion.

The cross section to be balanced and restored may be constructed in *Canvas*. The program functions like a sheet of graph paper and provides the necessary ruler and protractor functions needed for drafting. We typically input the data as an image and make the measurements or draw the section as an overlay on the image. The images may range from reduced well logs to a version of the cross section itself as, for example, a drawing, photograph, or seismic line. The output can range from as little as the required length and area measurements to final drafted cross sections. This paper is organized in the sequence in which we usually perform the analysis: image input, section construction, testing for balance, restoration, and printed output. Suitable digitizing and computer hardware is described in the Appendix.

Adapting general-market commercial software to structural purposes is an ongoing learning experience with ever-changing products. We have probably overlooked good programs and techniques and new products will be developed in the future. We will be grateful for suggestions for improvements.

Image Input

Images can be input from other programs or from a scanner. Orient the image to be scanned so that horizontal and vertical lines are exactly parallel to the scanner horizontal and vertical. This results in better scans and eliminates the need for rotations when piecing sections together from separate scans. *Canvas* allows rotations in one degree increments using the `Free Rotate` sub menu command in the `Effects` menu but this will probably not be accurate enough for perfect alignment. *Photoshop LE* has an `Arbitrary Rotate` function in the `Rotate` sub menu of the `Image` menu. This function allows the rotations of a fraction of a degree that are needed for accurate alignments. Place the image into *Canvas* without vertical exaggeration. Vertical exaggerations can be done in *Canvas* with `Scales` in the `Object` menu. Independent vertical and horizontal exaggerations are possible with this function.

Images from other programs must be saved in a format that can be read by *Canvas*. Some programs have a *Canvas* file option. Line drawings that are incompatible with *Canvas* can probably be saved in the original program as a CGM file, or as a bit-mapped image such as a paint file or a TIFF file which *Canvas* can read.

The simplest input procedure from a scanner is to scan as a `Copy` to the clipboard and then `Paste` into *Canvas* or into *Photoshop*. Large images may exceed the memory of the clipboard and cannot be imported this way. In this situation, save the image as a file in the scanning program and then open the file from within *Canvas*. It is also possible to scan directly into *Photoshop* via the Twain interface and bypass the clipboard.

A large cross section can be input with a small scanner. The section can be reduced before scanning, either photographically or by xerographic reduction (Fig. 1a). Large images can also be scanned as separate pieces and then fitted together inside *Canvas*. Be sure to have a common reference on each scan, such as a sea-level line for verifying that the pieces are reassembled properly (see Adams *et al.* in this volume). The `Place` command in the `File` menu makes it possible to add files to an already open *Canvas* file. After opening the destination file, choose `Place`, then choose and open the file to be added. The new file will be added as a new layer in the open document. It can be shifted to the first layer with the `Send To Layer` command in the `Arrange` submenu in the `Object` menu. Another approach to the large-image problem is to use a video camera as a scanner. The video board should come with the necessary image-capture software.

A video or photographic image will probably benefit from digital sharpening with the filters in *Photoshop LE*. The `Sharpen` and `Sharpen Edges` filters make significant improvements to most images. We have used these filters to improve a video image from unusable at 2× on-screen magnification to very good at 2× and still usable at 4× (Fig. 1b). Sharpening can make a scanned image better than the original for structural interpretation by, for example, making the bedding planes more distinct.

Cross Section Construction

Set Up

Begin by setting the working area. If an area greater than 8 1/2 × 14 in. is required, select additional pages in `Drawing Size` in the `Layout` menu. To orient the cross section along the long dimension of a page, choose the desired orientation under `Page Setup` in the `File` menu. Create a new layer for each separate aspect of the drawing. For example, if the section is to be drawn on top of a scanned image, make the image the bottom layer and the drawing an overlying layer. This precaution will prevent the drawing from changing the underlying image. `New` layers are created or the active layer changed in the `Layer Specs` submenu of the `Layout` menu. Quick access to the layer information is obtained by clicking on the `Layer#` in the information bar at the bottom of the window.

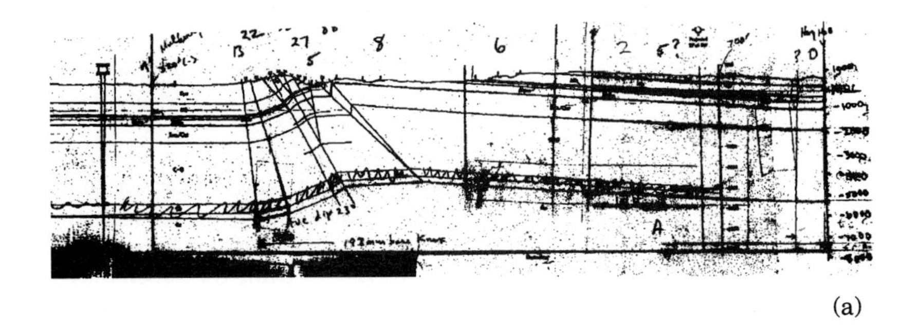

(a)

(b)

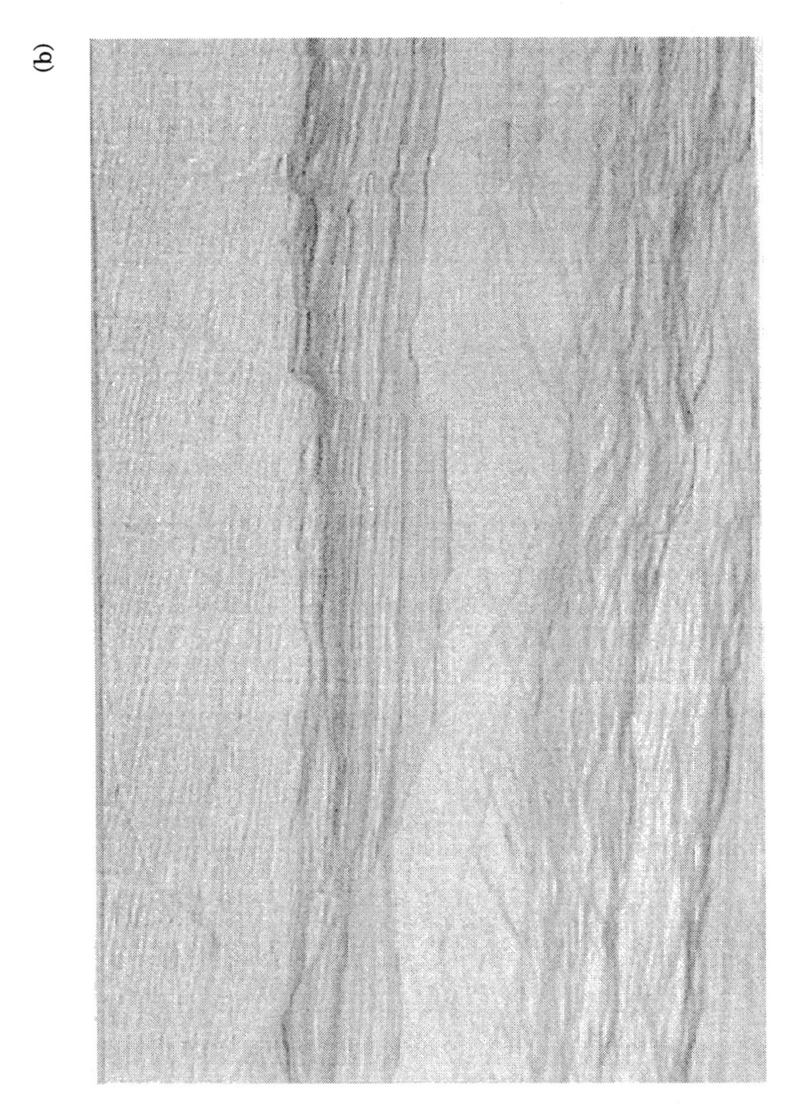

Set the scale of the cross section equal to the true scale. In the Show/Hide submenu of the Layout menu choose Rulers. In the Rulers submenu of the Layout menu, set the screen scale to the true scale of the cross section. A simple way to set the scale is from a bar scale in the imported image. Draw a line on the overlay in *Canvas* that matches the length of the bar and then adjust the scale ratio in the Rulers submenu until the indicated line length is exactly correct. The line-length measurement is in the Object Specs submenu listed under the Object menu. Double-click an object to go directly to the Object Specs. Line lengths are measured from the centers of the handles at the ends of the line, thus the measured length is always slightly less than the visible length. Use thin lines to reduce this difference to below the level of significance. Set line widths with the Pen Size palette icon in the on-screen toolbar.

A grid is useful for lining up the image with the horizontal and in fitting multiple images together. Choose Grid in the Show/Hide submenu. Alignment can be assisted by using the horizontal and vertical lines that can be dragged into view from the edges of the rulers.

Draw a Cross Section

Start a new layer for constructing the section so as not to disturb the original data. Graying the data image layer in the Layer Specs submenu may make the overlay easier to see. Drawing with colored lines may also enhance visibility. Group things on the cross section that will not be changed, such as section borders, data, *etc.*, using the Group command in the Object menu. Before drawing, consider the connectivity, or lack of it, that may be desirable in the lines. If the data on the section can be suitably grouped and moved, it may be possible to restore the cross section by manipulating the pieces of a copy of the original drawing, rather than being required to make a completely new drawing.

For a dip-domain cross section (method of Gill 1953), beds are planar between sharp hinges. Draw beds with the line or polygon tools from the tool bar. Make parallel lines with the Duplicate command in the Edit menu. Measure axial angles in folds by aligning a line on one limb and recording the orientation indicated in the bar at the bottom of the screen, then rotate the line to the other limb to find its orientation. For constant bed thickness structures, axial surfaces bisect the hinges. Draw the axial surfaces as construction lines and drag to the best-fit positions or locate the hinge positions by extending correlative stratigraphic horizons until they intersect (Fig. 2a).

Fig. 1 (opposite). Typical scanned images used as input for cross section analysis in *Canvas*. a) Cross section of southern Sequatchie anticline, Alabama (Cherry 1990). Original drawing (about 1 m long) was reduced xerographically to about 30 cm and scanned on a flatbed scanner. No effort is made to clean or improve image because cross section will be re-drafted using image as base. b) Sand peel, about 75 cm wide, of naturally deformed sand (peel by Dietmar Meier). Original peel was recorded as video image, input directly into computer, and sharpened in *Photoshop*.

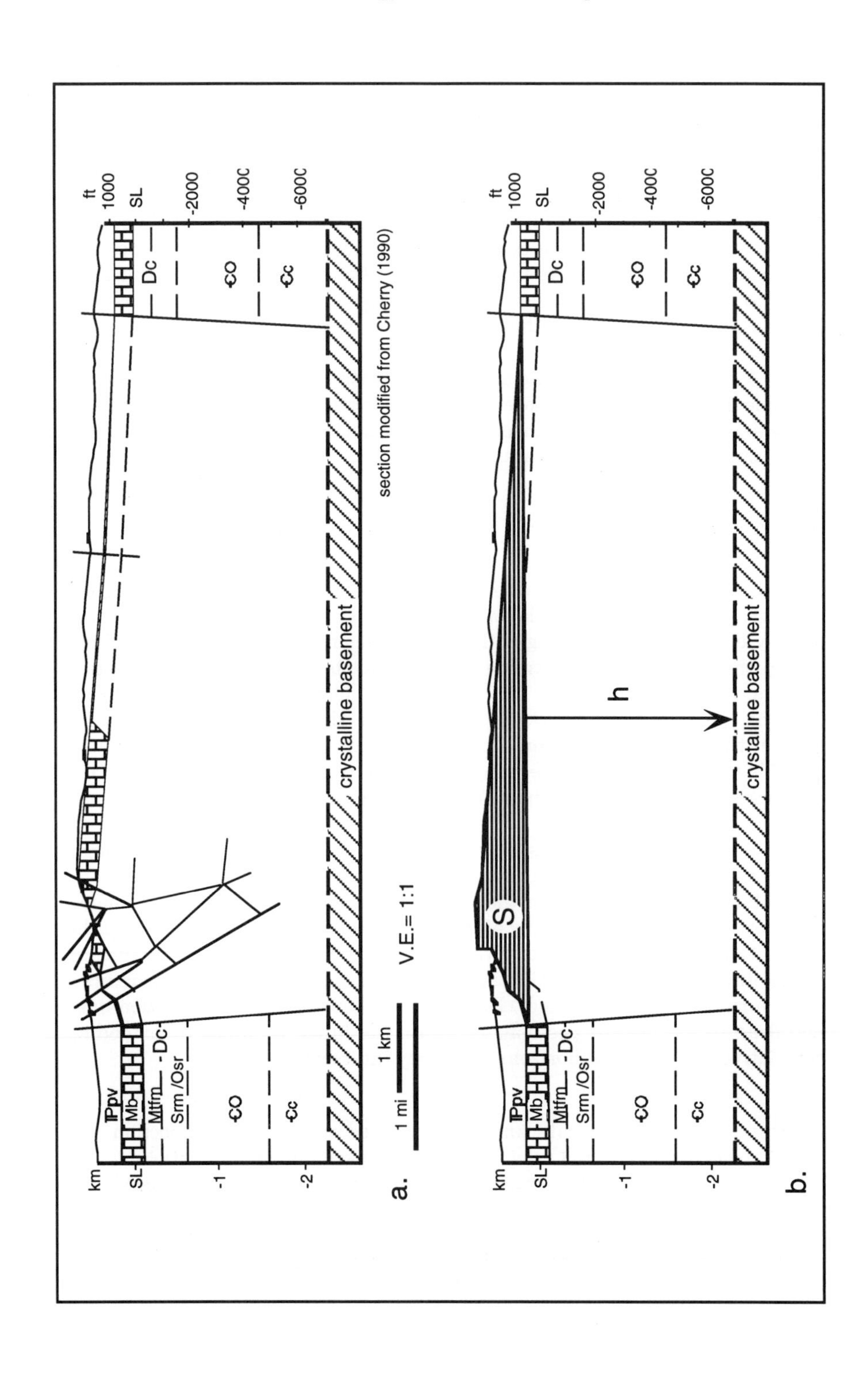
km
SL
-1
-2
ft
1000
SL
-2000
-4000
-6000
Mb
Mtfm
-Dc
Srm /Osr
Dc
crystalline basement
section modified from Cherry (1990)
1 mi
1 km
V.E.= 1:1
a.
S
h
b.

For a circular-arc cross section (method of Busk 1929), bed segments are portions of circular arcs which are tangent at their intersections. To create a circle around a specified center as a guide line, choose the `Oval` tool from the `tool bar`. With the cursor on the center of curvature, hold down the `shift` and `option` keys and drag the cursor to enlarge the circle until it has the desired diameter. Use the `split` tool, found as a `scissors` icon in the `Effects Tools` palette, to cut the circle at the segment boundaries, then delete the unneeded part of the circle. Smooth curves also can be drawn with either the `Bézier` tool or the `polygon` tool from the `tool bar`. To smooth a polygon, choose the `Smooth` command in the `Curves` submenu of the `Object` menu. The smooth curves so obtained yield Busk-style cross sections. Two curve segments can be combined into a single curve by selecting each segment while holding down the shift key, then selecting `Join` in `the Curves` submenu in the `Object` menu.

Polygons or curves can be transparent or opaque. Opaque objects may cover other important elements. Make objects transparent by selecting `Or` in the `Copy` box on the `tool bar` or send the object to the bottom of the stack of objects with the `Send to Back` command on the `Arrange` submenu in the `Object` menu. Move lines and polygons by dragging. Change lines by selecting an end point and dragging. Edit points in polygons and Bézier curves with the `edit` tools on the `tool bar`. Points can be added, deleted or dragged. The handles on Bézier and smooth curves can be manipulated to change the shape of the curve.

Testing for Balance

The primary criterion for section balance is that all bed lengths be the same. Lengths are determined from area measurements if there are tectonic thickness changes. Lengths and areas are measured from end points known as pin lines drawn on the cross section. These lines are commonly chosen to follow the axial surfaces of folds, or to be perpendicular to bedding or to be vertical. The pin lines in Fig. 2 are parallel to the outermost axial surfaces of the anticline where the beds return to regional dip. If the cross section is to be restored, one of the boundaries is designated as the pin line and is the start of the restoration. The other boundary is the loose line, the end line of the restoration, and may change shape in the restoration.

Fig. 2 (opposite). Sequatchie anticline construction and measurement techniques. a) Near-surface dip-domain geometry and stratigraphically controlled depth to basement. Axial surfaces appropriate for constant bed thickness shown in crestal region. Drawing produced as overlay on Fig. 1a. b) Illustration of excess area S measurement as area of filled polygon, and depth to reference level h as length of line from regional at top Bangor limestone Mb to top basement. Reference level (top basement) is deepest reasonable basal detachment position.

Length and Area Measurement

Use the `Object Specs` submenu in the `Object` menu to learn the length of a line, the perimeter of a polygon or curve, and the area enclosed by a polygon or curve. Polygons and curves need not be closed. The area enclosed by an object is the area between the object boundary and a straight line joining the end points of the polygon or curve. If the line joining the end points of a polygon or curve crosses the line that outlines the border of the area, the measured area will not be correct for use in area balancing. Correct this problem by drawing separate areas. Fill the object with a pattern to confirm the area being measured (Fig. 2b). To fill a polygon, select a background fill pattern from the box on the `tool bar`. The perimeter of the object is the length actually drawn and does not include the length of the imaginary line connecting the end points of an open polygon. Calculation of area and perimeter is automatic with the `line` and `polygon` tools. Choose `Calculate` in the `Object Specs` submenu to get the measurements for the Bézier and freehand drawing tools. Quick access to the measurements is obtained by clicking on the information manager (the space where the x, y coordinates of the pointer are shown) in the `information bar` at the bottom of the window after the object has been selected, or by double clicking the object to be measured. Length and area measurements are not available for objects that have been grouped; instead, the dimensions of the space occupied by the group are given. Choose `Ungroup` to separate the group.

Length Balance

Length balancing is applied where bed lengths and thicknesses are inferred to have remained constant during deformation. Bed lengths are measured and should all be the same. Bed lengths that show a consistent trend of length changes from the top to the bottom of the cross section may represent a systematic error due to an incorrect choice of pin line or loose line, not necessarily an error in the cross section.

Area Balance

Area balancing is applied where beds have changed length and/or thickness due to deformation. The bed area is measured and, assuming constant area, its length calculated by dividing the area by its original thickness determined in a location where the bed is undeformed. The calculated length is then compared to that of other beds as done in length balancing. In structures formed by pure vertical displacement, the original bed length is the straight-line distance between the pin lines located off the structure (Brewer & Groshong 1993); this length can be used to calculate the original bed thickness.

Excess Area / Lost Area Balance

'Excess area' (Figs. 2b, 3a) is the area displaced above regional as the result of deformation. 'Lost area' is the area dropped below regional as the result of

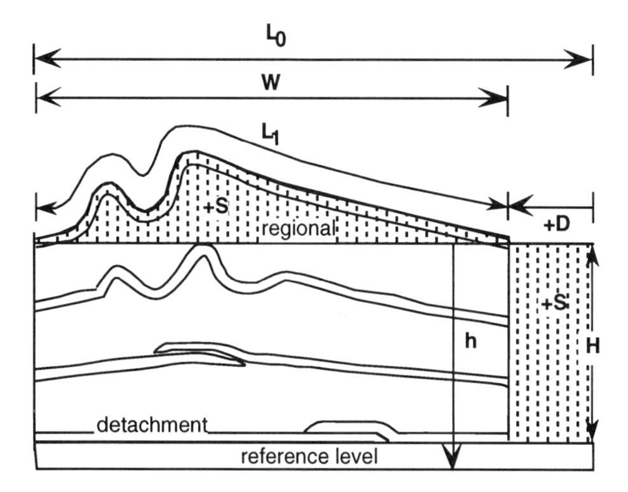

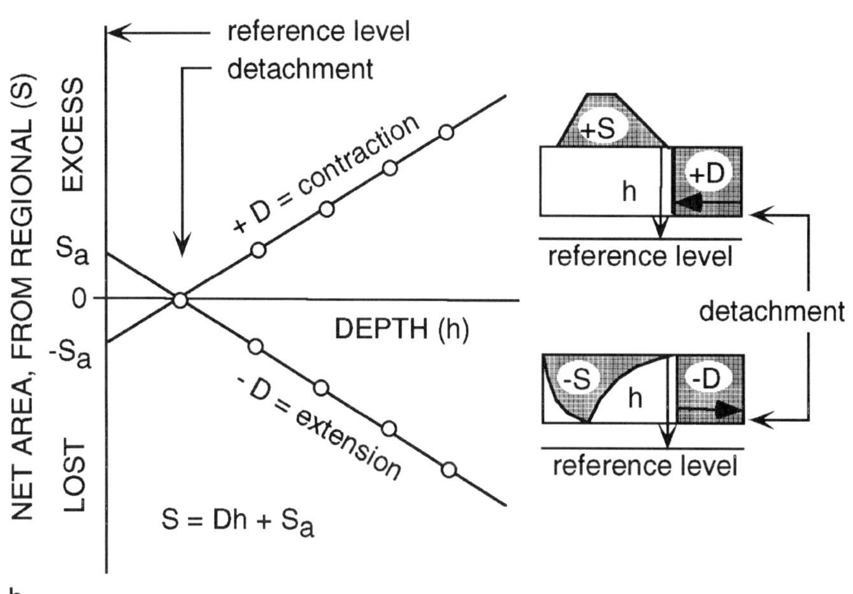

Fig. 3. Area-depth balancing technique. a) Terminology for contractional structure. Excess area is area uplifted above regional and is equal to area displaced along lower detachment. Reference level is parallel to regional but otherwise arbitrary. b) Area-depth graph for contraction or extension. Each point represents excess/lost area measured from regional for single horizon and its depth to reference level. All points fall on single line, slope D, that goes to zero area at lower detachment. D is displacement on lower detachment. Structures with more than one detachment horizon will have line segments of sequentially changing slopes that intersect at upper detachments (Epard & Groshong 1993).

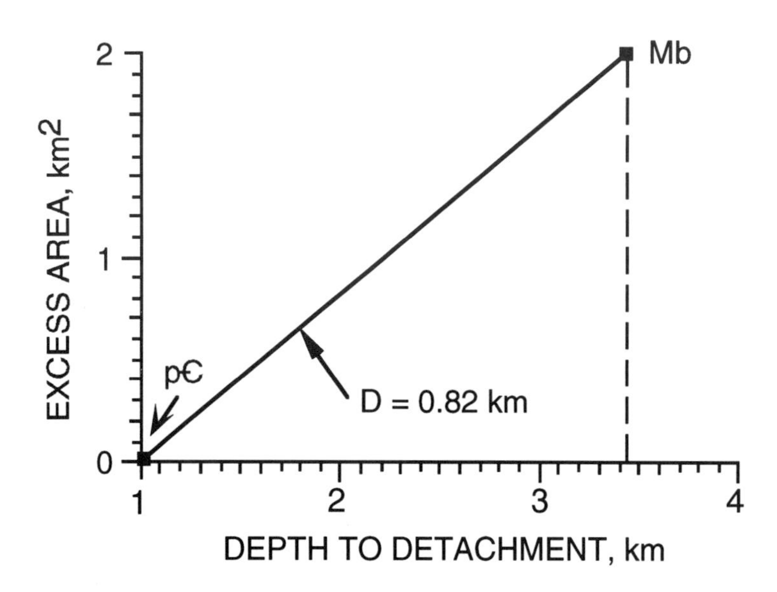

Fig. 4. Area-depth relationship for Sequatchie anticline interpreted as locally balanced fold. Measured excess area at top of Mississippian Bangor limestone *Mb* is 2 km^2 and at top of Precambrian crystalline basement *p*Є is assumed zero for this thin-skinned structure. Excess areas of all other units must fall along straight line between these end points: Devonian Chattanooga shale (*Dc*, depth = 2 km) must be 1.91 km^2, Cambrian Conasauga limestone (Є*c*, depth = 0.65 km) = 1.73 km^2. Required displacement on lower detachment is $D = 0.82$ km.

deformation. 'Regional' (McClay 1992) is the position of the bed before deformation. Excess area and lost area balancing (Epard & Groshong 1993, Groshong 1994) is done with an area-depth graph of the structure (Fig. 3b). The depth is the distance from the regional of a specific horizon to an arbitrary selected reference level that is the same for all measurements (Fig. 3). Create an area-depth graph and fit a line, line segments, or a curve to the data (Fig. 3b). Area-depth points that fall on lines, segmented lines, or smooth curves indicate an area-balanced cross section. The slope of the line (Fig. 3b) is the displacement on the lower detachment and the depth of the line at zero area is the position of the detachment. Lost areas and extensional displacements are negative numbers.

In addition to testing for balance, the area-depth relationship may be used to guide the construction of a balanced cross section. The Sequatchie anticline (Fig. 2) resembles a detachment fold, for which the excess area will decrease linearly to zero at the lower detachment (Epard & Groshong, 1993). The lower detachment for the Sequatchie anticline is probably just above the top of the basement. This constrains the anticline to have the area-depth properties shown in Fig. 4. A cross section that satisfies the area-depth relationship of Fig. 4 is shown in Fig. 5a.

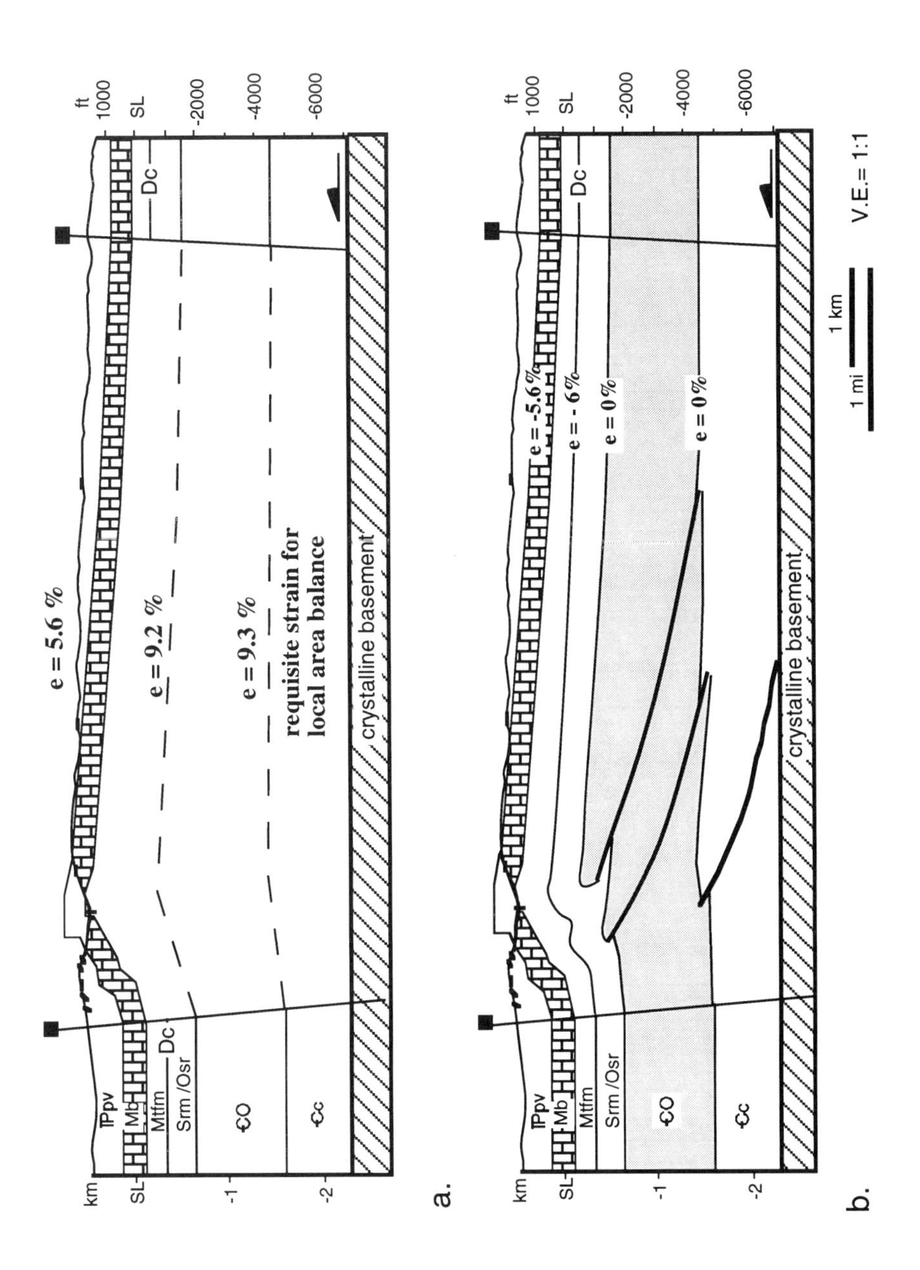

Fig. 5. Alternative interpretations of Sequatchie anticline. Both interpretations satisfy area-depth relationship in Fig. 4. a) Internally deformed detachment fold showing requisite layer-parallel strains. b) Duplex of Cambro-Ordovician carbonates showing requisite strains of zero (constant bed length) in duplex.

The requisite strain (Groshong & Epard, 1994) is an additional parameter for judging and constraining the cross section interpretation. The requisite strain is the layer-parallel strain required if the given cross section is area balanced. Unreasonable values of requisite strain suggest that the bed length on the cross section is incorrect. Requisite strain values of over -9% required by the cross section geometry of Fig. 5a are probably beyond the ductile limit of the Cambro-Ordovician dolomite deformed at low-temperature. The evaluation of alternative interpretations is facilitated by comparing undeformed and deformed bed lengths. The deformed bed length is that measured on the cross section. The undeformed bed length (Fig. 3a) is the sum of the width of the structure plus the displacement. The detachment-fold interpretation of the Sequatchie anticline (Fig. 5a) can be changed into an equivalent fold cored by a constant-bed-length duplex (Fig. 5b) by adding faults to the Cambro-Ordovician sequence while simultaneously increasing the bed lengths. The duplex has the same area-depth relationship as the detachment fold (Figs. 4, 5a) but maintains constant bed length (requisite strain = 0) in the Cambro-Ordovician carbonates. The duplex (Fig. 5b) is locally balanced, as is the detachment anticline (Fig. 5a). Local balance means that all the deformation is contained between the pin lines and that there is no upper-detachment displacement outside the pin lines.

Curve fitting, graphing, and strain calculation can be done with a spreadsheet program. A procedure for *Excel* 4.0 is given here. Measured data from each horizon are entered in rows with the highest stratigraphic level being in row 10. Let column `A` be the label for each stratigraphic unit, `B` the measured depth to the reference level, `C` the excess/lost area, `D` the measured bed length, and `E` the width of the fold at regional (Fig. 6). The next columns contain computed results. Let column `F` be the true depth to detachment based on the best-fit area-depth line, given by highlighting cell `F10` and typing

```
= B10 - $I$18
```
(1)

in the formula bar. Let column `G` contain the requisite strain as a fraction (equation 7, Groshong & Epard, 1994), given by typing

```
= ((D10*F10) / ((F10*E10) + C10)) - 1
```
(2)

in the formula bar. Equations can be copied to subsequent cells in a column using the `Edit Copy` and then `Edit Paste` commands. The results of the regression analysis are placed in the range of cells from `H10` to `I14` by highlighting the range and then selecting the `LINEST` function from the menu as follows:

`Formula` -> `Paste` Function -> Function Category: `statistical`, `LINEST`.

Be sure that the box, `Paste Arguments`, is selected. This enters the formula into the formula bar. Replace "known *y*'s" by the range of cells of the area in column `C`. Replace "known *x*'s" by the range of cells of the depth to reference level in column `B`. Replace both logical values "const" and "stats" by `TRUE`. Validate the formula by `Command+Enter`, not by a simple carriage return. The result is the regression line

$$y = mx + b, \tag{3}$$

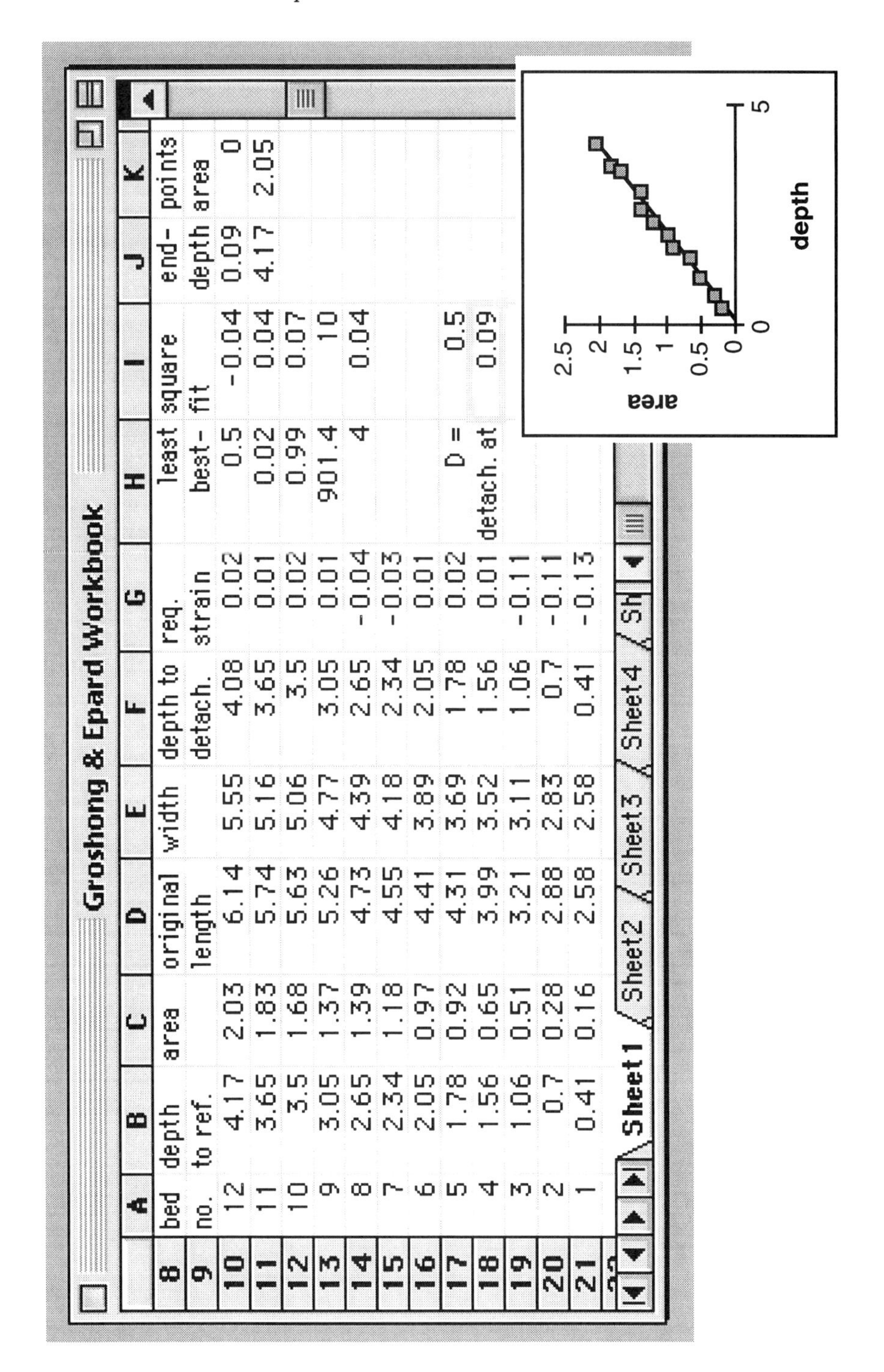

Groshong & Epard Workbook

	A	B	C	D	E	F	G	H	I	J	K
8	bed	depth	area	original	width	depth to	req.	least	square	end-	points
9	no.	to ref.		length		detach.	strain	best-	fit	depth	area
10	12	4.17	2.03	6.14	5.55	4.08	0.02	0.5	-0.04	0.09	0
11	11	3.65	1.83	5.74	5.16	3.65	0.01	0.02	0.04	4.17	2.05
12	10	3.5	1.68	5.63	5.06	3.5	0.02	0.99	0.07		
13	9	3.05	1.37	5.26	4.77	3.05	0.01	901.4	10		
14	8	2.65	1.39	4.73	4.39	2.65	-0.04	4	0.04		
15	7	2.34	1.18	4.55	4.18	2.34	-0.03				
16	6	2.05	0.97	4.41	3.89	2.05	0.01				
17	5	1.78	0.92	4.31	3.69	1.78	0.02	D =	0.5		
18	4	1.56	0.65	3.99	3.52	1.56	0.01	detach. at	0.09		
19	3	1.06	0.51	3.21	3.11	1.06	-0.11				
20	2	0.7	0.28	2.88	2.83	0.7	-0.11				
21	1	0.41	0.16	2.58	2.58	0.41	-0.13				

Fig. 6. Example of *Excel* spreadsheet for area-depth-strain analysis as described in the text.

where m is in cell H10, b in cell I10 and the coefficient of determination r^2 in cell H12. If the reference level is at the known detachment, the regression line can be forced through the origin by setting "const" equal to FALSE. Place the displacement in cell I17 with the formula

```
= H10,
```
(4)

and the detachment depth relative to the reference level in cell I18 with the formula

```
= -I16/H10.
```
(5)

Plot the regression line from two points. The first point is the detachment depth from I18, placed in cell J10 with

```
= I18.
```
(6)

The second point is at the maximum depth value which will be in cell B10 if data are always entered with the highest horizon in row 10; place the value in cell J11 with

```
= B10.
```
(7)

The corresponding area values are calculated and placed into cell K10 with

```
= (H10*J10) + I10
```
(8)

and K11 with

```
= (H10*J11) + I10.
```
(9)

To draw the graph, select the depth and area data range (columns B and C); click the chart button on the menu bar; drag across an area on the spreadsheet to open the chart window and follow the instructions of the ChartWizard. Select the *xy* scatter plot with no line joining the data points. The data are now plotted. To add the regression line, double click on the graph to activate the chart window. The regression line will be a new data series of two points joined by a line. Add this new data series to the chart with the steps:

```
Chart -> Edit Series... ->
```

Make the x values the addresses of column J with

```
('Worksheetname'! = $J$10:$J$11)
```
(10)

and the y values the addresses of column K with

```
('Worksheetname'! = $K$10:$K$11).
```
(11)

This causes the end points of the regression line to be plotted. Double click on the upper end point to activate the Patterns window, chose a line type but no marker for this series. The final result is a graph that is automatically adjusted when the

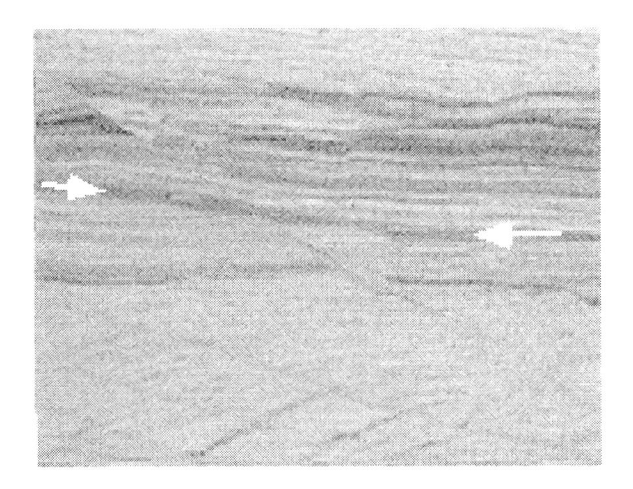

Fig. 7. Segment of left-side master fault from Fig. 1b restored by rigid-body translation along fault. Dark layer between arrows has been aligned across master fault and next few units above it are seen to line up as well. Displacement on other faults has not been restored.

input areas or depths in columns B and C are changed. The requisite strain will automatically be adjusted as the bed length or width of the structure are changed in columns D and E.

Restoring the Section

Rigid-Block Restoration

This method is suitable for deformation that occurred by rigid-block translation and rotation. In *Canvas*, copy the cross section to a new layer using the Copy to Layer command in the Arrange submenu of the Object menu or to a new file using the Copy/Paste commands in the Edit menu so that you can restore the section by changing a copy of the original, rather than by redrawing. Group all lines, curves and polygons in each fault block, then move the object to the restored position.

If the cross section to be restored is a paint object, it can be cut with the Lasso tool or the Marquee tool and then pasted into a new layer where it can be translated and rotated (Fig. 7). Pressing the Option key while selecting the outlined area copies the area instead of removing it. The shape of the area selected can only be changed by deselecting the Lasso tool and repeating the process. If the area selected is larger than needed and the object is copied and pasted to a new layer, then it can be trimmed to exact size with the eraser tool from the same submenu. A line in a line drawing may be cut with the Split tool found as a scissors icon in the Effect Tools palette. A line will be split into two lines at the point where the crosshairs of the tool are clicked on the line. If the line is in `Edit` mode, it can be split at a vertex point. It may be convenient to convert a complex line drawing into a `paint` object so that arbitrary pieces can be selected or erased before rearranging. Change a line drawing to a `paint` object by selecting `Object Specs` in the `Object` menu. In the dialog box, select the object type to see the list of possible object types and change the selection to the `paintbrush`. The `Object` menu then gives the default resolution of the new `paint` object. The resolution can be changed to one of the other values seen by clicking the `resolution` box. The resolution cannot be changed after the `Object Specs` menu has been closed once. A grouped object is converted by selecting `Group Specs` in the `Object` menu and selecting `Convert To Paint Object` in the dialog box. A `paint` object cannot be changed back into a `line` object, so be sure to perform this operation on a copy of a line drawing, not the original.

Constant Line Length Restoration

The section is redrawn with the folds straightened out and fault displacements removed. Bed segment lengths measured on the deformed-state cross section must be preserved on the restoration and bed thicknesses must be constant or their original thicknesses variations preserved. It may be possible to restore the section without completely redrawing by selecting line segments and moving them to their restored positions. Copy the cross section to a new layer or to a new file so that it can be restored without changing the original. To maintain bed thicknesses, either group bed segments before moving, or draw bed normals from one bedding top, group with the top, and move to the restored position. The bed normals provide the location of the base of the unit being restored (method of Brewer & Groshong 1993). Variable thickness units can be restored in this manner.

Area Restoration

Beds that have changed thickness due to the deformation can be restored if the original length or thickness can be determined. In key-bed area balancing (Mitra & Namson 1989), the original bed length is assumed to be equal to that of an adjacent unit that maintained constant thickness during deformation. Alternatively, the original thickness may be known from a location outside the structure. The measured area is divided by the known dimension to give the restored length and thickness and the unit added to the restoration.

Fig. 8 (opposite). Oblique simple shear restoration of hangingwall of extensional growth fault. a) Deformed-state cross section showing working lines parallel to assumed shear direction. Block displacement is D. Intersections of working lines with regional are marked by circles and intersections with bedding by squares. b) Footwall and regional copied to new location. Working lines copied and displaced to remove offset of bed 1. c) Working lines displaced into contact with fault. d) Restored beds drawn, construction lines shaded.

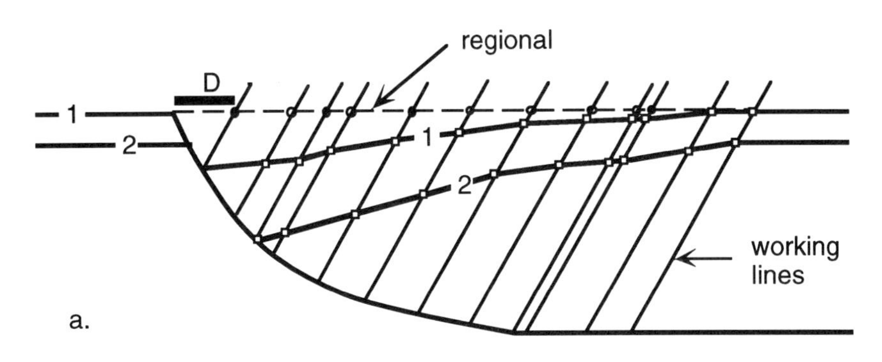
regional
D
1
2
1
2
working
lines
a.

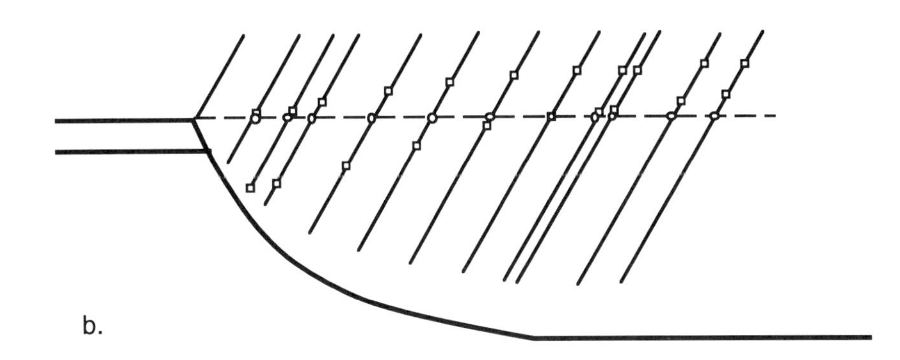
b.

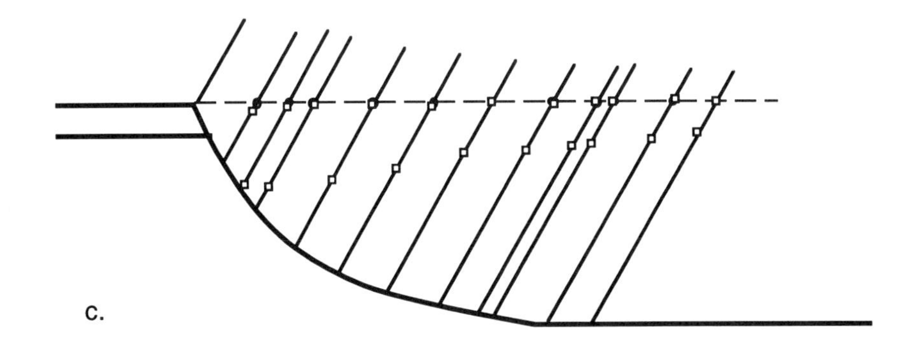
c.

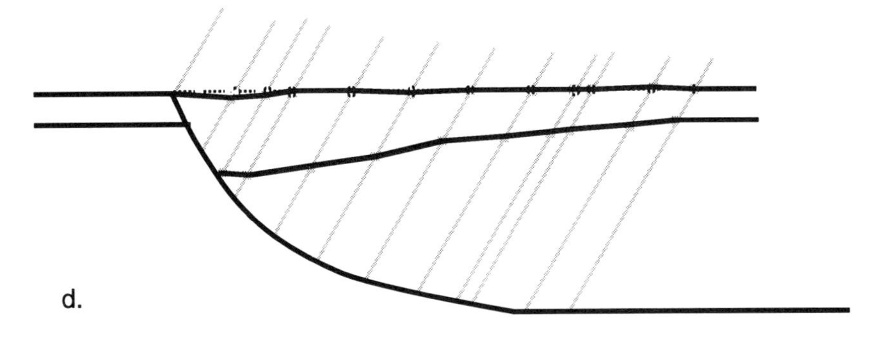
d.

Vertical and Oblique Shear Restoration

Simple-shear restorations (Williams & Vann 1987, Groshong 1990) are surprisingly easy. The restoration should be done on a new layer. Determine the key bed that is to be restored to its pre-deformation geometry along with its hangingwall and footwall cutoffs (bed 1 in the example of Fig. 8a). Draw the restored position of this bed (the regional). Draw a line in the shear direction (60° antithetic in Fig. 8) from the hangingwall cutoff of the key bed to the regional. The distance along the regional from the footwall cutoff to the intersection of this line through the hangingwall cutoff is the block displacement (*D*, Fig. 8a). Draw parallel working lines at increments of one block displacement (increments of *D*). Add other working lines at critical locations such as at the hangingwall cutoffs of all beds to be restored, where the fault flattens to regional, and where each bed flattens to regional. Terminate each working line at the master fault. Mark the intersections where the working lines cross the regional (circles in Fig. 8a) and `group` with the regional as a single unit. Then mark all intersections between beds to be restored and the working lines (squares in Fig. 8) and `group` each working line and its bed-intersection points. Copy the footwall to a new location (Fig. 8b) together with the regional line and its attached points. `Group` all the working lines and move (or copy) them to the new cross section so that the hangingwall cutoff of the key bed is restored to contact with the footwall cutoff (Fig. 8b). `Ungroup` the working lines and drag each one separately into contact with the fault, using the guide points along the regional to maintain the original spacing (Fig. 8c). Draw the restored beds by connecting bed intersection points (Fig 8d). The working lines and the regional may then be deleted to give the restored cross section.

Vertical simple shear restoration is demonstrated with a scanned seismic line (Fig. 9a). The original image was interpreted and then duplicated so that the copy could be manipulated. Vertical slices were deformed to flatten the interpreted horizon using the `Skew` command in the `Special Effects` submenu of the `Effects` menu. Shear the image by dragging the handle at the corner of the image. Each application of the `Skew` command to a `paint` object degrades the image slightly and so the correct amount of distortion should be accomplished in one step. Use the `Undo` command in the `Edit` menu to return to the original image if the restoration is not correct. The seismic line on each side of the fault was flattened separately and then moved to restore the fault displacement. The reflector correlation across the fault (Fig. 9b) looks more convincing on the restored profile.

Printed Output

A large drawing can be reduced so that it will print on a standard page by using the `Scale` submenu in the `Object` menu. If the original drawing is saved, then the reduction will not be permanent unless the reduced drawing is saved. Be sure to `group` the entire drawing prior to scaling, because each object is scaled independently from its center. Reductions can be selected on the `Page Set Up` submenu in the `File` menu. Reductions using the `Page Set Up` command are slower than the same operation done in *Canvas* and a large file may overtax the memory of the printer. Drastic scale reductions in *Canvas* may significantly

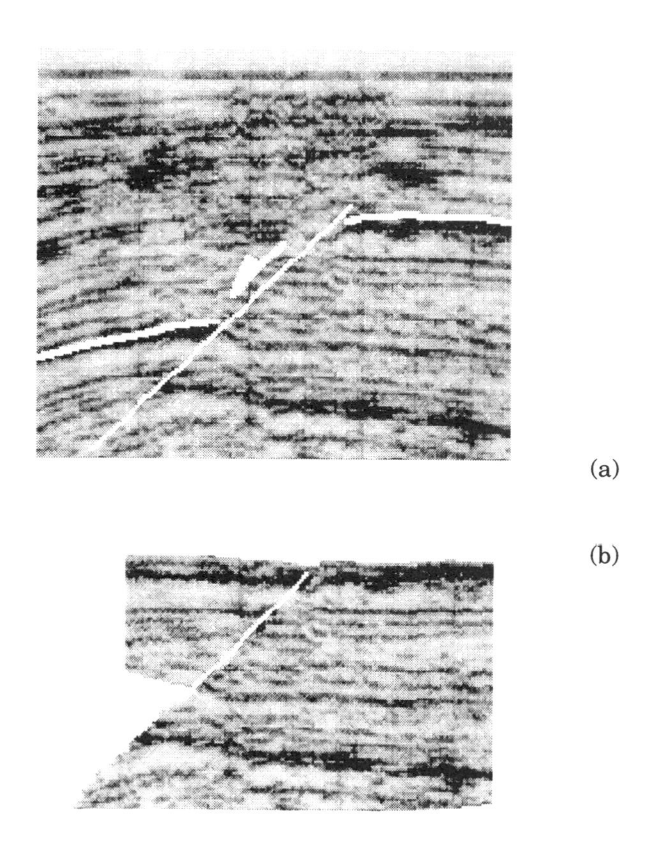

Fig. 9. Seismic line restored by vertical simple shear. a) Original seismic line scanned as a photograph on flatbed scanner, sharpened in *Photoshop* and interpreted in *Canvas*. b) Restoration by vertically shearing vertical slices bounded by marked reflector and fault. Slices were sheared to flatten marked reflector to horizontal, hangingwall and footwall slices pasted back together separately, and then displaced so that marked reflector matches across fault.

reduce the quality of the drawing. Multiple-page drawings can be printed. Set the page overlap in the `Drawing Size` submenu in the `Layout` menu to obtain enough overlap to assist in piecing the segments together.

The cutting and pasting of a `paint` object in *Canvas* required to produce results like Figs. 7 and 9 may result in a file that will not print properly on a Personal LaserWriter. Adding a `draw` object to a `paint` object will cause the file to be converted to `picture` objects which also may not print properly. Transferring an improperly printing image into *Photoshop* seems to fix the problem. The file can be returned to *Canvas*, if desired, and will print correctly.

Conclusions

After a little practice, the measurements and manipulations required to balance and restore cross sections by any of the standard techniques are relatively quick and easy using *Canvas*. Length and area measurements alone are very rapid when performed on a scanned cross section. *Photoshop* is valuable for improving the quality of scanned photos or of seismic lines scanned as photos. The combination of speed and convenience of measuring areas and lengths as required by the excess-area / lost-area balancing technique is our reason for choosing *Canvas*. The required calculations are instantaneous in an *Excel* spreadsheet.

Acknowledgements

We appreciate helpful discussions with Peter Clark, James Donahoe and Michael Lesher in setting up the system. Valuable suggestions on equipment or procedures have been made by William Dunne, Brown Hawkins, Diahn Hawkins, Martin Smith, Julia Smith, and John Spang. William Dunne and Kevin G. Stewart provided helpful reviews. Jean-Luc Epard gratefully acknowledges financial support from the Swiss National Science Foundation (grant 20-37470.93).

References

Adams, M. G. Mallard, L. D., Trupe, C. H., & Stewart, K. G. 1996. Computerized Geologic Map Compilation. In: De Paor, D. G. (ed.) *Structural Geology and Personal Computers*. Elsevier Science Ltd., London, pp. 457-469.

Brewer, R. C. & Groshong, R. H., Jr. 1993. Restoration of cross sections above intrusive salt domes. *Am. Ass. Petrol. Geol. Bull.* **77**: 1769-1780.

Busk, H. G. 1929. *Earth flexures*. Cambridge University Press, London.

Cherry, B. A. 1990. *Internal deformation and fold kinematics of part of the Sequatchie anticline, southern Appalachian fold and thrust belt, Blount County, Alabama*. MS. Thesis, Univ. of Alabama, Tuscaloosa, 78 pp.

Dahlstrom, C. D. A. 1969. Balanced cross sections. *Can. J. Earth Sci.* **6**: 743-757.

Dahlstrom, C. D. A. 1990. Geometric constraints derived from the law of conservation of volume and applied to evolutionary models of detachment folding. *Amer. Ass. Petrol. Geol. Bull.* **74**: 336-344.

De Paor, D. G. 1988. Balanced sections in thrust belts. Part I: Construction. *Amer. Ass. Petrol. Geol. Bull.* **72**: 73-90.

Epard, J. -L. & Groshong, R. H., Jr. 1993. Excess area and depth to detachment. *Amer. Ass. Petrol. Geol. Bull.* **77**: 1291-1302.

Geiser, P. A. 1988. The role of kinematics in the construction and analysis of geological cross sections in deformed terranes. In: Mitra, G. & Wojtal, S. (eds.) Geometries and Mechanisms of Thrusting. *Geol. Soc. Amer. Spec. Paper* **222**: 47-76.

Gill, W. D. 1953. Construction of geological sections of folds with steep-limb attenuation. *Amer. Ass. Petrol. Geol. Bull.* **37**: 2389-2406.

Groshong, R. H., Jr. 1990. Unique determination of normal fault shape from hanging-wall bed geometry in detached half grabens. *Eclogae geol. Helv.* **83**: 455-471.

Groshong, R. H., Jr. 1994. Area balance, depth to detachment and strain in extension. *Tectonics* **13**: 1488-1497.

Groshong, R. H., Jr. & Epard, J.-L. 1994. The role of strain in area-constant detachment folding. *J. Struct. Geol.* **16**: 613-618.

McClay, K. R. 1992. Glossary of thrust tectonics terms. In: *Thrust Tectonics* (edited by McClay, K. R.). Chapman and Hall, London, 419-433.

Mitra, S. & Namson, J. 1989. Equal-area balancing. *Amer. J. Sci.* **289**: 563-599.

Woodward, N. B., Boyer, S. E. & Suppe, J. 1989. Balanced geological cross sections: An essential technique in geological research and exploration. *Amer. Geophys. Union Short Course in Geology* **6**.

Williams, G. & Vann, I. 1987. The geometry of listric normal faults and deformation in their hangingwalls. *J. Struct. Geol.* **9**: 789-795.

Appendix: Hardware Considerations

The computer systems we have used for section balancing are described to suggest the minimum requirements for satisfactory operation. Any system that has sufficient speed and memory to work with photographs and to open more than one program at a time should be adequate. Most of our work is on a Macintosh IIci with 12 MB RAM and a 230 MB hard drive and we find the speed and memory to be sufficient. We have also successfully used a Macintosh IIfx with 8 MB RAM and a 200 MB hard drive and a PC-compatible 66 MHz 486DX2. A Macintosh Si with a 12 in. Apple color monitor and 5 MB memory is barely adequate and requires careful memory conservation to function. Image capture and manipulation requires speed and a large amount of memory, especially for high-resolution color. A snapshot-sized scanned color photograph can require about 2 MB of memory. Black and white photographs and line drawings require half or less memory. Manipulation of a 2 MB image in *Canvas* requires about 7.5 MB allocated to the program. *Photoshop* tends to require even more memory. A large color monitor is needed to best display the data, the higher the resolution, the better. An Apple 16 in. color monitor is satisfactory.

Images for section balancing may be input by a variety of devices. The quality and resolution of the input are critical factors in system design and performance. Color input can be vital for preserving the information on the image such as, for example, the color contrast between beds. We have used flat-bed scanners and a color video camera. An HP ScanJet IIcx provides excellent color images up to 8 1/2 ×14 inches in size. The slide scanner attachment for this scanner produces images of color transparencies that can be enlarged on screen up to about 25 cm wide without appreciable loss of resolution. Input via a home color video camera allows large images to be captured in one piece. Live video can be more convenient than still-image input because the camera can be continuously adjusted until the view is exactly correct, then the image captured. The RasterOps 24MxTV video card (Macintosh) is used because it was rated highest in resolution among low-priced video boards in a 1993 computer magazine article. We use a Sony CCD-TR81 home color video camera because it was top rated in a 1993 consumer magazine and has the high-resolution *s*-video output.

If memory is insufficient, it may be impossible to capture an image or to manipulate it. When the memory is overtaxed, *Canvas* is very slow and necessary functions (like `Undo`) are unavailable. Several steps can be taken to optimize memory use. *RAM Doubler*, a program by Connectix increases the effective use of memory. As much memory as possible needs to be allocated to *Canvas* and to *Photoshop*. To increase the memory allocation on a Macintosh system, close the program, select the application program file (*i.e., Canvas*) on the hard drive, choose `Get Info` from the `File` menu in the `Finder`, and then select and increase the preferred memory size. It may be impossible to have more than one program open at a time. Input the data in the most memory-conservative mode possible. For example, in the HP scanning program, scan line drawings in the drawing mode, not the photo mode or use the black and white photo mode rather than color.

The quality of the printed image depends on the output device. A Personal LaserWriter is the minimum suitable printer and will give passable grayscale output with *WonderPrint*™ by Delta Tao Software, Inc. A grayscale printer like the Apple LaserWriter Select, or a color printer like the HP Deskjet 1200C gives a better result.

Bitmap Rotation, Raster Shear, and Block Diagram Construction

Declan G. De Paor
Department of Earth & Planetary Sciences,
Harvard University, 20 Oxford Street,
Cambridge MA 02138, U.S.A. depaor@eps.harvard.edu

Abstract– This paper describes the mathematical theory required to construct block diagrams on a personal computer and presents output of a working Macintosh program written by the author. As is common in graphics-intensive applications, efficiency of computation is more important than succinct solution of equations. When applied to thousands of pixels, traditional methods of calculation are much too slow to permit interactive control of bitmapped images. Instead, the images on the sides of a block must be obtained by deformation. Despite appearances, no orthogonal rotation tensors are employed; instead, images are manipulated by shear displacement of rows or columns of pixels. This process has implications for strain studies; oblique superposition of simple shear zones of the same sense may lead to highly rotational deformation states with very little distortion.

Introduction

Block diagrams are attractive and effective tools for presenting three dimensional geological data and are important aids to the visualization of complex structures. In its simplest form, a block diagram combines an inclined view of a map surface, a cross section, and a long section of a region (generally, only three sides of the block are visible at once). Often the projection direction is chosen so as to present a 'down-plunge' view of cylindrically folded structures. Sophisticated diagrams may incorporate perspective and surface relief and may include serial slices or cutouts; however in this paper, we confine our attention to basic rectilinear blocks of arbitrary dimension and orientation.

The manual procedure for constructing a block diagram is tedious (Lisle 1980, Ragan 1985). First, the orientation and projected length of three reference axes are determined with the aid of an orthographic orientation net. Then these axes are used to transform the coordinates of key features from the map and two sections onto the visible sides of an obliquely viewed block. Finally, geological contacts are drawn freehand by interpolating between the key features. Clearly, this is an ideal candidate for automation on a personal computer. However, several significant computational problems have to be solved before a working application can be developed.

Rotation of Bitmapped Images

The classical approach to rotation about the origin in the plane of two reference axes, here labelled *x-y*, employs an orthogonal rotation tensor such as

$$\mathbf{R} = \begin{bmatrix} \cos\alpha & -\sin\alpha \\ \sin\alpha & \cos\alpha \end{bmatrix} \tag{1}$$

where α is the angle of rotation. Any point **P** in the *x-y* plane may be transformed into **P'** using the equation

$$\mathbf{P}' = \mathbf{RP} \tag{2}$$

or, in terms of vector components,

$$\begin{bmatrix} P'_x \\ P'_y \end{bmatrix} = \begin{bmatrix} \cos\alpha & -\sin\alpha \\ \sin\alpha & \cos\alpha \end{bmatrix} \begin{bmatrix} P_x \\ P_y \end{bmatrix}. \tag{3}$$

The first problem with this approach is one of computational efficiency; eqn. (3) requires four trigonometric functions, four multiplications, and two additions,

$$\begin{aligned} P'_x &= P_x\cos\alpha - P_y\sin\alpha \\ P'_y &= P_x\sin\alpha + P_y\cos\alpha \end{aligned}. \tag{4a,b}$$

A small angle approximation ($\alpha < 3°$) cuts the work in half,

$$\begin{aligned} P'_x &\approx P_x - P_y\sin\alpha \\ P'_y &\approx P'_x\sin\alpha + P_y \end{aligned} \tag{5a,b}$$

(*e.g.*, Foley *et al.* 1990, p. 213) and is applicable to frame-by-frame animation of rotations. Note that there is no typo – substitution of P'_x from eqn. (5a) into (5b) is intentional. However, errors accumulate when this approximation is applied repeatedly during the simulation of continuous rotational movements. Furthermore, the trigonometric functions in **R** require floating point calculations which are relatively slow to compute. One solution to this problem is to pre-calculate the sines of all angles from 0° to 90° in 1° steps, multiply them by a constant, and store them in an integer array or 'look-up table' here called Sint(). When sines are required in real-time, the tabulated values are recalled and divided by the chosen constant (see Ritter 1990). Maximum speed is achieved if the constant is chosen

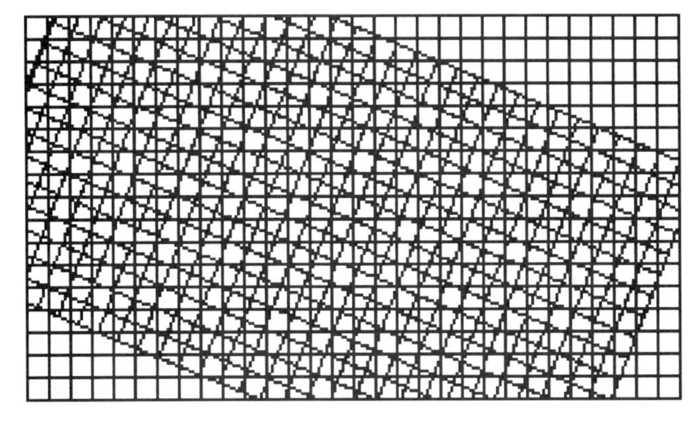

Fig. 1. Pixel overlap resulting from the free rotation of a bitmap. As many as six source pixels must be blended to define a destination pixel.

to be the highest power of two that can be accommodated by the compiler's integer variable type. For example, assuming two byte words, a 15-bit shift will convert from real numbers in the range [0,1] to integers in the range [0,32768]. The following *BASIC* routine illustrates the procedure;

```
Pi! = ATN(1)<<2
Rads! = Pi!/180
FOR   a = 0 to 90
    Sint(a) = SIN(a*Rads!)>>15
NEXT a
.
.
INPUT a
output! = Sint(a)<<15
```

where << and >> denote left and right bit shifts. A separate table for cosines is not required because the values are simply read from the other end of the sine table. However, even these clever tricks are too slow when thousands or millions of pixels have to be manipulated in real time. And when the analysis is expanded to three dimensions, it is impossible to avoid floating point calculations because products of two or three trigonometric functions are required.

An additional consideration concerns the nature of raster image display devices (Fig. 1). A raster device is any monitor, scanner, or printer that represents images by rows and columns of pixels. When the rotation tensor of eqn. (1) is applied to an array of pixels, the source and destination arrays overlap in such a way that from one to six pixels in the source image contribute to the color of one pixel in the destination image. To find which elements contribute to a given destination pixel, the rotational transformation must be inverted so that rows of the destination bitmap, not the source bitmap, are scanned sequentially. To avoid loss of image quality, the contributions of the source pixels must be scaled by their areas of overlap with the destination pixel and then these area-weighted contributions must be blended so as to define the color of the destination pixel. Even if

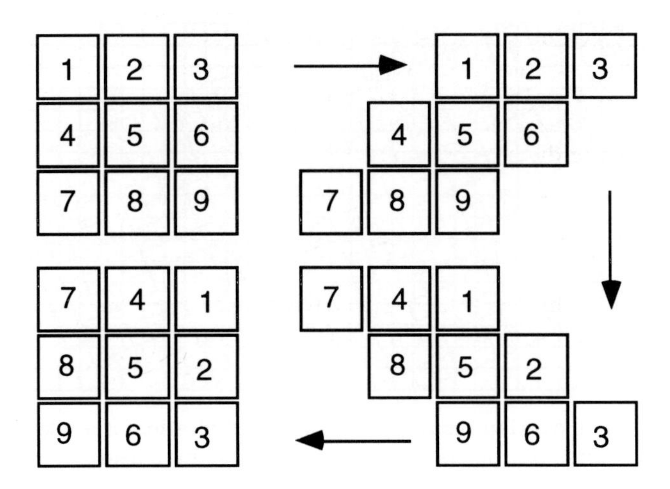

Fig. 2. Rotation of a pixel array through 90° by means of three successive 45° simple shears. Arrows indicate sequence of shears.

the image color depth is only eight bits, the amount of computation involved for, let us say, a 600 × 400 pixel image is prohibitive. To permit bitmap manipulation in interactive applications, we must abandon the rotation tensor approach to the computation of rotations and adopt a new approach (De Paor 1995).

90° Rotations

For simplicity, let us begin with the special case of a 90° rotation (Fig. 2). An arbitrary point **P** in the source bitmap is transformed into **P'** by eqns. (4),

$$P'_x = P_y \qquad P'_y = -P_x \tag{6a,b}$$

or, in matrix format,

$$\begin{bmatrix} P'_x \\ P'_y \end{bmatrix} = \begin{bmatrix} 0 & 1 \\ -1 & 0 \end{bmatrix} \begin{bmatrix} P_x \\ P_y \end{bmatrix}. \tag{7}$$

The 90° rotation tensor in eqn. (7) may be replaced by three 45° simple shears,

$$\begin{bmatrix} P'_x \\ P'_y \end{bmatrix} = \begin{bmatrix} 1 & 1 \\ 0 & 1 \end{bmatrix} \begin{bmatrix} 1 & 0 \\ -1 & 1 \end{bmatrix} \begin{bmatrix} 1 & 1 \\ 0 & 1 \end{bmatrix} \begin{bmatrix} P_x \\ P_y \end{bmatrix}. \tag{8}$$

This would be a perverse substitution to make but for one overriding advantage: Simple shears may be applied to whole rows or columns of pixels at one time and so three simple shears are much more efficient computationally than the pixel-by-pixel transformation of eqns. (6a,b), especially where there are thousands of pixels per row or column. The price that must be paid for efficiency is memory allocation; from Fig. 2 it is clear that the width of the bitmap must equal the width plus the height of the image. For a non-square image, the bitmap height must equal the greater dimension of the image.

Arbitrary Rotations

We now extend the discussion to include rotation through any angle α. Paeth (1986, 1990) found a solution using the following three sequential simple shears,

$$\begin{bmatrix} P'_x \\ P'_y \end{bmatrix} = \begin{bmatrix} 1 & \tan\alpha/2 \\ 0 & 1 \end{bmatrix} \begin{bmatrix} 1 & 0 \\ -\sin\alpha & 1 \end{bmatrix} \begin{bmatrix} 1 & \tan\alpha/2 \\ 0 & 1 \end{bmatrix} \begin{bmatrix} P_x \\ P_y \end{bmatrix}. \quad (9)$$

Equation (8) is clearly the special case where $\alpha = \pi/2$. The first and third tensors represent dextral simple shear parallel to the x-axis and the middle tensor is a dextral simple shear parallel to y. As Paeth (1990) has demonstrated, the programming steps for solving eqn. (9) can be reduced to three by substitution of interim results,

$$\begin{aligned} P'_x &\leftarrow P_x + P_y \tan\alpha/2 \\ P'_y &\leftarrow P_y - P'_x \sin\alpha \\ P'_x &\leftarrow P'_x + P'_y \tan\alpha/2 \end{aligned} \quad (10)$$

Figure 3a is a bitmapped image of Grand Canyon topography which has been rotated through an arbitrarily chosen angle of 34° using the three-pass simple shear technique of eqn. (9). The first horizontal simple shear is through an angular shear of 17° (equivalent to a shear strain of 0.3057 - Fig. 3b). The vertical shear strain of Fig. 3c equals the sine of 34°. The third shear (Fig. 3d.) is again a horizontal simple shear through an angular shear of 17°. No pixel blending or anti-aliasing procedures were used in this simple demonstration so there is a slight reduction in image quality but this is quite within the limits of tolerance for block diagram construction purposes. The main deterioration is due to the inevitable Moiré effect on certain map patterns.

Smith (1987) gives an alternative solution which involves superposing two general shear deformations,

$$\begin{bmatrix} P'_x \\ P'_y \end{bmatrix} = \begin{bmatrix} 1 & 0 \\ \tan\alpha & \sec\alpha \end{bmatrix} \begin{bmatrix} \cos\alpha & -\sin\alpha \\ 0 & 1 \end{bmatrix} \begin{bmatrix} P_x \\ P_y \end{bmatrix}. \quad (11)$$

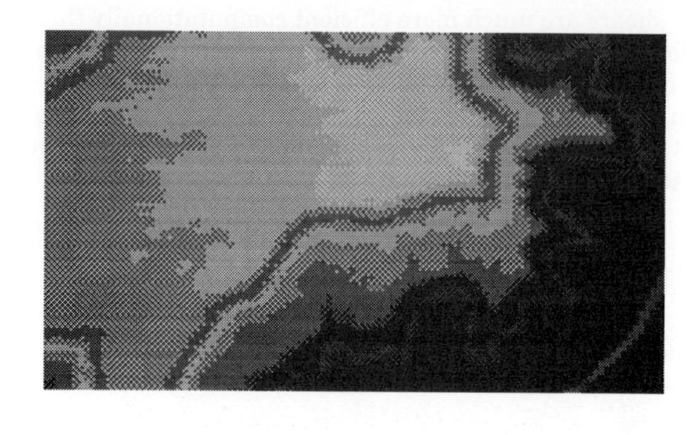

(a)

(d)

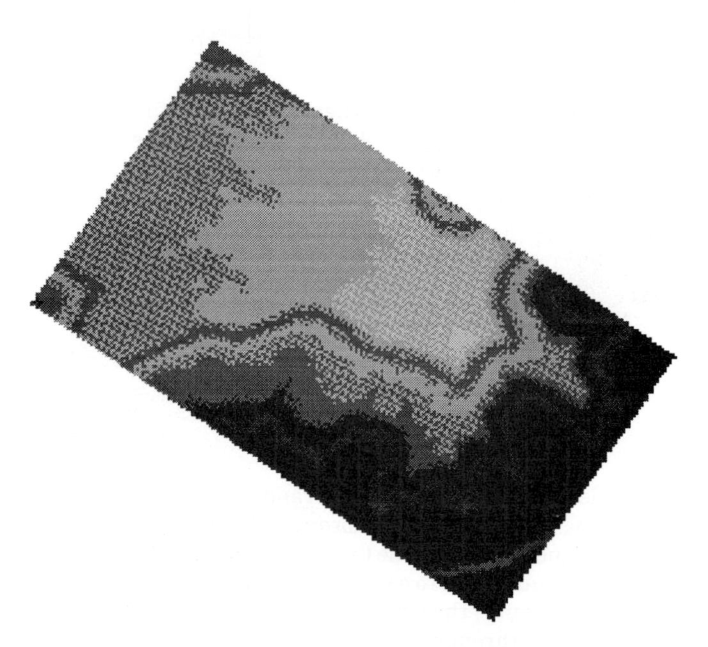

Fig. 3. Image rotation through 34°. (a) Source bitmap, (d) after third shear (horizontal).

The first tensor (counting from right to left, as usual) is a combination of simple shear and shortening parallel to the x-axis with an area loss of magnitude cos α. (as indicated by its determinant). The second is a shear and elongation parallel to the y-axis with an area gain of magnitude sec α. Because of these area changes, more than one pixel in the source bitmap may be mapped into a single location in the intermediate bitmap. The second shear tensor involves an area increase, so there are more pixels in the final destination bitmap than in the intermediate bitmap. So, although the net effect is area-conserving, the image will suffer from the developments of overlaps and gaps during its two-stage transformation. Therefore, Paeth's solution is preferable.

(b)

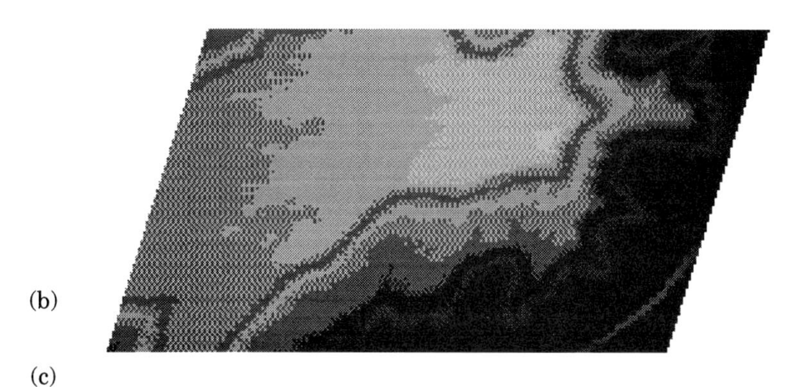

(c)

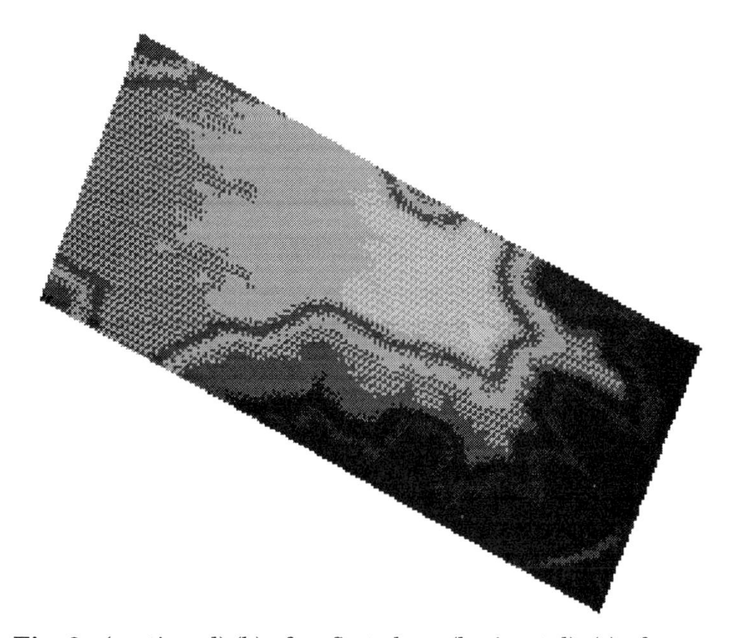

Fig. 3. (continued) (b) after first shear (horizontal), (c) after second shear (vertical).

Breene & Bryant (1993) extended the 3-pass solution from rigid rotations to all area-conserving deformations in two dimensions (*i.e.*, all deformation matrices with unit determinants, $D_{11}D_{22}\text{-}D_{12}D_{21} = 1$),

$$\begin{bmatrix} D_{11} & D_{12} \\ D_{21} & D_{22} \end{bmatrix} = \begin{bmatrix} 1 & 0 \\ G_2 & 1 \end{bmatrix} \begin{bmatrix} 1 & D_{12} \\ 0 & 1 \end{bmatrix} \begin{bmatrix} 1 & 0 \\ G_1 & 1 \end{bmatrix} \tag{12}$$

where

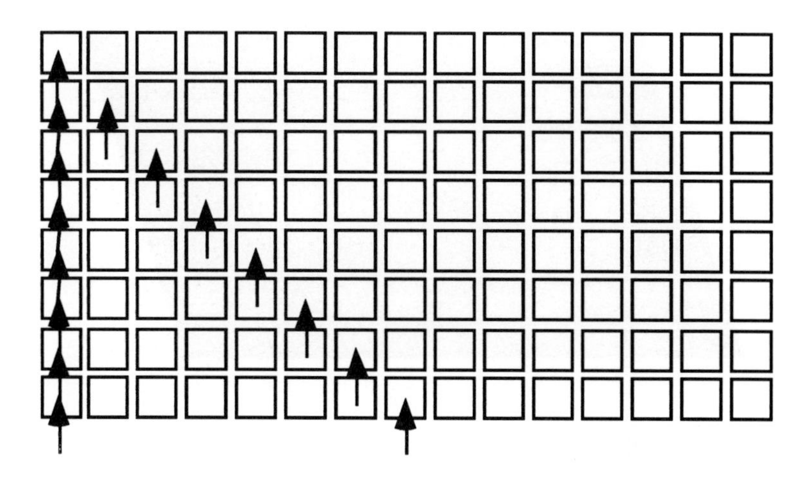

Fig. 4. Dextral bitmap shear by apparently sinstral bit shifting of row addresses.

$$G_1 = \frac{D_{11}-1}{D_{12}};\ G_2 = \frac{D_{22}-1}{D_{12}}. \qquad (13a,b)$$

D_{21} is not an independent variable because of the unit determinant requirement (commonly, $D_{21} = 0$). Clearly, there are solutions for all but the diagonal tensor, $D_{12} = D_{21} = 0$. In the block diagram application that follows, we will need to create foreshortened views using diagonal tensors of non-unit determinant,

$$\begin{bmatrix} P_x' \\ P_y' \end{bmatrix} = \begin{bmatrix} D_{11} & 0 \\ 0 & D_{22} \end{bmatrix} \begin{bmatrix} P_x \\ P_y \end{bmatrix} \qquad (14)$$

where D_{11} and D_{22} lie in the range [0,1] such that $D_{11}D_{22} \neq 1$. Intuitively, it is not possible to use raster shears for these operations.

The 3-pass simple shear approach outlined above greatly speeds the process of bitmap rotation because rows or columns of pixels are subject to identical transformations. Thus in Fig. 4, to apply a 45° dextral shear the bitmap's row addresses (shaded arrows) are shifted in an apparently sinistral sense. For a bitmap of width w and height h, this requires h bit-shift operations – an enormous saving compared to recalculating and storing $w \times h$ pixel values. For rotation angles other than 45° a majority of row or column offsets involve fractions of a pixel width. To maintain maximum image quality, it is necessary to blend the source pixels that overlap a given destination pixel horizontally for row shear or vertically for column shear. However, blending ratios based on overlap areas are constant for every pixel in a given row or column and recur periodically, so the blending process is much faster than that previously discussed and illustrated in Fig. 1.

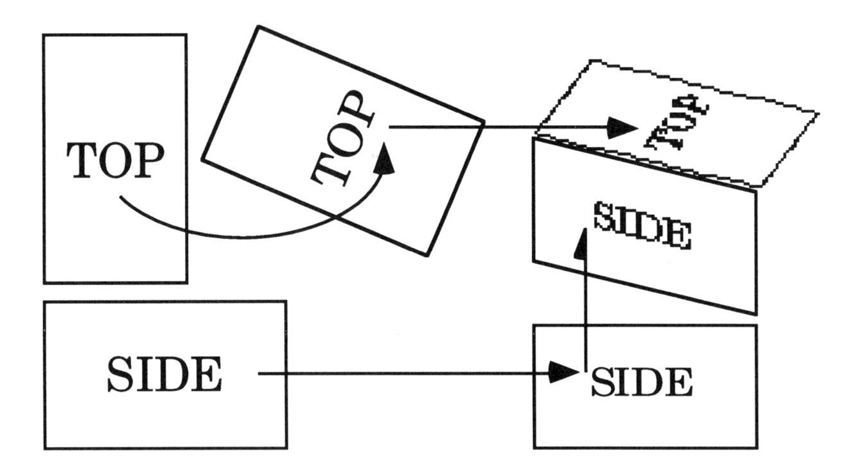

Fig. 5. Transformations used to create the sides of a block.

Implementation

We now address the practical problem of constructing a three dimensional view of a block given bitmapped images of the visible sides (Fig. 5). Let us suppose that we wish to view a block in a direction of plunge $\phi = 25°$, trend $\theta = 60°$, and twist $\tau = 0°$. Because the twist is zero, one edge of the block is constrained to remain vertical which simplifies the routines required to draw the sides; only the top can rotate freely in all three dimensions. Beginning with a view direction of $\phi = 90$, $\theta = 000$, the top bitmap is rotated through -60° using the 3-pass procedure of eqn. (9). Then this rotated image is deformed by

$$\mathbf{D} = \begin{bmatrix} 1 & 0 \\ 0 & \sin\phi \end{bmatrix}. \tag{15}$$

Next, the left side of the block is shortened horizontally to match the width of the transformed top panel and also vertically to match the desired inclination of the view direction using the deformation

$$\mathbf{D} = \begin{bmatrix} \sin\theta & 0 \\ \cos\theta\sin\phi & \cos\phi \end{bmatrix}. \qquad (16).$$

The front side of the block (not shown in Fig. 5) is treated in a similar fashion.

The above theory is implemented in a Macintosh application called *Block Diagram* version 2.2 written by the author and distributed by Earth'nWare Inc. The Macintosh operating system provides certain advantages for bitmap ma-

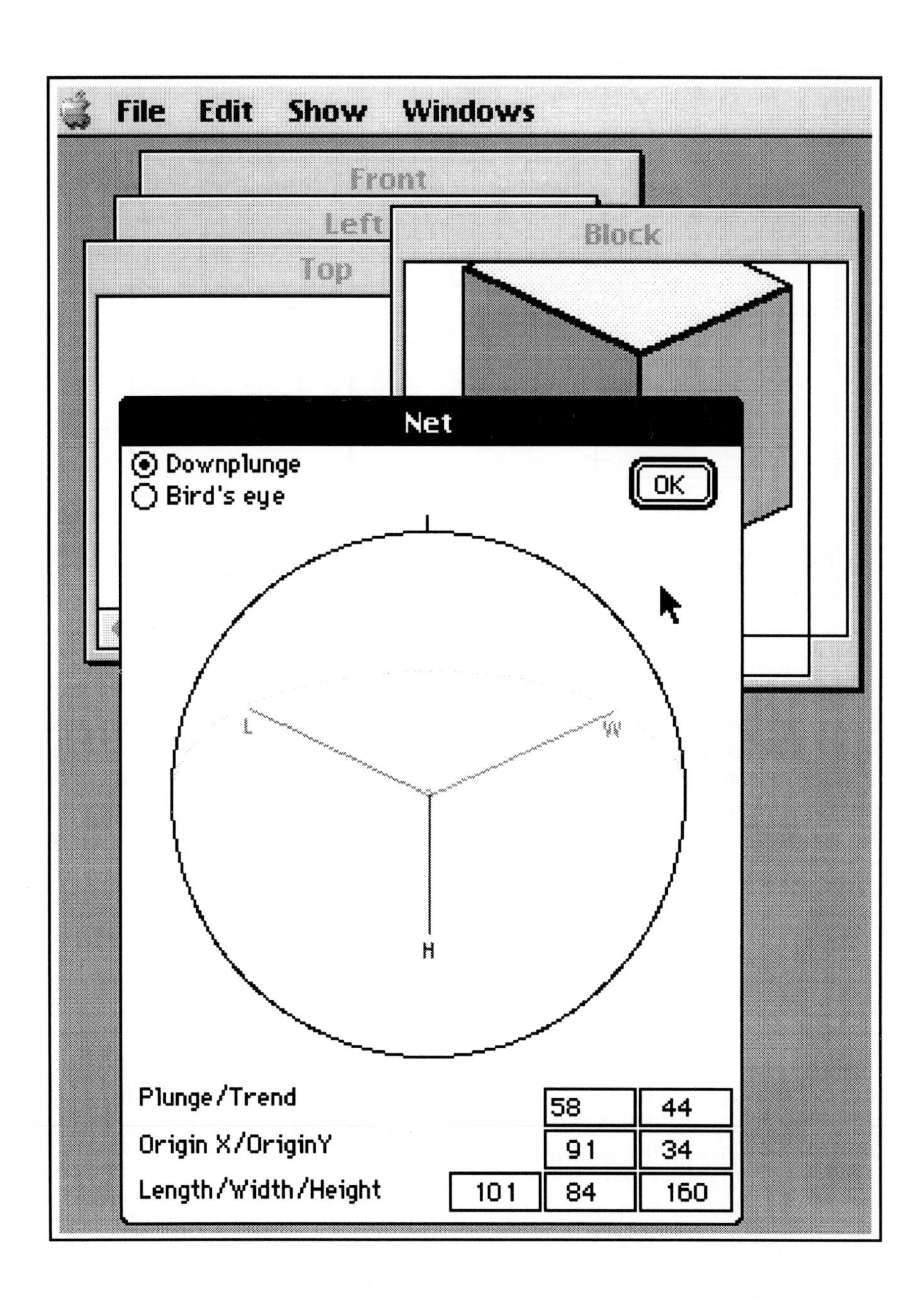

Fig. 6 Graphical user interface of the program *Block Diagram*. Block orientation is controlled by the orthographic orientation net (front window). Unit vectors parallel to the block's length, width, and height are labelled l, w, and h, respectively.

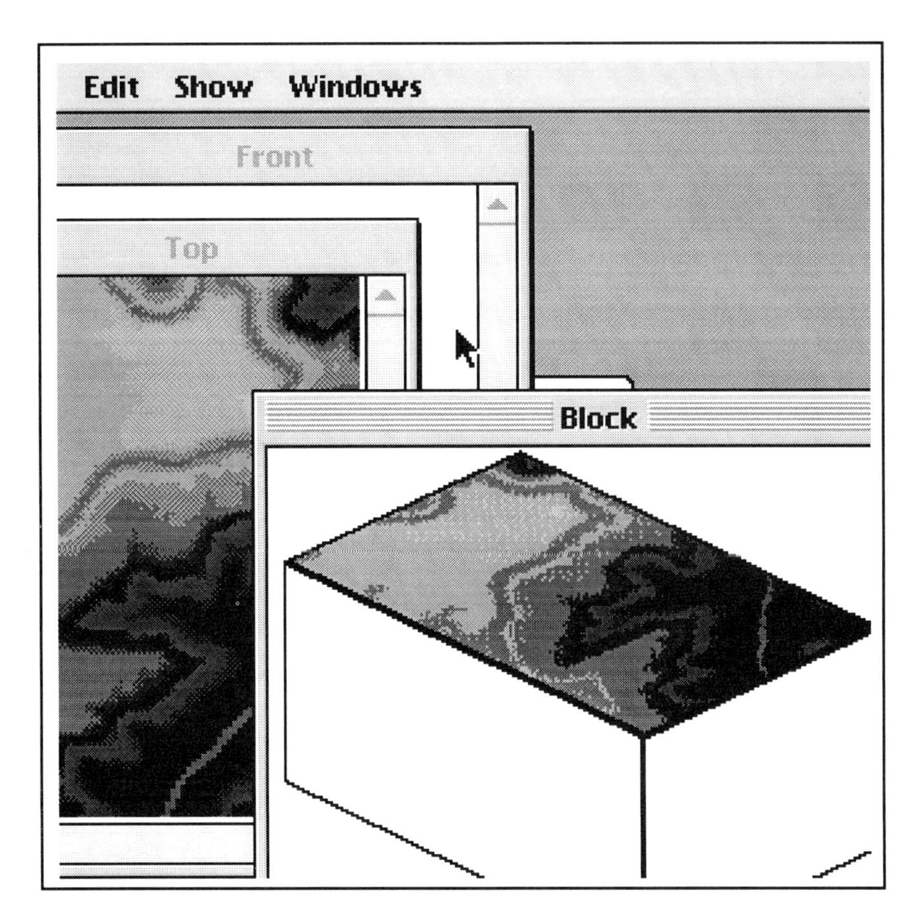

Fig. 7. Rotation of bitmapped image from the orthogonal view in the `Top` window to the inclined view of the top of the block in the `Block` window.

nipulators. First and foremost, it gives the application developer access to *Quickdraw*™ graphics routines such as `Copybits()` which maps a bitmap from a source rectangle to a destination rectangle with breathtaking speed. Also, the Macintosh memory management system permits six images to be opened in separate windows in an application whose RAM allocation is capable of storing only one. Images are stored as 'purgeable' PICT resources which are written to disk and recalled into RAM by the memory manager when they are needed.

As illustrated in Fig. 6, the location of the origin and the length , width, and height of the block can be entered into edit fields in the `Net` window. Alternatively, the block displayed in `Block` window may be relocated and resized by clicking and dragging the mouse. The block's orientation is controlled by entering values for the plunge and trend of the viewer's line of sight in edit fields or by clicking and

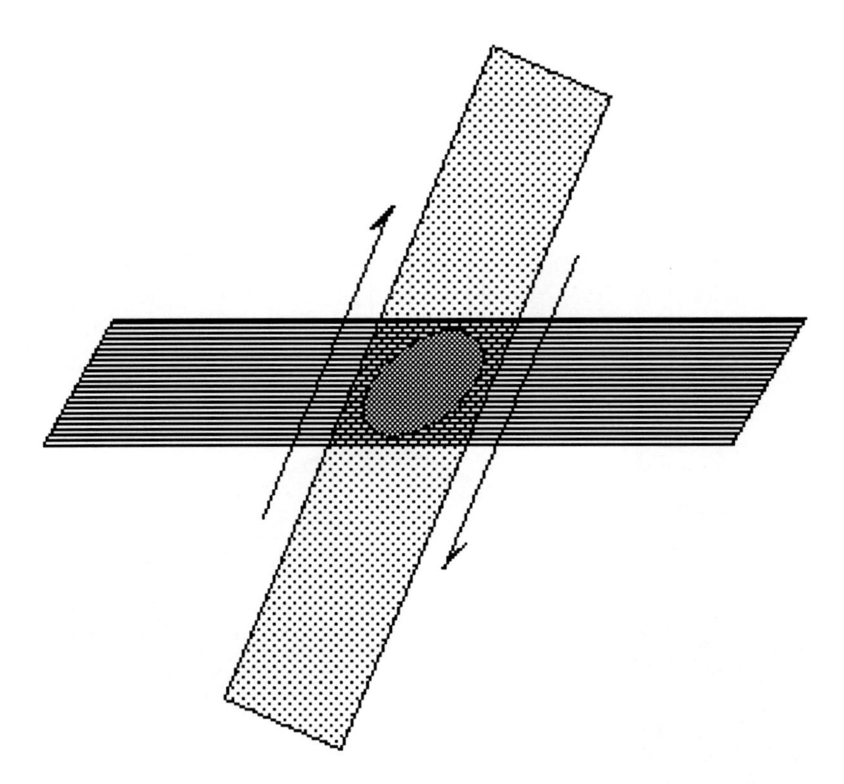

Fig. 8. Implications of superposition of shears of the same sense but different orientations. Shaded strain ellipse records strain state resulting from dextral shear in a horizontal zone (lined). Inclined zone of dextral shear (dotted) has not yet been applied.

dragging the mouse on points labelled l, w, and h on an orientation net. Figure 7 shows sample output.

Implications for Shear Structures

We have used the theory of shear deformation for the practical task of rotating bitmapped images. However, the multiple shear method of rotation has important implications for the study of strain in rocks also. It is unlikely that natural shear zones will be oriented exactly orthogonal to one another or that they will move in a sequence that mimics eqn. (9). However, intersecting shear zones of like sense are occasionally seen in nature (personal observation). Figure 8 illustrates the state of strain in a shear zone after the imposition of a dextral horizontal simple shear upon which an inclined zone of dextral simple shear is about to be superimposed. The long axis of the strain ellipse resulting from the first shear zone is oriented in the field of shortening with respect to the second zone and so will undergo shortening accompanied by clockwise rotation. Conversely, the strain ellipse's short axis is in the field of elongation of the second shear zone and will lengthen while it rotates clockwise. The result of this superposition of simple shears is a strain state with a high degree of rotation but low distortion.

Discussion

The mathematical tools traditionally used for applying rigid rotations – namely, orthogonal tensors – were developed by theoreticians who were not concerned with such issues as computational efficiency. They are not well suited to the task of modeling rotation in 3-D on a computer monitor. It doesn't make sense to expend time calculating places of decimals for pixel locations which are effectively truncated by the resolution of a raster device. A more efficient approach is to represent a rotation as a sequence of shear deformations. Every student of structural geology learns that a simple shear deformation can be decomposed into a pure shear and a rigid rotation. Conversely, it follows that a rigid rotation may be thought of as a combination of pure and simple shear. Although two general shear deformations can be used to generate a net rigid rotation, the best combination of speed and maintenance of image quality is obtained from the three-pass simple shear approach of eqn. (9). As implemented in the author's program *Block Diagram*, this method permits reorientation and refreshing of three 10,000-pixel images in a couple of seconds.

References

Breene, L. A. & Bryant, J. 1993. Image warping by scanline operations. *Computers & Graphics* **17**: 127-130.

De Paor, D. G. 1995. Quaternions, raster shears, and the modeling of rotation in structural studies. *Geol. Soc. Amer. Program with Abtracts.*

Lisle, R. J. 1980. A simplified work scheme for using block diagrams with the orthographic net. *J. Geol. Ed.* **29**: 381-395.

Paeth, A. W. 1986. A fast algorithm for general raster rotation. *Proceedings Graphics Interface '86.* Canadian Information Processing Society, Vancouver. pp. 77-81.

Paeth, A. W. 1990. A fast algorithm for general raster rotation. In: *Graphics Gems* I (ed. A.S. Glassner). Academic Press, Boston. pp. 179-195.

Ragan, D. M. 1985. *Structural Geology: An Introduction to Geometrical Techniques.* (3rd. ed). J. Wiley & Sons, New York.

Smith, A. R. 1987. Planar 2-pass texture mapping and warping. *Siggraph* **21**: 263-272.

Subject Index

3-D analysis 155
3-D Strain 75, 94
  analysis 155
  structure 156
  teaching applications 94

A.V.A. 129–131
  diagram 123–124, 129, 155
Access 99
Achsenverteilungsanalyse
    diagram. *See* A.V.A. diagram
Adobe Photoshop. *See Photoshop*
aerial photograph 135, 424
After Dark 99
Airy stress function 363
Alps 167–168
alteration halo 313
angular annulus 200
animation 14, 18, 26, 34, 43, 46, 299
anisotropy 153, 167, 173
anticline 483, 486
antiformal duplex 422
apparent dip 51–53
Applescan 18
ArcView 474
area
  estimation 105, 152
  -depth relationship 477, 486
ArtWorks 478
ASCII 170, 182, 184–187, 189,
  191, 242, 291
ASI 183–184
aspect ratio 75, 89
attitude 186
  data 185
  symbol 181, 184–186
audio cassette 16
Authorware 34
  Professional 18
AutoCAD 181–186, 189, 191–192,
  471, 474–475
autocorrelation 148, 153
AutoLISP 182–183, 192–193
automation 422, 477, 500
axial plane 442
axial trend 187
azimuth 169–170, 196, 203, 205,
  207, 212, 218–224, 227, 241–242

balance
  area 484
  excess area 484, 496
  key bed area 492
  length 484
  lost area 484, 496
barycentric 399
  combination 395, 399
  coordinates 397, 399–400, 405
base map 462
  scanning 459
BASIC 325, 501
basin 359
beam test 344
Beatrice 46
  HyperCard Stack 46
bed length 477, 483–484, 492
  deformed 488
  undeformed 488
bed thickness 359, 370–372, 477,
  481, 484, 492
bedding 308
  surface 308
Bézier curves 389, 391, 393
  affine invariance 393
  arclength 402
  bi-cubic patches 409
  continuity 393
  convex hull 393

CAD 389, 392, 474, 478.
  See also Computer Aided Design
  program 125, 127
  /CAM 391
CAL6 *See* Computer Aided Learning
CALCGRID 170
California 184, 372
Canadian Cordillera 421–422
Canvas 71, 291, 295, 457, 459
  –461, 477, 479, 481, 491, 494–498
  for *Windows* 459
CarDec 404, 416
  teaching aid 416
CARIS-DTM 426, 428–429, 440
  -V3D 429
cartridge drive 16
Catmulls 392
CD-ROM drive 16

CGM file 479
charge coupled device 138
ClarisWorks 291
clast 84, 89
  rotation 75, 84, 89
cleavage 42, 218, 220, 224, 226, 229, 239, 247, 307–308
  crenulation 247
  growth 247
  plane 135
CNTR file type 52, 54, 56
coaxial 94
COCORP 168
color mapping 114
Color OneScanner 458
Color StyleWriter 458
Compact Pro 469
complex
  conjugate 451
  number 447, 451
  strain 58
  structure 181
  visualization 499
complexly deformed terrain 422, 444
compressional stress 369
Computer Aided Design 471
Computer Aided Learning 13
  delivery area 17
  development area 15
  hardware environment 14–15
  management area 17
  software environment 18–20
computer
  assisted tomography 159
  graphics program 57–58
  image projector 20
  integrated polarization 155
  laboratory 17
  presentation panel 17
  programming 243
  simulation 343
  -generated images
  direct display 20
concepts maps 13
continuity 26
continuum mechanics 75
contour 51–52
  generation 428
  data 426, 434
Contour File Maker 52, 56
contouring 203
contraction 81
coordinates 473
CorelDraw 208, 440
correlation 149, 151, 153
courseware 13, 39
  geological 23
  multimedia 16
cross section 46, 51–53, 56, 119–120, 305, 391, 397, 422, 457, 465–466, 469, 499
  balance 469, 477–478, 483–490, 496
  circular arc 483
  construction 51, 479
  dip domain 481
  drawing 481
  extrapolating 397
  interpolating 397
  interpretation 488
  line-length balancing 403
  restoration 477–478, 491–494, 496
  scale 481
crystal
  class 173
  defect 98
  face 218
  growth 101
  -melt mixtures 124
  structure 100
  symmetry 171
crystalline aggregate 167
crystallization
  history 100
  pre-, syn-, post-tectonic 100
crystallographic
  direction 168
  fabric 173
  orientation 135, 155, 168
crystallographic
  a-axis 227
  angle 220
  axes 220
  b-axis 223
  c-axis 220, 227
  distribution 130
  fabric 155
  orientation 123–124, 155
  plunge, trend 130
crystallography 230
cubic 173
Cullen Granite 313
curve fitting 488

cyclic rupturing 343, 344, 351
cylindrical fold
  modeling 406

data storage 168
database 17, 44, 168, 181, 183–185, 189, 193, 238, 243, 471, 473
  design 472
  operations 473
  relational 425
  manager 46
  table 185–186
Datashow 20
dBASE 183, 185, 189
  dBASE III 184, 193
  dBASE III Plus 183
de Casteljeau algorithm 397
DECORP 168
deformation 39, 68, 81, 105, 123–124, 160, 285–286, 289, 311, 448, 477, 484, 499
  analysis 259
  apparent inhomogeneity 64
  computer simulation 76
  ductile 22
  experiment 33
  experimental 123, 160
  fabric 97
  field 40
  general shear 511
  geological 78
  geometry 135
  heterogeneous 83, 389
  history 113, 303
  homogeneous 57, 59–61, 70, 260, 262
  increment 266, 269, 272
  inhomogeneous 61–62, 70, 72
  kinematics 248
  matrix 75, 83, 85, 87, 94
  mechanism 100
  microscopic 97
  microstructure 98
  multiple episodes of 87
  particle movement 83
  path 40, 259, 287
  plane isochoric 262
  plane strain 79
  processes 24
  progressive 259, 262, 264
  pure shear component 78, 84
  quantifying 76
  rotational 447, 499
  shear 503, 511
  simple shear component 84
  simulation 100
  steady-state 76, 78, 83–84
  tensor 82
  transtensional 94
  velocity field 76
  visualization of 75–76
deformation field 123, 125
  map 129
Delauney Triangulation 428
DEM. *See* digital elevation model
Design Jet 650C 469
design tools 389
deviatoric stress 40
dextral shearing 87
Dicer 157
diffraction pattern 135
digital
  camera 136
  Chisel 18, 34
  courseware 14
  elevation model 422
  image 125, 136–137, 144, 153
  sharpening 479
  Terrain Model 421–422, 425, 428–431, 434, 440, 444
  applications 428
  data 424, 426
  facets 425, 428
  model 429
  orthographic visualization 440
  packages 426
  regularly gridded 425, 428
  surface 428–429, 442–443
  theory 422
  triangulated 428
  visualization 443
digitization 131
  on-screen 125, 129, 131
  tablet 129
digitizer 473
digitizing 293
  field stop locations 181
  pad 471
  tablet 125, 424
dike 360
dilation 22

dilution 146
dip 51, 205–206, 448
directional data 325
Director 34
dislocation 101
  climb 100
  glide 98, 100
dislocation tangles 135
displacement 348, 359
  algorithm 400
  component 363
  equations 306, 308, 313
  field 344, 348, 352, 359, 361, 363, 373-375
  vector 359
displacement maps 123
dissolution 125
distortion 22, 434, 511
DOS 71, 182, 195, 204, 207, 212, 242, 478
Douglas-Peucker filter algorithms 426
drafting 477–478
drawing 18
  applications 389
  program 52, 478
  software 169
DTM. *See* Digital Terrain Model
ductile
  deformation 42, 118
  shear 413
  shear zones 42
DXF file 472–475
dykes 307
dynamic history 135

earthquake 344, 352, 360, 374
  displacement fields 373–374
  Loma Prieta 359
ECORS 168
elastic
  medium 360
  stress 344, 352
  homogeneous bodies 359
Elastic Reality 18
elasticity equation 343–344, 347–348, 352
electron channelling 155
electronic
  atlas 99
  classroom 13–14, 17
  courseware 14, 17, 34
elevation 52, 221, 223, 423, 425, 429, 442
ellipticity 89
elongation 504
encapsulated postscript 299
Endnote 34
Euler angle 447–449
  limitations 448–449
Euler's theorem 447, 452
Excel 183, 233, 238, 477–478, 488, 496
  Macro 238, 240
experimental deformation 24
extension 75, 81
extensional displacement 486
extinction 130
extrapolation 305, 416

Fabfile 203–207, 212, 214–215
  downloading 215
Fabric 203–207, 212–215
  downloading 215
fabric 24–25, 101
  analysis 152
  element orientation 433
failure theory
  Mohr Coulomb 352
Fast Fourier Transform 148
fault 307, 327, 359–361
  displacement 402–403
  geometry 24, 374–375
  plane orientation 325, 327
  rupture strength 343
  slip 374
  strike-slip 359, 366, 369, 374
  system evolution 343–344, 346, 348, 351–352
  terminology 24
  throw 307
  tips 370
  traces 135
faulting 344
faults 40
  normal 40
  reverse 40
  strike-slip 40
  thrust 40
Fetch 18
FFT. *See* Fast Fourier Transform
fiber
  curvature change 275
fiber growth 270
  antitaxial 102, 271–274, 280

syntaxial 102, 270, 272, 274, 277, 280
field data 457, 462, 469
management 471
plotting 471, 474
recording 471, 474
field strength
calculation 309
field-scale structures 24
Fieldlog 471–475
AutoCAD interface module 471, 474
data entry 473
database design 472
database module 471–472
interface module 471
map generation 473–474
report generation 474
system requirements 471
file transfer program 20
filters 139
convolution 141
high-pass 146
low-pass 141
morphological 146
smoothing 141
Finger tensor 83
finite strain 76, 78, 85, 92, 247–250, 256, 259–260, 274–275, 281
axes 83, 94
calculations 75
ellipse 83
evolution 87
parameters 75, 82
particle paths 75
principal axes 83
quantities 82, 87
strain ellipses 75
values 84
flexural slip 42
Flinn diagram 40, 285–288, 291
plotting program 289–290
Ramsay's logarithmic 288
values 94
flood risk 428
flow. *See also* deformation
apophyses 78–82, 87
coaxial 130
field 83
heterogeneous 76
homogeneous 260
lines 75, 84–87
non-coaxial 130, 261
pure shear 261
simple shear 261
flow types
pure shear 261–262, 266
simple shear 261–262, 264
sub-simple shear 261–262
super-simple shear 261, 264
fold 181, 184, 187, 307, 360, 433
asymmetry 189
axial angle 481
axial surface 483
axis 186, 421, 440
domain 421, 433–434, 440
geometry 25
interference pattern 440
isoclinal 398
morphology 25
nomenclature 25
orientation 481
plunging 443
fold classification 42
folding 42, 303, 447
folds 24
foliation 131, 181, 186, 248
external 247
internal 247
footwall 494
FORTRAN 84, 167, 173, 364
Forward modelling 303–305, 310, 313, 391
Fourier transform 146, 151
continuous function 146
fractal analysis 111, 113, 115
box-counting 111
mathematics 115
Richardson 111
fracture 24, 360
analysis 159
mechanics 344, 352, 359–360, 368
constraints 344, 352
elastic 375
surfaces 157
tip 369
frame grabber 16, 123, 138
Freehand 18, 22–34, 71–72
freeware 471
Fry 293–295
Fry strain analysis 293–295
FutureBASIC 55

General
  Flow Lines 75, 85–86
  Shear 75
  Shear Box 85
geodetic methods 373
Geographic Information System 303, 389, 391, 421–422, 425, 447, 455, 471, 474
geologic
  computerized map 458
  contact 182, 474, 499
  data compiling 462–464
  map 457–458, 465–466, 469
  map compilation, presentation 457
  three dimensional data 499
geological
  history 311
  structures 39
geomorphometric parameters 429
geophysical response 303
geophysics course
  teaching aid 310
geothermal fluid 360
gimbal lock 448, 452
GIS. *See* Geographic Information System
Glide 100, 101
Global Positioning System 373–374
gneiss 422
Golden Dyke 311
  model 313
goniometry
  neutron 170
  texture 168–169
  x-ray 155, 170
GPS. *See* Global Positioning System
grain boundary 130–131, 143–144, 146, 153, 163
  detection 135
  migration 102
  sliding 102
grain size 135
grain-shape preferred orientation 105
Grand Canyon 503
graphics 14, 45–46, 359, 499
  adapter 196, 207
  image display 21, 440
  program 57, 202
  routine 365
  window 235
graphical
  digitization 424
  output 243
  presentation 106
  search 325
graphing 299, 488, 490
  three dimensional 299
  two dimensional 299
gravity anomaly 303, 309, 311, 313
grayscale 118, 125, 130, 136, 295
  filters 109
  table 108
great circle 52, 195, 200, 204, 206 209, 212–213, 221, 233, 450
  diagram 203
GREYDENS 170
ground survey 423
growth fibers
  antitaxial 25
  syntaxial 25
  syntectonic 24

hangingwall 494
harmonics
  linearly independent 171
  spherical 171
  symmetry 171
Hermite spline 392
heterogeneous general shear 413
  convergent 413
  divergent 413
heterogeneous strain 409
Hewlett Packard
  HP7474 197
  LaserJet 197
  Laserjet III 197, 200
  Graphics Language 197, 202, 208
hill-shading 428, 431
homogeneity 26
homogeneous strain 307
Hooke's law 344, 347
hornblende 220
hydrocarbon exploration 390
HyperCard 43–46, 49
  environment 44
  stack 43–45, 49
  teaching aid 44
Hyperstudio 18, 34
hypertext 39
hypothesis testing 375

IBM PC 97, 195, 204
illustration 457
Illustrator 22, 71, 299
Image. See NIH Image
image 46, 130, 135, 477–478, 499
  acquisition 136
  analysis 106, 113, 135–136, 138, 143, 153, 160, 163
  input 478–479
  measurement 138
  operations 125
  processing 106, 113, 136-138, 144, 148, 151-153, 155-156, 160, 163, 167-168
  program 131
image
  capture 479
  file 52
  manipulation 18
  splicing 460–461
ImageFractal 105, 111, 118
  spurious results 112
ImageTool 138
imaging 125
  system 123–124
IMGEO 138
inclination 169
inclusion trail 119, 248, 253
  simulation 249
inclusions 247
information management system 49
instantaneous strain 76, 78, 82–83, 92
  flow apophyses 75, 86
  kinematic vorticity 75
  parameters 75
  velocity fields 75
instantaneous strain
  axes. *See* ISA
  quantities 82, 86, 94
  strain axes 86
instantaneous stretching axes. *See* ISA
interactive
  graphics 391
  multimedia packages 18
interference colors 114
internal vorticity 261
interpolate 169
interpolation 305, 422, 424–425, 447–451, 453–455, 500
  cubic 416
  functions 425
  grid-based 425
  linear 416
  path 449
  pointwise 425
intracrystalline strain 113
introductory geology courseware 33
intrusive events 303
inversion 159, 168, 170–174, 304
  geophysical field 304
INVPOLE 167–168, 170, 173
ISA 80, 82, 87
island-arc terrane 422
isoclinal trough 443
isotropy 26

joint 359–361
  propagation 370
  spacing 359, 370–372

k values 40
kernels 139
kinematic
  analysis 57, 75, 447
  constraints 314
  history 135, 256
  indicator 42, 97
  modeling animation 22
  process modeling 22
  vorticity 75, 81–82
kinematics 389
kink-fold
  construction 390
  cross section 390
Kodak PhotoCD 97, 99, 102
Koolpin Formation 313

LabView 167
landscape
  visibility 428
Lanzo massif 167–168, 174
laser printer 19, 203, 213
LaserWriter 495
latitude-longitude 474
lattice preferred orientation 105, 113, 115, 168
LEFM. *See* linear elastic fracture mechanics
lherzolite 174
linear elastic fracture mechanics 359–361, 369, 375

linear elastic medium 360
linear interpolation structure 425
lineation 135, 181, 307
Lisle's stress inversion method 326, 327–329, 333
  assumptions 327
  implementation 327
LISP 183
lithologic contact 464, 469
lithosphere 344, 352
loading condition 359
Local Area Network 23
look-up table 107–109, 112, 139, 148
  customizing 115
LPO. *See* lattice preferrec orientation
LUT. *See* look-up table

Mac Plus 19, 100
Macintosh 13, 18, 20, 43, 51–52, 59, 84, 97, 100, 124, 182, 233, 235, 238, 293, 299, 458–459, 469, 478, 499
Maclaren Glacier Metamorphic Belt 256
MacPaint 45, 105–107, 291
Macromedia 34
  Director 18
magnetic anomaly 303, 309, 311, 313
mantle texture 168
map 51–53, 56
  data 444
  surface 499
  symbol 473
Maple 299, 301, 363
mapping tool 474
markers 75, 83, 87, 89, 123, 125, 127, 129–160
  digitizing 123, 125, 127
  displacement 124
  rotation rate 248
  strain 40
material behaviour 26
mathematic co-processor 208
Mathematica 34, 301, 363, 366, 368, 374
mathematical theory 499
MATLAB 363
matrix 75
  homogeneous deformation 75
maximum elongation 80
maximum shear stress
  direction 326
MacWrite 291
mean normal stress field
  compressive 368
  tensile 368
mean stress 369
MetaFile 299
metamorphic event 313
metamorphism 390
microcomputer 14–15, 299, 360
micrograph 135
microscope 97, 123, 135–136, 218
  stage 125, 218
  -video camera 123
microscopy 97
microstructure 97, 100, 113, 123, 131, 135–136, 153, 156, 160, 163
Microstructures CD ROM 99, 102
Miller indices 168
model
  dynamic 305
  evolutionary 351
  geometric 305
  geophysical 314
  geophysical response 304
  integrated 306, 310, 313
  kinematic 305–306, 477
  magnetic 314
  potential-field 304, 309
  spring network 344, 346–347, 352
  three-dimensional 305–306
  topological data 425
  two dimensional 361
module 39
Mohr
  circle diagram 22, 40, 42
  Stress circle 26
Moiré effect 503
Monashee
  Complex 421–422
  Mountains 421
monoclinic 173–174, 220
Monterey Formation 372
morphing 449–450
morphological
  features 218
  filter 146
Mount Bonnie Formation 313
multimedia 13–14
  development facility 15
  hardware 15, 17
  presentation software 34
  storage 18

multiply deformed terrains 303, 305

neighborhood operators 139
neotectonic structure 360
neotectonics 373
Net 195, 196, 200
nets
  Lambert equal
    area 195, 199, 203
  orientation 455
  orthographic 195, 199, 499
  polar 196
  stereographic 195, 199, 450. *See also* stereonet
network 14, 15, 17
  management 17
neutral surface
  fold 42, 414
  modelling 414
Newtonian viscous
  flow 260
  fluid 248
  matrix 249, 264
NIH Image 105–107, 111, 113–115, 120, 123, 125, 127, 130–132, 138, 152, 157
  analyze menu 109
  applications 113
  downloading 106, 125
  enhance menu 109
  importing images 107
  options menu 108
  special menu 110
  stacks menu 110
  system requirements 106, 125
  using 108
Noddy 306, 313, 317–318
noncoaxial 94
norcamphor 124
normal stress 40, 363, 373
North Sea 43, 44
  HyperCard stack 46
  structural geology 44, 46
notebook computer 17
numerical
  simulation 269, 280
  techniques 168
NURBs 392
NuSplines 392

object centers 293
oceanic terrane 422
octachloropropane 124
octahedral shear stress 94
ODF. *See* Orientation Distribution Function
Ofoto 459–461
oil 44
olivine 174
Omineca Belt 421–422
Omnipage 18
ooids 293
optic
  axis 219
  principal directions 220
optical
  character recognition 18
  directions 217
  disk drive 125
  indicatrix 220
  mineralogy 97
oral presentation 457
orientation 60, 167–168, 184, 203, 217–218, 220, 224, 239
  data 181, 195, 203–206, 208, 215, 233, 237–238
  diagram 203, 205–209, 212, 215
  measurement 184
  notation 238
  data 308, 447, 455, 471
  data density counting 169
  data density diagram 168
  distribution function 167–174
  distribution function coefficients 168
Orocopia
  Mountains 184, 187
orthoamphibole 239
orthogonal tensors 447
orthographic view 429
orthonet 438, 440
orthorhombic 174, 176
outcrop
  pattern 53, 308
  scale 360
overgrowth 247
overhead transparencies 19

P-wave 174
paleocontinent 422
paleostress field 325
Paradox 183
particle

motion 85
path 264
Pascal 125
pascal triangle 399
Passive Clasts 75, 89
pattern matching 135
PC 458, 471
peridotite 167–168
peripheral equipment 16
personal computer 13, 15
perspective 434
petrofabric 167–168, 171
petroleum 44
exploration 46
phase boundary
migration 125
PhotoCD. *See* Kodak PhotoCD
photogrammetry 159, 423–424
photograph 496
photometry 124, 155
photomicrograph 97, 99
Photoshop 18, 99, 107, 114–115, 169, 295, 477–479, 495–498
PICS 105–107
PICT 45, 52, 54, 56, 105–107, 114, 169, 294, 299, 404, 457, 460, 509
Pine Creek Geosyncline 311
Pingston Fold 421–422, 431, 434, 438, 440, 442–443
pitch 51, 448, 450
pixel 501
planar
data 205
structures 51–53, 56
surface idealization 361
plane
isochoric flow 261, 264–265, 280
-elastostatic problem 362
plastic
behavior 167
flow 167
plot
contour 369
three-dimensional surface 369
plotter
ink jet 469
plotting program 474
plunge 51, 185, 196, 205–206, 212, 448
direction 507
point
counting 105, 113
operation. *See* point transformation
transformation 139
polarized light
cross, plane 218
pole
interpolated 454
figure 167–174, 176, 203, 447
contoured 203
polycrystal 173
polynomial 391
expression 426
polyphase 440
porosity 153, 159
Porphyroblast 100, 101
porphyroblast 247–249, 253, 256, 454
delta tails 454
growth models 247–250, 252
growth pattern 253
rotation 119, 247, 252
synkinematic 120
texture 253
Portfolio 99
position gradient. *See* deformation tensor
position gradient tensor 262, 269
Postscript 19, 235
potential-field
information 306
response 304
gravity survey 303
magnetic survey 303
power spectrum 149
PowerMac 100, 235, 299
PowerPC 299
Powerpoint 18, 19, 20–22, 24, 26, 33–34
PowerWave 100
presentation
images 18
package 34
software 14, 18
pressure fringe 259–260, 269–275, 277, 280–282
solution 259, 369
soultion surface 359
PREVIEW 428
principal directions 57, 65–67, 70–71
principal stress 40, 369
orientation 325
printer 195, 197, 208, 233, 458

printing 494
profilometer 157
programming environment 43
projected image 159
projection 434, 442
  isometric 438
  orthographic 421, 438
  parallel/orthographic 429
  perspective 429
projections 120
  brightest point 120
  mean value 120
  nearest-point 120
propagation path 369
pull-apart basin 369
punch cards 168
pure shear 40, 57–58, 64–68, 71, 76, 79, 81, 83, 85, 87, 89, 248, 250–252, 256, 413, 511
  homogeneous 57, 68
  principal directions 68
  rotation 65
  strain rate 84
  visualization 58
pyrrhotite 313

Quadra 650 458
quadrangle 458, 469
quadratic interpolation 398
quantitative analysis 106
quaternion 447, 451–455
  operation 455
  theory 451
Quattro Pro 182, 183
QuickBASIC 84
QuickDraw 235, 509
QuickTime 46

RAM Doubler 458
raster 306, 424
raster device 501, 511
reconstruction
  structural 303
reference sphere 195–196, 200, 203–204, 218
regional
  structural geology 181
  structure 303
  modeling 311
regression 490
relative motion 75
relaxation method 352
relief 428
remote load 359
remote sensing 135
remote stress 367
Reports 46
requisite strain 488
restoration 477
  area 492
  constant line length 492
  oblique shear 494
  rigid block 491
  vertical simple shear 494
"RGN " resource type 54
rheology 24, 26
ridge 359
rigid
  clast rotation 84, 89
  fiber growth 269
rock deformation 105
Rock Deformation and Geologic Structures
  availability 42
  brittle deformation 40
  courseware 39
  ductile deformation 42
  strain 40
  stress 40
  system requirements 39
rock
  fabrics 42, 100
  fracture mechanics 343
  mechanics 26
  texture 167
ROMSA. *See* Lisle's stress inversion method
Rotating Clasts 75, 84, 89
rotation 22, 75, 89, 212, 260, 269, 447, 450, 452–453, 500, 511
  bitmap 503–506, 510
  dynamics 89
  multiple shear method 510
  progressive 266
  rigid 511
  shear-induced 81
rotation axes 217
  east-west 217
  inner vertical 217
  north-south 217
rotational operations 455
rupture strength 344, 352

S-wave 174
satellite image 424
scale limitations 313
scaling laws 118
scanner 16, 477–479
  flatbed 457
scanning 18, 52, 136, 496
  electron microscopy (SEM) 155
  programs 52
SCANSTEREO 167–170, 174, 177
Schmidt grid 169
scientific data
  display quality 18
  illustration quality 18
screw dislocation 100
  migration 100
section 308
sedimentary clasts 286
segmentation 144, 146
seismic
  anisotropy 167–168
  displacement 352
  line 494, 496
  phase velocity 174
  properties 168, 174
  reflection profile 168
separation-arc diagrams 189
Sequatchie anticline 486
serial section 105, 156
Shape Preferred Orientation 135, 153
shareware 195, 234, 237
shear
  displacement 499
  failure 40
  stress 40
  structures 510
  zone 510
Shear Box 84–86
shear sense 167, 260
shear
  vertical strain 503
  stress 363
  wave splitting 167
  zone, active 303, 307
shortening 75, 504, 507
shrinking 146
simple shear 40, 58, 64, 67, 70,–71, 76, 79, 81-89, 248-251, 256, 260, 413, 477, 499, 502-503, 510-511
  dextral 503
  heterogeneous 413
  homogeneous 57, 68, 415
  inclined 402
  principal directions 65, 68
  progressive 65
  rotation 65
  strain 100
  strain rate 76, 84
  visualization 58
SimpleText 291
simulation 26
  algorithm 344, 346, 348
single crystals 167
sinistral shearing 87
site licence 18, 20
skeletonising 146
SLERP 454–455
Slicer 157
slickenlines 326
slip 373
  direction 325
  systems 102, 124
small circle 195, 200
sound 46
SPARC Station 5 351
spatial
  data 451
  orientation 447
  resolution 159
special effects 43
SpheriCAD 181–189, 191, 193
  downloading 191
spherical
  harmonics 169
  projection 195, 199, 212, 217–218
  See also nets, stereonet
  equal area 203
  triangle 221
spiral trail 247
splay crack 359, 369
SPO. *See* Shape Preferred Orientation
spreadsheet 238, 243, 474, 488, 490, 496
spring
  element 347–348
  network model 343
SQL. See Structured Query Language
stacks 125
statistical analysis 187, 229
stereogram 326
stereographic projection net 238
StereoNet 233–234

system requirements 233
stereonet 51–53, 56, 130, 181, 185–186, 233, 325, 327, 455, 474
contouring 187
StereoPlot 233–235
system requirements 235
stiffness 344, 348
after rupture 343, 345, 351–352
before rupture 343, 345, 351–352
stop location 184–185
digitizing 184
symbol 184
strain 24, 26, 39, 57, 59, 64, 75, 161
analysis 57–59, 105, 113, 135, 153, 288, 447
axes 175
calculation 488
compatibility 363
computer programs 58
conceptualizing 39
ellipse 40, 65, 250, 286, 408, 510–511
ellipse principal axes 286
grid 412, 414–416
heterogeneous 40, 415
history 75, 81, 89, 248–249, 252
homogeneous 40, 415
increment 260, 272
instantaneous 175
layer-parallel 477
map 123
ovoid 408
path 248, 288
principal directions 57, 65
rate 78, 269
state 510–511
studies 499
superposition 76
theory 75, 85–87
strain ellipsoid 40, 285–286, 288
oblate 286
principal axes 286
prolate 286
Strain Grid 412, 414–416
Strain-o-Matic 75, 87
stratigraphy 303, 307
stress 24, 26, 39, 359–360
axes 40, 87
circumferential 369–370
components 40, 363
conceptualizing 39
ellipse 40
ellipsoid 40
equilibrium equations 362
function 363, 366, 372, 375
hydrostatic 40
inversion 327
pattern 359
shadow 372
stress field 359, 363, 366, 375
fault-tip 369
Stress vs. Strain 75, 87
stress-strain curve 26
stretching
shear-induced 81
striations 326
strike 206, 448
line 51–52
structural
analysis 181
data 464
elements 131, 442
feature 308, 434
history 303, 307
information 306, 313
interpretation 440, 477
map 471
mapping 303
measurements 181
modeling 411
orientation data 181
structural geology course 39, 97
advanced 13, 285
advanced lecture modules 33
electronic lectures 23
fieldwork program 13
graduate 76
history 256
introductory 13, 285
laboratory 13, 51, 56, 58
lecture linking 23
lecture modules 24–26
lectures 13
microstructure laboratory 101
presentation software 18–20
teaching aid 97, 310
teaching collection 97
undergraduate 13, 39, 76
workshop 13
structure
three dimensional 421, 440
Structure Lab 1 51, 56

Structured Query Language 183
Stylolite 102
sub
-Bézier curves 400
-grain formation 98, 113
-simple shear 76, 78–79, 81, 85
subsidence 368, 369
SuperPaint 22, 59, 71, 291, 293–295
slant command 59, 64, 71
stretch command 59, 64, 71
Suppe construction 399
surface 159
characterisation 157
reconstruction 157
symbolic math 299, 359–360, 363–366, 368, 375
symmetry 173
synkinematic microscopy 123–124, 131

tacheometers 423
tape drive 125
technical presentation 19
tectonic
information 360
interpretation 167
teleseismic 176
television 16
TEM images 135
tensile
fracture 40
stress 359, 369
tensor 325, 449, 453, 503–504
orthogonal 511
orthogonal rotation 499–500
rotation 453, 501
terrain 303, 426, 428
features 424
text display 21
text file 473–474
texture analysis 168
theodolites 423
Theorist 299–301, 363
importing data 300
mathematical capabilities 299
system requirements 299
Thiessen polygons 426, 428
thin section 97, 135, 138, 156, 218
Thor-Odin 421–422, 440
three dimensional
display 359
structure 303–304, 311, 314
visualizations 119
three-point problems 51
thresholding 139, 157
thrust duplex 422
thrust sheet
modeling 412
TIFF file type 105–107, 113–114, 118, 169, 479
tilt 307, 447
Tiltwalls 102
time sequence 138
time-lapse 125
tomography 155, 159
topographic
base map 457, 469
features 421
image 422
map 423, 431
profile 466
relief 421–422, 434
resolution 457
surface 425
topographic map
scanning 469
topography 52, 182, 308, 421, 434, 443, 503
representation 423
topological
relationships 425
structure 426
transformation
rotational 455, 501
transformer displacement sensor 159
translation 22, 89
transport direction 189
trend 51, 448
Triangulated Irregular Network 421, 425–426, 428
triangulation 428
triaxial ellipsoid 285
triclinic 173–174
true dip 51
Turbo Pascal 195, 204, 217, 223
tutorial 34
twin plane 218
two dimensional display 359

U.K.E.S. Courseware Consortium 39, 42
unconformity 307
uniaxial 123, 217, 219
uniaxial compression 344

universal stage 124, 130, 155, 167–168, 171, 174, 217 –220, 224, 226, 229, 237, 239
  measurements 217
UNIX 138, 157, 182
uplift 368–369
upper crust 360
upper mantle 167, 176
Urai-type press 124
urban planning 428
UTM 474
Vacancy Migration 102
vantage point 430
vector 306, 424, 447, 451, 500
  data set 307
veins 24, 360
velocity
  field 78, 82, 83, 87
  gradient tensor 80, 82–83, 261, 264
vertical exaggeration 430
video 46
  camera 125, 136, 138, 477, 479
  cassette recorder 16
  input 137
Visual BASIC 182
visualization 51, 56
  three dimensions 51
volcanic hazard 360
volume loss
  anisotropic 75, 78–81, 86–87
  isotropic 75, 86–87
vortical flow 447
voxel 306–307, 309
  data set 307
voxels 157

wavelet 151
  directional 152
  theory 151
  transform 151
weighted
  average 425
  sum 395
Westergaard stress functions 366–367
Windows 182, 233, 238, 299
  Windows NT 299
Word 18, 291
  for *Windows* 191
word processing 18
WordPerfect 208

x-ray density 159
XDataSlice 156